Stabilization and Degradation of Polymers

David L. Allara and
Walter L. Hawkins, EDITORS
Bell Laboratories

Based on a symposium
sponsored by the Division
of Polymer Chemistry at
the 173rd Meeting of the
American Chemical Society,
New Orleans, Louisiana,
March 21–25, 1977.

ADVANCES IN CHEMISTRY SERIES **169**

AMERICAN CHEMICAL SOCIETY
WASHINGTON, D. C. 1978

Library of Congress CIP Data

Stabilization and degradation of polymers.
(Advances in chemistry series; 169 ISSN 0065-2393)

Includes bibliographical references and index.

1. Polymers and polymerization—Deterioration—Congresses. 2. Stabilizing agents—Congresses.
I. Allara, David L., 1937- . II. Hawkins, W. Lincoln, 1911- . III. American Chemical Society. Division of Polymer Chemistry. IV. American Chemical Society. V. Series.

QD1.A355 no. 169 [QD381.8] 540'.8s [547'.84]
ISBN 0-8412-0381-4 ADCSAJ 169 1-445 78-10600

PRINTED IN THE UNITED STATES OF AMERICA

Advances in Chemistry Series

FOREWORD

ADVANCES IN CHEMISTRY SERIES was founded in 1949 by the American Chemical Society as an outlet for symposia and collections of data in special areas of topical interest that could not be accommodated in the Society's journals. It provides a medium for symposia that would otherwise be fragmented, their papers, distributed among several journals or not published at all. Papers are reviewed critically according to ACS editorial standards and receive the careful attention and processing characteristic of ACS publications. Volumes in the ADVANCES IN CHEMISTRY SERIES maintain the integrity of the symposia on which they are based; however, verbatim reproductions of previously published papers are not accepted. Papers may include reports of research as well as reviews since symposia may embrace both types of presentation.

CONTENTS

Preface ix

1. **Fundamental Processes in the Photodegradation of Polyolefins** 1
   James Guillet

2. **Spectroscopic Study of Photosensitized Oxidation of 1,4-Polybutadiene** 11
   Morton A. Golub, Robert V. Gemmer, and Mark L. Rosenberg

3. **Low-Temperature Chemiluminescence from *cis*-1,4-Polybutadiene, 1,2-Polybutadiene, and *trans*-Polypentenamer at Temperatures Near Ambient** 19
   Richard A. Nathan, G. David Mendenhall, Michelle A. Birts, Craig A. Ogle, and Morton A. Golub

4. **The Effect of Thermal Processing History on the Photooxidation of Polymers** 30
   Gerald Scott

5. **Some Effects of Production Conditions on the Photosensitivity of Polypropylene Fibers** 56
   D. J. Carlsson, A. Garton, and D. M. Wiles

6. **Luminescence and Photooxidation of Light-Sensitive Commercial Polyolefins** 68
   N. S. Allen, J. F. McKellar, and G. O. Phillips

7. **Oxidative Degradation of Polymers. III. Photooxidation of Poly(vinyl alcohol) in Aqueous Solution** 78
   Etsuo Niki, Yorihiro Yamamoto, and Yoshio Kamiya

8. **Photoyellowing of Polystyrene and Poly(styrene-*Alt*-Methyl Methacrylate)** 96
   Robert B. Fox and Thomas R. Price

9. **Thermooxidative and Photooxidative Aging of Polypropylene: Separation of Heptane-Soluble and -Insoluble Fractions** 109
   H. P. Frank and H. Lehner

10. **Ultraviolet Stabilization of Polymers: Development with Hindered-Amine Light Stabilizers** 116
    A. R. Patel and J. J. Usilton

11. **^{13}C NMR Observation of the Effects of High Energy Radiation and Oxidation on Polyethylene and Model Paraffins** 133
    F. A. Bovey, F. C. Schilling, H. N. Cheng

12. **Stability of γ-Irradiated Polypropylene. I. Mechanical Properties** . 142
    J. L. Williams, T. S. Dunn, H. Sugg, and V. T. Stannett

13. **Stability of γ-Irradiated Polypropylene. II. Electron Spin Resonance Studies** 151
    T. S. Dunn, J. L. Williams, H. Sugg, and V. T. Stannett

14. **The Effect of Transition Metal Compounds on the Thermal Oxidative Degradation of Polypropylene in Solution** 159
Zenjiro Osawa and Takashi Saito

15. **Pyrolysis and Oxidative Pyrolysis of Polypropylene** 175
James C. W. Chien and Joseph K. Y. Kiang

16. **Kinetic Studies on Degradation in Polyimide Precursor Resins** 198
David E. Kranbuehl, Jean Takeuchi, Deborah Gibbs, and George Tsahakis

17. **The Effects of Some Structural Variations on the Biodegradability of Step-Growth Polymers** 205
S. J. Huang, M. Bitritto, K. W. Leong, J. Paulisko, M. Roby, and J. R. Knox

18. **Stabilization Fundamentals in Thermal Autoxidation of Polymers** . 215
J. Reid Shelton

19. **The Role of Certain Organic Sulfur Compounds as Preventive Antioxidants. III. Reactions of *tert*-Butyl *tert*-Butanethiolsulfinate and Hydroperoxide** 226
Donald M. Kulich and J. Reid Shelton

20. **Inhibited Autoxidation of Polypropylene** 237
Dale E. Van Sickle and David M. Pond

21. **Nonmigrating Antioxidants via Sulfonyl Azide Intermediates** 253
Stephen E. Cantor

22. **The Distribution of Additives and Impurities in Isotactic Polypropylene** 261
T. G. Ryan, P. D. Calvert, and N. C. Billingham

23. **Microscopic Mechanisms of Oxidative Degradation and Its Inhibition at a Copper–Polyethylene Interface** 273
D. L. Allara and C. W. White

24. **Stabilization of Poly(fluoroalkoxyphosphazene) (PFAP) Elastomer against Thermal Degradation** 293
G. S. Kyker and J. K. Valaitis

25. **Recent Fundamental Developments in the Chemistry of Poly(vinyl chloride) Degradation and Stabilization** 309
W. H. Starnes, Jr.

26. **Reductive Dehalogenation with Tri-*n*-butyltin Hydride: A Powerful New Technique for Use in Poly(vinyl chloride) Microstructure Investigations** 324
W. H. Starnes, Jr., R. L. Hartless, F. C. Schilling, and F. A. Bovey

27. **Molecular Orbital Theory of Polyenes Implicated in the Dehydrochlorination of Poly(vinyl chloride). II. Electronic Structures, Equilibrium Geometries, and Energetics of the Ground States of Polyenyl Cations and Neutral Polyenes** 333
Robert C. Haddon and William H. Starnes, Jr.

28. **The Ligand Exchange Reaction of Some Dialkyltin Dimercaptides and Dicarboxylates with Dialkyltin Dichlorides: Observation by ^{1}H and ^{13}C NMR Implications for Poly(vinyl chloride) Stabilization** 363
Richard G. Parker and Charles J. Carman

29. Stabilizer Consumption during Processing of Poly(vinyl chloride) . 374
Tran Van Hoang, Alain Michel, André Revillon,
Michel Bert, Alain Douillard, and Alain Guyot

30. Synergistic Mechanisms between Zinc, Calcium Soaps, and Organo Compounds in Poly(vinyl chloride) Stabilization 386
Alain Michel, Tran Van Hoang, and Alain Guyot

31. Antioxidative Properties of Phenyl-Substituted Phenols: Their Role in Synergistic Combinations with β-Activated Thioethers 399
C. R. H. I. de Jonge, E. A. Giezen, F. P. B. van der Maeden,
W. G. B. Huysmans, W. J. de Klein, and W. J. Mijs

Index 427

PREFACE

The chapters in this book present a review of current active areas of research in polymer stabilization and degradation and represent a blend of fundamental research on both well-defined laboratory systems and practical commercial systems. The intent of this volume was to explore the extent to which fundamental research is contributing to the current study of polymer degradation and stabilization. As demonstrated in the papers presented, modern scientific techniques and concepts are making important contributions to the fundamental understanding of the highly complex mechanisms by which polymers degrade and are stabilized. The papers are grouped into three main sections. In the first, Guillet provides a review of the field of photodegradation of polyolefins and presents new data with conclusions on the significant mechanisms. Several chapters follow on photodegradation and photostabilization in a variety of polymeric systems.

The second section deals with degradation induced by thermal processes, high-energy radiation, metal catalysis, and biological processes. The section begins with the chapter by Bovey and Schilling on the application of the powerful tool of carbon-13 NMR to studies of products in polymer degradation. The final portion of this section leads off with a review paper by Shelton on fundamentals in stabilization against thermal oxidation of polymers. Several papers which are concerned with stabilization mechanisms of sulfur compounds and phenols, preparation of nonmigrating antioxidants by backbone grafting, and the physical as well as chemical aspects of stabilization in bulk polymers and at metal–polymer interfaces. The last paper presents work on stabilization in a semi-inorganic polymer.

The last section deals with stabilization and degradation of poly(vinyl chloride). A detailed review is presented by Starnes. Papers follow on current approaches for mechanisms in stabilization and degradation including molecular orbital calculations. The final chapters are concerned with stabilization during processing conditions.

We certainly conclude that fresh, solid scientific approaches to the old problems are evolving in both industrial and academic laboratories.

Bell Laboratories
Murray Hill, New Jersey
September 1, 1978

DAVID L. ALLARA
WALTER L. HAWKINS

1

Fundamental Processes in the Photodegradation of Polyolefins

JAMES GUILLET

Department of Chemistry, University of Toronto, Toronto, Canada M5S 1A1

Mechanisms for the photooxidation of polyolefins are reviewed briefly. It is shown that although hydroperoxides are the main carriers of the photooxidative chain, their decomposition does not represent an efficient mechanism for main chain scission. In the photolysis of cis-*polyisoprene hydroperoxide, for example, the quantum yield for chain scission is only ca. 2% of that for hydroperoxide decomposition. On the other hand, ketonic and aldehyde groups continue to build up in the polymer and soon absorb most of the light. They contribute to chain scission by (1) direct photolysis (Norrish Type II), (2) formation of radicals (Norrish Type I), and (3) energy transfer from the carbonyl to hydroperoxy groups. Thus, their effect becomes increasingly important as oxidation continues, and depending on the structure of the polymer, they may make significant contributions to the degradation of polymer properties even at low degrees of oxidation. Direct experimental evidence for this is given from studies of the photodecomposition of* tert-*butyl hydroperoxide and* cis-*polyisoprene hydroperoxide.*

Polyolefins represent the largest single class of thermoplastic materials in commercial production at the present time. Their degradation under external environments is a subject of both practical and theoretical importance and has been studied extensively during the past decade. The relatively rapid deterioration of such materials when exposed to the outdoors seems to be related in nearly all cases to chain scission caused by photooxidation reactions induced by the UV light of the sun. Therefore, the inhibition of degradation or the control of the lifetime of these materials is a subject of considerable importance.

0-8412-0381-4/78/33-169-001$05.00/1

Because of the ozone layer in the upper atmosphere, most of the short-wave UV radiation is filtered out and for all practical purposes the UV light reaching the surface of the earth has wavelengths longer than 290 nm. There are usually only two chemical groups present normally in polymers that absorb in this region, namely the ketone or aldehyde carbonyl and the hydroperoxide group. Both are produced by oxidation of the polymer, either thermally or photochemically during processing, or even by exposure to atmospheric oxygen at room temperature. Recently there has been some controversy about the relative role of these two groups in the photooxidation of polyolefins and this paper is an attempt to review recent work relating to that problem.

Recent work by Carlsson and Wiles (*1*) on the photooxidative degradation of polypropylene has demonstrated conclusively the importance of hydroperoxides in initiating photodecomposition and maintaining the oxidative chain reaction leading to polymer degradation. Although keto and aldehyde groups are usually the major UV absorbing groups in photooxidizing polyolefins, their role in the process has been considered minimal by these authors because of the relatively low quantum yields for the Norrish Type I process leading to the formation of free radicals in the polymer matrix. The purpose of this chapter is to review recent results relating to this problem and to elucidate further the relative roles of the two chromophores in photooxidizing systems.

It is accepted generally that photooxidation follows the usual mechanism for the oxidation of hydrocarbons, that is, a chain process initiated by free radicals which are formed after the absorption of a quantum of UV light. The primary initiating processes are indicated in Reactions 1 and 2 below.

$$ROOH + h\nu \rightarrow RO\cdot + \cdot OH \qquad (1)$$

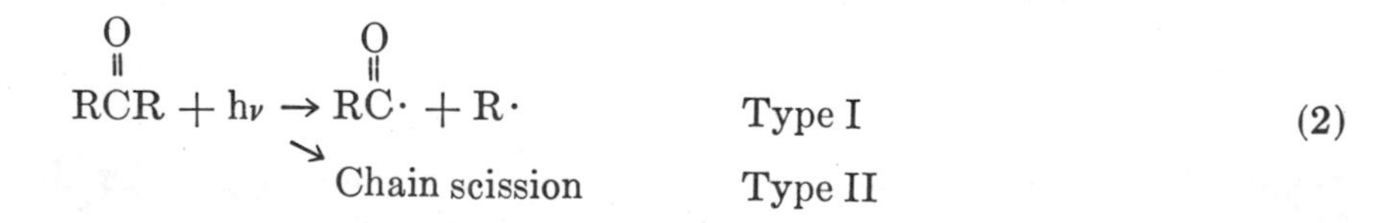

In addition to these two primary processes, there are also possibilities of sensitization of the photooxidation by a variety of mechanisms. The first of these is the formation of an excited state of a ketone or a polynuclear aromatic compound by absorption of light followed by quenching with oxygen to form singlet oxygen ($^1O_2$) which later adds to a double bond in the polymer to form an allylic hydroperoxide by the scheme shown below. This reaction has been demonstrated to occur quantitatively in *cis*-polyisoprene (2) but its importance in the photodegradation of saturated polyolefins such as polyethylene and polypropylene remains to be demonstrated.

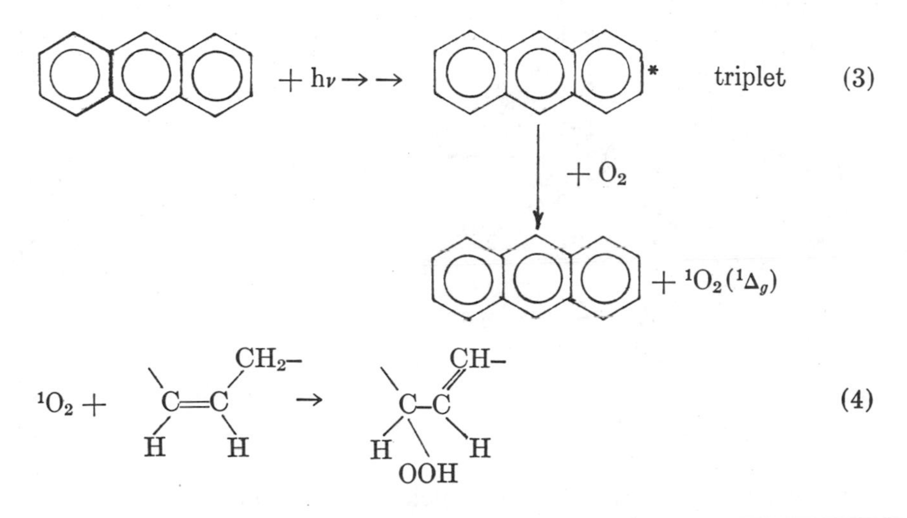

A second possible mechanism of sensitization involves the induced decomposition of hydroperoxides caused by the direct photolysis of groups such as peroxides and ketones to give radicals which later attack hydroperoxide groups to produce peroxy radicals.

$$R\cdot + R'OOH \rightarrow RH + R'OO\cdot \qquad (5)$$

A third mechanism of sensitization involves the formation of an excited state of a ketone by absorption of light followed by energy transfer to a hydroperoxy group and subsequent decomposition of the hydroperoxide. In complex polymer systems it is possible that all three of these mechanisms play an important role.

After the initial radicals are formed, a hydrocarbon polymer follows a typical radical-chain oxidation mechanism as described by the reactions listed below where *P* represents an aliphatic polymer and R· can be any radical.

$$R\cdot + PH \rightarrow P\cdot + RH \qquad (6)$$

$$P\cdot + O_2 \rightarrow POO\cdot \qquad (7)$$

$$POO\cdot + PH \longrightarrow POOH + P\cdot \qquad (8)$$

$$POOH \xrightarrow[\text{or thermal}]{h\nu} PO\cdot + \cdot OH \qquad (9)$$

$$\mathrm{POOH} + \mathrm{R}\cdot \longrightarrow \mathrm{RH} + P\mathrm{OO}\cdot \quad (10)$$

$$\cdot\mathrm{OH} + P\mathrm{H} \longrightarrow P\cdot + \mathrm{H_2O} \quad (11)$$

$$P\mathrm{O}\cdot + P\mathrm{H} \longrightarrow P\mathrm{OH} + P\cdot \quad (12)$$

$$P\mathrm{O}\cdot \xrightarrow{\text{disproportionation}} P\cdot + \text{olefin} + \text{ketone or aldehyde} \quad (13)$$

$$P\cdot + P\cdot \longrightarrow P\text{–}P \quad (14)$$

$$P\mathrm{OO}\cdot + P\cdot \longrightarrow P\mathrm{OO}P \quad (15)$$

$$P\mathrm{OO}\cdot + P\mathrm{OO}\cdot \longrightarrow P\text{–}\overset{\overset{\displaystyle \mathrm{O}}{\|}}{\mathrm{C}}\text{–}P + P\mathrm{OH} + \mathrm{O_2} \quad (16)$$

From kinetic considerations of such a mechanism it is expected that for each radical formed in the primary photolysis of a hydroperoxide or ketone (Reactions 1 or 2) a number of oxidation steps will occur. In particular, a number of hydroperoxy groups are expected to be decomposed by an induced-radical mechanism (Reaction 10) for each radical produced in the primary photolysis step, and in fact this is observed in practice. Termination of the radical chain occurs by radical combination (Reactions 14 and 15) or disproportionation of two peroxy radicals by a complex mechanism to give keto and alcohol compounds (Reaction 16). The stable products of the photooxidative chain are polymeric or small molecule compounds containing keto, aldehyde, and alcohol groups. However, the aldehyde groups are oxidized readily to acid or ester under these conditions and the carbonyl groups monitored in polymers as a measure of the extent of degradation usually represents the sum of the absorptions from all four types of carbonyl structures. The alcohol groups are observed less readily because of the relatively low intensity of their absorption in the ir.

The mechanism described above does not require the intervention of keto groups in order to carry out a photooxidative chain reaction and many investigators suggest that their role is not important (*1, 3, 4*). However, it is well known that the concentration of ketone groups increases as the photooxidation proceeds and that they absorb a much greater proportion of the solar radiation than do hydroperoxides. In addition, it is well established that after absorption of light by a ketone, radicals are formed by the Norrish Type I, Type II, and other processes

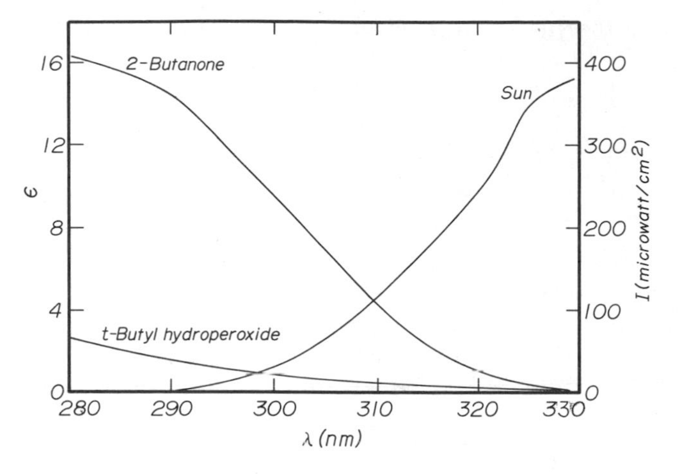

Figure 1. Absorption spectra of 2-butanone and tert-*butyl hydroperoxide compared with terrestrial UV intensity*

(*5*). Furthermore, in many polymer samples the concentration of hydroperoxy groups is negligible under photooxidating conditions. Figure 1 shows the molar extinction coefficients for 2-butanone and *tert*-butyl hydroperoxide compared with the emission spectrum of the sun at terrestrial levels. It is obvious from this figure that keto groups present in equimolar quantities with hydroperoxide will absorb about four times as much radiation. If the quantum yields of radical production were the same for the two compounds, then one would expect four times the yield of radicals from ketone photolysis than from hydroperoxide photolysis. Furthermore, as the degradation proceeds and the keto groups build up in concentration, one would expect to observe an even greater effect. For this reason it becomes important to determine the quantum yields for photolysis of the two different species under comparable conditions.

Table I shows the quantum yields for photolysis of *tert*-butyl hydroperoxide in hexane at 36°C (*6*). It is evident that the quantum yield for photolysis is greater than unity, which is typical for a radical-chain process, and depends on the hydroperoxide concentration, ranging from 3.5 at 1% hydroperoxide to 10 at 5%. Similar results are shown in

Table I. Quantum Yields for Photolysis of *tert*-Butyl Hydroperoxide in Hexane, 36°C, λ = 313 nm

$C_{\text{tert}BHP}$, %	ϕ
1	3.5 ± 0.2
3	5.2 ± 0.3
5	9.9 ± 0.5

Table II for the photodecomposition of the hydroperoxide of *cis*-polyisoprene formed from singlet-oxygen addition (*7*). Similar high quantum yields for hydroperoxide decomposition have been reported by Carlsson and Wiles for polypropylene hydroperoxide (*1*). On the other hand, quantum yields for formation of radical products from polymeric ketones have been previously believed to be considerably less than this—of the order of 0.002 (*5*). However, these results were obtained with keto groups which were in the backbone of a polyethylene chain and hence were hindered both sterically and by the mobility of the free radical produced, from having a high quantum yield for radical production. Recently, it has been shown that in the photolysis of polyethylene-containing side-chain ketone groups, the quantum yield of free radical formation can be as high as 0.3 (*8, 9*). By including such ketone groups in polyethylene chains, accelerated photodegradation is observed (*10*) and the photodegradation of pure polyethylene is sensitized by blending

Table II. Quantum Yield of Photodecomposition of *cis*-1,4-Polyisoprene Hydroperoxide at 313 nm; Light Intensity = 7.2×10^{-7} Einstein/min.

Sample	*Conc. (–OOH) mol/l*	*T, °C*	ϕ_{-OOH}
PIPH 1	0.00104	30	2.5
PIPH 2	0.00215	30	5.3
PIPH 2	0.00215	40	5.1
PIPH 2	0.00215	22.3	5.2
PIPH 3	0.00205	12.9	5.1
PIPH 4	0.00601	30	7.5

with minor amounts of such ketone copolymers (*11*). The mechanism is presumably a radical-chain oxidation induced by the polymeric ketones.

This prompted us to review the work on the decomposition of hydroperoxides and in particular to determine what portion of the hydroperoxides that decomposes leads ultimately to bond scission. It is important to point out in this context that it is possible in principle for a polymer to oxidize extensively without losing its physical properties. The most important process in the breakdown of physical properties is bond scission in the backbone of the polymer chain. This can be best determined by measurements of the molecular weight of a photooxidizing or photodegrading polymer. In our laboratories we have studied this process by automatic viscometry which permits a very precise measurement of bond scission in polymers: the apparatus has been described previously (*12*). For these studies we prepared a singlet-oxygen adduct of *cis*-polyisoprene by the following scheme:

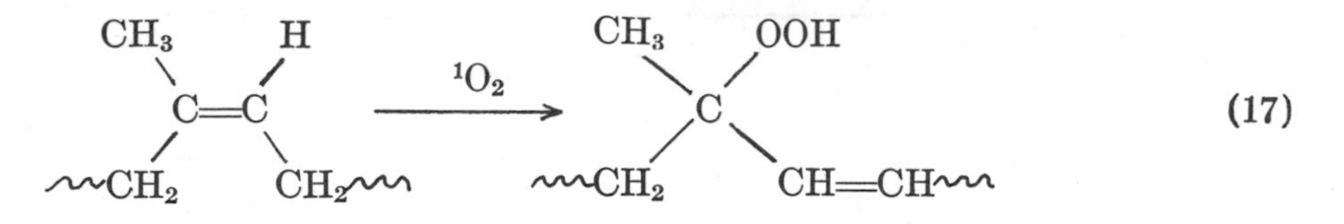

The hydroperoxide formed by singlet-oxygen addition is photolyzed then in the automatic viscometer; typical results are shown in Figure 2 (7). By monitoring the light absorbed, it is possible to calculate the quantum yields as shown in Figure 3. It is obvious that the quantum yield for chain scission increases as the wavelength of the exciting light decreases. It is affected also by the solvent used, as indicated in Table III. Using the same apparatus the quantum yield of disappearance of hydroperoxide was determined by assaying the peroxide using the triphenylphosphine method; typical results are shown in Figure 4. Comparisons of the quantum yields for hydroperoxide decomposition and chain scission are shown in Table IV. In all cases, the quantum yield for scission is only 2–3% of that for hydroperoxide decomposition. That is consistent with the hypothesis that most of the hydroperoxide that decomposes on exposure to UV light is decomposed by an induced-radical mechanism rather than by primary photolysis of the peroxy bond. Therefore, it is suggested that chain rupture results from the disproportionation of $PO\cdot$ radicals (Reaction 13) which are formed only in the primary photolytic step (Reaction 9) and not in the induced-decomposition step (Reaction 10). This would explain the large difference between the rates of these

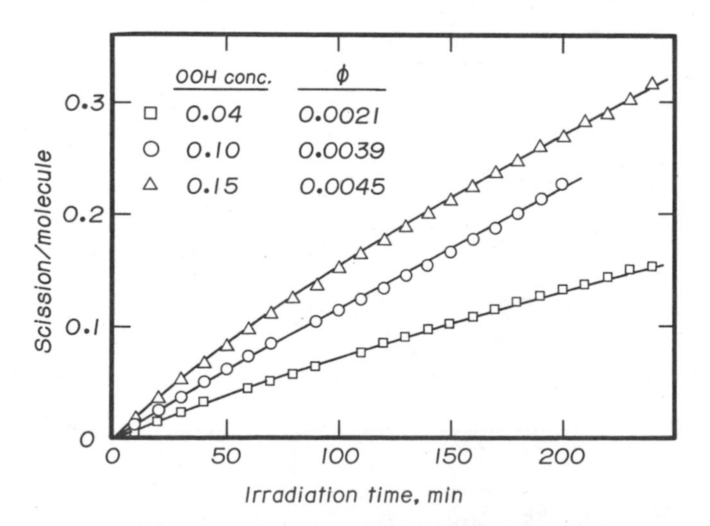

Figure 2. Photolysis of polyisoprene hydroperoxide in 1,2-dichloroethane solution, 30°C, $\lambda = 365$ nm

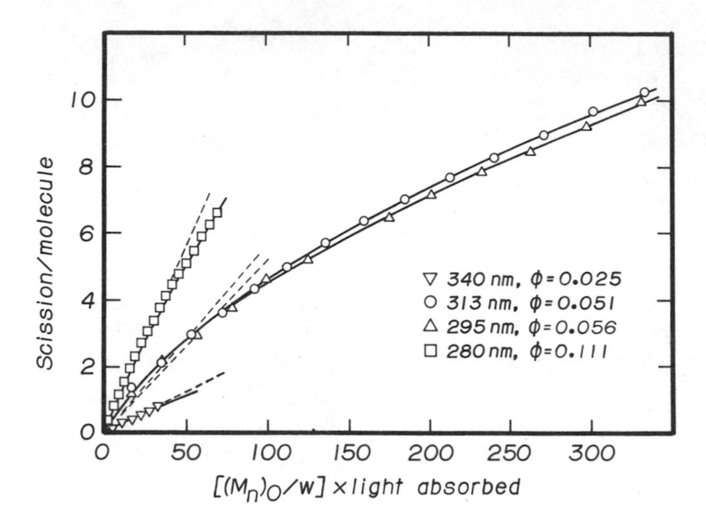

Figure 3. Photolysis of cis-*polyisoprene hydroperoxide (0.06% –OOH) in 1,2-dichloroethane at 30°C; determination of quantum yields at various wavelengths*

Table III. Effect of Solvent on ϕ_s for Polyisoprene at 30°C[a]

λ	*DCE*	*Cyclohexane*	*Benzene*
280	0.111	0.024	0.006
313	0.051	0.031	0.004

[a] –OOH = 0.06%.

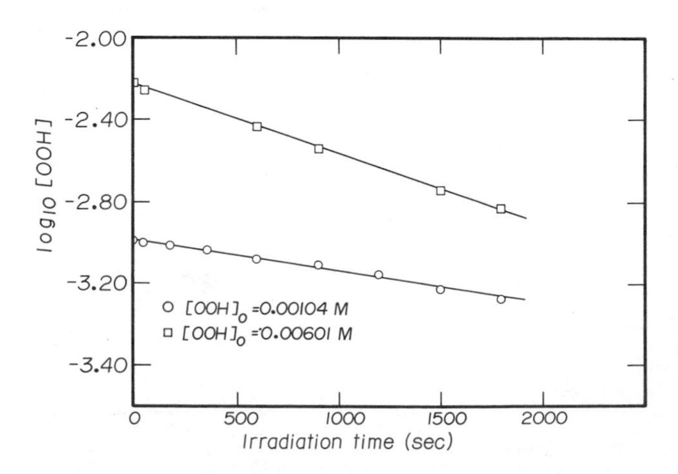

Figure 4. Photolysis of cis-*polyisoprene hydroperoxide estimated by loss of peroxy groups, λ = 313 nm, dichloroethane solvent*

two processes in hydrocarbon polymers and the relative inefficiency of the hydroperoxide decomposition reaction in causing chain scission.

In a series of further experiments, the photolysis of *tert*-butyl hydroperoxide and also the *cis*-polyisoprene hydroperoxide was studied in the presence of a variety of ketones. The results of the photodecomposition of *tert*-butyl hydroperoxide in hexane in the presence of methyl isobutyl-

Table IV. Efficiency of Polymer Chain Scission Relative to Photodecomposition of Hydroperoxide at 313 nm

Sample	*Conc. (–OOH) mol/l*	*Mol wt* $\overline{M}_{n0}$	*Temp., °C*	*No. scissions / No. –OOH decomposed*
PIPH 1	0.00104	126,000	30	0.018
PIPH 2	0.00215	40,000	30	0.028
PIPH 2	0.00215	40,000	40	0.028
PIPH 2	0.00215	40,000	22.3	0.026
PIPH 3	0.00205	33,000	12.9	0.028
PIPH 4	0.00601	3,000	30	0.017

ketone is shown in Table V (*6*). The apparent quantum yield for photolysis of the hydroperoxide increases in the presence of the ketone, confirming the presence of an efficient energy transfer or induced decomposition or both. This confirms the earlier results of Walling and Gibian (*13*) showing the induced decomposition of hydroperoxides catalyzed by ketones.

Table V. Apparent Quantum Yields for *tert*-Butyl Hydroperoxide Photolysis in the Presence of Methyl *tert*-Butyl Ketone; 0.31*M* Hydroperoxide in Hexane at 36°C, λ = 322 nm

Ketone, wt %	*Light Absorbed by Ketone, %*	$\phi_{\text{-OOH}}$
0	0	5.2
1	32	8.4
3	59	13
10	80	21

In conclusion, it is suggested that keto groups can and do play a central role in catalyzing the photooxidative degradation of polyolefins by a variety of mechanisms, including (1) direct photolysis causing bond breaking by the Norrish Type II scission, (2) chain scission and initiation of photooxidative chains by the formation of radicals by Norrish Type I and other processes, and (3) induced decomposition of hydroperoxides caused by energy transfer from the ketone to the hydroperoxide. The relative importance of these processes will depend on the structure of the olefin polymer and its thermal history, as well as the extent of the degradation process itself, and the conditions of exposure to UV light. While hydroperoxides are undoubtedly key intermediates in the formation of highly oxidized products, their role in causing chain scission and consequent deterioration of physical properties may well be secondary to that of the ketonic groups derived from their decomposition.

Acknowledgment

The author wishes to express his appreciation to the National Research Council of Canada for financial support of this research.

Literature Cited

1. Carlsson, D. J., Wiles, D. M., *J. Macromol. Sci., Rev. Macromol. Chem.* (1976) **C14**, 65.
2. Ng, H. C., Guillet, J. E., *Proc. EUCHEM Conf. on Singlet Oxygen Reactions with Polymers, Stockholm, Sept. 2–4, 1976.*
3. Amin, M. U., Scott, G., Tillekevatne, L. M. K., *Eur. Polym. J.* (1975) **11,** 85.
4. Harper, D. J., McKellar, J. F., *Chem. Ind.* (1972) 848.
5. Hartley, G. H., Guillet, J. E., *Macromolecules* (1968) **1**, 165, 413.
6. Guillet, J. E., Sherwin, N. (manuscript in preparation).
7. Ng, H. C., Guillet, J. E. (manuscript in preparation).
8. Sitek, F., Guillet, J. E., Heskins, M., *J. Polym. Sci., Polym. Symp.* (1976) **57,** 343.
9. Li, L., Guillet, J. E. (manuscript in preparation).
10. Guillet, J. E., U.S. Patent **3,753,952,** Aug. 21, 1973.
11. Guillet, J. E., Troth, H. G., Canadian Patent **1,000,000,** Nov. 16, 1976.
12. Kilp, T., Houvenaghel–Defoort, B., Panning, W., Guillet, J. E., *Rev. Sci. Instrum.* (1976) **47,** 1496.
13. Walling, C., Gibian, M. J., *J. Am. Chem. Soc.* (1965) **87,** 3413.

RECEIVED July 1, 1977.

Spectroscopic Study of Photosensitized Oxidation of 1,4-Polybutadiene

MORTON A. GOLUB, ROBERT V. GEMMER, and
MARK L. ROSENBERG

Ames Research Center, National Aeronautics and Space Administration,
Moffett Field, CA 94035

The microstructural changes which occur in cis- *and* trans-*1,4-polybutadienes during photosensitized oxidation using visible light were studied by means of infrared, proton, and carbon-13 NMR spectroscopy. The singlet oxygenation of the 1,4-polybutadienes yielded the expected allylic hydroperoxides with shifted double bonds, with the new double bonds having a nearly all*-trans *configuration. A convenient infrared measure of the extent of hydroperoxide formation is given by the absorbance ratio,* $A_{2.9}/A_{6.9} \equiv A'$. *In the case of methylene blue and chlorophyll as sensitizers* A′ *gave a smooth correlation with oxygen uptake; Rose Bengal, on the other hand, gave erratic* A′*-oxygen uptake plots. The singlet oxygenation followed zero-order kinetics, the relative rates for* cis- *and* trans-*1,4-polybutadienes being approximately 6:1. The spectroscopic results are interpreted in terms of an "ene"-type photooxygenation of the 1,4-polybutadienes.*

Considerable attention has been given to the reaction of *cis*- and *trans*-1,4-polybutadienes with singlet oxygen ($^1O_2$), whether produced by microwave discharge, by photosensitization, or by chemical means in situ (*1–6*). The major finding, based on ir spectroscopy, was that the reaction of 1,4-polybutadiene with $^1O_2$ leads to the formation of allylic hydroperoxides (as evidenced by the growth of a 2.9-μm band), presumably accompanied by double bond shifts, according to an "ene"-type process (*7, 8*):

$$-\overset{|}{C_1}=\overset{|}{C_2}-\underset{H}{\overset{|}{\underset{|}{C_3}}}- \quad + \quad {}^1O_2 \quad \rightarrow \quad -\underset{O_2H}{\overset{|}{\underset{|}{C_1}}}-\overset{|}{C_2}=\overset{|}{C_3}- \qquad (1)$$

Since the nature and extent of the shifted double bonds in the case of the $^{1}O_2$-reacted 1,4-polybutadienes have not been previously demonstrated, there is a need to reexamine this question using not only ir spectroscopy but proton and carbon-13 NMR spectroscopy as well. This paper, which is a follow-up to our prior ir–NMR studies on both the photosensitized oxidation of 1,4-polyisoprene (*9*) and the thermal oxidation of 1,4-polybutadiene (*10*), presents a detailed spectroscopic study of the microstructural changes occurring in *cis*- and *trans*-1,4-polybutadienes during photosensitized oxidation (or singlet oxygenation). In this paper we show that the "ene"-type Reaction 1 indeed occurs in 1,4-polybutadiene and that the new shifted double bonds have a nearly all-trans configuration and are formed at the same rate as the hydroperoxide groups.

Experimental

The *cis*- and *trans*-1,4-polybutadienes used in this work were obtained from The B. F. Goodrich Research and Development Center, Brecksville, OH, and had initial cis/trans ratios of ~98/2 and 2/98, respectively. The procedures for the photosensitized oxidation of purified samples of the 1,4-polybutadienes, using visible light and methylene blue, chlorophyll and Rose Bengal as sensitizers, the procedures for the reduction of the hydroperoxidized polymers to the corresponding alcohols, as well as the procedures for the spectroscopic analysis of the reaction products, were similar to those described previously (*8*).

Paramagnetic shift reagent experiments were carried out using *tris*-(6,6,7,7,8,8,8-heptafluoro-2,2-dimethyloctanedionato) europium (Eu(fod)$_3$). Portions of the Eu(fod)$_3$ were added to a $CDCl_3$ solution of the reduced product obtained from the singlet oxygenation of *cis*-1,4-polybutadiene having about 0.2 O_2/monomer unit. After each incremental addition of the reagent, a ^{13}C NMR spectrum was obtained. The changes in chemical shift were recorded as a function of added Eu(fod)$_3$. Assignments were made assuming that the rate of change of chemical shift is greatest for carbons closest to the oxygen functional group.

Results and Discussion

Figure 1 shows the changes in the ir spectrum of *cis*-1,4-polybutadiene (CB) after extensive singlet oxygenation using methylene blue (MB) as photosensitizer. Normally the use of MB as sensitizer results in a minor additional absorption at ~6.2 μm, but this may be removed by washing a film of the $^{1}O_2$-reacted CB (supported on an NaCl plate) with methanol. Spectra similar to Figure 1 were obtained with chlorophyll (CL) as sensitizer, but Rose Bengal (RB) tended to give extraneous peaks caused by autoxidative side reactions just as in the case of 1,4-polyisoprene (*8*). The significant changes observed in Figure 1 are the growth of the 2.9-μm band (O_2H), the development of a moderately strong band

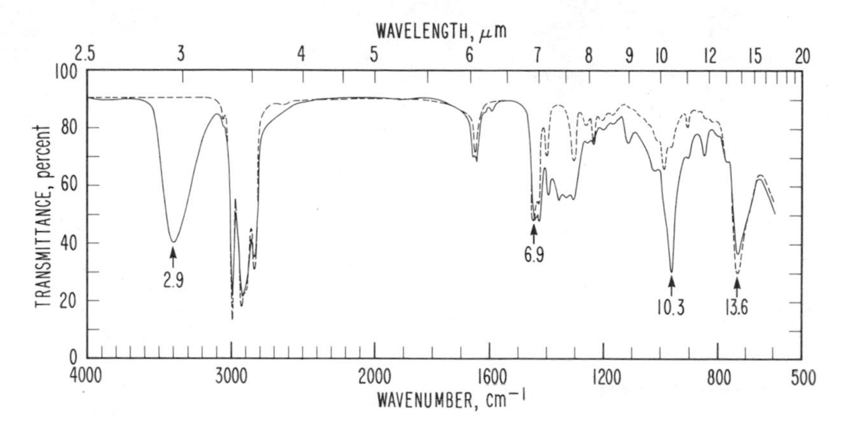

Figure 1. Ir spectra of CB before (- - -) and after (——) singlet oxygenation with methylene blue as photosensitizer

at 10.3 μm (*trans*—CH═CH—), and the corresponding diminutive of the 13.6-μm band (*cis* —CH═CH—), all of which point to the occurrence of Reaction 1.

The ir spectrum of *trans*-1,4-polybutadiene (TB) after photosensitized oxidation (Figure 2) likewise displays a growth of the 2.9-μm band. However, there was no significant change in intensity of the 10.3-μm trans band, nor was there any development of a new *cis* band at ca. 14.0 μm. Thus any shifted double bonds in TB as in Reaction 1 are essentially all-trans. An interesting feature of Figure 2 is the growth of the 6.0-μm absorption which is clearly not caused by C═C stretch of cis double

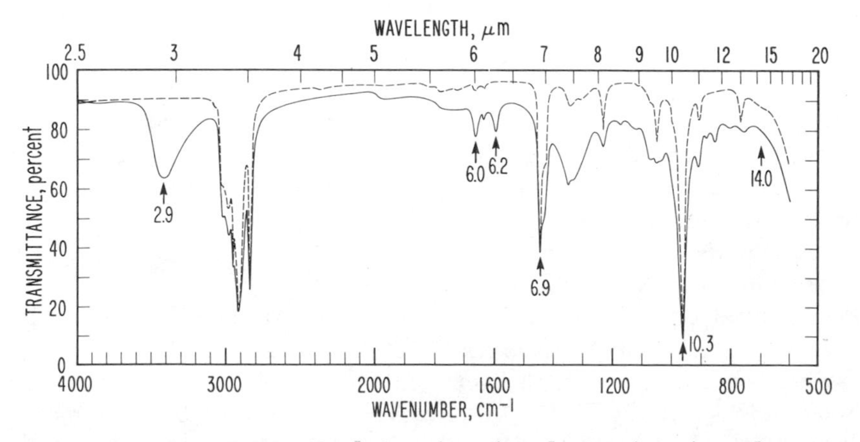

Figure 2. Ir spectra of TB before (- - -) and after (——) singlet oxygenation with methylene blue as photosensitizer

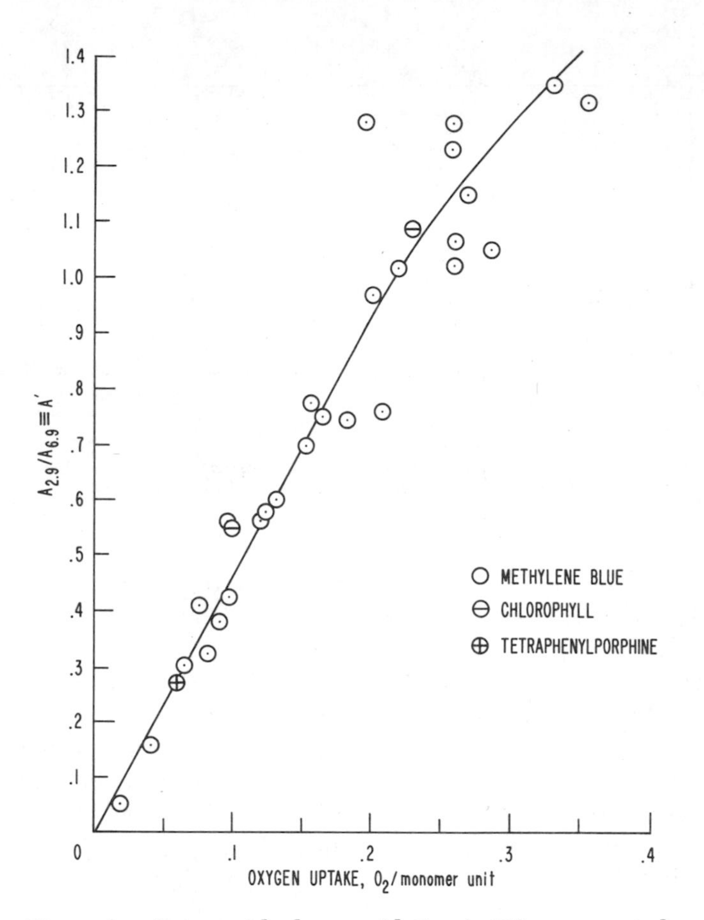

Figure 3. Extent of hydroperoxidation in CB, represented by ir absorbance parameter, A′, plotted vs. oxygen uptake

bonds but must be associated with the analogous stretching vibration of the new *trans* —CH=CH—, the symmetry of which is perturbed by an O_2H group attached to the alpha carbon. The presence of the 6.2-μm peak is caused by residual photosensitizer (MB) and not by new conjugated double bonds.

A suitable spectroscopic means of following the degree of hydroperoxidation is provided by the absorbance ratio $A_{2.9}/A_{6.9} \equiv A'$ (*8*). A smooth correlation of *A′* with oxygen uptake for MB- and CL-photosensitized oxidation of CB is given in Figure 3. Data for RB, which were erratic and tended to fall to the right of the curve because of autoxidative side reactions, are not shown. Interestingly, the use of tetraphenylporphine as photosensitizer for the $^1O_2$–CB reaction, examined elsewhere (*11*), yielded a data point which fit well on the curve shown in Figure 3.

Before consideration of the other spectroscopic data, it is pertinent to mention that the singlet oxygenation exhibited zero-order kinetics as expected, the relative rates for CB and TB being ~ 6:1. This ratio is substantially the same as that obtained (~6.1:1.0) for the relative rates of $^1O_2$ reaction with cis and trans forms of 4-methyl-2-pentene (*12*). The rates of photosensitized oxidation of CB were not significantly affected by the presence of an antioxidant such as 2,6-di-*tert*-butylphenol. On the other hand, the rates were greatly decreased by the presence of a singlet oxygen quencher such as diazabicyclooctane. It is worth noting that cis and trans polypentenamer (homologs of CB and TB with an additional CH_2 unit between successive double bonds) undergo singlet oxygenation with a similar ratio of relative rates (~5:1), suggesting that this kinetic behavior is general for disubstituted olefins. Moreover, CB and cis polypentenamer display comparable hydroperoxidation rates when expressed in terms of oxygen uptake per monomer unit per unit time.

The ^1H NMR spectrum of $^1O_2$-reacted CB shown in Figure 4 provides complementary microstructural information to support Reaction 1 (or 1′).

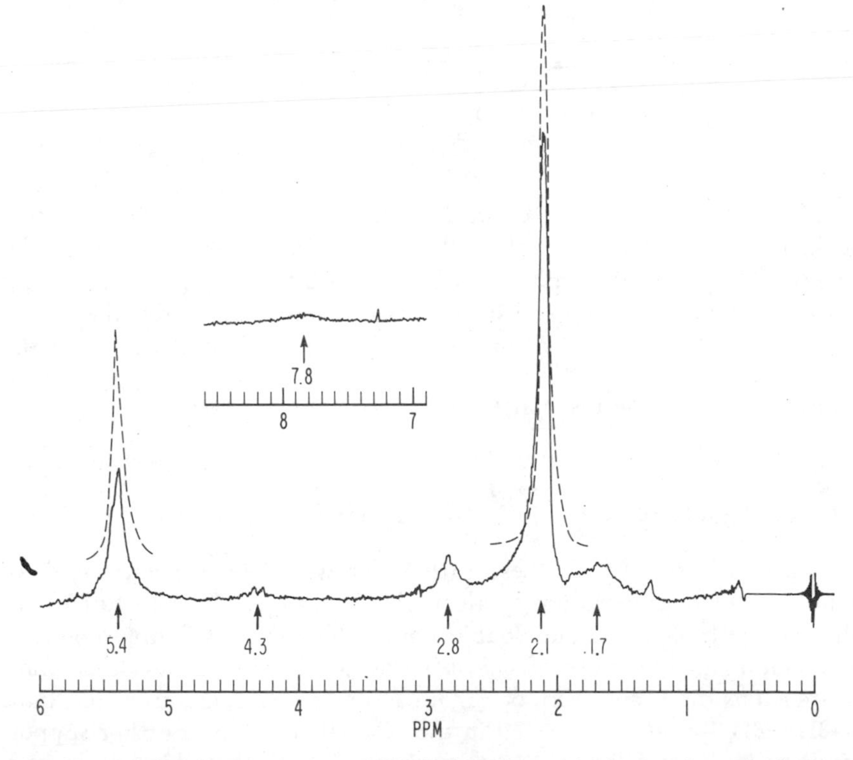

Figure 4. ^1H NMR spectra of CB before (- - -) and after (——) photosensitized oxidation

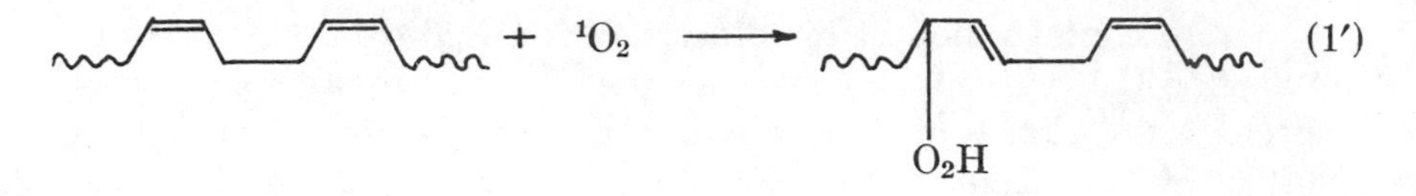

Thus in addition to the 2.1 (—CH_2C=) and 5.4 δ (—CH=) resonances of the original polymer, this spectrum shows a broad resonance centered at 1.7 δ (—CH_2—C—O—) and other new resonances at 2.8 (=C—CH_2—C=), 4.3 (—CH—O—), and 7.8 δ (—C—O_2H). Characteristically, the last peak can be removed by washing the ^{1}H NMR sample with D_2O.

Typical ^{13}C spectra of $^1O_2$-reacted CB before and after reduction are shown in Figure 5, with the following assignments of peaks based on paramagnetic shift reagent experiments and supported by analogies in the literature:

2 δ ε
1 α γ β
O_2H

Of immediate interest are the prominent displacements of the δ and γ peaks (to δ′ and γ′) as a result of the substitution of –O_2H by –OH: the δ peak at 86 ppm shifts upfield by 13 ppm, while the γ peak at 34 ppm moves downfield by 5 ppm. These displacements reinforce the above assignments. Similar upfield shifts of ~12–14 ppm were observed for the analogous —C—O— resonances in $^1O_2$-reacted *cis*- and *trans*-1,4-polyisoprenes (*8*). Significantly, Figure 5 shows only one resonance for —C—O_2H and only one for —C—OH, not two resonances (separated by ~5 ppm) for each of these carbons as would be expected if there were present hydroperoxide or alcohol (*13*) groups attached to carbons alpha to both cis and trans double bonds. The shifted double bonds in Reaction 1′ are therefore of one isomeric form, and this must be trans inasmuch as the ir spectrum of Figure 1 shows unambiguously that *trans* —CH=CH— units are formed in the $^1O_2$-CB reaction. Further support for Reaction 1′ and the trans nature of the shifted double bonds, regardless of whether the double bonds are cis or trans, prior to shifting was

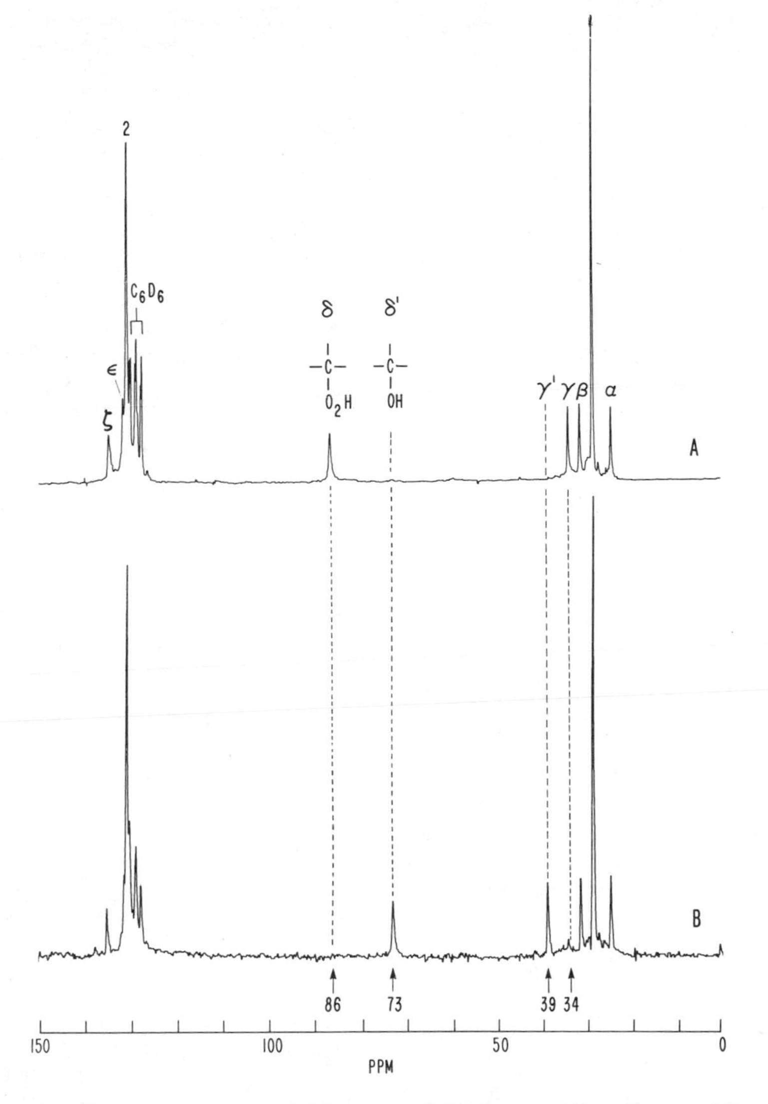

Figure 5. ^{13}C *NMR spectra of* $^{1}O_2$*-reacted CB before (A) and after (B) reduction with* $NaBH_x$

provided by the ^{13}C NMR spectra of $^{1}O_2$-reacted TB, before and after reduction. These spectra (not shown here), in common with Figure 5, display a single δ peak at 86 ppm and a single δ' peak at 72 ppm. The conclusion reached here concerning the trans nature of the shifted double bonds agrees with the fact that the singlet oxygenation of simple low molecular weight olefins is known to yield almost exclusively trans-shifted double bonds (*14*).

Finally, it should be noted that the cis/trans ratio in the $^1O_2$-reacted CB cannot be calculated in the ordinary way (*15*) since the new *trans* —CH=CH— units, being alpha to the hydroperoxide groups, exhibit modified (and as yet undetermined) extinction coefficients at 10.3 μ.

In summary, this study has demonstrated that singlet oxygenation of *cis*- and *trans*-1,4-polybutadienes follows an "ene"-type reaction yielding allylic hydroperoxides in 1:1 correspondence with new shifted double bonds, and the latter are essentially all-trans. The ratio of relative rates for the $^1O_2$ reaction with CB and TB ($\sim$6:1) is approximately the same as that for *cis*- and *trans*-4-methyl-2-pentene. Evidently there are no fundamental differences in singlet oxygenation mechanism between the unsaturated polymers, CB and TB, and their low molecular weight counterparts.

Literature Cited

1. Kaplan, M. L., Kelleher, P. G., *Science* (1970) **169,** 1206.
2. Kaplan, M. L., Kelleher, P. G., *J. Polym. Sci. Part A-1* (1970) **8,** 3163.
3. Kaplan, M. L., Kelleher, P. G., *Rubber Chem. Technol.* (1971) **44,** 642.
4. Kaplan, M. L., Kelleher, P. G., *Rubber Chem. Technol.* (1972) **45,** 423.
5. Breck, A. K., Taylor, C. L., Russell, K. E., Wan, J. K. S., *J. Polym. Sci. Polym. Chem. Ed.* (1974) **12,** 1505.
6. Rabek, J. F., Rånby, B., *J. Polym. Sci. Polym. Chem. Ed.* (1976) **14,** 1463.
7. Kearns, D. R., *Chem. Rev.* (1971) **71,** 395.
8. Foote, C. S., *Pure Appl. Chem.* (1971) **27,** 635.
9. Golub, M. A., Rosenberg, M. L., Gemmer, R. V., *Amer. Chem. Soc., Div. Polym. Chem., Prepr.* **17,** 699 (San Francisco, September, 1976).
10. Gemmer, R. V., Golub, M. A., *Amer. Chem. Soc., Div. Polym. Chem., Prepr.* **17,** 676 (San Francisco, September, 1976).
11. Mendenhall, G. D., personal communication, July, 1976.
12. Higgins, R., Foote, C. S., Cheng, H., ADV. CHEM. SER. (1968) **77,** 102.
13. Stothers, J. B., "Carbon-13 NMR Spectroscopy, p. 188, Academic, New York, 1972.
14. Gollnick, K., ADV. CHEM. SER. (1968) **77,** 78.
15. Golub, M. A., *J. Polym. Sci., Polym. Ltt. Ed.* (1974) **12,** 295.

RECEIVED May 12, 1977. R. V. Gemmer was a National Research Council-National Aeronautics and Space Administration Resident Research Associate, 1975-77. M. L. Rosenberg is a Research Associate, San Jose State University, San Jose, CA 95192.

3

Low-Temperature Chemiluminescence from *cis*-1,4-Polybutadiene, 1,2-Polybutadiene, and *trans*-Polypentenamer at Temperatures near Ambient

RICHARD A. NATHAN, G. DAVID MENDENHALL,
MICHELLE A. BIRTS, and CRAIG A. OGLE

Battelle-Columbus Laboratories, Columbus, OH 43201

MORTON A. GOLUB

Ames Research Center, NASA, Moffett Field, CA 94035

The chemiluminescence emission at 25–60°C was measured from films of cis-*1,4-polybutadiene, 1,2-polybutadiene, and* trans-*polypentenamer. The polymers were autoxidized previously in air 100°C, or allowed to react with singlet molecular oxygen in solution, and then cast into films. Values of β (or $k_d(^1O_2 \rightarrow {}^3O_2)/k_r(^1O_2 + \text{polymer} \rightarrow \text{products})$) were determined in benzene for* cis-*1,4-polybutadiene and* cis-*1,4-polyisoprene, and for model compounds* cis-*3-hexene and* cis-*3-methyl-3-hexene by independent methods. The chemiluminescence emission from irradiated films of the polymers containing a dye sensitizer showed a complicated time dependence, and the results depended on the length of irradiation.*

The light associated with combustion is a commonplace phenomenon, and the luminescence from cool flames in the gas phase is well known to specialists in the oxidation field. A much weaker light emission is associated with the autoxidation process in organic materials and can be observed frequently in samples at temperatures near ambient with sufficiently sensitive photomultipliers. In terms of efficiency of production, this emission does not comprise a significant part of the oxidation, but the chemiluminescence reflects degradation rates in materials that are often too slow to measure easily by more conventional methods.

0-8412-0381-4/78/33-169-019$05.00/1

Recently, we reported on the weak chemiluminescence emitted in the autoxidation of thin films of *cis*-1,4-polyisoprene and also from a film of *cis*-1,4-polyisoprene containing hydroperoxide groups that were introduced through singlet oxygenation (*1*). Activation energies of 20–25 kcal/mol were obtained for chemiluminescence emission over the temperature range 25–66°C from both partially autoxidized and hydroperoxidized samples of *cis*-1,4-polyisoprene. These values were taken to represent the activation energy for decomposition of hydroperoxide groups in either of those polymer systems. The weak emission itself was thought to arise from electronically excited species (carbonyl-containing structures) formed in termination reactions involving peroxy radicals (*2–6*), whether generated in the autoxidation or by hydroperoxide decomposition. This assumption is consistent with the thermodynamics of the autoxidation process steps and with kinetic evidence.

Following Ashby's first observation (*7*) of chemiluminescence from the oxidative degradation of polymers, a number of papers have appeared dealing with oxidative chemiluminescence from a variety of polymers (*8–16*). In this chapter we continue the 1,4-polyisoprene work with a study of the low-temperature chemiluminescence emitted in the autoxidation of three additional elastomers, *cis*-1,4-polybutadiene, amorphous 1,2-polybutadiene, and *trans*-polypentenamer. We also report the chemiluminescence obtained from singlet-oxygenated samples of *cis*-1,4-polybutadiene and *trans*-polypentenamer, as well as rate data for singlet oxygen reactions with the 1,4-polyisoprene, 1,4-polybutadiene, and model compounds in solution.

Experimental

The B. F. Goodrich Research and Development Center provided the *cis*-1,4-polybutadiene; the 1,2-polybutadiene was obtained from the Institut Francais du Pétrole; and the *trans*-polypentenamer was obtained from the Farbenfabriken Bayer GmbH. These polymers were purified by three reprecipitations from benzene solution with methanol as precipitant. The solvents were reagent grade and were degassed by flushing with argon before use. Operations with the solutions were carried out in an argon or nitrogen atmosphere, and the final benzene stock solutions ($\sim$ 15 g/l) were stored in the dark under positive argon pressure. Polymer films for autoxidation were prepared by adding portions of the stock solutions to petri dishes (9.5-cm diameter) and allowing the solvent to evaporate. The weights of the resulting films were determined, and they were used immediately for chemiluminescence experiments.

Triphenyl phosphite ozonide was prepared from base-washed, redistilled triphenyl phosphite (Eastman) and ozone in redistilled fluorotrichloromethane at -78°C. Cold aliquots of the ozonide (1.0 mL) were injected into solutions of the other reactants at room temperature in 10 mL of benzene containing 0.30 mL of 1:1 v/v pyridine–methanol catalyst

(*17*). The consumption of rubrene (Eastman, recrystallized from benzene–methanol) in solution was followed by the decrease in optical density at 492.5 nm. The photosensitizers, dibenzanthrone (K & K Laboratories) and *meso*-tetraphenylporphine (Strem Chemicals), were used as received. The olefins were pure (> 98%) commercial samples and were passed through a short column of aluminum before use.

The irradiations for the competitive photosensitized oxidation (method (*b*), *see* below) were conducted wtih light from a 150-watt Eimac high-pressure xenon lamp in conjunction wtih a Bausch and Lomb monochromator centered at 580 nm. The exit slit width was set at 3.5 mm. Under these conditions virtually all of the light was absorbed by the dibenzanthrone.

The photosensitized oxidations in solution for determination of β by the slope–intercept method were carried out in a constant temperature bath with light from a 250-watt projector lamp with a UV-cutoff filter. Oxygen uptake was monitored with a conventional mercury burette. Polymer films also were irradiated with the same lamp containing a Corning CS 3–69 filter.

The apparatus and procedures for carrying out the chemiluminescence measurements, as well as two techniques for autoxidation in air at 100°C and for singlet oxygenation, were otherwise similar to those described previously (*1, 2*). The weights of the autoxidized and hydroperoxidized films varied between 0.1 and 0.25 g.

Results and Discussion

Low-Temperature Chemiluminescence from Polymers. The activation energies for chemiluminescence observed at 25–60°C from various autoxidized and singlet-oxygenated polymers are recorded in Table I. Exemplary Arrhenius plots for the chemiluminescence from two polymers are shown in Figure 1. The E_a values range from a low of 9.6 kcal/mol for a sample of singlet-oxygenated *trans*-polypentenamer to a high of about 32 kcal/mol for a sample of autoxidized 1,2-polybutadiene. The latter sample, however, also gave a lower value of 23 ± 2 kcal/mol from a second experiment, so that the initial high value may have been an artifact caused by morphological changes in the film on warming. Elucidation of this point will require additional work. However, reproducible results were obtained from the other oxidized polymer samples. The oxygen content of the autoxidized samples varied 2–17%; as expected, 1,2-polybutadiene autoxidized at the slowest rate .

Except for the autoxidized 1,2-polybutadiene values, all of the E_a values in Table I were equal to or less than those found earlier for autoxidized or singlet-oxygenated *cis*-1,4-polyisoprene (*1*).

The data in Table I reveal a trend towards higher E_a values with increasing extent of singlet oxygenation or time of autoxidation. A possible interpretation of this observation may be sought in a consideration of the autoxidation–chemiluminescence mechanism (*1–17*). Since the rate of free-radical initiation equals the rate of termination, and if $d(h\nu)/dt = \Phi R_{term} = \Phi R_{init}$ where Φ is the quantum yield for light emission (per termination event), which is assumed to be nearly temperature-independent (2), it follows that: $d(h\nu)/dt = \Phi[\text{initiator}] \cdot k_{init} = \Phi[\text{initiator}] \cdot A_{init}e^{-E_{init}/RT}$.

In our system, E_a is approximately E_{init}. An increase in E_{init} with extent of oxidation may correspond to a change in mode of initiation to one with higher activation energy. This is reasonable, since the concentrations of initiating species (principally peroxides and hydroperoxides) are expected to increase, at least initially, in the samples during the course of the oxidation. The nature of the medium changes also during the oxidations. The influence of the medium in determining decomposition rates and E_a's of hydroperoxides is well established (*18*).

Photosensitized Oxidation of Polymer Films. In the singlet-oxygenated polymer samples prepared in solution, the oxygen is expected to be present only as hydroperoxide groups. The increase in E_a with an increase in oxygen uptake in singlet-oxygenated films suggests that a change in mechanism from induced-unimolecular ($ROOH + M \rightarrow$ radicals) to a bimolecular ($2ROOH \rightarrow RO\cdot + RO_2\cdot + H_2O$) mode of initiation may account for this trend. Since the intensity of chemiluminescence from films cast from singlet-oxygenated polymer solutions was not very

Table I. Activation Energies for Chemiluminescence from Autoxidized[a] and Singlet-Oxygenated Polymers

Polymer	*Autoxidized (A, min) or Singlet-Oxygenated (S)*	*O_2 per 100 Monomer Units*	*Percent Oxygen in Sample*[b]	*E_a, kcal/mol*
cis-1,4-Polyisoprene	A, 65	—	28.4	24 ± 3[c]; 23 ± 2[c]
	S	8.93	4.0	23.7 ± 2
	S	18.9	8.1	24.9
cis-1,4-Polybutadiene	A, 22	—	—	18 ± 2
	A, 40	—	6.6	21 ± 2
	S	5.46	3.1	11.5
	S	6.67	3.8	16 ± 1
	S (another film)	6.67	3.8	15 ± 1
1,2-Polybutadiene	A, 74	—	2.3	32 ± 4; 23 ± 2
trans-Polypentenamer	A	—	17.1	20 ± 1
	S	1.89	0.9	9.6 ± 2
	S	4.55	2.1	16 ± 2

[a] In air at 100°C.
[b] A, from combustion analysis; S, calculated from O_2 uptake.
[c] Previous work (*1*).

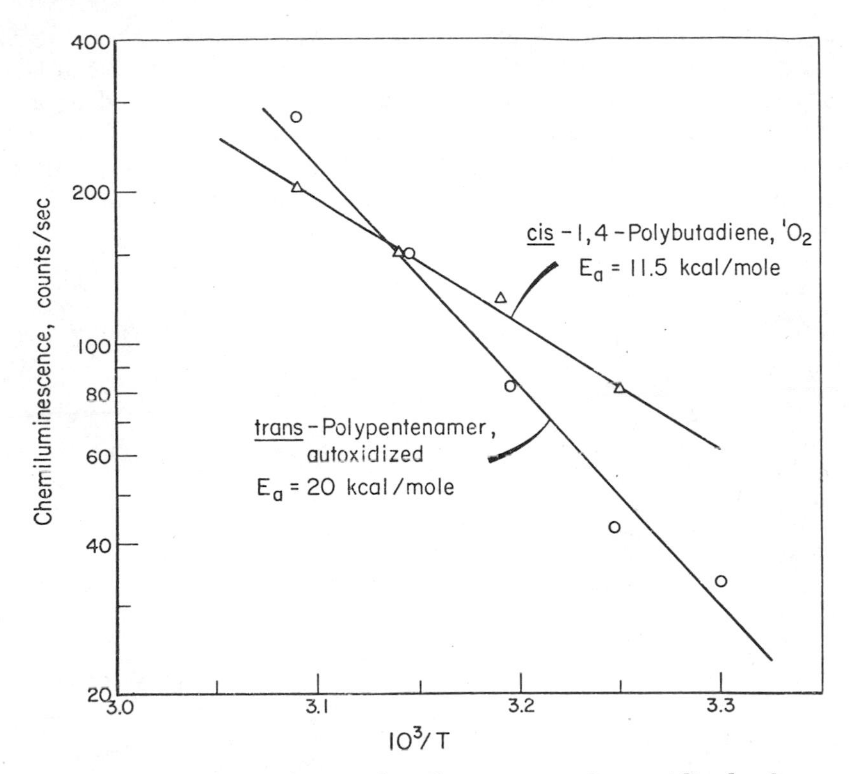

Figure 1. Arrhenius plots of chemiluminescence from oxidized polymer films (cf. Table I)

reproducible, we attempted to photooxidize polymer films to different extents and measure both chemiluminescence emission and hydroperoxide content in individual films. Since the photosensitized production of singlet oxygen is expected to occur at a constant rate under uniform illumination, an induced-unimolecular mechanism should show chemiluminescence directly proportional to the time of irradiation. For a bimolecular mode of initiation the chemiluminescence should increase as the square of the irradiation time.

Films of *cis*-1,4-polyisoprene, *cis*-1,4-polybutadiene, or *trans*-polypentenamer containing dissolved tetraphenylporphine or dibenzanthrone were prepared from solutions by evaporation and irradiated with filtered incandescent light. For all samples studied, the chemiluminescence after a short time of irradiation (< 60 sec) showed a rapid decrease in intensity to the background level after a few minutes. Longer irradiation times that were sufficient to show changes in the ir spectra of the polymers gave samples with greater chemiluminescence emission, but the emission rates were not stable and decreased, or increased and then decreased with time. Typical results for a film of *trans*-polypentenamer are shown in Figure 2. Similar results were obtained from *cis*-1,4-polybutadiene films containing dibenzanthrone that were irradiated and examined at 0°C. Fresh polymer films with or without sensitizers gave no appreciable chemiluminescence at room temperature, and a film without sensitizer did not show chemiluminescence under the same conditions after 10-sec irradiation with unfiltered incandescent light.

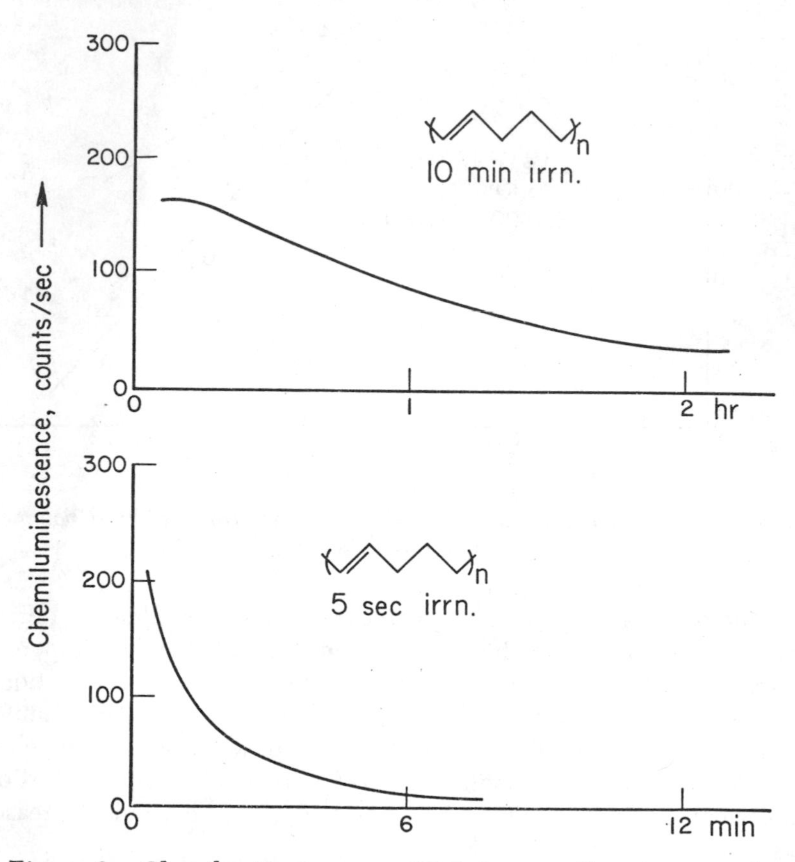

Figure 2. Chemiluminescence at 25°C from a film of trans-*poly-pentenamer (0.0152 g) containing 5.0 × 10^{-5} g of dibenzanthrone after different irradiation times*

Although we cannot give a detailed interpretation of these results at the present, the short-lived chemiluminescence seen after short-term exposure of polymer–sensitizer films suggested that we were observing the decay of free radicals formed during the irradiation process. The ESR spectrum of a benzene solution containing *trans*-polypentenamer and dibenzanthrone and a film prepared from the solution, both displayed a weak, unresolved signal consistent with a peroxy radical, but in neither case did it increase upon irradiation with visible light.

The longer-lived chemiluminescence seen in films irradiated for longer periods (Figure 2) is ascribed to thermal decomposition of a second labile species. Simple first-order kinetics are not observed in either decay curve in Figure 2.

After 10-min irradiation, the ir spectrum of the *trans*-polypentenamer film showed the expected weak 2.9 μ peak (O–H) without the 5.8 μ absorption (C=O), by analogy to $^1O_2$-reacted 1,4-polybutadiene (*20*). Only minor changes in the spectrum were noted after decay of the chemiluminescence. In contrast to this result, the chemiluminescence from polymer films oxidized at 100°C or singlet-oxygenated in solution and cast into films usually was invariant at 25°C for several hours.

β-Values for Polymer and Model Compounds. 1,4-Polyisoprene and 1,4-polybutadiene are so sufficiently reactive toward singlet oxygen that we can conveniently obtain β-values by three methods: (*a*) from oxygen-uptake rates by photosensitized oxidation at different polymer concentrations; (*b*) by the initial disappearance rate of rubrene on photooxygenation of solutions with and without polymer (or model olefin); and (*c*) for 1,4-polyisoprene, by the amount of rubrene consumed upon addition of aliquots of triphenyl phosphite ozonide in the presence and absence of olefin or polymer. The treatment of data from methods (*b*) and (*c*) was modified to give β-values directly from the following equations:
For method (*b*),

$$\beta = \frac{k_d[A]}{k_r[R]_o + k_d[(m/m_A) - 1]}$$

$$= \frac{[A]}{(k_r/k_d)[R]_o + (m/m_A) - 1}$$

In these equations, $[A]$ and $[R]_o$ are the respective initial concentrations of polymer (or olefin) and rubrene, k_r and k_d are the respective rate constants for reaction of rubrene with singlet molecular oxygen and for the unimolecular decay of the latter to triplet oxygen. The quantities m and m_A are the initial rates of rubrene consumption in the absence and presence of A, respectively. Under our experimental conditions the first term in the denominator is small, so we have:

$$\beta \simeq \frac{[A]}{(m/m_A) - 1}$$

For method (*c*) we derive:

$$\beta = \frac{k_d}{k_a} = \frac{k_r\left[\frac{[PO_3]_o - [R]_o + [R]_T}{\ln [R]_o/[R]_T}\right] - \frac{k_d}{k_r}}{\frac{k_r}{[A]}\left[\frac{[PO_3]_o - [R]_o + [R]_T{}^A}{\ln [R]_o/[R]_T{}^A}\right] - \frac{k_d}{k_r}}$$

where k_a refers to the rate constant for reaction of A with $^1O_2$. Under our experimental conditions this reduces to:

$$\beta \simeq [A]\left[\frac{\ln ([R]_o/[R]_T{}^A)}{\ln ([R]_o/[R]_T)}\right]$$

In these equations $[PO_3]_o$ is the initial concentration of triphenyl phosphite ozonide in the reacting solution, $[R]_T{}^A$ and $[R]_T$ are the final concentrations of rubrene in the solutions with and without olefin acceptor, respectively. The other symbols have the same significance as indicated earlier.

Table II. Values of β for *cis*-1,4-Polybutadiene and *cis*-1,4-Polyisoprene

Competitive Photooxidation Method[a]

Hydrocarbon, M	*Rubrene* *10*^{5}M	$-$m, *10*^{5}M *sec*$^{-1}$	$-$m$_A$, *10*^{5}M *sec*$^{-1}$	β, M
cis-1,4-Polyisoprene				
.144	4.13	8.7	4.6	.16
.086	4.13	8.2	6.1	.25
cis-3-Methyl-3-hexene				
.164	4.13	9.1	3.6	.11
.598	4.13	7.0	1.2	.12
cis-3 Hexene				
.536	4.13	7.9	4.9	.88
.592	4.13	19.8	8.1	.41
cis-1,4-Polybutadiene				
.825	4.13	20.5	15.4	2.50
.660	4.13	23.3	17.7	2.09

Phosphite Ozonide Method[b]

Hydrocarbon, M	R_o, *10*^{5}M	R_t, *10*^{6}M	$R_T{}^A$, *10*^{6}M	β, M
cis-1,4-Polyisoprene				
0.127	8.89	6.3	10.6	0.10
cis-3-Methyl-3-hexene				
0.218	8.89	6.3	10.3	0.18

[a] Dibenzanthrone sensitizer ($1.53 \times 10^{-4}M$).
[b] Initial ozonide concentration $2.28 \times 10^{-3}M$.

Values of β determined by the various techniques are summarized in Table II. A plot of the oxygen-uptake data and β-values obtained by the slope–intercept method (*21*) appear in Figure 3. The reactivity of *cis*-3-hexane ($\beta \simeq 0.6 \pm 0.2M$) was found to be about a factor of four less than that of *cis*-1,4-polybutadiene ($\beta \simeq 2.3 \pm 0.2M$), whereas 1,4-polyisoprene ($\beta \simeq 0.16 \pm 0.05M$) is almost comparable in reactivity to its model olefin 3-methyl-3-hexene ($\beta \simeq 0.13 \pm 0.04M$).

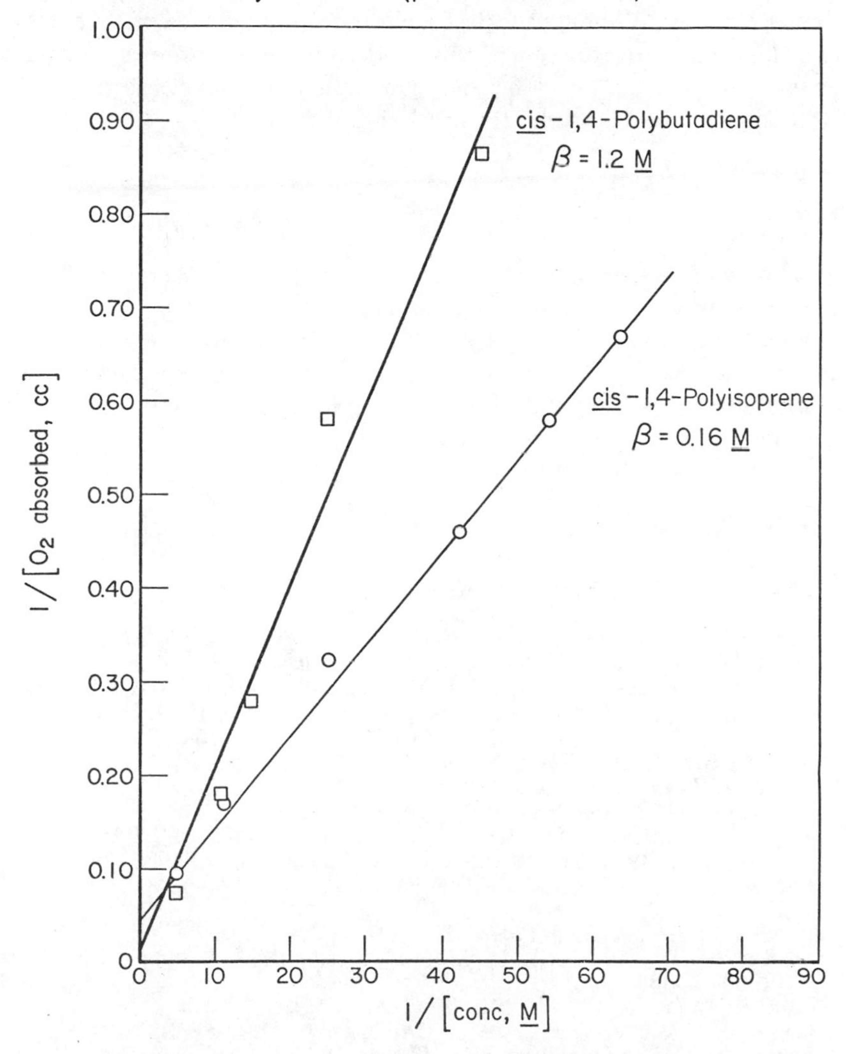

Figure 3. Double reciprocal plot for determination of polymer β-values (benzene, 6°C, 4 × 10⁻⁶M porphyrin sensitizer). For 1,4-polybutadiene, ordinate calculated from oxygen absorbed in 37 mL of solution in 95 min. For 1,4-polyisoprene, ordinate from oxygen absorbed in 40 mL of soltuion in 45 min.

Thus, the differences between polymer and model olefin reactivities are small. This is reasonable because the small size of the singlet oxygen molecule should allow it to interact with the olefin groups in a polymer without appreciable steric interference from the rest of the polymer chain.

The variability in chemiluminescence from the photooxidation of solid polymer films prevented measurements of β-values in solid media, where the results would be most pertinent to the general question of polymer stability. We might expect a close correspondence of β-values to those determined in solution in this work, since the lifetimes of singlet oxygen in benzene and cyclohexane, which should bracket the properties of the polymers in this study, are nearly identical (*22*).

Acknowledgment

This work was supported in part by the National Aeronautics and Space Administration under Contract No. NAS2-8195 and by a group program for chemiluminescence studies at Battelle supported by several industrial companies. We wish to thank Frank Huber for carrying out the combustion analyses and Miles Chedekel for assistance with the EPR experiments.

Literature Cited

1. Mendenhall, G. D., Nathan, R. A., Golub, M. A., *Polym. Prepr., Am. Chem. Soc., Div. Polym.* (1976) **17**, 726.
2. Vassil'ev, R. F., in "Progress in Reaction Kinetics," Vol. 4, pp. 305–352.
3. Shlyapintokh, V. Ya., Karpukhin, O. N., Postnikov, L. M., Tsepalov, C. F., Vichutinskii, A. A., Zakharov, I. V., "Chemiluminescence Techniques in Chemical Reactions," Consultants Bureau, New York, 1968.
4. Vassil'ev, R. F., *Russ. Chem. Rev. (Eng. Transl.)* (1970) **39**, 529.
5. Kellogg, R. E., *J. Am. Chem. Soc.* (1969) **91**, 5433.
6. Beutel, J., *J. Am. Chem. Soc.* (1971) **93**, 2615.
7. Ashby, G. E., *J. Polym. Sci.* (1961) **50**, 99.
8. Stauff, J., Schmidkunz, H., Hartmann, G., *Nature* (1963) **198**, 281.
9. Schard, M. P., Russell, C. A., *J. Appl. Polym. Sci.* (1964) **9**, 985, 997.
10. Barker, Jr., R. E., Daane, J. H., Rentzepis, P. M., *J. Polym. Sci., A* (1965) **3**, 2033.
11. Reich, L., Stivala, S. S., *J. Polym. Sci.* (1965) **3**, 4299.
12. de Kock, R. J., Hol, P. A. H. M., *Int. Synth. Rubber Symp., Lect., 4th* (1969) **2**, 53.
13. de Kock, R. J., Hol, P. A. H. M., *Chem. Abstr.* (1971) **74**, 4425.
14. Isacsson, U., Wettermark, G., *Anal. Chim. Acta* (1974) **68**, 339.
15. Pokholok, T. V., Karpukhin, O. N., Shlyapintokh, V. Ya., *J. Polym. Sci., A1* (1975) **13**, 525.
16. Mendenhall, G. D., *Angew. Chem., Int. Ed. Engl.* (1977) **16**, 225.
17. Mendenhall, G. D., Barlett, P. D., unpublished results.
18. Hiatt, R., in "Organic Peroxides," Chapter 1, D. Swern, Ed., Wiley-Interscience, N.Y., 1971.

19. Golub, M. A., Gemmer, R. V., Rosenberg, M. L., ADV. CHEM. SER. (1978) **169**, 11.
20. Carlsson, D. J., Mendenhall, G. D., Suprunchuk, T., Wiles, D. M., *J. Chem. Amer. Soc.* (1972) **94**, 8060.
21. Higgins, R., Foote, C. S., Cheng, H., ADV. CHEM. SER. (1968) **77**, 102.
22. Merkel, P. B., Kearns, D. R., *J. Amer. Chem. Soc.* (1972) **94**, 7244.

RECEIVED May 12, 1977.

4

The Effect of Thermal Processing History on the Photooxidation of Polymers

GERALD SCOTT

University of Aston in Birmingham, Gosta Green,
Birmingham, B4 7ET, England

The processing of several polymers involves an initial mechanochemical stage resulting in damage to the polymer chain and sensitization to subsequent photooxidation. In the case of PVC, unsaturation and peroxides are produced during the first minute of processing and a tin stabilizer that acts by retarding the formation of peroxides and unsaturation is an effective photostabilizer. Polyolefins are stabilized against UV light by the removal of peroxides produced during processing. The latter are effective photosensitizers in combination with unsaturation present initially in the polymer; a similar effect can be achieved by removing peroxides during processing by means of a peroxide-decomposing antioxidant (zinc dialkyldithiocarbamate). The latter interacts synergistically with a benzophenone UV absorber.

There is a substantial amount of evidence suggesting that the photoinitiating groups responsible for the rapid photooxidation of most of the thermoplastic polymers are oxygen-containing species introduced either during the manufacturing process or during conversion of the polymers to fabricated products. Unsaturated groups frequently formed during the polymerization process are not chromophores but appear to be implicated closely in the formation of hydroperoxides and derived-carbonyl compounds by adventitious contact of the polymer with oxygen (*1, 2*). Although oxidation is possible, during manufacturing (e.g., during the drying of emulsion-polymerized polymers) the most favorable conditions are during the processing operation, since this involves not only high temperatures and the presence of air but also conditions of high shear during the melting of the polymer. The latter leads to mechanical breakage of the polymer chain yielding alkyl radicals (*3*).

0-8412-0381-4/78/33-160-030$06.50/1

Mechanooxidation of Polymers During Processing

Mechanooxidation of rubber to produce a product of lower viscosity has been practiced by the rubber industry as part of the polymer-fabrication process since the origins of the industry. Without such treatment it would be impossible to incorporate compounding ingredients which are essential to the final performance of the fabricated rubber products.

The pioneering work of Busse (*4*) showed the sensitivity of natural rubber mastication to temperature. The rate of mechanooxidative degradation was high at low temperatures (cold mastication) and decreased to a minimum at about 100°C. The rate of molecular weight decrease was found to increase rapidly above 100°C. The low-temperature stage is dominated primarily by the mechanical shear of the polymer (*5, 6*). As the plasticity of the rubber increases, the rate of this process decreases and becomes minimal at 100°C. Above 100°C thermal oxidative degradation initiated by peroxides introduced by mechanochemical chain scission becomes the dominant degradation mechanism. Ceresa and Watson (*7*) extended these studies to plastics. They showed that most polymers, particularly in plasticized form, undergo a rapid reduction in molecular weight on cold mastication and that the macroradicals produced by shear initiate conventional free radical processes. In the presence of air the macroradicals are converted rapidly to alkylperoxy radicals (*3, 8*).

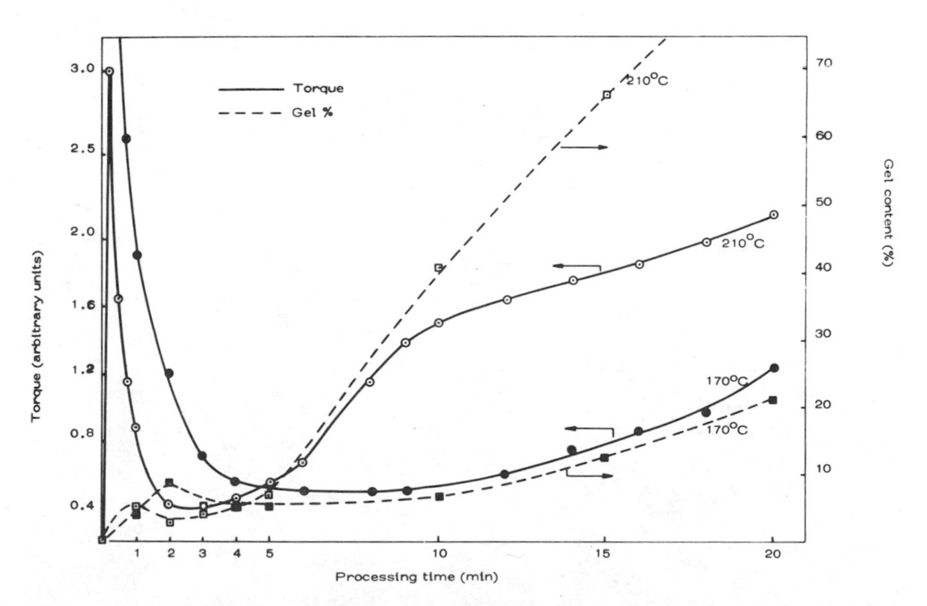

Figure 1. Relationship between the applied torque during the processing of PVC and the formation of solvent-insoluble gel

Mechanooxidation, Thermal Oxidation, and Photooxidation of Poly(vinyl chloride) (PVC)

The change in applied torque with time during the processing of PVC in a typical mixer (RAPRA torque rheometer) at two temperatures is shown in Figure 1. Oxygen ingress was excluded formally by means of a hydraulic ram; the only oxygen present was that initially trapped in the polymer particles. Figure 1 also relates the formation of solvent-insoluble gel in the polymer to the applied torque in the mixer. Two distinct stages in the growth of both of these parameters can be observed.

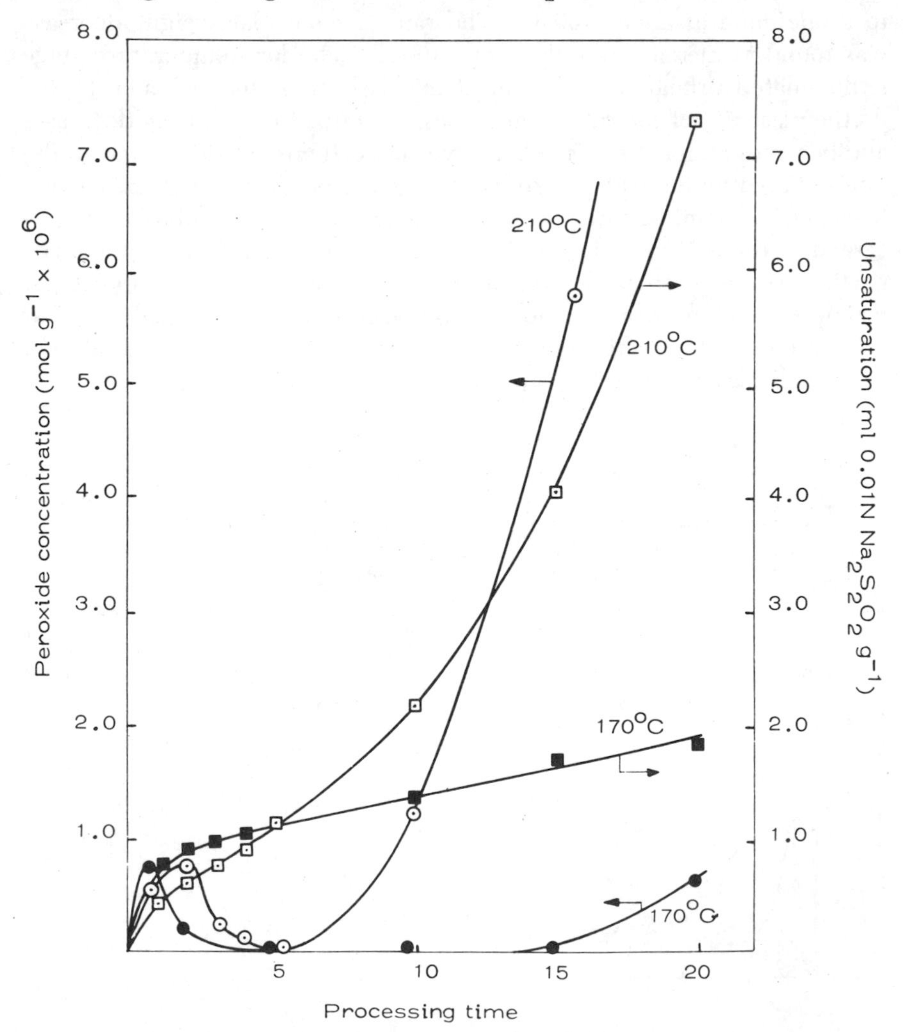

Figure 2. Development of peroxide and olefinic unsaturation with time during the processing of PVC

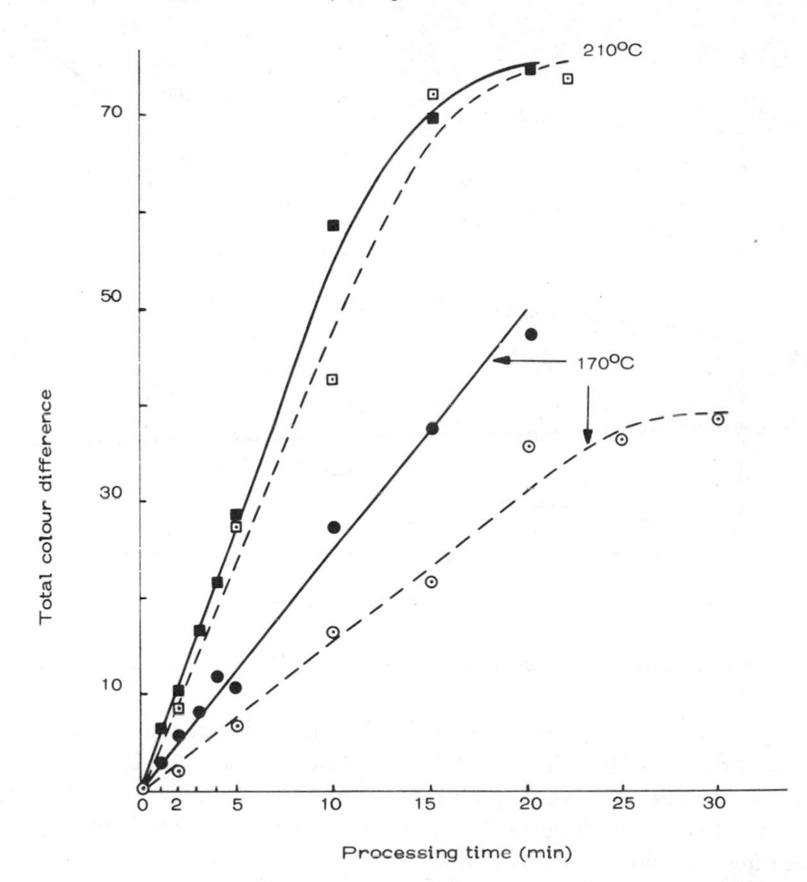

Figure 3. Effect of processing on color formation in PVC: (——) without additive; (- - -) with Wax E lubricant

Both reach a higher level in the first stage at 170°C than at 210°C and both show a faster increase rate in the second stage at 210°C than at 170°C. Both chemically determined unsaturation and peroxide increase very rapidly during the first minute of processing at both temperatures (Figure 2). The rate of growth of both species in the initial phase is higher at 170°C than at 210°C; but the reverse is true in the second stage. The rate of color development is linear at both temperatures, but is lower at 170°C than at 210°C (Figure 3). However, the conjugated species that are the precursors of visible color show a rapid rise during the first minute of processing, followed by an induction period before further increase (Figure 4).

It is clear from these results that the formation of ethylenic unsaturation, peroxides, and cross-linked gel are primary processes associated with the shear-induced mechanochemical reactions occuring during the first minute of processing. The lower the temperature, the higher the

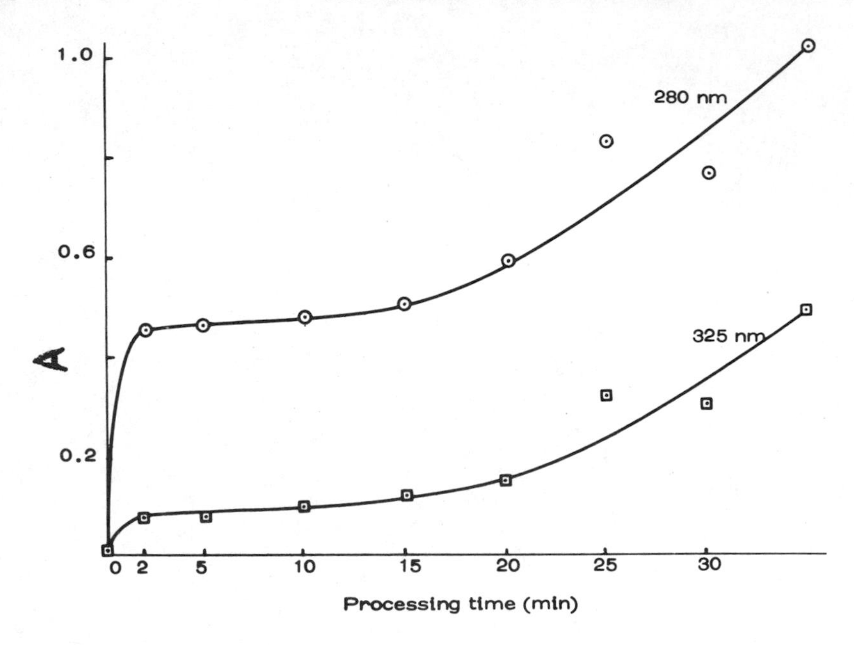

Figure 4. Effect of processing at 170°C on UV absorbance maxima

applied torque and the more severe the damage inflicted on the polymer. It is important to note that the level of unsaturation introduced by the mechanochemical process is much higher than that initially present in the commercial polymer (Figure 2). The latter is barely detectable by the techniques used in the present study.

The reactions occuring during the initial stage involve the scission of the polymer chain by mechanical shear (Scheme 1). The radicals

Scheme 1. Reactions resulting from mechanoscission of the PVC chain

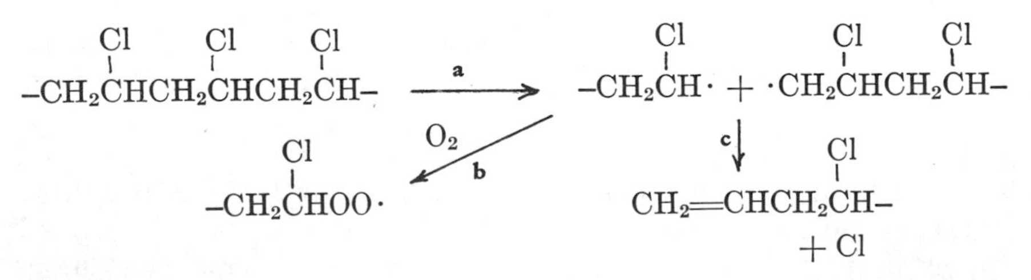

formed originally can undergo two alternative reactions; they can either react with available oxygen to give alkylperoxy radicals (Scheme 1(b)) and ultimately hydroperoxides, or they can form the site for dehydrochlorination by loss of a chlorine atom (Scheme 1(c)). The chlorine atom or the radicals formed from the peroxide species may initiate the zipper reaction leading to conjugated unsaturation (Scheme 2).

Scheme 2. Reactions of allylic groups in PVC

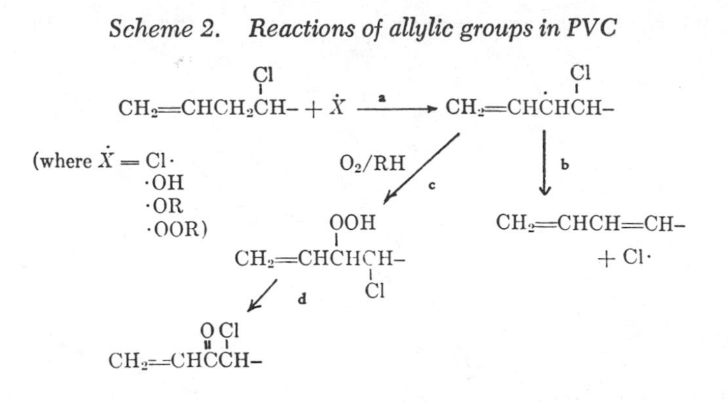

In the presence of residual oxygen, attack on the allylic radical may also occur (Scheme 2(c)) leading to the formation of conjugated carbonyl by thermolysis.

The reactions of vinyl compounds (particularly conjugated vinyl compounds) with oxygen to give alternating copolymers are well known (9) and could be responsible for the thermally unstable gel formed in the present instance (Scheme 3).

Scheme 3. Peroxy gel formation in PVC

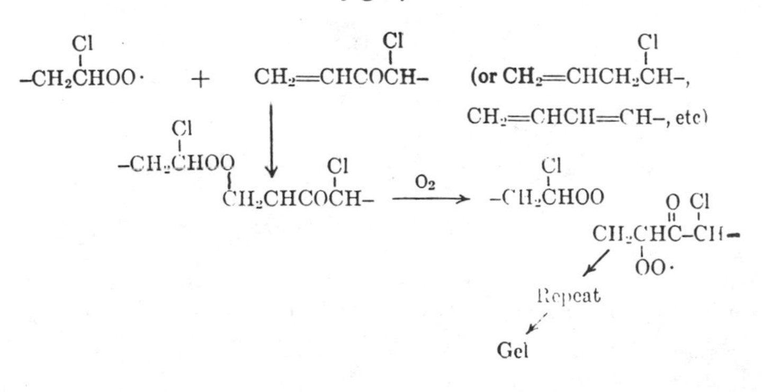

PVC films made by compression-molding the polymer after processing for increasing times at 170°C were exposed to UV light (S/B lamp; > 290 nm) and the rate of formation of carbonyl and hydroxyl in the polymer by ir also was measured. The results for hydroxyl growth are shown in Figure 5; the behavior of carbonyl growth was very similar. The effect of 1 min of processing is striking; longer processing times have relatively less effect. The activating effect is associated with the formation of both reactive allylic groups and peroxides. Both appear to be necessary to the photoactivating effect observed.

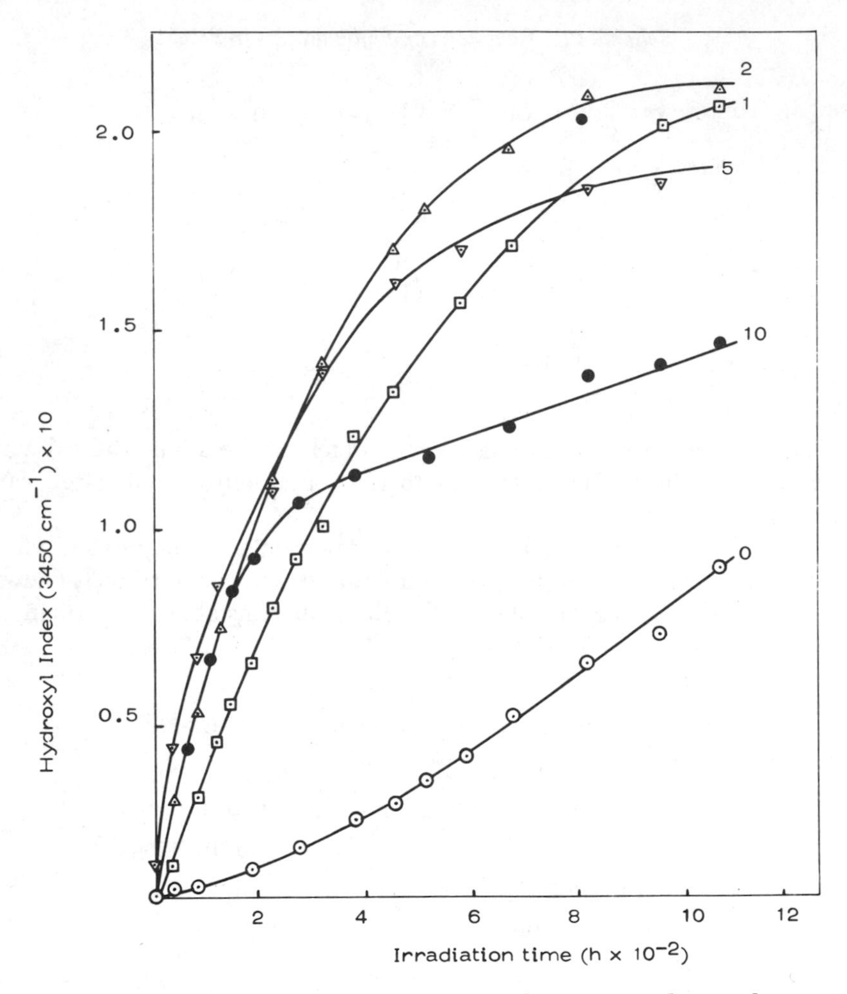

Figure 5. Effect of processing time on the rate of photooxidation (hydroxyl formation) of PVC (numbers on curves are processing times at 170°C)

Photooxidation of Rubber-Modified Polymers (10)

Rapid peroxide formation also occurs in rubber-modified polystyrene (HIPS) on processing (Figure 6). The unsaturation present initially was found to decrease rapidly with carbonyl increase as a secondary breakdown product. Similar behavior was observed when heating extruded films of HIPS at 98°C for longer time periods; but here the induction period to carbonyl and hydroxyl formation is much more evident (Figure 7). The effect of the prior heating process on the photooxidation of HIPS is very apparent from Figure 8 which shows that the autoaccelerating period evident in the control is removed effectively by the presence of peroxide in the polymer. The reduced extent of photooxidation in the more extensively oxidized samples appears to be associated with the depletion of the unsaturation. Figures 7 and 8 illustrate very clearly the dependence of the photooxidation rate on the hydroperoxide concentration rather than on carbonyl concentration. The former reaches a maximum in about 30 hr whereas the latter shows an induction period of about 20 hr. Before this time, the photooxidation autoaccelerating period has been almost eliminated owing to the increasing peroxide concentration. The evidence suggests that thermally produced peroxides in conjunction with initial unsaturation present in the polymer from the beginning are responsible for the extreme photosensitivity of processed HIPS.

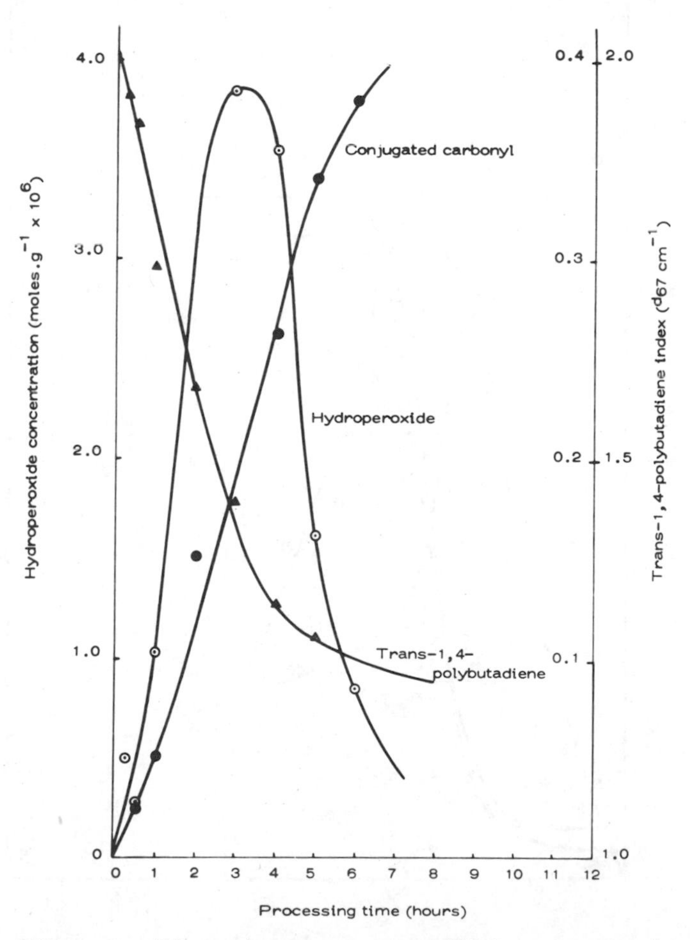

Figure 6. Effect of processing at 200°C on functional group formation in HIPS

Effect of Processing on the Polyolefins

The activating effect of prior processing on the UV stability of LDPE appears to be less than in the case of HDPE and PP (*11*). Nevertheless, both physical and chemical changes occur during the processing of the former.

In the presence of excess air, the melt viscosity of the polymer (as measured by melt-flow index (MFI)) remained constant at 150°C (Figure 9), but carbonyl concentration increased rapidly after an induction period (Figure 10). When air was excluded by using a sealed mixer,

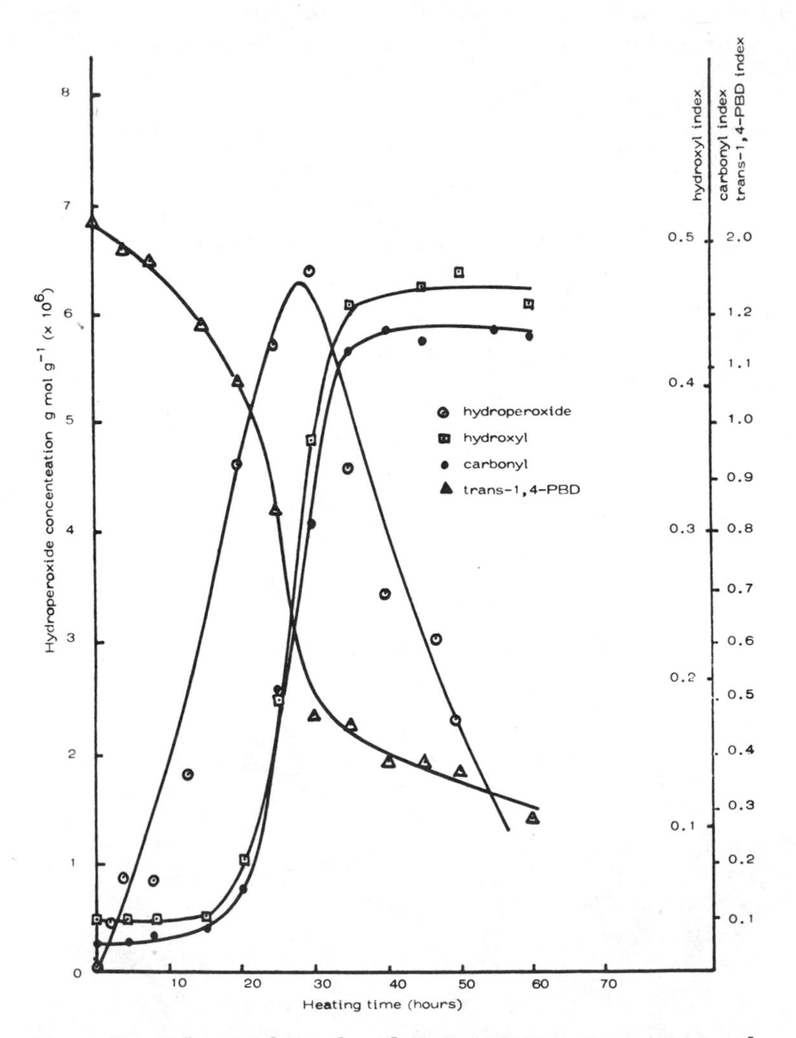

Figure 7. Effect of thermal oxidation at 98°C in air on functional group formation in HIPS

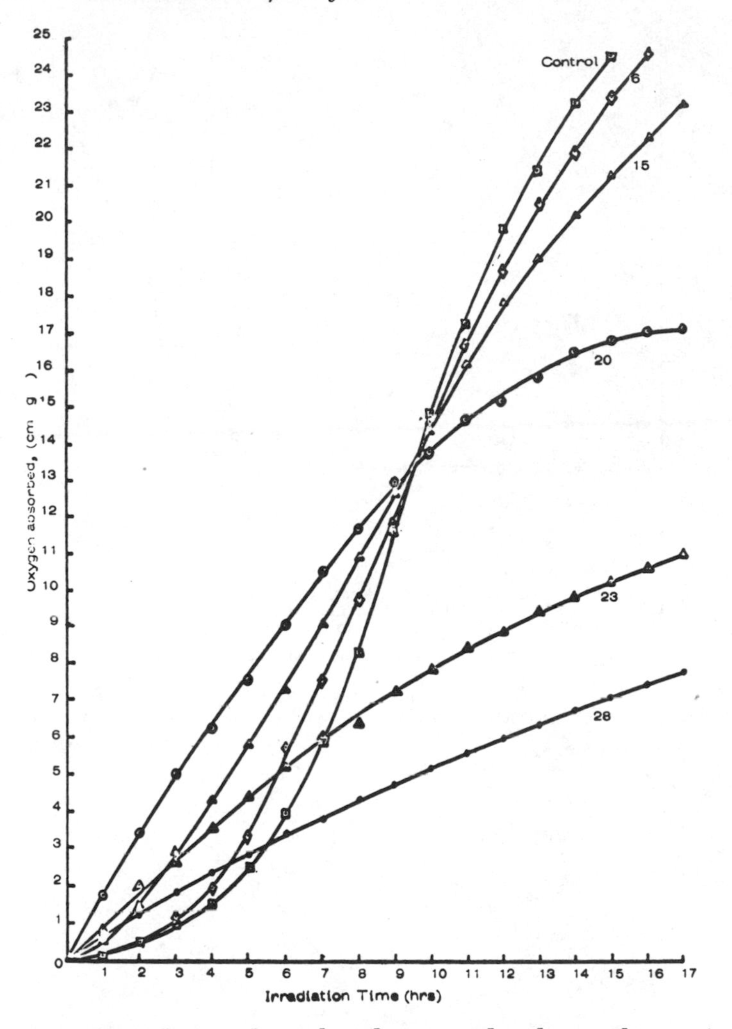

Figure 8. Effect of thermal oxidation on the photooxidation of HIPS (numbers on curves indicate the times for which the films were heated in air at 98°C)

there was an immediate decrease in MFI (Figure 9) but only a slight increase in carbonyl concentration (Figure 10). Prior displacement of air by argon eliminated all changes. The MFI changes were paralleled by similar changes in polymer molecular weight. In the presence of excess air there was a decrease in molecular weight after an induction period (Figure 11). In a deficiency of air (sealed mixer) the molecular weight increased with processing (Figure 12). In both cases hydroperoxides were formed in the polymer from the beginning of processing (Figure 13) but as expected, this reaction was restricted in the sealed mixer.

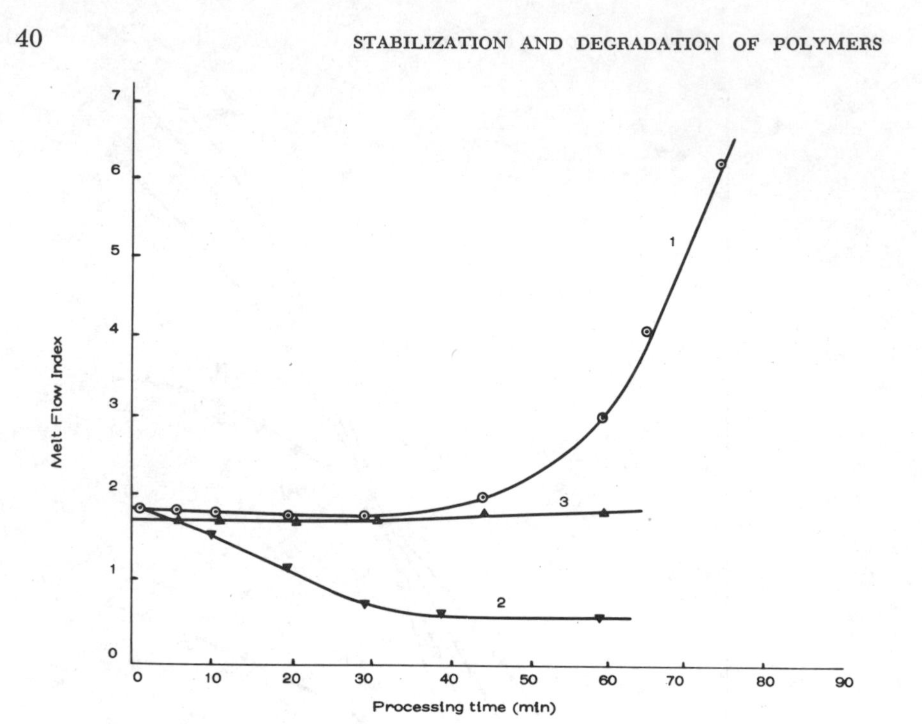

Figure 9. Effect of processing time on the melt-flow index (MFI) of LDPE. (1) open chamber (excess air); (2) closed chamber (limited oxygen access); (3) closed chamber purged with argon (inert atmosphere).

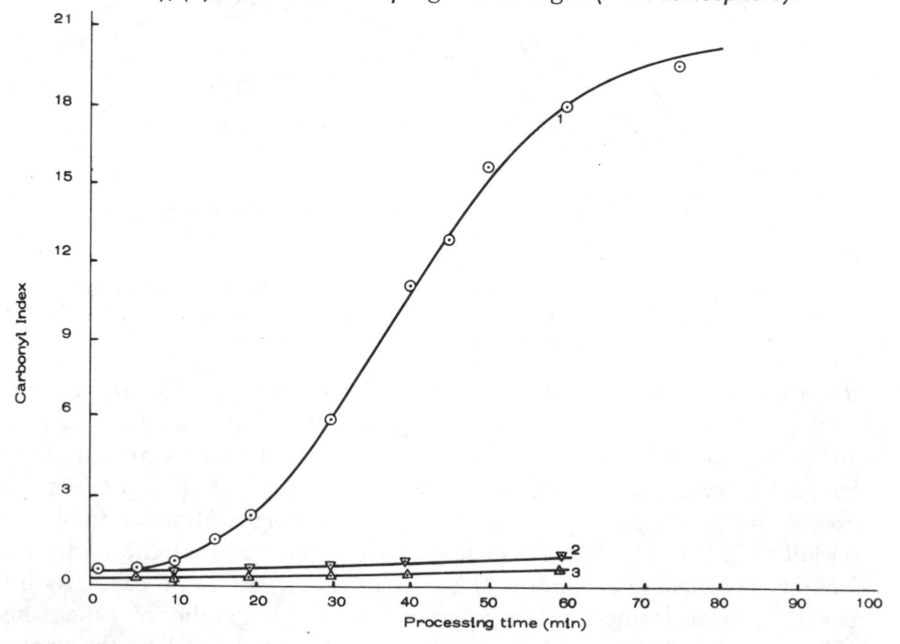

Figure 10. Effect of processing time on carbonyl formation in LDPE. (1) open chamber; (2) closed chamber; (3) closed chamber purged with argon.

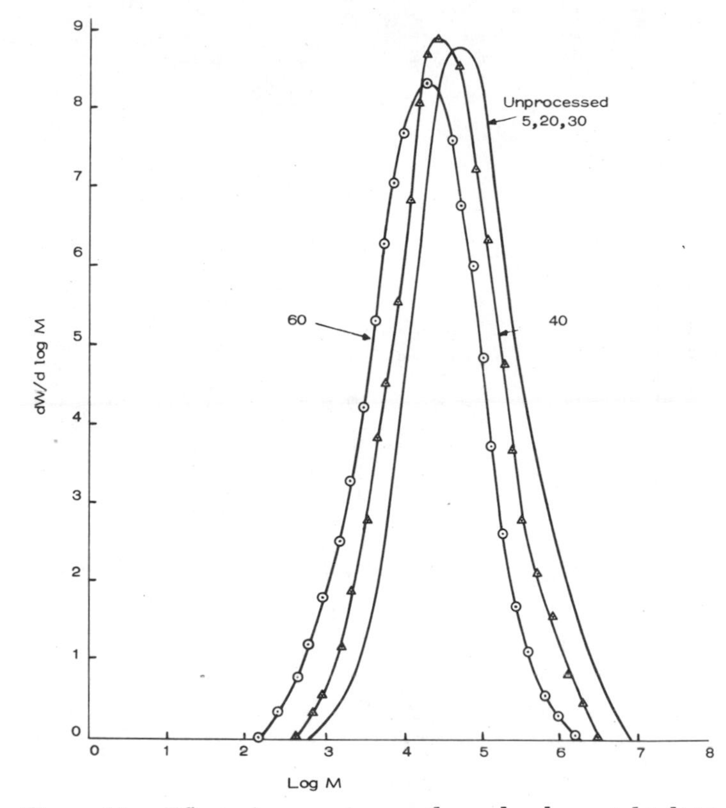

Figure 11. Effect of processing on the molecular weight distribution of LDPE (numbers on curves indicate processing times at 150°C; min in an open chamber)

As in the case of HIPS the formation of peroxides in LDPE was associated with the destruction of the initial unsaturation (1,1-dialkylethylene) after an induction period (*2*). Similarly, on exposure to UV light, the rate of photooxidation as measured by the formation of carbonyl compounds was related to the initial hydroperoxide concentration (*12*). In oxidized LDPE, the UV carbonyl growth curve showed a rapid rate initially (Figure 14, Curve c); it was associated with the presence of hydroperoxide by heating a similar film in argon at 110°C until hydroperoxide could no longer be detected. The initial carbonyl concentration increased but the UV carbonyl growth rate showed an initial drop followed by a rate of photooxidation similar to that of an unprocessed control (Figure 14, Curves a and d).

The rate of decay of the vinylidene group on photooxidation also showed a marked increase wtih the processing time (Figure 15). The induction period disappeared after 30-min processing; this corresponded to a maximum hydroperoxide concentration (Figure 13).

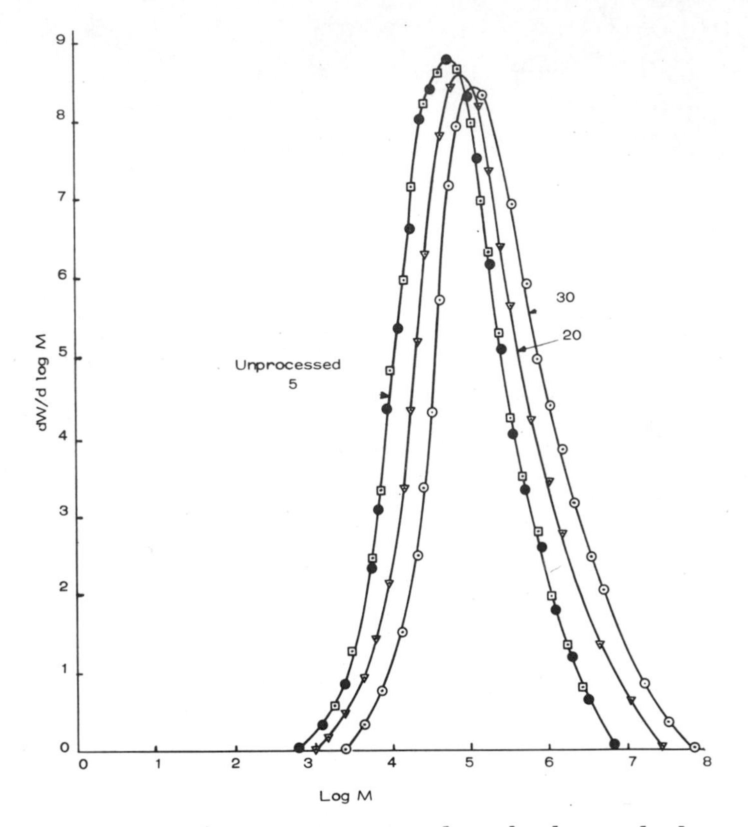

Figure 12. Effect of processing on the molecular weight distribtuion of LDPE (numbers on curves indicate processing times at 150°C; min in a closed chamber)

The chemistry of the thermal oxidative and photooxidative reactions (which is very similar to that taking place in photooxidizing HIPS) occuring is summarized in Scheme 4.

Studies of model compounds have confirmed that both unsaturation and peroxides are important in the photo-initiation process. Figure 16 shows that diamylethylene(**1**) autoxidizes in the presence of UV light in

the classical autoaccelerating mode. The introduction of a hydroperoxide removes the induction period completely while the addition of diamyl ketone (**2**), although undergoing photolysis as evidenced by the initial

Scheme 4. Thermal and photooxidative reactions of vinylidene groups in LDPE

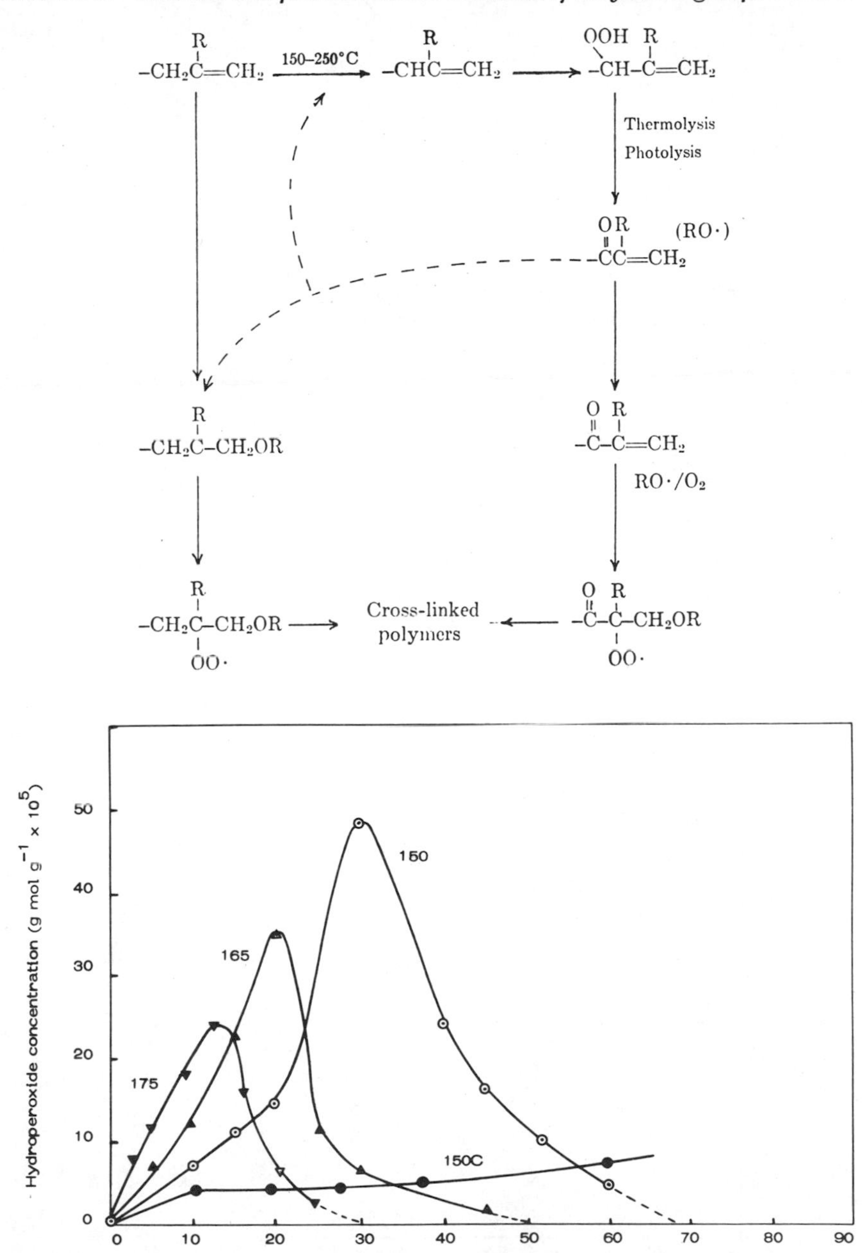

Figure 13. Effect of processing time and temperature on the formation of peroxides in LDPE (numbers on curves indicate temperatures (°C) in open chamber, 150C indicates 150° in a closed chamber)

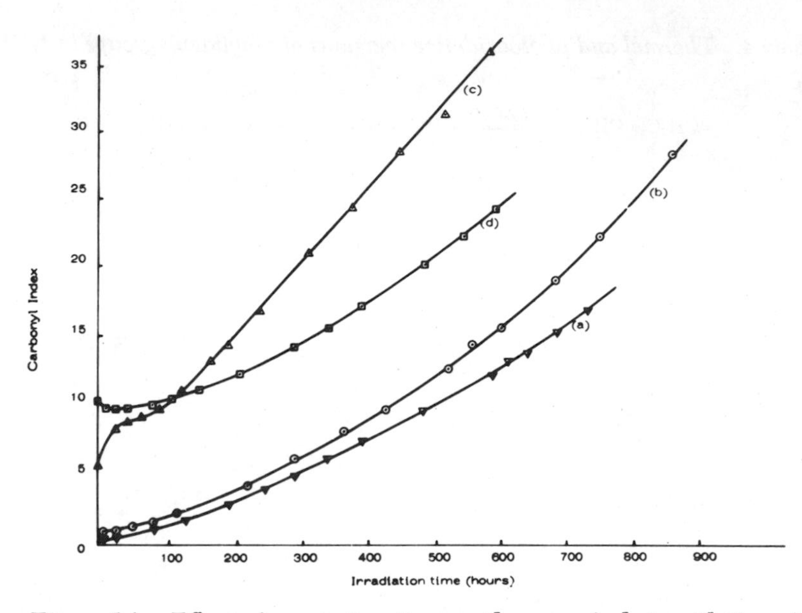

Figure 14. Effect of processing time on the rate of photooxidation of LDPE. (a) compression-molded control; (b) processed for 5 min; (c) processed for 30 min; (d) processed for 30 min and heated in argon for 8 hr at 110°C.

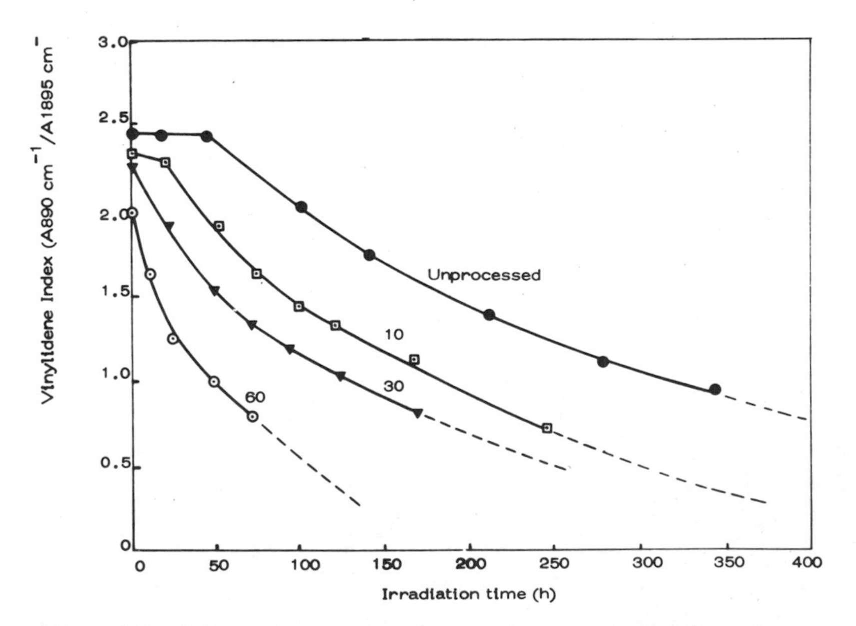

Figure 15. Effect of processing time on the rate of vinylidene decay in LDPE during photooxidation (numbers on curves indicate processing time in min)

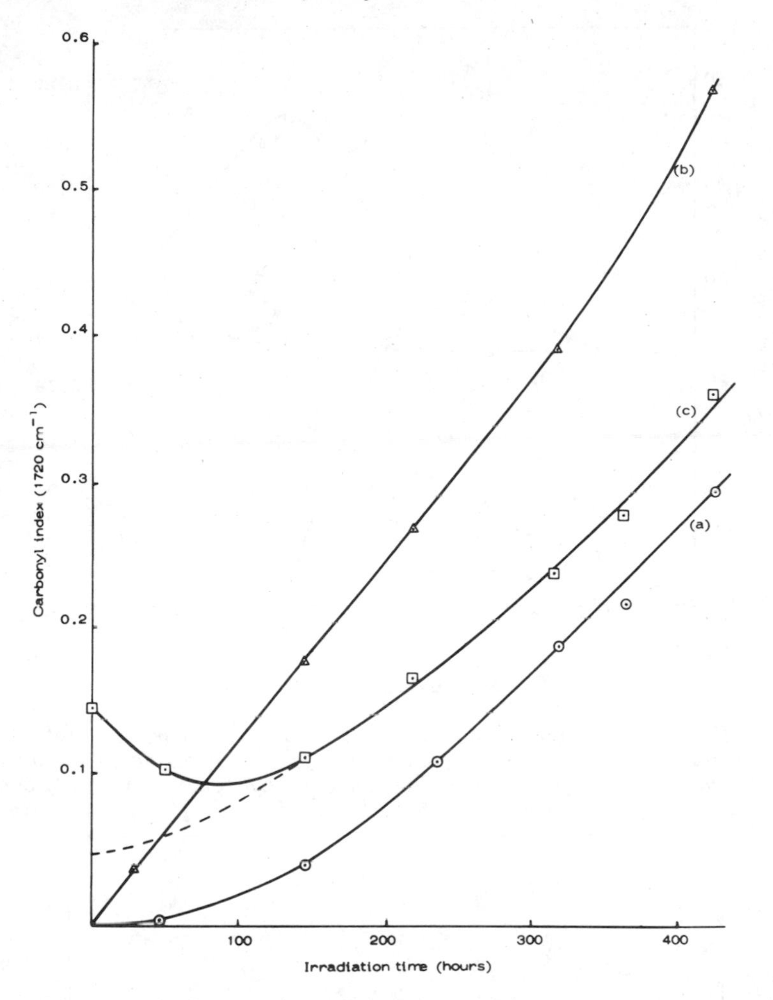

Figure 16. Effect of photo-initiators in the photooxidation of 1,1-diamylethylene. (a) No additive; (b) tert-*butyl hydroperoxide (2 × 10^{-4} mol g^{-1}) (c) diamyl ketone (2 × 10^{-4} mol g^{-1}).*

reduction in carbonyl concentration, has a negligible effect on the induction period or on the rate of photooxidation. Similar conclusions have been reached in the case of HIPS (*10, 13*).

PP shows a similar behavior to polyethylene when photooxidized. Conjugated carbonyl is present initially in the polymer (*14*) and it is interesting to compare its growth and disappearance upon processing with that of hydroperoxide. It can be seen from Figure 17 that the rate of growth of carbonyl shows an induction period and is slower both in excess oxygen and in a deficiency of oxygen than is the growth of hydroperoxide. Furthermore, although the rate of photooxidation as measured by car-

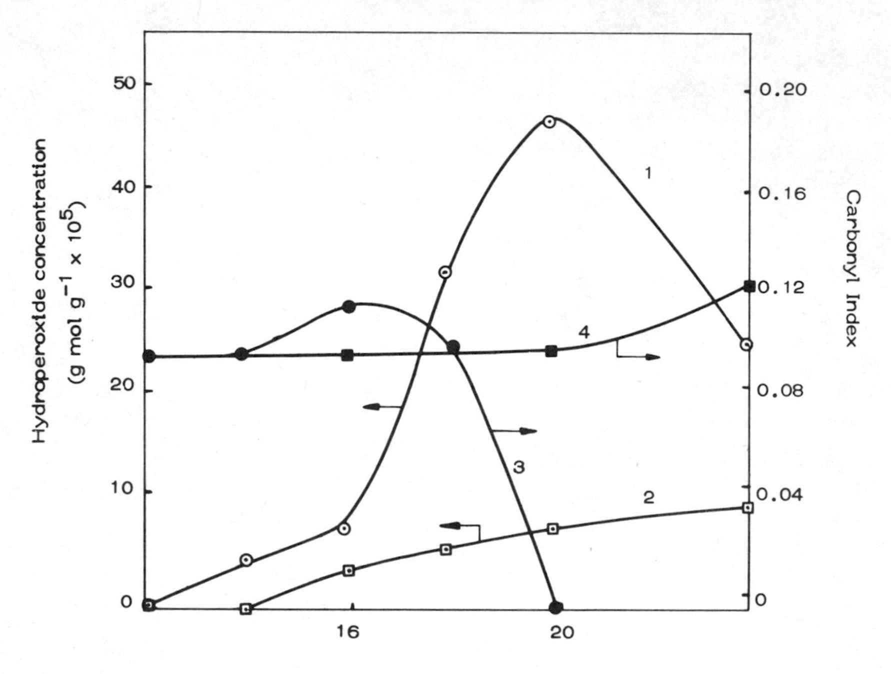

Figure 17. Effect of processing time at 180°C on the peroxide and conjugated carbonyl concentration in polypropylene. (1) peroxide (open mixer); (2) peroxide (closed mixer); (3) conjugated carbonyl (open mixer); (4) conjugated carbonyl (closed mixer).

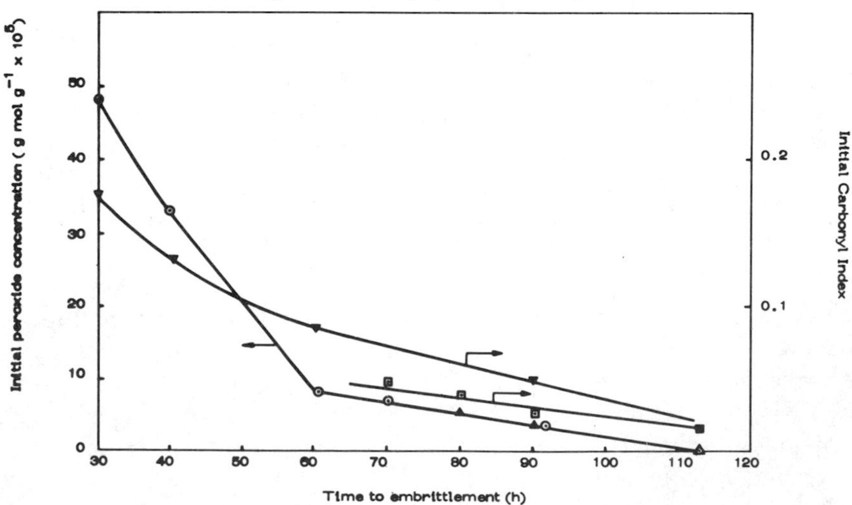

Figure 18. Relationship between peroxide and carbonyl concentrations and embrittlement time for polypropylene. (—⊙—) peroxide formation, open chamber; (—▼—) carbonyl formation, open chamber; (—△—) peroxide formation, closed chamber; (—□—) carbonyl formation, closed chamber.

bonyl formation can be related directly to the concentration of peroxide, it bears no relation to the concentration of conjugated carbonyl (*15*). Figure 18 relates the embrittlement time of PP to both peroxide concentration and carbonyl concentration and although embrittlement time is a

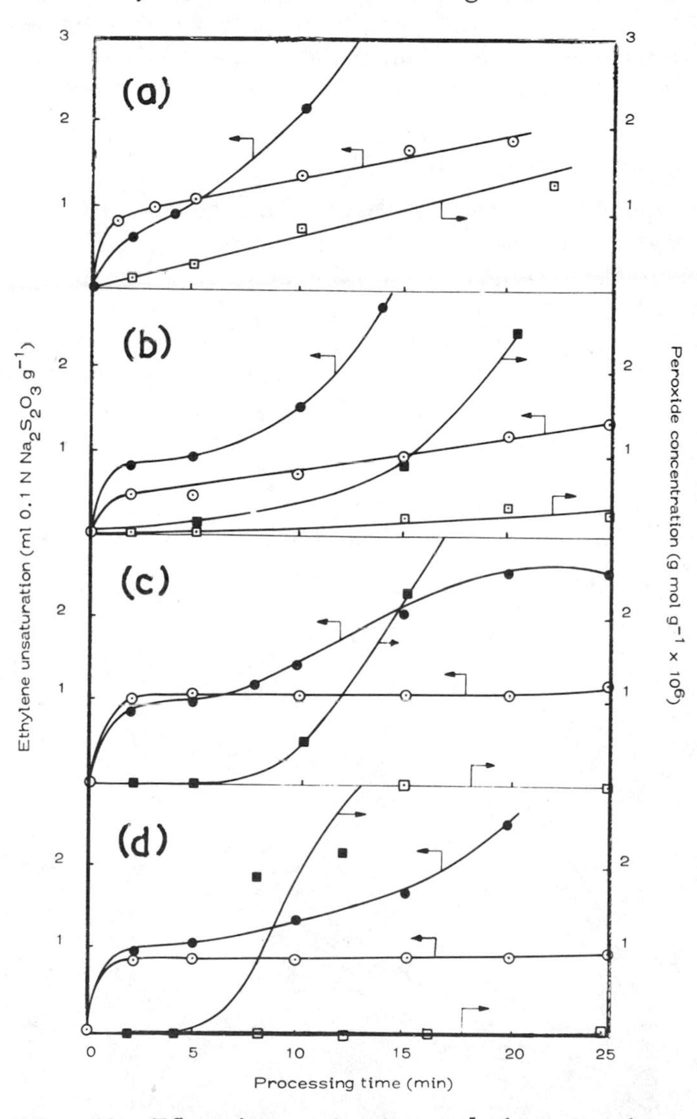

Figure 19. Effect of processing time on the formation of peroxides and ethylenic unsaturation in PVC containing additives. (a) Wax E (0.6%); (b) calcium stearate (0.8%); (c) tin maleate (2.5%); (d) three-component synergistic mixture of (a), (b), and (c). (—□—) peroxides at 170°C; (—○—) unsaturation at 170°C; (—■—) peroxides at 210°C; (—●—) unsaturation at 210°C.

direct function of peroxide concentration in samples thermally processed both in excess and in deficiency of oxygen, the embrittlement time–carbonyl concentration relationship is quite different for the two modes of thermal treatment indicating that carbonyl concentration, although related to the concentration of peroxide from which it is formed, is not involved directly in the photo-initiation process—at least during the early stages of thermal treatment.

Effects of Stabilizers and Antioxidants

The effect of an efficient tin maleate stabilizer on the formation of unsaturation and peroxides in PVC is unexpected. First, the additive has no inhibitory effect on the initial formation of unsaturation in the polymer (Figure 19(c)) and relative to the control without additive (Figure 2), the unsaturation actually increases. However it markedly retards the formation of further unsaturation and it completely eliminates the initial formation of peroxide. In the latter respect it is more effective than a typical ester lubricant (Wax E), and calcium stearate which do not inhibit peroxide formation although they retard its initial rate of formation. Both lubricants reduce the initial formation of unsaturation; Wax E is more effective at 170°C than at 210°C, whereas the reverse is true with

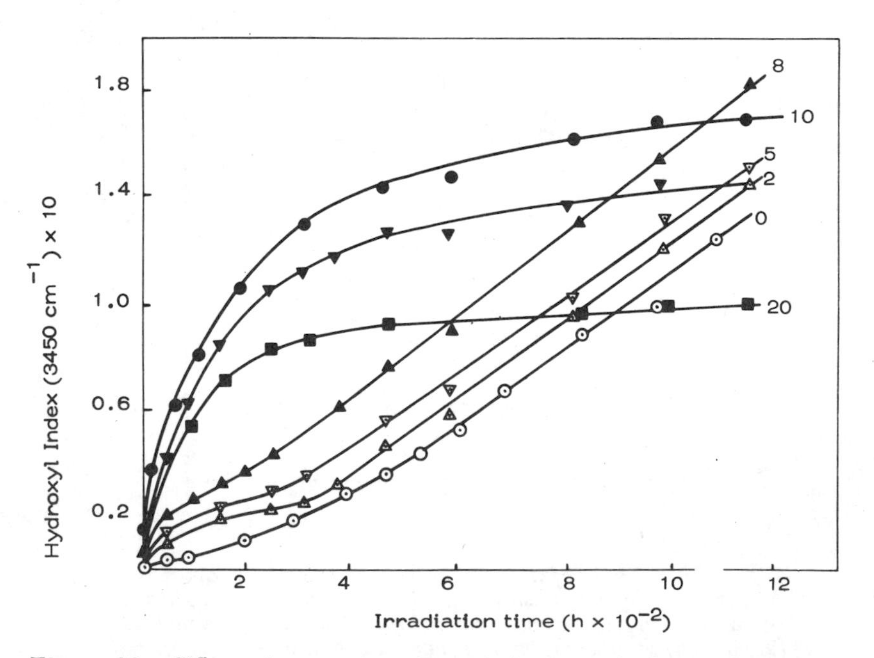

Figure 20. Effect of processing time on the rate of photooxidation (hydroxyl formation) of PVC containing a tin maleate stabilizer (2.5 g g^{-2}) (numbers on curves indicate processing times at 210°C in min)

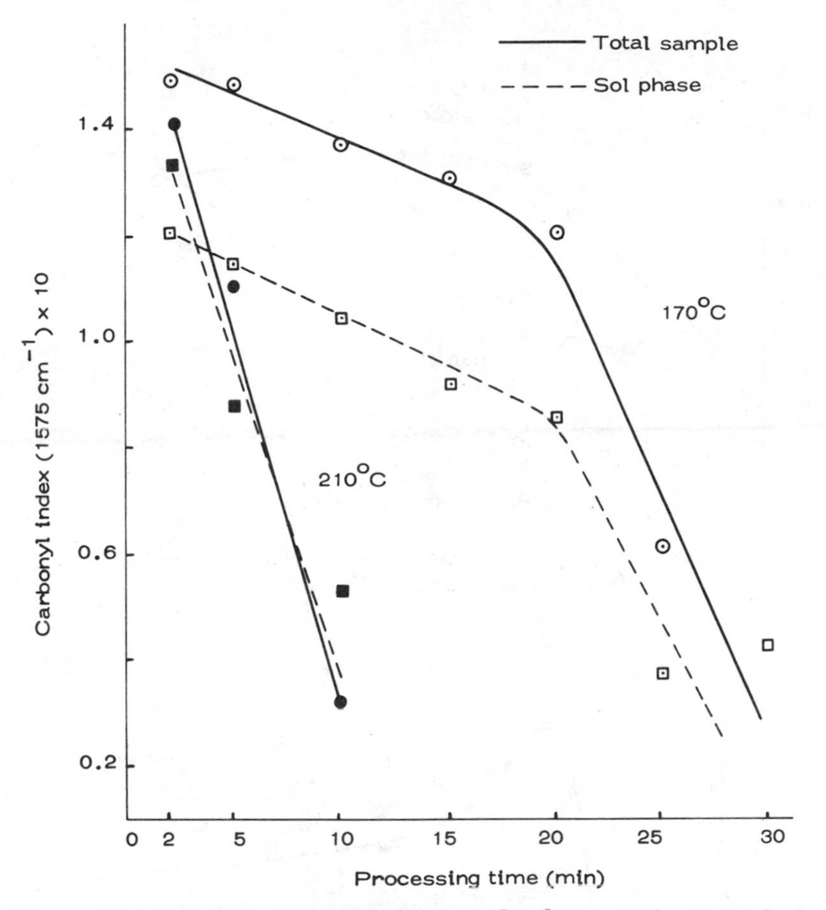

Figure 21. Effect of processing time on the decay of the tin maleate carboxylate ir absorption (1575 cm⁻¹)

calcium stearate. The effect of Wax E is also evident from the decreased rate of thermal color formation at 170°C relative to the control (Figure 3).

The delay in the formation of both unsaturation and peroxides by the tin maleate stabilizer is reflected in an induction period to photooxidation (Figure 20). It is significant that the sharp change-over from an autoaccelerating to an auto-retarding mode of hydroxyl formation in the photooxidizing polymer samples coincides with the end of the thermal induction period to peroxide formation at 210°C (Figure 19). This clearly implicates peroxide in the photo-initiation step. Unsaturation alone does not lead to rapid photooxidation (*see* also Figure 16) and an appreciable rate of photodegradation is not achieved until peroxides are present in the system.

The disappearance of the tin maleate stabilizer during the induction period can be followed by means of the ir carboxylate peak at 1575 cm^{-1}.

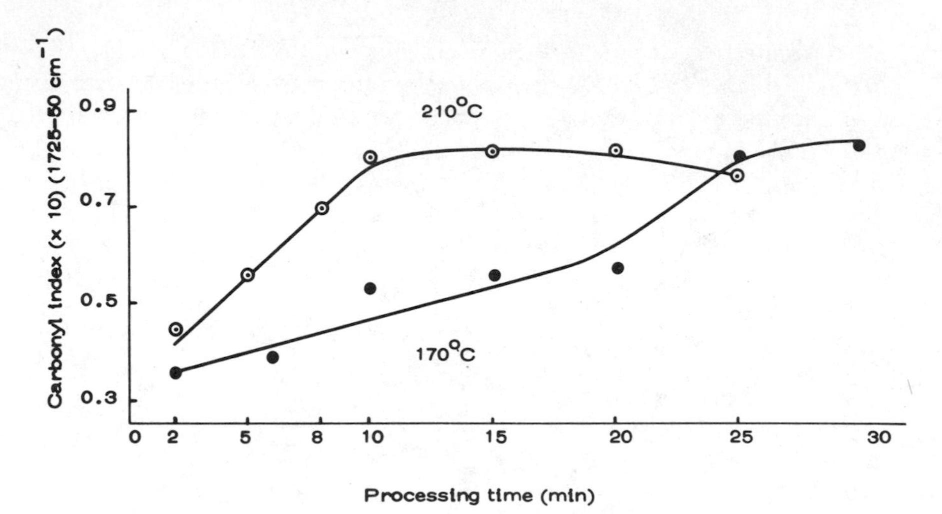

Figure 22. Effect of processing time on the growth of ester in PVC containing a tin maleate stabilizer

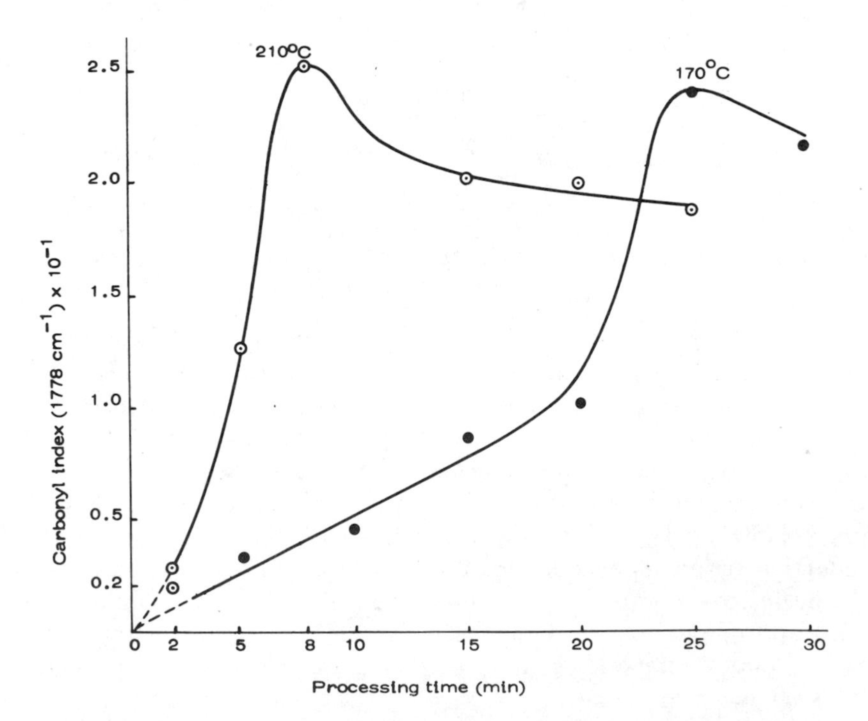

Figure 23. Effect of processing time on the growth of maleic anhydride UV absorption at 1778 cm^{-1} in PVC containing a tin maleate stabilizer

The concentration decays to 0 in $\simeq$ 12 min at 210°C (Figure 21). At 170°C the decay is a two-stage process; the rate increases rapidly after 20 min. The decay of carboxylate is associated with the parallel formation of ester at 1725–50 cm^{-1} (Figure 22) and maleic anhydride at 1775 and 1848 cm^{-1} Figure 23). The two-stage process is evident again at 170°C in both cases and the maximum extent of formation corresponds again to the end of the induction period for peroxide formation. Ester carbonyl absorption shifts from initial unsaturated carbonyl (1725 cm^{-1}) at the start of processing to saturated carbonyl (1750 cm^{-1}) towards the end of the induction period, indicating reaction of the maleate ester to conjugated dienes and polyenes. This accounts for complete growth inhibition of further unsaturation during the thermal induction period. The change in photooxidation mode (Figure 20) clearly is coincident with the end of the induction period. The chemical reactions occuring during processing are summarized in Scheme 5.

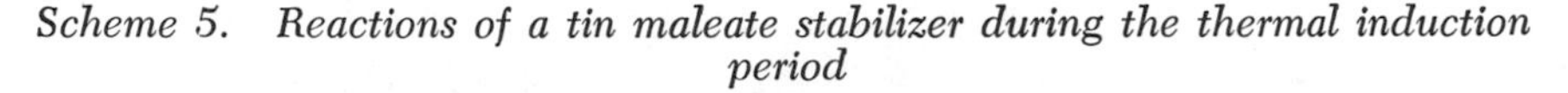

Scheme 5. Reactions of a tin maleate stabilizer during the thermal induction period

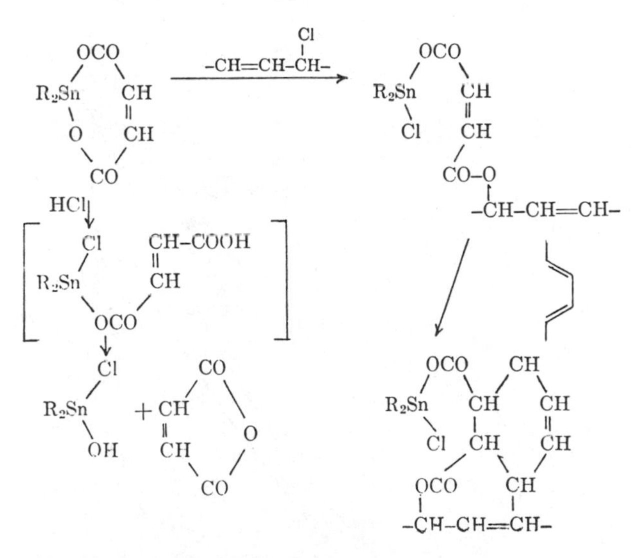

Peroxide Decomposers as UV Stabilizers

The implication of peroxides in the initiation step during the photooxidation of PVC suggests that additives that destroy peroxides in polymers can behave as UV stabilizers. The tin maleate may fulfill this function in addition to removing polyconjugated unsaturation but evidence is lacking at present. However, several examples of this behavior have been reported in other polymers and peroxide decomposition now is accepted generally as an important mechanism of UV stabilization. How-

Table I. Induction Periods to Onset of Carbonyl Formation in the Thermal Oxidation of LDPE Films (110°C) [a]

Formulation	*Induction Period*	*Calculation on Additive Basis*
No additive	10	—
HOBP	10	—
1076	105	—
ZnDEC	110	—
1076 + ZnDEC	330	215
HOBP + ZnDEC	130	120
NiDEC	105	—
1076 + NiDEC	300	210

[a] All concentrations are 3×10^{-4} mol/100 g.

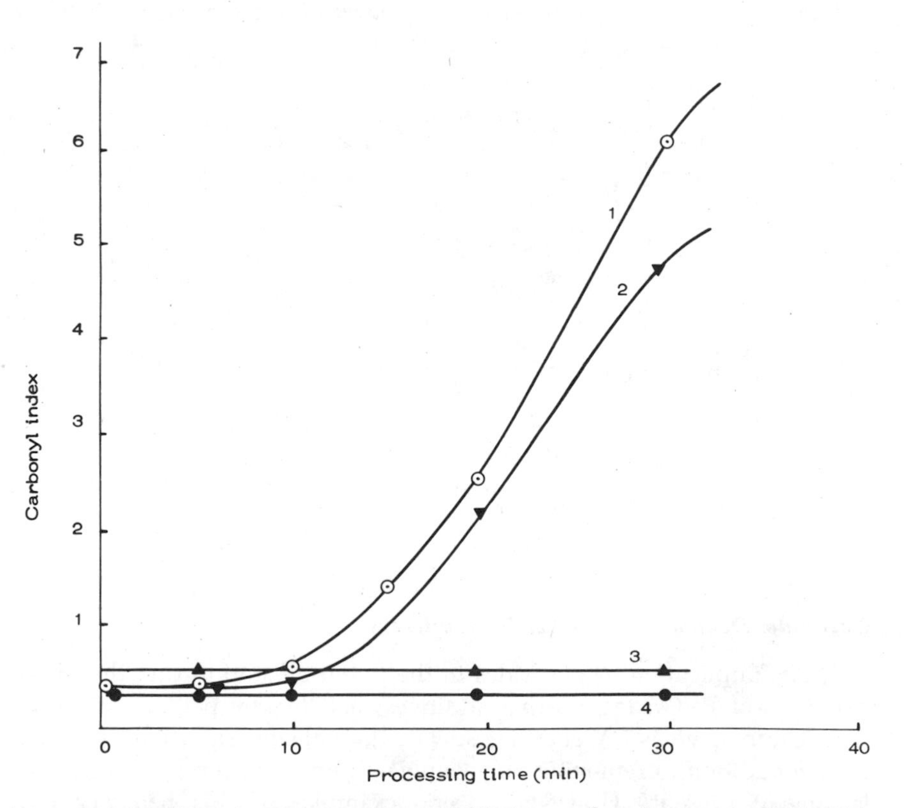

Figure 24. Effect of processing time at 150°C on carbonyl formation in LDPE-containing antioxidants and stabilizers (3×10^{-4} g g^{-2}). (1) control (no antioxidant); (2) HOBP; (3) 1076; (4) ZnDEC.

ever, different types of stabilizers may behave quite differently in thermal and UV stabilization of polymers. Figure 24 and Table I show that a typical UV absorber, 2-hydroxy-4-octyloxybenzophenone (HOBP), is only slightly active as a thermal antioxidant. Hydroperoxides and carbonyl groups are formed in the polymer after an induction period that differs only slightly from the control without antioxidant. A typical high molecular weight phenolic antioxidant (1076) and two typical peroxide

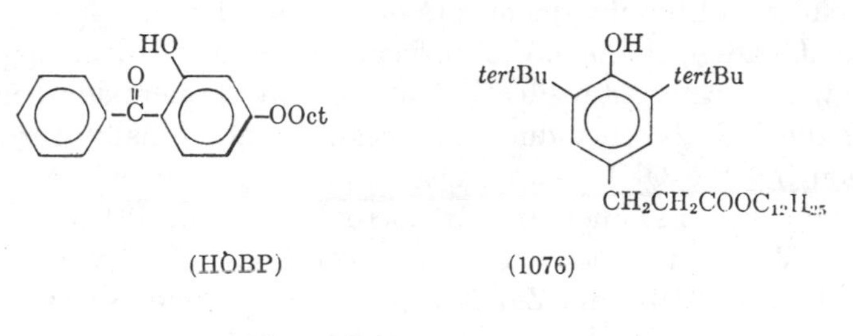

$Et_2NC(=S)S_2MS_2C(=S)NEt_2$ (M = Zn, ZnDEC, M = Ni, NiDEC)

decomposers (zinc diethyldithiocarbamate (ZnDEC) and nickel diethylthiocarbamate (NiDEC), were much more effective as thermal stabilizers (Figure 24 and Table I). Furthermore, 1076 and ZnDEC showed a synergistic interaction whereas HOBP and ZnDEC did not.

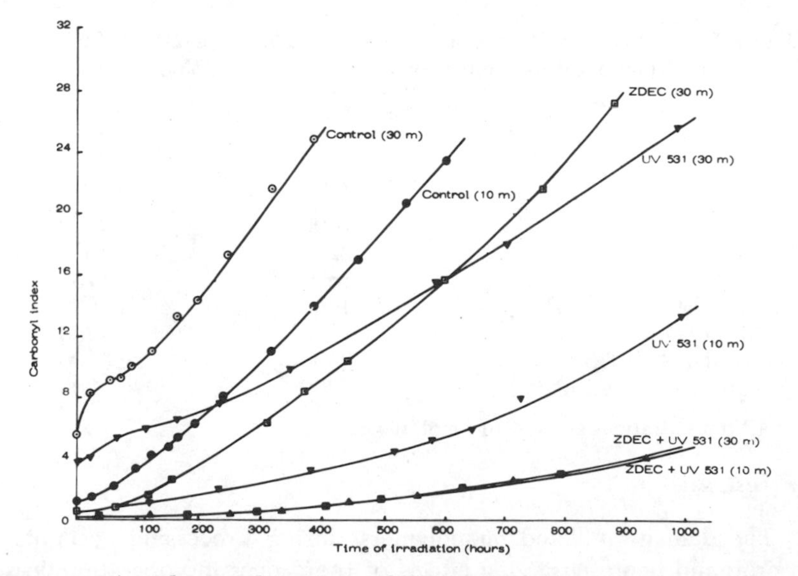

Figure 25. Photooxidation of mildly (10 min) and severely (30 min) processed LDPD containing a peroxide decomposer (ZnDEC), a UV absorber (HOBP), and a synergistic combination of the two

The behaviors of HOBP and the dithiocarbamates were quite different in UV stabilization. HOBP, as anticipated, retarded photooxidation even in a severely oxidized film although not as effectively as it does in a mildly oxidized film (Figure 25). ZnDEC, on the other hand, eliminated the peroxide-initiated photooxidation completely and the rate of photooxidation is independent of the severity of processing. ZnDEC caused an induction period relative to the control but it had only a minor effect on the subsequent rate of photooxidation.

Combination of the UV absorber and peroxide decomposers gave very powerful synergistic effects. Not only do the peroxide decomposers protect the UV absorber during processing and initially during photooxidation but the UV absorber also appears to protect the metal complexes by a process other than UV screening. Embrittlement times are listed in Table II which shows the very powerful synergistic effect obtained with HOBP and ZnDEC, particularly under severe processing conditions. The 1076, although reasonably effective as a UV stabilizer, is antagonistic towards the peroxide decomposer in spite of the good thermal stabilization achieved with this combination. Further studies of these synergistic and antagonistic effects under conditions of UV irradiation are in progress and preliminary evidence suggests that the UV absorber protects the metal complex by deactivating a photo-excited species formed from it, whereas oxidation products of the phenolic antioxidant sensitize the photolysis of the dithiocarbamates.

Table II. Effect of Processing Conditions on the UV Lifetime of Polyethylene Films Containing Synergistic UV Stabilizers[a]

	Time to Embrittlement (hr)	
Stabilizer	*10 Min Processed*	*30 Min Processed*
ZnDEC	1400	1400
HOBP	2200	1600
1076	1800	1750
ZnDEC + HOBP	> 4000	> 4000
ZnDEC + 1076	1250	1250
HOBP + 1076	≃ 3500	≃ 3500
Control (no additive)	1200	900

[a] All concentrations are 3×10^{-4} mol/100 g.

Conclusions

The damage inflicted on polymers during processing accounts for the profound photosensitizing effects of the processing operation on subsequent photooxidation. In particular, the mechanochemical production of free radicals in the presence of small amounts of air is the cause of subsequent thermal oxidative damage to the polymer during processing and the reason for the presence of photoinitiating peroxide groups in the polymer. The presence of peroxides appears to be mandatory for rapid photooxidation and antioxidants that destroy peroxides during processing and in light are potential photo-stabilizers.

Acknowledgments

The contribution made to these studies by my co-workers is gratefully acknowledged. In particular I wish to thank J. Vyvoda who carried out the study of PVC, A. Scott for the work on the photodegradation of HIPS, K. B. Chakraborty for his studies of the photooxidation and stabilization of the polyolefins and M. Humphrey for his studies of model compound autoxidation.

Literature Cited

1. Mellor, D. C., Moir, A. B., Scott, G., *Eur. Polym. J.* (1973) **9**, 219.
2. Amin, M. U., Scott, G., Tillekeratne, L. M. K., *Eur. Polym. J.* (1975) **11**, 85.
3. Bresler, J. E., Zhurkov, S. N., Kuzbekov, F. M., Saminkii, E. M., Tomashevskii, E. E., *Rubber Chem. Technol.* (1960) **33**, 462.
4. Busse, W. F., Cunningham, E. N., *Proc. Rubber Technol. Conf.* (1938) 288.
5. Pike, M., Watson, W. F., *J. Polym. Sci.* (1952) **9**, 229.
6. Watson, W. F., *Trans., Inst. Rubber Ind.* (1953) **29**, 32.
7. Ceresa, R. J., Watson, W. F., *J. Appl. Polym. Sci.* (1959) **1**, 101.
8. Potter, W. D., Scott, G., *Eur. Polym. J.* (1971) **7**, 489.
9. Scott, G., "Atmospheric Oxidation and Antioxidants," p. 24, Elsevier, London, 1965.
10. Ghaffar, A., Scott, A., Scott, G., *Eur. Polym. J.* (1976) **12**, 615.
11. Scott, G., *Resour. Recovery Conserv.* (1976) **1**, 381.
12. Scott, G., *ACS Symp. Ser.* (1976) **25**, 340.
13. Ghaffar, A., Scott, A., Scott, G., *Eur. Polym. J.* (1977) **13**, 83.
14. Allen, N. S., Cundall, R. B., Jones, M. W., McKellar, J. F., *Chem. Ind. (London)* (1976) 110.
15. Chakraborty, K. B., Scott, G., *Polymer* (1977) **18**, 98.

RECEIVED May 12, 1977.

5

Some Effects of Production Conditions on the Photosensitivity of Polypropylene Fibers

D. J. CARLSSON, A. GARTON, and D. M. WILES

Division of Chemistry, National Research Council of Canada, Ottawa, Canada K1A 0R9

The sensitivity of polypropylene monofilaments to photooxidation by near UV is enhanced by certain processing conditions. Melt oxidation resulting from high extruder temperatures and the presence of oxygen promotes photooxidation, whereas spin-line oxidation appears to be unimportant. Dramatic increases in photosensitivity result from fiber drawn under high shear conditions (low temperatures and high draw rates) when backbone scission by mechanical action is followed by chain oxidation. Hydroperoxide formation in this process is prevented only partially by the presence of stabilizers. Draw-induced oxidation also results in a rapid increase in embrittlement on UV exposure, although the formation of photooxidation products and embrittlement do not correlate. Morphological effects on photooxidation for the series of fibers used were assumed unimportant from γ-initiated oxidation studies.

The weathering of unstabilized polypropylene (PPH) fibers results in a catastrophic drop in elongation at break together with a build-up of hydroperoxide (–OOH) and carbonyl (>C=O) products after relatively short exposures (*1*). Deterioration of mechanical properties stems from –OOH photolysis, the –OOH itself resulting from free radical oxidation chains initiated by the photocleavage of UV-absorbing impurities (chromophores) such as catalyst residues or thermal oxidation products introduced during processing (*2*). The nature, concentration, and location of these oxidation products resulting from processing may then be major factors in determining the subsequent resistance to weathering of the filaments. To establish the key processing steps responsible for chromophore introduction, the extrusion and draw conditions for PPH mono-

0-8412-0381-4/78/33-169-056$05.00/1

filament production were systematically varied, and the photooxidative stability of the produced fibers determined.

As well as influencing chromophore types and levels, variations in production conditions result in fibers of varying morphology (orientation, crystallinity, and supramolecular order) (*3*). Because morphology itself may markedly effect photooxidative sensitivity, the extent of this effect was explored by first studying the susceptibility of the various filaments to γ-initiated oxidation. The γ-initiation is believed to be completely random and it is not dependent on trace quantities of chromophores or the build-up of oxidation products, in contrast to photooxidation, yet it is followed by the same propagation and termination steps as photo-initiated oxidations (*2, 4*).

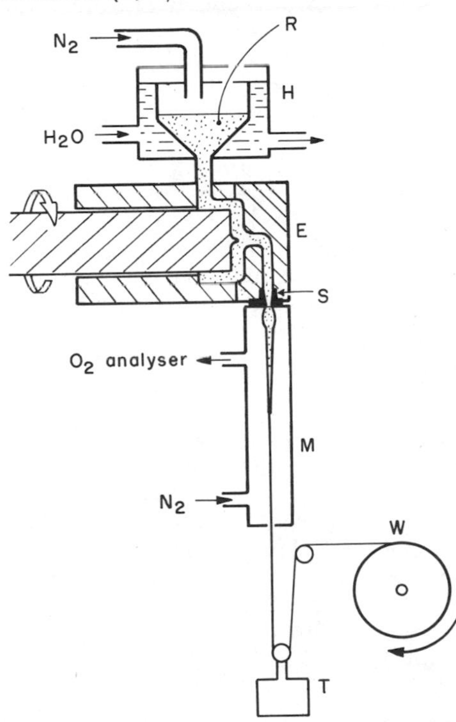

Figure 1. Small-scale extrusion system. R = resin powder; H = water-cooled hopper; E = modified Maxwell extruder; S = spinneret; M = spin line manifold; T = tension indicator; W = variable speed wind-up assembly.

Experimental

Materials. Unprocessed, commercial, additive-free, isotactic PPH (Monomer Polymer Labs. $M_v \sim 400{,}000$) (*5*) was extruded into 50–

200-μm monofilaments on a Maxwell type screwless extruder with a 0.125-cm diameter, single-hole spinneret (Figure 1). Extruder temperature, wind-up speed, and spin-line tension could be varied and continuously monitored. The PPH delivery hopper was water-cooled and could be N_2-purged. Directly on emerging from the spinneret, the filament was N_2- or air-blanketed for 60 cm before wind-up. Off-line hot drawing was carried out by passage through an air oven at 180°C (nominal draw rate ~ 2500% min^{-1}, based on the residence time), or by drawing in fixed-temperature water baths at 12 in. min^{-1} (400% min^{-1}). Fiber crystallinities were measured by x-ray diffraction with a powder camera.

Some measurements were made on an unoriented, commercially extruded, isotactic PPH film (25 μm, Enjay chill roll cast) and on an oriented commercial PPH monofilament [150 μm, Amtech Inc., Odenton, U.S.A., 90% residual elongation at break, stabilized with 0.24 wt % octadecyl 3-(3′,5′-di-*tert*-butyl-4′-hydroxyphenyl)propionate]. The unprocessed PPH powder also was pressed into film (~ 70 μm). All samples were Soxhlet-extracted with acetone (48 hr) and vacuum dried before use. The stabilizer in the commercial fiber was not removed by this treatment, although additives were extracted rapidly from low orientation samples (*5*).

Oxidations. Photo- and γ-oxidations were carried out on single filament lengths rigidly located on formers in an Atlas 6000W Weather-Ometer (Pyrex inner and outer filters 30% relative humidity, 34°C silver panel temperature) and Gammacell 220 sources respectively. Oxidation product build-up was monitored at 3400 cm^{-1} (predominantly OOH) and ~ 1715 cm^{-1} (C=O) (*6*) by ir spectroscopy using the single fiber technique (*7*). OOH levels also were measured iodometrically after solution and precipitation of ~ 0.1-g fiber samples in pure Decalin (heating at 140°C for $<$ 1 min) by addition of NaI in acetic acid/2-propanol (*6*).

Scanning Electron Microscopy (SEM). Samples intended for SEM analysis normally were mounted directly on polished aluminum microscope stubs before irradiation. Several filaments were peeled before mounting to expose the sample interior to UV irradiation as described previously (*8*). The mounted filament samples were gold coated before examination to minimize charging in the electron beam. A Cambridge S-2 scanning microscope was used for all studies.

Mechanical Testing. An Instron tensile tester, Model TTC was used to draw filaments in air at 25°C and 200% min^{-1}.

Table I. Role of Sample Thickness in PPH γ-Initiated Oxidation[a]

PPH Sample	*Thickness*	*Product Yields (M)* [OOH]	*Product Yields (M)* [>C=O]
Commercial film (Enjay)	25 μm	0.21	0.030
Pressed film	~ 70 μm	0.21	0.040
Undrawn lab. fiber	114 μm	0.23	0.054
Undrawn lab. fiber	178 μm	0.14	0.036

[a] 36 hr at 1000 rad min^{-1}.

Results and Discussion

Morphological Effects. A series of PPH samples were γ-irradiated in air at both high and low dose rates, and product build-up was monitored by ir spectroscopy. From the observed product levels shown in Table I, sample thickness affects the oxidation rate only above 150 μm. This is in good agreement with the γ-results of Decker, et al. (9). All samples studied subsequently in our work were < 120 μm.

The γ-initiated oxidation of PPH filaments of widely differing draw history (Table II) showed that oxidation was not influenced by molecular orientation. The residual elongation at break of the filaments studied implied that the filaments ranged from the dominantly spherulitic (> 1000% residual elongation) to the highly fibrillar (< 100% residual elongation) (3). The fibrillar nature of the low residual elongation filaments was shown also by SEM examination of peeled sections (Figure 2).

Table II. γ-Initiated Oxidation of PPH Fiber Samples[a]

Filament Sample	*Diameter (μm)*	*Residual Elongation (%)*	*Product Yields (M)*	
			[OOH]	*[>C=O]*
Undrawn	114	1100	0.23	0.054
Increasing Orientation ↓	116	500	0.22	0.048
	112	380	0.24	0.055
	113	160	0.24	0.047
	103	90	0.22	0.038
	65	~ 70	0.21	0.036

[a] After 36 hr at 1000 rad^{-1} min.

A dependence of γ-oxidation rate on production conditions might be expected to result from variations in O_2 solubility, permeability, and accessibility with morphology. Apparently for the series of samples discussed here, these effects are insignificant (or fortuitously opposing). The samples do, in fact, all have similar overall crystallinities (Table III). PPH γ-oxidation and O_2 permeability have been shown previously to be relatively insensitive to crystallinity (9) and orientation (*10*), respectively. When fiber diameters were increased beyond 150 μm, γ- and photooxidation rates became thickness-dependent, presumably because of O_2 depletion in the sample interior.

Effects of Production Conditions

Melt Oxidation. In the absence of a morphological effect on oxidation of our series of PPH filaments, differences in their photostabilities must depend upon differences in the concentrations of UV absorbing impurities

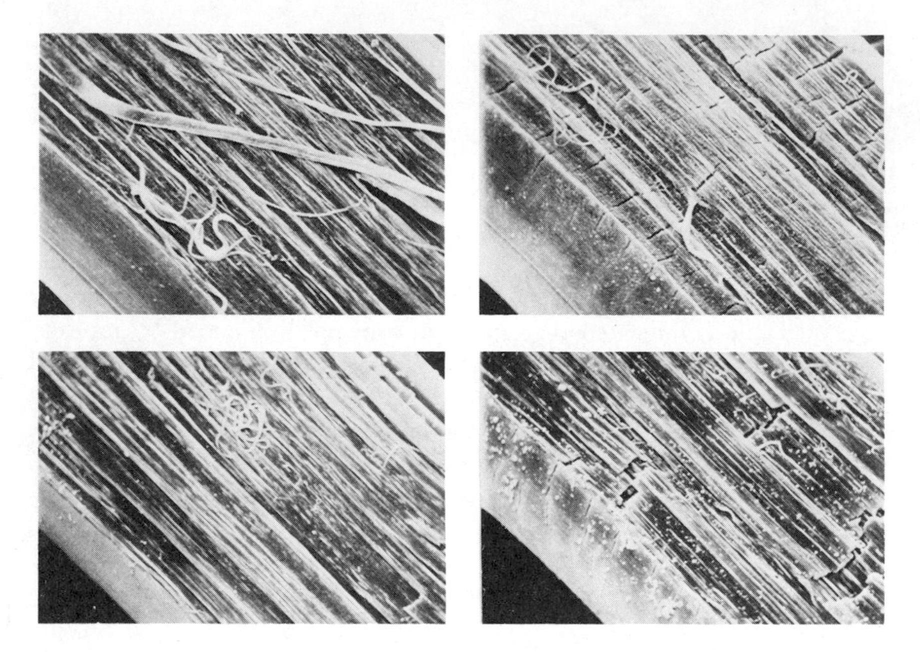

Figure 2. Micrographs of photooxidized filaments (at 10-μ filaments peeled before UV exposure). Partially oriented filament (110 μm), Xe irradiation; 0 hr. (upper left); ~ 300 hr (upper right). Highly oriented filament (70 μm), Xe irradiation; 0 hr. (lower left); ~ 300 hr (lower right).

Table III. γ-Initiated Oxidations of Polypropylene

Sample	Crystallinity (%)	Thickness (μm)	Initial Residual Elongation[a] (%)	Oxidation Product Conc. (M)	
				[OOH][b]	[>C=O][b]
Laboratory filament (high hot draw)	51	113	160	0.41	0.081
Laboratory filament (medium hot draw)	56	112	380	0.40	0.093
Laboratory filament (low hot draw)	49	116	500	0.35	0.079
Laboratory filament (no hot draw)	54	114	1100	0.39	0.090
Laboratory filament (no hot draw, cold drawn × 3)	—	~ 75	210	0.38	0.063
Commercial Film (Enjay)	~ 40	25	—	0.43	0.093

[a] Drawn at $200\%^{-1}$ min in air at 25°C.
[b] From ir data, 60 hr γ-irradiation at 1000 rad^{-1} min.
[c] Calculated from density measurements. All other crystallinities from x-ray data (±5%).

(photo-initiators) within them (2). Production conditions were varied systematically to determine at which stage significant thermal oxidation effects occurred, significant from the viewpoint of affecting filament photostability. The criteria for photo-susceptibility were [OOH] or [OOH] + [C=O] build-up, brittle failure usually occurring at ~ 0.1*M* [OOH].

As might be expected from the work of Burch (*11*), increased extrusion temperature or the presence of O_2 in the PPH feed both increased photosensitivity (Figure 3). To minimize melt degradation, the remainder of the work was performed at an extruder temperature of 225°C and with rigorous N_2 purging of the feed hopper. A comparison of filaments prepared under identical extrusion and draw conditions showed that N_2-blanketing of the spin line did not enhance filament photostability over that of filaments extruded into air. Presumably under our conditions, cooling of the melt is too rapid to allow significant thermal oxidation at this stage. Similarly, increasing the wind-up speed from 20 cm sec^{-1} to 50 cm sec^{-1} had no effect on the subsequent photooxidation of the filaments.

Draw Oxidation. Hot drawn filaments (×3 at ~ 2500% min^{-1} at 180°C) were found to photooxidize significantly more rapidly than undrawn fiber, or undrawn fiber exposed to oven heat at 180°C for the normal draw duration. Solely, thermal oxidation in the oven appeared unimportant. Consequently the effects of draw parameters (temperature and speed) were investigated further.

Figure 4a shows that photosensitivity increases enormously for fibers drawn ×3 at a series of decreasing temperatures but at a constant draw rate from a common parent filament (i.e. the converse of a thermal oxidation effect). Furthermore, for another series drawn ×3 at a fixed temperature but at varying speeds, photosensitivity decreased as the draw speed decreased (Figure 4b).

The photosensitivity increase on decreasing draw temperature or increasing draw speed probably results from free radical generation under high shear conditions (as observed by ESR (*12, 13*)), followed by a long-chain thermal oxidation of the PPH. This was confirmed when fiber was drawn rapidly at low temperature in the absence of O_2 (Figure 5) when photosensitivity was only slightly increased. That the shear-induced photosensitivity was not related to the final fiber morphology was confirmed further by γ-oxidation of fibers that had been rapidly cold drawn. From Table III, this fiber γ-oxidized at a rate very similar to that of the other hot drawn or undrawn filaments, which were in turn similar to an unoriented commercial film (Enjay). Furthermore, the fiber after rapid cold drawing was shown by iodometry to contain ~ $1 \times 10^{-3}M$ OOH, as compared with undetectable levels ($< 1 \times 10^{-4}M$) in the filament before drawing.

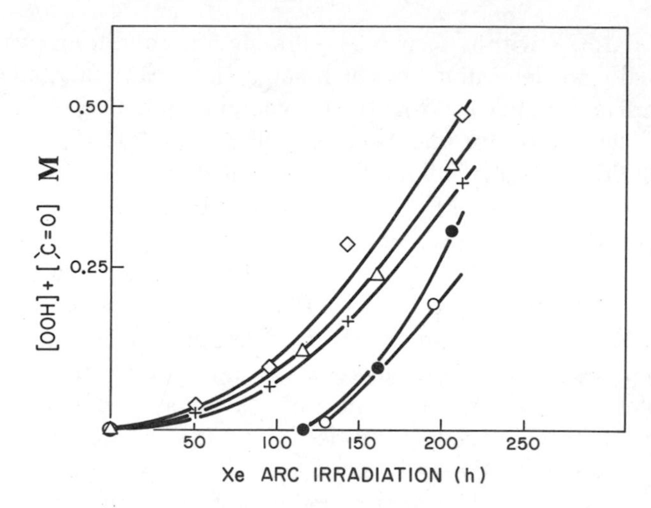

Figure 3. Effects of hopper blanketing and extrusion temperature on PPH photostability. Filaments extrude into air with 20 m min⁻¹ wind-up rate. Extruded temperature (hopper atmosphere): ○, 200°C (N_2); ●, 225°C (N_2); +, 280°C (N_2); △, 225°C (air); ◇, 280°C (air).

The overall process for strain-induced photosensitivity is summarized by Reaction 1. As well as producing —OOH chromophores, this burst

$$\begin{array}{ccccc} \text{PPH filament} & \xrightarrow[\text{draw}]{\text{Fast cold}} & \text{PP}\cdot & \xrightarrow{O_2} & \text{PPO}_2\cdot \\ & & \uparrow & & \downarrow \text{PPH} \\ & & \text{PP}\cdot & + & \text{PPOOH} \xrightarrow{h\nu} \text{Photo-initiation} \end{array} \quad (1)$$

of thermal oxidation on drawing must also partly consume any antioxidant in the system (discussed later). The hydroperoxide yields observed after the rapid cold drawing of PPH fibers are consistent with the free radical yields reported from ESR spectroscopy. Thus, fracturing polyethylene at 25°C produced $\sim 5 \times 10^{15}$ radicals cm^{-3} (i.e. $\sim 1 \times 10^{-5}M$) (*13*) while oxidation chain lengths easily can exceed 100 for $PPO_2\cdot$ propagation (*4, 9*). Breaks probably occur throughout the drawn volume (*12, 13*). Although most of these chain scissions will have a trivial contribution to fiber tensile properties, all potentially can initiate thermal oxidation and so yield light sensitive –OOH groups.

Tensile Changes. Although filaments drawn under high shear conditions (low temperature, high draw speed) oxidize rapidly on exposure to near UV irradiation, it is of great practical importance to study also the changes in tensile properties in these filaments. Embrittlement is in

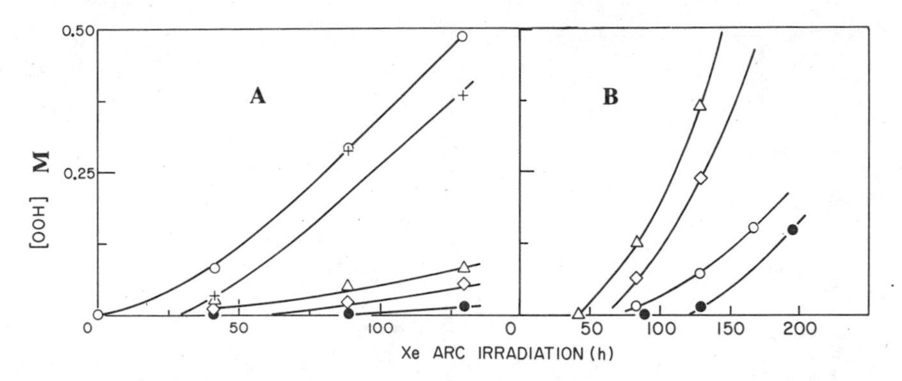

Figure 4. Effects of draw temperature and draw speed on PPH photostability. (a) Drawn 400% min^{-1} at 20°C (○), 55°C (+), 75°C (△) and 95°C (◇), or undrawn (●). (b) Drawn at 25°C with draw speeds 700% min^{-1} (△), 70% min^{-1} (◇), and 0.7% min^{-1} (○) or undrawn (●).

fact the usual practical criterion of PPH photodegradation, yield load and load-at-break being affected only slightly (*7, 14*).

Changes in the residual elongation at break for four filaments of varying draw history are shown in Figure 6. From Figure 6a, which is based on the time taken to halve the initial elongation at break, it is obvious that photosensitivity as measured by embrittlement increases in the sequence: highly oriented (hot drawn ×6 at 180°), << undrawn ≃ partially oriented (hot drawn ×2.5 at 180°) << cold drawn (×3 at 25°C).

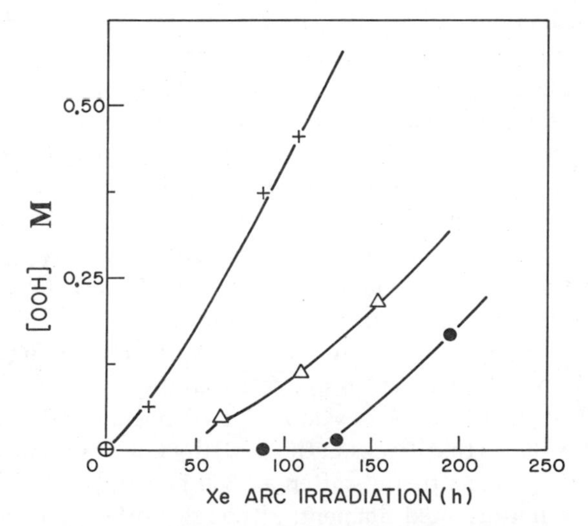

Figure 5. Effect of draw atmosphere on PPH photostability. Filaments undrawn (●) or drawn ×3 at ~ 1000% min^{-1} at 25°C in air (+) and vacuum (△).

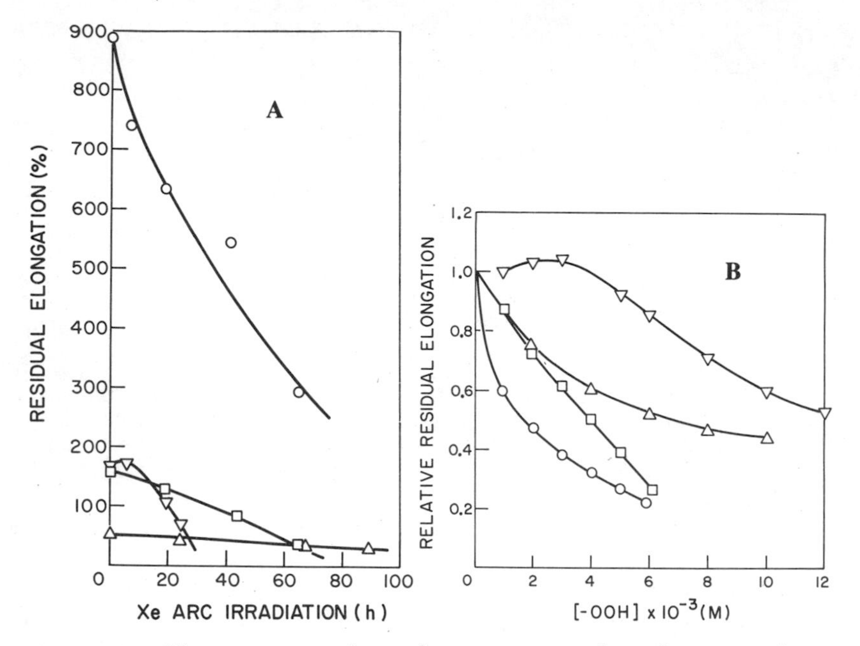

Figure 6. Changes in mechanical properties and oxidation products during PPH photooxidation. [OOH] by iodometry. Filaments undrawn (○), partially oriented ×3 at 180°C (□), highly oriented ×6 at 180°C (△), or cold drawn ×3 at 25°C (▽). All drawn in air.

However, when the relative elongation at break of each filament is plotted as a function of the associated hydroperoxide build-up (Figure 6b), this order is changed dramatically. From Figure 6b, filament embrittlement at a fixed degree of photooxidation (say $5 \times 10^{-3}M$) increases in the sequence cold drawn < highly hot drawn < partially hot drawn < undrawn. A possible explanation for this dependence on product accumulation is that the highly voided filament from cold drawing undergoes a uniform photooxidation throughout its cross-section, as the result of uniform photo-initiation by scission-generated hydroperoxide groups. The other three filaments probably photooxidize predominantly in the surface regions as found for PPH film.

All hot drawn filaments show extensive surface cracking after photooxidation (*8*), and these cracks propagate into the filament upon stressing and lead to brittle failure. An examination of peeled and then photooxidized filaments shows that cracks in the partially hot drawn filament extend deep into the fibrillar structure after ~ 300-hr irradiation (Figure 2B). For the highly hot-oriented filament, although fibrils have been cleaved, the cracks are shallow (Figure 2D). The poor interfibrillar cohesion in this fiber has allowed the spontaneous crystallization of the degraded surface to occur by slippage of the surface fibrils, without crack extension into the interior. The highly oriented filament behaves as if the central fibrillar core is unaffected by the photooxidation-induced surface restructuring.

For the partially oriented filament, interfibrillar cohesion is sufficiently high to cause relatively deep spontaneous cracking as the surface restructures by photooxidative cleavage, as reported for PPH films (*14*). Surface–core cohesion is sufficiently high to allow propagation of the surface cracks through the fiber core under tensile stress.

The uniform oxidation of the cold-drawn filament results in a high overall degree of oxidation and associated backbone scission before surface deterioration reaches the level at which spontaneous restructuring and crack formation can occur. Lateral cohesion of these fibers was much greater than for the highly oriented filaments. For example, the latter could be easily peeled, whereas the cold-drawn fiber stretched rather than fibrillated after notching.

Effects of Stabilizers. Breakage of polymer chains during draw will generate macro-alkyl radicals that can combine with oxygen to give peroxy radicals and cause some hydroperoxide formation by a thermal oxidation of the polymer (Scheme 1). The propagation of peroxy radicals in PPH can be prevented by free radical scavengers such as the hindered phenols (*4*). To study the effectiveness of additives in preventing oxidation during draw with high shear (and so enhancing UV stability), filaments containing light stable phenols were cold drawn. Simple phenols are destroyed rapidly during photooxidation, and may yield sensitizing products (*15*). The commercial stabilizers 2,4-di-*tert*-butylphenyl(4′-hydroxy-3′,5′-di-*tert*-butylbenzoate) (P-1) and octadecyl 3-(3′,5′-di-*tert*-butyl-4′-hydroxyphenyl)propionate (P-2) are excellent and fair UV stabilizers for PPH (*15*), respectively.

The incorporation of P-1 into monofilament by compounding with PPH powder and extrusion was effective only partially in reducing draw-initiated oxidation. After $\times 3$ cold draw (at 25°C) the observed –OOH level was $0.3 \times 10^{-3}M$, in the presence of 0.1 wt % P-1, in comparison with $\sim 1.1 \times 10^{-3}M$ for the stabilizer-free fiber. Furthermore, this lower level of draw-induced degradation still had a large photosensitizing effect, even in the presence of the stabilizer. After 200 hr of UV exposure in the xenon arc, the P-1 concentration was reduced by more than 20 fold in the cold-drawn filament and photooxidation was accelerating rapidly, whereas for the undrawn filament no P-1 loss was visible, and photooxidation was undetectable.

Slow cold drawing ($\times 2$ at 50% min^{-1}) of commercial (Amtech) monofilament that already was highly oriented had a similar effect on photostability. After cold drawing, filament UV lifetime was $\sim 60\%$ of that of the commercial filament, despite the presence of 0.24 wt % of P-2.

P-2 is expected to have a $PPO_2\cdot$ scavenging effectiveness similar to 2,6-di-*tert*-butyl-4-methylphenol used by Niki et al. (*4*). These workers found that the hindered phenol efficiently scavenged propagating $PPO_2\cdot$ radicals in the solid polymer at ~ 45°C. However, although P-1 and P-2 were only partially effective in preventing draw-induced oxidation, other radical scavengers may be more effective alkyl- or peroxy-radical traps under cold draw conditions. These scavengers, either used alone or in conjunction with other UV stabilizers, might minimize the draw sensitization of PPH monofilament photooxidation.

Conclusions

- For the range of samples tested, the oxidation of PPH films and filaments is independent of morphological differences, but the concentrations of UV-absorbing photo-initiators vary with production conditions.
- Photosensitivity is enhanced by melt oxidation because of dissolved O_2 and high melt temperatures, but spin-line oxidation is insignificant.
- Photo-initiator formation by bond rupture occurs during draw at low temperatures and/or high draw rates.
- Photooxidation product build-up does not correlate with filament embrittlement because of differing loci of oxidation and differing internal cohesion of filaments.
- The sensitizing effects of cold drawing can be reduced only partially by the presence of free radical scavenging phenolic additives.
- Photo-initiator formation at points of PPH cold deformation during fabrication and use will enhance the photosensitivity of these points and contribute to premature weathering failure of PPH knots, tight loops, hinges, and impact zones.

Literature Cited

1. Carlsson, D. J., Clark, F. R. S., Wiles, D. M., *Text. Res. J.* (1976) **46**, 590.
2. Carlsson, D. J., Wiles, D. M., *J. Macromol. Sci., Rev. Macromol. Chem.* (1976) **14**, 65.
3. Peterlin, A., *Polym. Eng. Sci.* (1974) **9**, 627.
4. Niki, E., Decker, C., Mayo, F. R., *J. Polym. Sci., Polym. Chem. Ed.* (1973) **11**, 2813.
5. Carlsson, D. J., Garton, A., Wiles, D. M., *J. Appl. Polym. Sci.* (1977) **21**, 2963.
6. Carlsson, D. J., Wiles, D. M., *Macromolecules* (1969) **2**, 587, 597.
7. Carlsson, D. J., Clark, F. R. S., Wiles, D. M., *Text. Res. J.* (1976) **46**, 318.
8. Blais, P., Carlsson, D. J., Clark, F. R. S., Sturgeon, P. Z., Wiles, D. M., *Text. Res. J.* (1976) **46**, 641.
9. Decker, C., Mayo, F. R., Richardson, H., *J. Polym. Sci., Polym. Chem. Ed.* (1973) **11**, 2879.
10. Connor, W. P., Schertz, G. L., *SPE Trans.* (1963) 186.
11. Burch, G. M., *The Chemical Engineer* (July, 1971) 264.

12. Peterlin, A., *J. Magn. Resonance* (1975) **19**, 83.
13. Peterlin, A., "Frontiers of Polymer Science," L. E. Murr, C. Stein, Eds., Marcel Dekker Inc., New York, 1976.
14. Blais, P., Carlsson, D. J., Wiles, D. M., *J. Polym. Sci., Polym. Chem. Ed.* (1972) **10**, 1077.
15. Carlsson, D. J., Wiles, D. M., *J. Macromol. Sci., Rev. Macromol. Chem.* (1976) **14**, 155.

RECEIVED May 12, 1977. Issued as NRCC # 16163.

6

Luminescence and Photooxidation of Light-Sensitive Commercial Polyolefins

N. S. ALLEN and J. F. McKELLAR

Department of Chemistry and Applied Chemistry, University of Salford, Salford, Lancs. M5 4WT, U.K.

G. O. PHILLIPS

Kelsterton College, North Wales Institute, Connah's Quay, Deeside, Clwyd, N. Wales, U.K.

The fluorescence and phosphorescence excitation and emission spectra of commercial polypropylene and poly(4-methylpent-1-ene) are examined using a fully compensated spectrofluorimeter. The excitation spectra of the polymers are compared with the absorption spectra of model chromophores of those believed to be present in the polymers. The fluorescence emission is associated primarily with the presence of enone and the phosphorescence is associated with dienone impurity chromophoric units. Bromination of cold hexane extracts of the polymers reduces significantly the intensity of the fluorescence confirming the presence of ethylenic unsaturation. The behavior of the luminescent enone and dienone groups during irradiation under sunlight-simulated conditions is examined also. Possible mechanisms for the participation of these chromophoric units in the photooxidation of the polymers are discussed.

Luminescence studies of commercial polymers have provided valuable information on the nature of some of the light absorbing chromophoric impurities believed to be responsible for sunlight-induced oxidation (*1–13*). The luminescence (fluorescence and phosphorescence) from commercial polyolefins has been attributed to the presence of impurity carbonyl groups (*1, 2, 5, 6, 8*), and recent work on polypropylene has indicated that these groups are conjugated with ethylenic unsaturations

0-8412-0381-4/78/33-169-068$05.00/1

(*9, 10*). In this paper we report on the identification of the luminescent impurity carbonyl species in two of the most light-sensitive polyolefins, polypropylene and poly(4-methylpent-1-ene) and examine the behavior of these groups during irradiation under sunlight-simulated conditions.

Experimental

Materials. Commercial polypropylene and poly(4-methylpent-1-ene) powders containing no commercial additives were supplied by I.C.I. (Plastics Division) Ltd. The powders were vacuum pressed into film of 200-μ thickness at 190° and 280°C respectively, for 1 min. The *n*-hexane was of spectroscopic quality.

Hexane Extraction and Bromination. The polymer powders (20 g) were shaken with *n*-hexane (100 mL) for 48 hr. The hexane extracts then were filtered through glass wool. A few micro-liters (10 μL) of liquid bromine were added to a portion of each extract (20 mL) and were allowed to stand for 30 min. The excess unreacted bromine was removed by bubbling nitrogen through the extracts for 15 min until shaking with an aqueous silver nitrate solution gave no precipitate. The solutions then were allowed to equilibrate with air and compensation was made for the evaporation of solvent.

Luminescence Measurements. Corrected fluorescence and phosphorescence spectra were obtained using a Hitachi Perkin–Elmer MPF-4 spectrofluorimeter. Fully corrected excitation spectra were obtained also using the newly developed spectrofluorimeter of Cundall, et al. (*see* Ref. 9, instrumental details to be published).

Photooxidation. Polymer films were irradiated in a Xenotest-150 weatherometer (supplied by Original Hanau, Quartzlampen, G.m.b.H.) set up for natural sunlight-simulated conditions (45°C; 50% relative humidity). The rate of photooxidation of the polymer films was measured by monitoring the build-up in the nonvolatile carbonylic oxidation products absorbing at 1710 cm^{-1} using a Perkin–Elmer 157G ir spectrophotometer (*4, 6*). The carbonyl index = $[\log_{10} I_O/I_t)/d] \times 100$.

Results and Discussion

Polymer Luminescence Spectra. Figure 1 shows typical fluorescence and phosphorescence excitation and emission spectra obtained from commercial polypropylene film (or powder). Poly(4-methylpent-1-ene) exhibits similar spectra to those of polypropylene. The excitation spectrum for the fluorescence has two distinct maxima at 230 and 285 nm while that of the phosphorescence has only one distinct maximum at 270 nm with rather weak and diffuse structure above 300 nm. It is clear from these results that the fluorescent and phosphorescent chromophoric species cannot be the same. This, of course, does not rule out the fact that both may arise from carbonyl emitting species, as will be shown later, since these chromophoric groups when linked to ethylenic unsaturation can have quite distinct absorption (*14*) and emission spectra (*15, 16, 17*).

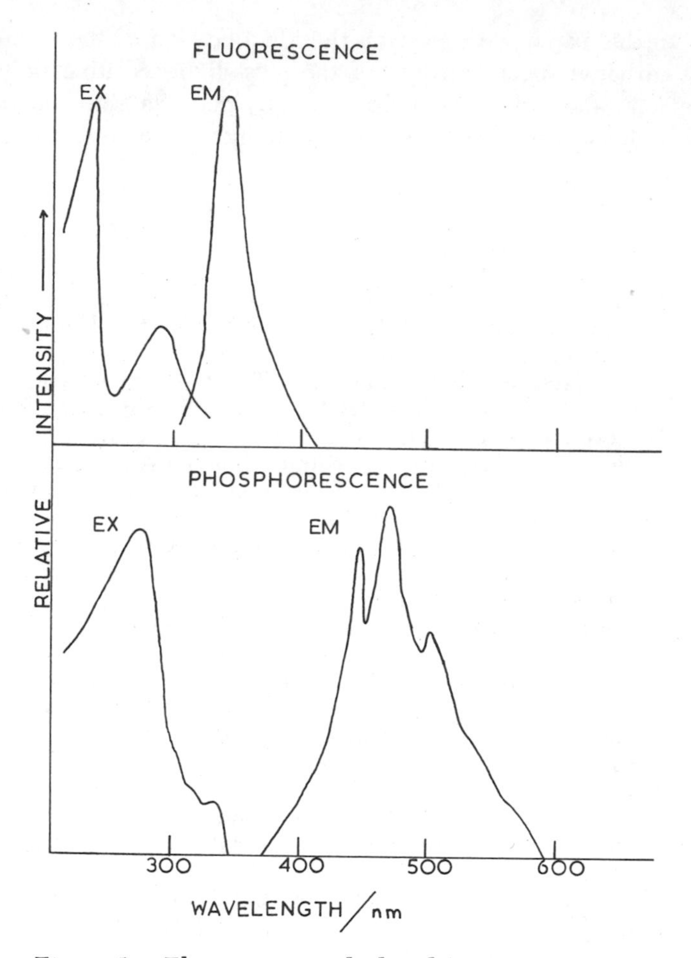

Figure 1. Fluorescence and phosphorescence excitation (EX) and emission (EM) spectra of polypropylene film

Identification of the Fluorescent Species. Figure 2 compares the fluorescence excitation spectra of the polymers with the absorption spectrum of a simple α,β-unsaturated carbonyl compound (pent-3-ene-2-one) (*13*). The three spectra are very similar. Figure 2 shows also that the fluorescence from the polymers in the region 300–400 nm cannot be caused by the presence of polynuclear aromatic hydrocarbons such as naphthalene as postulated earlier by Carlsson and Wiles (*13*). Furthermore, as shown below, the excitation spectrum also differs significantly from that of a fully saturated aldehyde or ketone.

Identification of the Phosphorescent Species. A comparison of the fluorescence and phosphorescence excitation spectra clearly shows that they cannot be caused by the same chromophoric species in the polymer.

This, therefore, diminishes the possibility of a simple α,β-unsaturated carbonyl chromophore being responsible for the majority of the phosphorescence emission. However, these groups may indeed phosphoresce (*16, 17*) but they are overlaid evidently by a much stronger emitting impurity species.

In Figure 3 we compare the polymer phosphorescence excitation spectra with other possible phosphorescent carbonyl chromophores known to be present in the polymer (*1, 8, 9, 10*). The absorption of a typical dienone (*14*) or -al), *trans,trans*-hexa-2,4-dienal, more closely matches that of the polymer excitation than does a typical long chain aliphatic aldehyde or ketone (*14*).

A further interesting feature of the phosphorescence emission from the polymers is that it is long-lived (*5*). For the polypropylene and poly(4-methylpent-1-ene) samples examined in this study the emission lifetimes were 1.2 and 0.7 sec, respectively. Long-lived phosphorescence

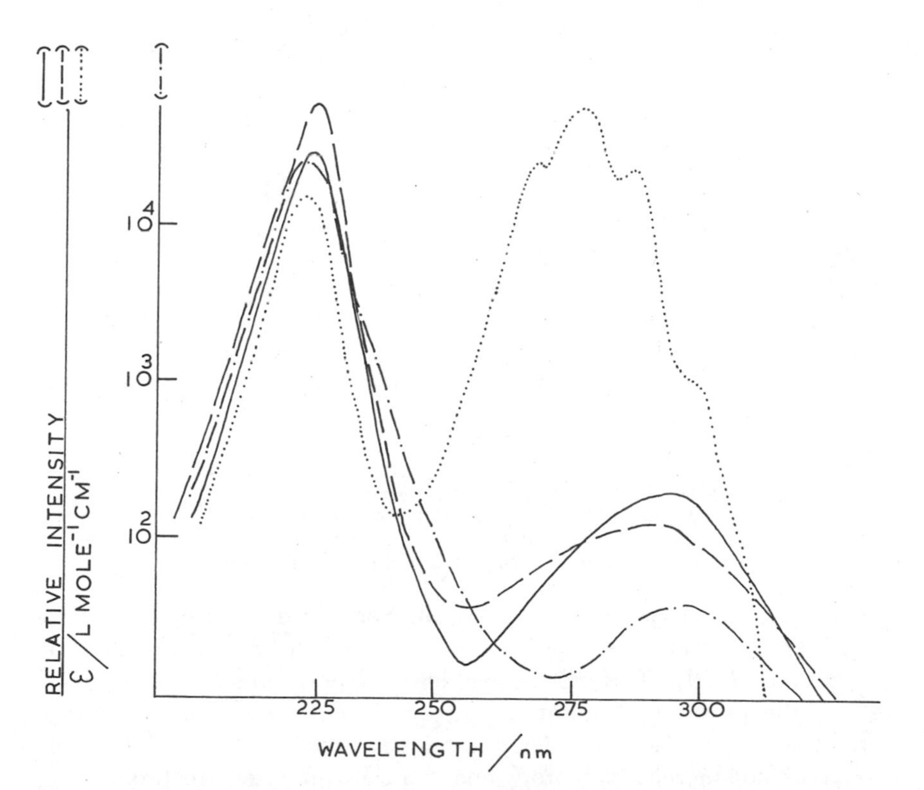

Figure 2. Comparison of the fluorescence excitation spectra of polypropylene (——) and poly(4-methylpent-1-ene) (– – –) films with the absorption spectrum of pent-3-ene-2-one (–·–·) and the fluorescence excitation spectrum of naphthalene (····) in n*-hexane* (10^{-5}M)

emission ($\tau_p > 0.1$ sec) from rigid, cyclic α,β-unsaturated carbonyl compounds is not uncommon (*15, 16, 17*). In fact, when present in the polymer matrix, deactivation of these groups by rotation around the C=C bond will be inhibited resulting in a relatively long-lived, excited triplet state. Interestingly, long-lived phosphorescence emission of a similar type has been reported from other commercial polymers such as the polyamides (*4*) and polybutadiene (*11, 12*).

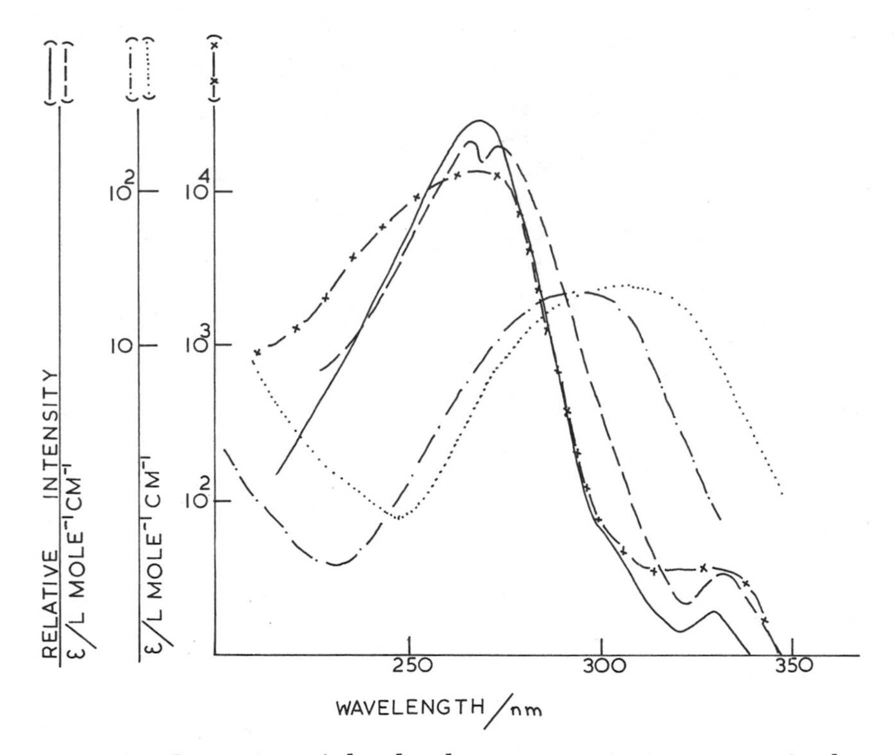

Figure 3. Comparison of the phosphorescence excitation spectra of polypropylene (——) and poly(4-methylpent-1-ene) (– – –) with the absorption spectra of trans,trans-*2,4-hexadienal (–×–×), 2,2,4,4-tetramethylpentan-3-one (–·–·), and 3-methylpentanal (····) in* n-*hexane*

Hexane Extraction and Bromination. Some of the fluorescent enones are extractable from the amorphous regions of the polymer matrix by cold hexane (*1, 13*). This process enables us to perform a simple chemical test for the presence of unsaturation, i.e., bromination (*18*). Figure 4 shows the fluorescence excitation and emission spectra of the cold hexane extract of commercial polypropylene before and after treatment with bromine. After bromination there is a significant reduction in the fluorescence intensity, confirming the presence of unsaturation. A similar result was obtained on brominating a cold hexane extract of poly-4-

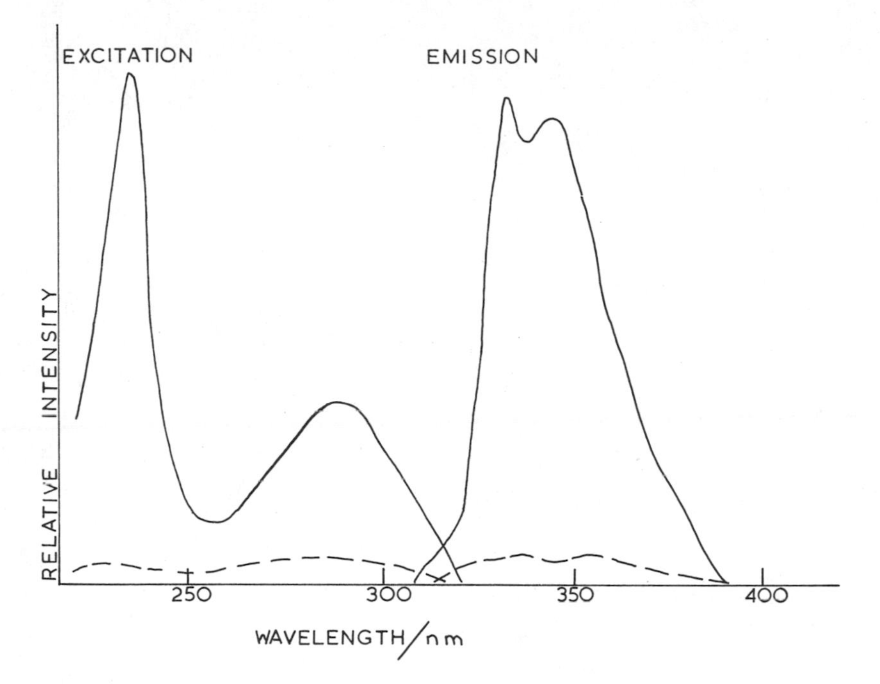

Figure 4. Fluorescence excitation and emission spectra of a cold hexane extract of polypropylene powder before (——) and after (- - -) bromination

methylpent-1-ene). Excess, unreacted bromine was removed carefully to eliminate quenching of the fluorescence by heavy-atom, spin orbit coupling (*19*) (*see* Experimental). The phosphorescent dienones were not extractable from the polymers.

Effect of Photooxidation on the Luminescent α,β-Unsaturated Carbonyl Groups. The longest wavelength $n - \pi^*$ absorption bands (*15, 16, 17*) of both the enone and dienone chromophoric impurities overlap the high energy end of the spectrum of natural sunlight known to be harmful to the commercial polyolefins (*20, 21*) (i.e., 300–350 nm).

During irradiation in a Xenotest-150 the fluorescent enones were consumed gradually as shown by a gradual reduction in the intensity of the fluorescence excitation spectrum (Figure 5). The phosphorescent dienones were consumed also during irradiation. However, at the onset of embrittlement there was a bathochromic shift in the phosphorescence excitation wavelength maximum. Figure 6 shows that this bathochromic shift was caused by the conversion of the dienones to saturated carbonyl groups that absorb at longer wavelengths and have a much smaller extinction coefficient (*14*). During the early stages of irradiation the bathochromic shift was not observed because of the much stronger

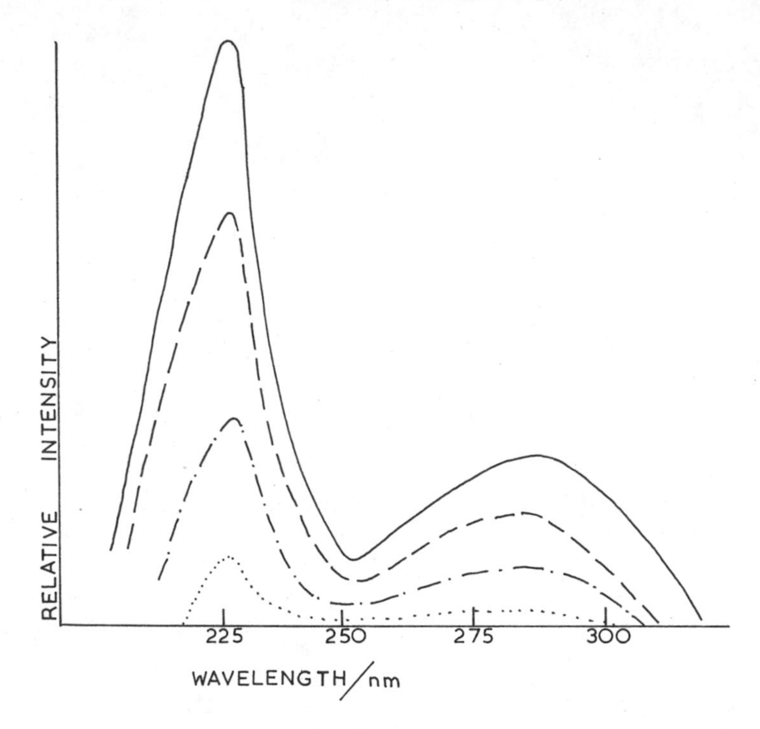

Figure 5. Intensity of the fluorescence excitation spectrum of polypropylene film before (——) and after 75-hr (- - -), 150-hr (-·-·), and 250-hr (····) irradiation in a Xenotest-150 weather-ometer

absorption by the dienones compared with the saturated carbonyls (Figure 6). The fluorescence excitation spectrum showed no such corresponding shift since simple saturated aliphatic carbonyl groups do not fluoresce (*19*).

Further evidence supporting the participation of α,β-unsaturated carbonyl impurities in the sunlight-induced oxidation of the light-sensitive polyolefins is provided in the following experiment for polypropylene. Four samples of polypropylene powder that exhibited different fluorescence intensities at 340 nm were selected from different batches of commercial polymer. The samples did not contain antioxidants, or any other stabilizing additives, and had similar MFI values of about 20 and *Ti* residues of about 25 ppm. Thus, the possible effects of differing polymer molecular weights and catalyst residue contents on the rate of photooxidation were minimized. Each sample of polymer powder was vacuum-pressed into film of 100-μ thickness to eliminate as many differences as possible owing to their processing (oxidative) history. Figure 7 shows that the rate of photooxidation was greatest with those samples showing the higher initial intensity of fluorescence emission at 340 nm.

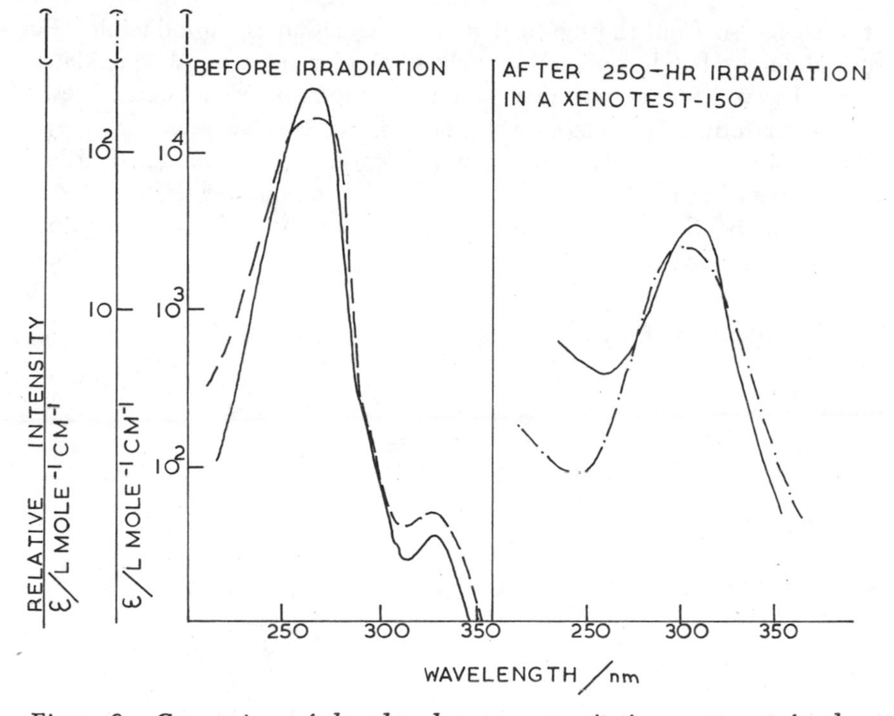

Figure 6. Comparison of the phosphorescence excitation spectrum of polypropylene film (——) before and after irradiation for 250 hr in a Xenotest-150 weatherometer with the absorption spectra of trans,trans-*hexa-2,4-dienal (– – –) and 2,2,4,4-tetramethylpentan-3-one (–·–·) in* n-*hexane, respectively*

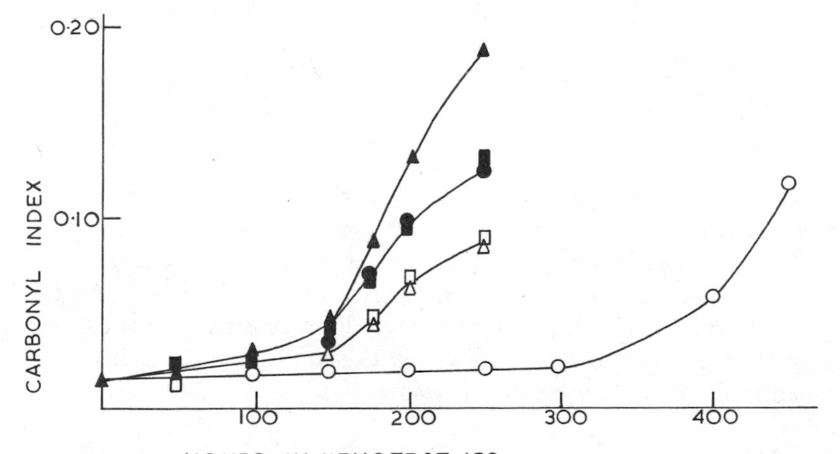

Figure 7. Variation of carbonyl index vs. time of irradiation in a Xenotest-150 weatherometer (50% RH; 45°C) for different batches of commercial polypropylene. Before irradiation the samples had relative fluorescence intensities at 340 nm of (▲), 1.0; (■, ●), 0.73; (□, △), 0.44; (○), 0.25. EX λ_{max} = 230 and 285 nm.

It is seen also from the figure that the correlation is reproducible. For four different batches of polypropylene, similar rates of photooxidation were observed for samples that have the same fluorescence intensities.

Light-Induced Mechanism Involving α,β-Unsaturated Carbonyls. The photochemistry of α-β-unsaturated carbonyl compounds is documented well in the literature (*11, 12, 22*). If we consider the enone case for simplicity, these chromophoric units generally undergo two photoreactions. These are:

1. The formation of β,γ-carbonyls

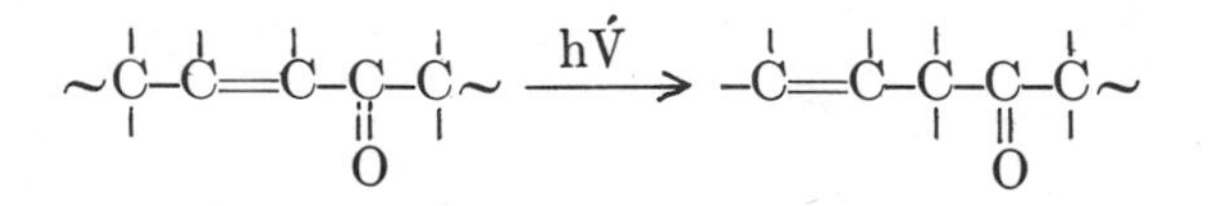

2. Cross-linking between adjacent α,β-unsaturated carbonyls

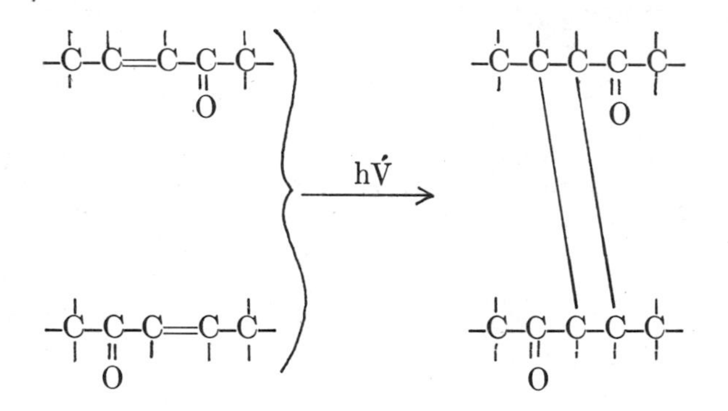

Our results in Figure 6 show that during photooxidation of the polymers the α,β-unsaturated carbonyl groups are converted into saturated ketonic/aldehydic groups that can themselves be converted subsequently to nonluminescent products by Norrish Type I and II processes, e.g., carboxylic acids (*2*). For the two light-sensitive polyolefins considered here, Reaction 1 is likely to be the more important, because the concentration of the species is very low.

Acknowledgments

The authors thank D. G. M. Wood, ICI (Plastics Division) Ltd., for helpful discussions and for supplying the polymer samples used in this study.

Literature Cited

1. Charlesby, A., Partridge, R. H., *Proc. R. Soc. London* (1965) **A283**, 312, 329.
2. Allen, N. S., McKellar, J. F., Phillips, G. O., *J. Polym. Sci., Polym. Chem. Ed.* (1974) **12**, 2647.
3. Allen, N. S., McKellar, J. F., Phillips, G. O., *J. Polym. Sci., Polym. Chem. Ed.* (1974) **12**, 1233, 2623.
4. Allen, N. S., McKellar, J. F., Phillips, G. O., *J. Polym. Sci., Polym. Chem. Ed.* (1975) **13**, 2857.
5. Allen, N. S., Homer, J., McKellar, J. F., Phillips, G. O., *Brit. Polym. J.* (1975) **7**, 11.
6. Allen, N. S., Homer, J., McKellar, J. F., *J. Appl. Polym. Sci.* (1976) **20**, 2553.
7. Allen, N. S., McKellar, J. F., *J. Appl. Polym. Sci.* (1976) **20**, 1441.
8. Briggs, P. J., McKellar, J. F., *J. Appl. Polym. Sci.* (1968) **12**, 1825.
9. Allen, N. S., Cundall, R. B., Jones, M. W., McKellar, J. F., *Chem. Ind. London* (1976) 110.
10. Allen, N. S., Homer, J., McKellar, J. F., *Chem. Ind. London* (1976) 692.
11. Beavan, S. W., Phillips, D., *J. Photochem.* (1974) **3**, 349.
12. Beavan, S. W., Phillips, D., *Eur. Polym. J.* (1974) **10**, 593.
13. Carlsson, D. J., Wiles, D. M., *J. Polym. Sci., Polym. Lett. Ed.* (1973) **11**, 759.
14. "Directory of Molecular Spectra," U.V. Atlas of Organic Compounds, Vol. II, Butterworths, 1966.
15. Zimmerman, H. E., Binkley, R. W., McCullough, J. J., Zimmerman, G. A., *J. Am. Chem. Soc.* (1967) **89**, 6589.
16. Cargill, R. L., Bundy, W. A., Pond, O. M., Sears, A. B., Saltiel, J., Winterle, J., *Mol. Photochem.* (1971) **3**(2), 123.
17. Loutfy, R. O., Morris, J. M., *Chem. Phys. Lett.* (1973) **19**(3), 377.
18. Mann, F. G., Saunders, B. C., "Practical Organic Chemistry," 4th ed., p. 84, Longmanns, London, 1960.
19. Calvert, J. G., Pits, J. N., "Photochemistry," John Wiley & Sons, Interscience, New York, 1966.
20. Allen, N. S., McKellar, J. F., *Chem. Soc. Rev.* (1975) **4**(4), 533.
21. Ranby, B., Rabek, J. F., "Photodegradation, Photooxidation, and Photostabilization of Polymers," John Wiley & Sons, Interscience, New York, 1976.
22. Bellus, D., Kearns, D. R., Schaffner, K. S., *Helv. Chim. Acta* (1969) **52**, 971.

RECEIVED May 12, 1977.

7

Oxidative Degradation of Polymers. III.

Photooxidation of Poly(vinyl alcohol) in Aqueous Solution

ETSUO NIKI, YORIHIRO YAMAMOTO, and YOSHIO KAMIYA

Department of Reaction Chemistry, Faculty of Engineering,
University of Tokyo, Hongo, Tokyo 113, Japan

The rates and products of poly(vinyl alcohol) (PVA) photooxidation were measured over a temperature range of 30°–90°C. Oxidation was initiated with 253.7-nm light, and several model compounds were included. The PVA photooxidation rate was autocatalytic and relatively insensitive to temperature with the major products being carbon dioxide, acids, peroxides, polymeric β-hydroxyketone, and hydrogen peroxide. Acids were mostly formic acid and carboxylic acids at the polymer chain ends. Several molecules of carbon dioxide and acid were formed per statistical chain scission. The mechanism of the photooxidation of PVA is discussed.

PVA is one of the most widely used polymers with the characteristic property that it is soluble in water but insoluble in most organic solvents. PVA oxidation by molecular oxygen in aqueous solution is interesting fundamentally and practically. From a fundamental point of view, it gives information on the chemistry and reactions of oxy radicals in aqueous solution. Practically, it is important in suppressing oxidative deterioration and such ecological problems as wastewater treatment and chemical recovery.

PVA oxidation has been studied by several groups, and it has been observed that the degree of polymerization decreases when PVA is oxidized with hydrochloric acid (*1*), potassium permanganate (*2–4*) potassium bichromate (*3, 5, 6*), hydrogen peroxide (*7, 8*), periodic acid, and ozone (*9*). However, only a limited number of studies have been performed on the oxidation of PVA by molecular oxygen. Shiraishi and Matsumoto (*10*) observed that it is oxidized by molecular oxygen in alkaline aqueous solution, which causes discoloration and a decrease in

0-8412-0381-4/78/33-169-078$05.00/1

the degree of polymerization. Dulog, Kern, and Kern (*11*) studied the photosensitized oxidation of PVA in water at 30°C using a high pressure mercury lamp as a light source. They found that PVA was cleaved statistically, and that one carboxylic group accompanied each main chain scission. They also observed hydrogen peroxide and carbon dioxide as oxidation products. Photolysis of PVA preoxidized with sodium hypochlorite or periodic acid was also studied by using ESR spectrometry (*12*). The photooxidative cross-linking of PVA was also studied by several investigators in the presence of alkali dichromates from the viewpoint of photoengraving processes (*13, 14*). Recently, Indicator, Auerbach, and co-workers (*15–18*) studied the rate of PVA oxidation and cleavage in the presence of *tert*-butyl hydroperoxide and various metal salts.

However, the mechanism and products of PVA oxidation by molecular oxygen are not clear. The primary objective of this study is to measure rates and products of photooxidation of PVA and some model compounds in aqueous solution and to elucidate the oxidation mechanism. Such a fundamental understanding is needed to learn how to retard, accelerate, and direct photooxidation.

The autoxidation of an alcohol gives the corresponding ketone or aldehyde and hydrogen peroxide, and the yield of hydrogen peroxide in the autoxidation of PVA is of interest. It is also expected that photochemical cleavage of ketone will reduce the molecular weight. Another product of interest is acetaldehyde, formed by the reverse aldol-type reaction, (*19, 20*) where β-hydroxyaldehyde or ketone is cleaved in alkaline aqueous solution.

Experimental

Materials. PVA (kindly supplied by Dr. Yasui of Kuraray Co. Ltd.) was purified by the conventional method using water and methanol as solvent and precipitant respectively. The dried PVA was then saponified completely by slow addition to 0.1*N* $NaOH–CH_3OH$ solution. This mixture was kept at 40°C for 2 hr and at room temperature for 2 days. The polymer was then washed with methanol several times and dried in the oven under vacuum. The degree of saponification, intrinsic viscosity at 45°C in water, and number average degree of polymerization of the purified PVA were determined as 99.49%, 0.692 dl/g and 1750 respectively. Model compounds were the highest grade available commercially. Water was purified by passage through an ion exchange resin.

Oxidation Procedures. Ten mL of 4% aqueous solution of PVA was put into a 50-mL quartz vessel, which was evacuated by the freeze and thaw method, and oxygen was introduced. For the low molecular weight compounds, 1 mL of substrate was oxidized in 10 mL of water. The solution was stirred vigorously with a magnetic stirrer. Oxidation was started by irradiation of light after thermal equilibrium was reached. A 30-W low pressure mercury lamp was used as a light source. Initial and final gases were determined by measuring the total amount of gases noncondensable at −80°C with a Toepler pump and by gas chromatographic analysis. Oxygen, hydrogen, carbon monoxide, and methane were analyzed with a molecular sieve 13X column; carbon dioxide was analyzed with an active charcoal column. A Teflon valve equipped with Swagelok fittings, type 6VD of Nupro Co., was used to connect and to disconnect the vessel to the Toepler pump.

The oxidation and photochemical decomposition of simple model compounds were conducted under similar conditions.

Analytical Methods. Peroxides were determined by iodometric titration. Hydrogen peroxide was determined from the difference between the titers before and after treatment with catalase that decomposes hydrogen peroxide selectively. It was ascertained by separate experiments that hydrogen peroxide could be decomposed quantitatively by catalase over the pH range 2.8–8.0.

Acids were measured by conductmetric titration (*21*) and isotachophoresis analysis. An aqueous solution of 0.01*M* glutamic acid was used as a terminal solution and an aqueous solution containing 0.01*M* L-histidine and 0.01*M* L-histidine hydrochloride was used as a leading solution.

Low boiling products were analyzed by GLC equipped with FID. For most analyses, Porapak Q and poly(ethylene glycol) columns were used.

Total carbonyl groups in the polymer chain were determined by the method of Imai and Kazusa (*22*) using 2,4-dinitrophenylhydrazine. The ketone associated with double bonds, $-C(O)-(C{=}C)_2-$, was measured by UV spectra at $\lambda_{max} = 270$ nm with $\epsilon = 2.7 \times 10^3 M^{-1}$ cm^{-1} (*23*). β-Hydroxyketones, peroxides, and 1,2-glycols and α-hydroxyketones were decomposed respectively with dilute alkali (reverse aldol reaction), potassium iodide in sulfuric acid aqueous solution, and periodic acid in sulfuric acid aqueous solution and then measured by GLC.

After the oxidation, a portion of the reaction mixture was placed on a polystyrene sheet and dried at 60°C overnight to obtain a film, that was analyzed by ir spectroscopy. The viscosities of unoxidized and oxidized PVA were determined at 45°C in water with a modified Ostwald viscosimeter.

Results

The results of PVA photooxidations in aqueous solution are summarized in Table I. The rate of PVA thermal autoxidation in the dark initiated with water-soluble initiators was quite low, and the contribution of nonphotolytic oxidations was small. Figure 1 shows the rate of oxygen uptake and product formation at 70°C and shows that the oxidation is slightly autocatalytic. The major oxidation products are carbon dioxide, acids, and peroxides along with hydrogen peroxide. Formation of carbon dioxide, acids, and peroxides increased with increasing oxygen uptake. The decrease in hydrogen peroxide is attributed to its photolysis (*see* later text), which may be responsible for part of the autocatalysis.

The acids were mainly formic acid and carboxylic acid at the polymer chain ends. Although formic acid could not be determined directly by

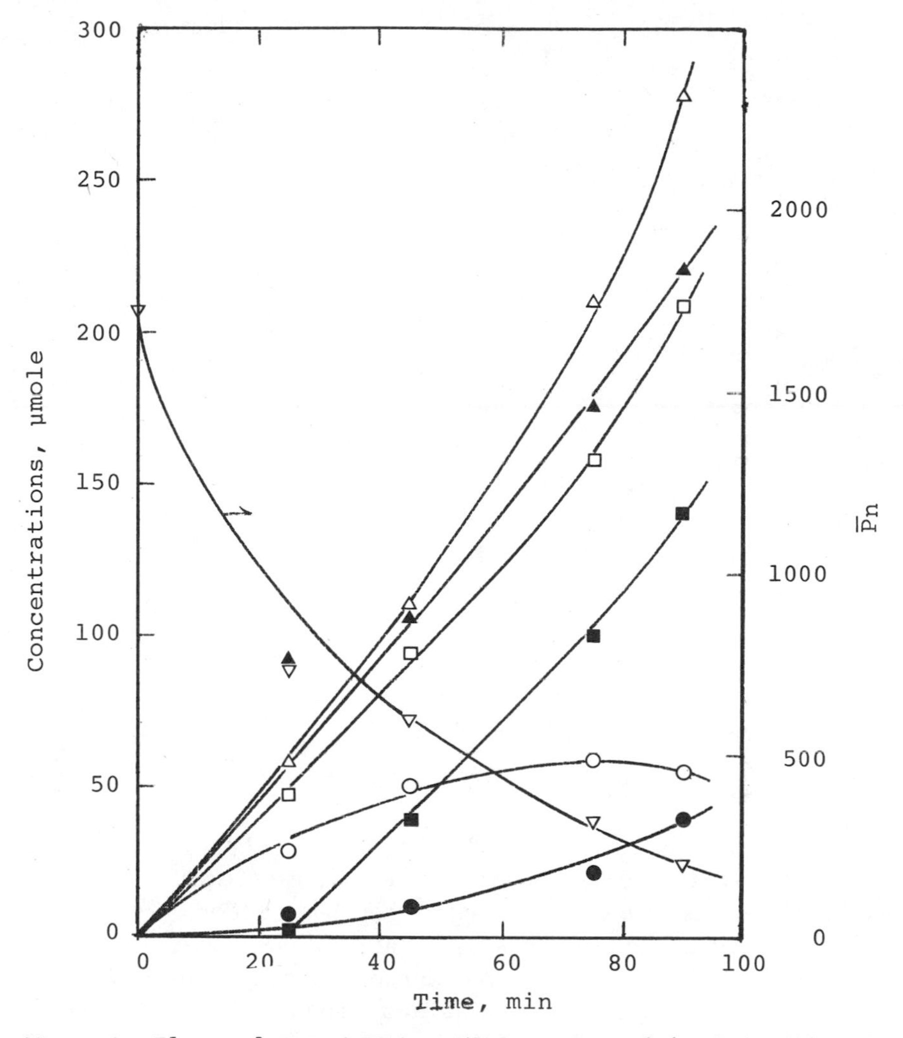

Figure 1. Photooxidation of PVA at 70°C in water. (△) $\Delta O_2/3$, (▲) acids, (□) CO_2, (■) peroxides, (○) H_2O_2, (●) total number of statistical chain scissions, (▽) Pn.

GLC with an FID detector, it was confirmed by isotacophoresis analysis and by the formation of methyl formate when the product solution was treated with methanol containing sulfuric acid. No other low boiling carboxylic acid was observed by GLC except a trace amount of acetic acid. Therefore, the formation of formic acid was estimated as indicated in Table I. This estimate agreed fairly well with that determined by isotacophoresis analysis. Little formation of other low boiling products was observed.

Table I. Photooxidation of

Run No.	*1*	*2*	*3*
Temperature, °C	*70*	*70*	*70*
Time (min)	*25*	*45*	*45*
PVA (μmol)	*8870*	*8870*	*8370*
O_2 uptake (μmol)	*174.0*	*329.5*	*329.8*
Products in μmol			
CO	3.2	2.1	trace
CO_2	46.6	93.6	102.5
peroxides[a]	2.3	38.6	44.0
H_2O_2	29.5	49.9	50.5
acids	90.4	105.4	135.7
HCOOH[b]	40.9	58.0	88.5
$-C(=O)-(C=C)_2-$	1.4	3.0	2.8
$[\eta]$[c]	0.428	0.380	0.362
$\bar{P}n$[d]	740	599	551
S_o (μmol)[e]	6.9	9.7	10.4
CO_2/S_o	6.8	9.6	9.9
$HCOOH/S_o$	5.9	6.0	8.5

[a] This value does not include hydrogen peroxide.
[b] HCOOH = acids − (−OAc remaining) − $2S_o$. The remaining acetoxy group was estimated from infrared spectrum.

The total number of statistical chain scissions during oxidation, S_o, was calculated from the following equation,

$$S_o = [\text{PVA}](1/\bar{P}n - 1/\bar{P}n_o) \quad (1)$$

where Pn_o and Pn are respectively the degrees of polymerization of initial and oxidized PVA, determined by intrinsic viscosity. Table I shows that intrinsic viscosity and degree of polymerization decrease considerably, whereas the number of total statistical chain scissions is small compared with the amount of carbon dioxide and acids formed. The significance of this relation is discussed later. These results contrast with those of Dulog et al. (*11*), where the acid formed and number of statistical chain scissions were approximately the same and only a small amount of carbon dioxide was observed.

Figure 2 shows the change in PVA ir absorption spectra. The bands at 1715 and 1560 cm^{-1} of initial PVA may be ascribed to the remaining acetoxy groups. The bands at 1560 cm^{-1} decreased whereas the band at 1715 cm^{-1} increased as the oxidation proceeded. This suggests that the remaining acetoxy group was removed from the main chain and that the band at 1715 cm^{-1} for the oxidized PVA must be mainly the result of new carbonyl and carboxylic groups in the polymer chain.

The total amount of carbonyl group was determined as described in the experimental section, and 22.4 μmol of carbonyl group were found

Poly(vinyl alcohol) in 10 mL Water

4	*5*	*6*	*7*	*8*
70	*70*	*90*	*50*	*30*
75	*90*	*45*	*45*	*45*
8870	*8870*	*8870*	*8630*	*8630*
630.4	*832.1*	*433.9*	*257.4*	*147.6*
1.4	trace	4.3	trace	trace
158.1	208.1	89.9	106.4	66.3
99.9	140.7	54.5	12.6	20.1
59.0	54.5	56.8	46.8	trace
175.7	220.9	125.5	105.4	95.4
115.1	132.9	73.1	62.9	58.5
13.0	20.9	5.1	1.7	0.5
0.270	0.206	0.348	0.411	0.474
328	203	513	689	888
22.0	38.6	12.2	7.6	4.8
7.2	5.4	7.4	14.0	13.8
5.2	3.4	6.0	8.3	12.2

[c] Intrinsic viscosity after photooxidation. Initial $[\eta]_o = 0.692$.
[d] Degree of polymerization. Initial $\overline{Pn}_o = 1750$.
[e] Number of statistical chain scissions calculated from Equation 1.

in Run 3. Figure 3 shows the change in UV spectra. It can be seen that the absorption at 270 nm increases with time. This band is caused by –C(=O)–(C=C)$_2$–, and from its reported extinction coefficient the amount of this conjugated ketone was estimated as 2.8 μmol. The amount of non-conjugated ketone was estimated as 3.0 μmol from the formation of acetone by the treatment of oxidized PVA with sodium hydroxide. Since little absorption was observed at 330 nm, which was caused by –C(=O)–(C=C)$_3$– (*23*), and at longer wavelength, the amount of –C(=O)–C=C– was estimated as 22.4–2.8–3.0 = 16.6 μmol. Formation of –C(=O)–(C=C)$_2$– in each experiment, as measured from ultraviolet absorption spectra, is shown in Table I.

The effects of temperature are also summarized in Table I, which shows that the rate of oxidation increases only moderately with increasing temperature and the apparent activation energy for chain scission is 2.9 kcal/mol. Such a low value is characteristic of photooxidations. Figure 4 shows the effect of temperature on the UV spectra of photooxidized PVA. It can be seen that the formation of diene conjugated with ketone increased with increasing temperature.

Tables II and III summarize representative photooxidations of model compounds.

Methanol. One characteristic of PVA photooxidation is the formation of formic acid with high yield. Formic acid was believed to arise

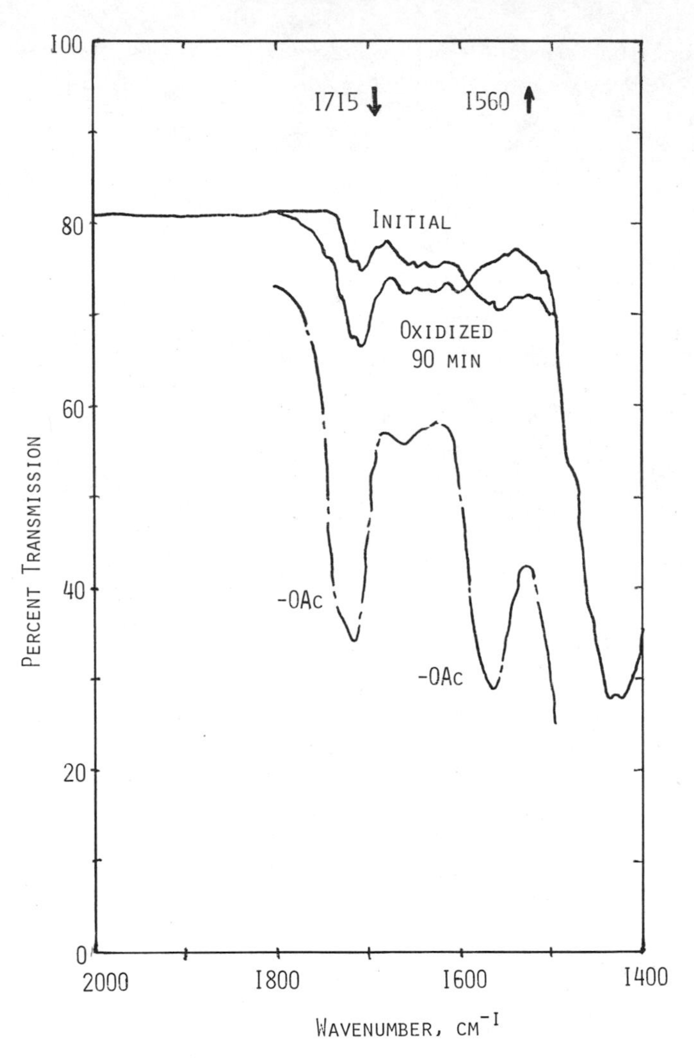

Figure 2. IR spectra of initial and oxidized PVA

from formaldehyde. To prove this assumption, methanol was photooxidized at 70°C in water, and the result in Table II shows that formaldehyde and formic acid are indeed the major products from methanol. Carbon dioxide, hydrogen peroxide, carbon monoxide, hydrogen, and methane were also observed as minor products. In this experiment, the remaining oxygen was small. If enough oxygen had been supplied, formation of hydrogen and methane would have been smaller.

Acetone. As described later, the reactions of acyl and acylperoxy radicals play an important role in the photooxidation of PVA. Therefore, the photochemical reactions of acetone were studied. The primary reaction of acetone must be the photochemical cleavage of carbonyl group to give acetyl and methyl radicals. Table II shows that in the absence of oxygen most of the acetyl radicals give carbon monoxide and methyl radicals, but approximately 20% of acetyl radicals combine to yield biacetyl, and fewer abstract hydrogen atoms to give acetaldehyde. Methyl ethyl ketone must arise from the combination of methyl and acetonyl radicals which must be formed by the hydrogen atom abstraction from acetone. The material balance for methyl radicals is poor. Most of the missing methyl radicals must be present as ethane which was not determined by our analytical procedure.

Under oxygen, the product distribution is quite different from that under vacuum. Since little methane, biacetyl, and acetaldehyde are formed, substantially all of the acetyl and methyl radicals react with oxygen to give acetylperoxy and methylperoxy radicals. Acetylperoxy radicals can abstract hydrogen to give peracetic acid or react with each other or with other peroxy radicals. Most of the biomolecular interactions of acetylperoxy radicals give methyl radicals and carbon dioxide (*24, 25*). Methylperoxy radicals can abstract hydrogen to give methyl hydroperoxide or react with themselves or with other peroxy radicals to form formaldehyde and methanol. Formaldehyde is reactive toward both light and heat and gives carbon monoxide, carbon dioxide, and formic acid.

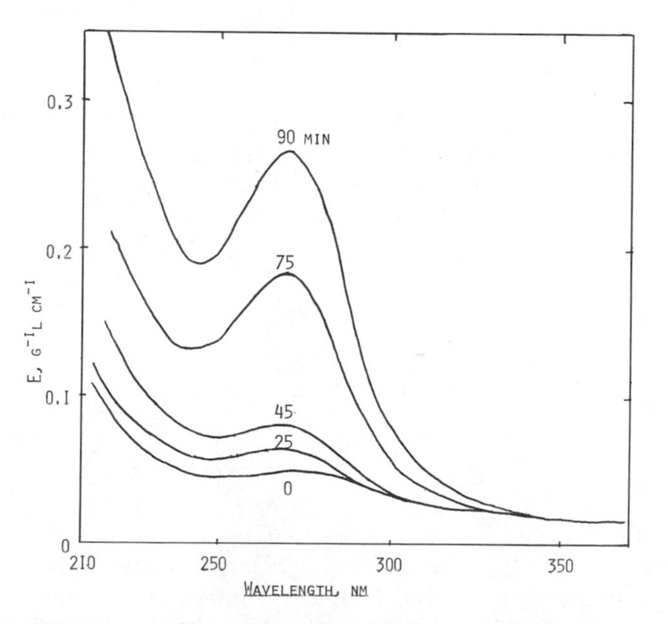

Figure 3. Effect of time on UV spectra of photooxidized PVA

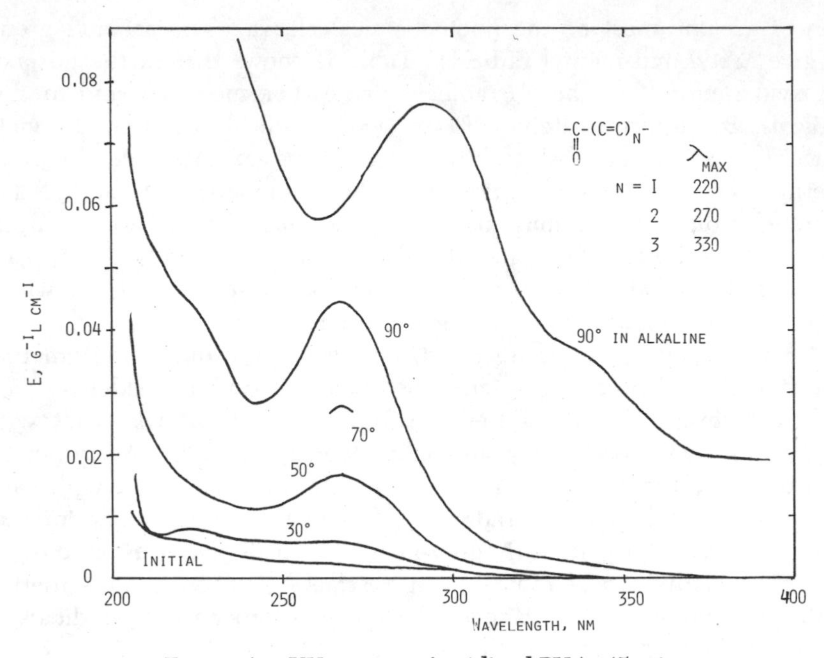

Figure 4. UV spectra of oxidized PVA, 45 min

***tert*-Butyl alcohol.** This model compound was studied to obtain information on the reactions of the α-hydroxy primary alkyl radical which is one of the important transient radicals in the photooxidation of PVA. The photolysis of *tert*-butyl alcohol in vacuo gave acetone, methane, hydrogen, and smaller amounts of carbon monoxide. Acetone is formed by any of the following pathways (*26*).

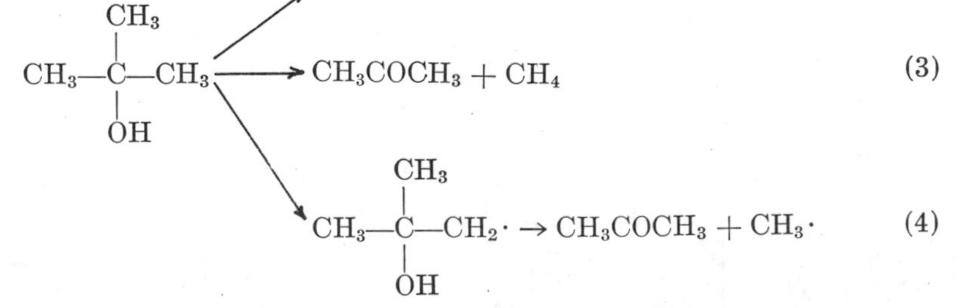

$$H\cdot + (CH_3)_3CO^{\ddagger} \rightarrow CH_3\cdot + CH_3COCH_3 \quad (2)$$

$$(CH_3)_3COH \rightarrow CH_3COCH_3 + CH_4 \quad (3)$$

$$CH_3-C(CH_3)(OH)-CH_2\cdot \rightarrow CH_3COCH_3 + CH_3\cdot \quad (4)$$

The marked decrease in the formation of hydrogen and methane under oxygen implies the presence of hydrogen atoms and methyl radicals. Most of the products and their distribution are similar to those of acetone photolysis. The two exceptions are 2-methylpropylene glycol and 2-hydroxy-2-methylpropanal, both of which arise from the 2-hydroxy-

Table II. Photooxidations of Model Compounds at 70°C in Water[a]

Run No.	*9*	*10*	*11*	*12*	*13*
Substrate	*MeOH*	*Acetone*		tert-*BuOH*	
Atmosphere	O_2	*vac*	O_2	*vac*	O_2
Time (min)	*60*	*120*	*20*	*120*	*120*
Oxygen uptake (μmol)	*1174*[b]		*613*		*1195*[b]
Products in μmol					
H_2	91	0	0	347	77
H_2O_2	14		11		19
CO	67	401	92	14	62
CO_2	86		205		95
CH_4	4	123	tr	257	29
CH_3OH			63		334
CH_3OOH			35		30
HCHO	found[c]		tr		tr
HCOOH	608		130		218
CH_3CHO		6	tr		tr
CH_3COOH			tr		tr
CH_3COOOH			60		97
CH_3COCH_3				299	612
biacetyl		50	0		
$(CH_3)_2C(OH)CH_2OH$					65
$(CH_3)_2C(OH)CHO$					45
$CH_3COC_2H_5$		10			
$HCOOCH_3$	s				
peroxides	20		97		125
acids	608		193		315

[a] vac: in vacuo, s: small, tr: trace.
[b] Remaining oxygen was small.
[c] HCHO was found but not determined quantitatively.

2-methylpropylperoxy radical. About twice as much acetone was observed under oxygen as under vacuum. There may be several possible routes for acetone formation in the presence of oxygen in addition to Reactions 2–4. One is the β-scission of 2-hydroxy-2-methylpropyloxy radical to give formaldehyde and 2-hydroxy-2-propyl radical, which eventually gives acetone and hydrogen peroxide as in the autoxidation of 2-propanol. The other possible route may be the oxidation of 2-hydroxy-2-methylpropanal which yields carbon dioxide, acetone, and hydrogen peroxide. Still another route may be the intramolecular hydrogen shift of the 2-hydroxy-2-methylpropylperoxy radical through a six-membered transition state. Although such an intramolecular hydrogen abstraction is not important in thermal oxidations of alkanes (*27*), it may not be excluded in the photooxidations since it resembles the Norrish type II reaction and since peroxidic bond is cleaved photochemically by 254-nm light.

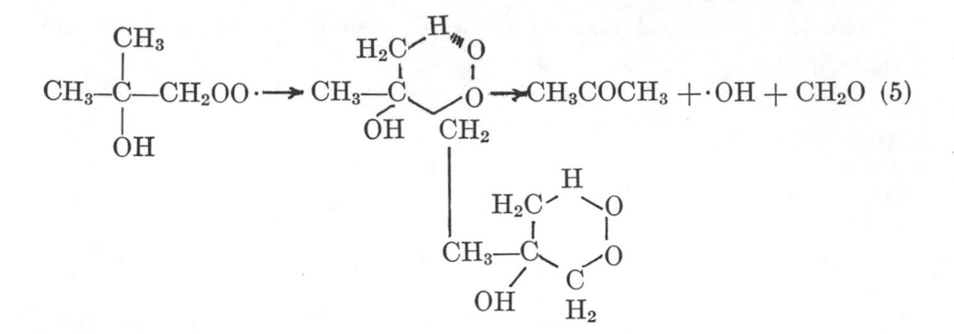

2-Propanol. 2-Propanol was chosen as a model for a simple alcohol. Irradiation under vacuum gave hydrogen and acetone as major products along with smaller amounts of methane, acetaldehyde, carbon monoxide, *tert*-butyl alcohol, and ethanol. These minor products must arise mainly from acetone. Acetaldehyde may be formed either by direct photolysis of 2-propanol, hydrogen atom abstraction by acetyl radical, or β-scission of the 2-propyloxy radical. The formation of *tert*-butyl alcohol implies the presence of methyl and 2-hydroxy-2-propyl radicals.

No hydrogen was observed in the presence of oxygen, suggesting that the hydrogen molecule is formed not by concerted mechanism but through hydrogen atoms. Acids must be the oxidation products from aldehydes. The absence of peroxide excludes the presence of per-acids and 2-hydroperoxy-2-propanol.

1,3-Butanediol. The oxidation products from 1,3-butanediol are complicated, but the primary products must be 4-hydroxy-2-butanone, 3-hydroxybutanal, and hydrogen peroxide.

2,4-Pentanediol. This is a good model for PVA. As shown in Table III, the major products are 4-hydroxy-2-pentanone and its photooxidized products. No 2,4-pentanedione was observed. A detailed discussion is given below.

Hydrogen Peroxide. Hydrogen peroxide was decomposed completely in one hour when its aqueous solution (1.02 mmol in 10 mL water) was irradiated with 253.7-nm light at 70°C, giving approximately half an equivalent amount of oxygen molecule (0.46 mmol). This photolysis must proceed by the initial cleavage of the oxygen–oxygen bond and subsequent hydroxy radical-induced decomposition of hydrogen peroxide. This photolysis explains why the formation of hydrogen peroxide from PVA levels off as the oxidation proceeds (Figure 1).

Discussion

Scheme 1 summarizes the proposed major reaction pathways in PVA photooxidations. It is not clear how the initiation proceeds. The photolysis of methanol, *tert*-butyl alcohol, and 2-propanol by 254-nm light is puzzling. Mercury was eliminated by placing a liquid nitrogen trap between the reaction vessel and vacuum line equipped with mercury diffusion pump and Toepler pump. The contribution of mercury-sensitized initiation must be negligible. It was shown that the photolysis of *tert*-butyl alcohol proceeded when mercury had no chance to be intro-

Table III. Photooxidations of Model Compounds at 70°C in Water[a]

Run No.	*14*	*15*	*16*	*17*	*18*
Substrate	*2-Propanol*			*BD*	*PD*
Atmosphere	*vac*	O_2	O_2	O_2	O_2
Time (min)	*120*	*30*	*52*	*45*	*45*
Oxygen uptake (μmol)		*333*	*907*	*722*	*483*
Products in μmol					
H_2	655	0	0	0	0
H_2O_2		11	6	136	30
CO	34	9	38	3	4
CO_2		13	38	149	103
CH_4	65	11	36	7	tr
CH_3OH		14	43	7	15
CH_3OOH				tr	50
$HCHO$		tr	tr	tr	tr
$HCOOH$		91	160	60	88
C_2H_5OH	s				
CH_3CHO	69	15	30	61	12
CH_3COOH		tr	tr	0	36
CH_3COOOH				93	80
CH_2OHCH_2OH				130	0
$CH_3CHOHCH_2OH$				20	10
CH_3COCH_3	291	468	1017	15	15
$(CH_3)_3COH$	51				
crotonaldehyde				10	
CH_3—CH(OH)—CH_2COOH				30	
CH_3—C(=O)—CH_2—CH_2OH				20	
CH_3—C(=O)—CH_2—CH(OH)—CH_3					94
CH_3—C(O)—CH=CH—CH_3					60
acetylacetone					0
peroxides		0	0	86	130
acids		91	160	182	204

[a] BD: 1,3-butanediol, PD: 2,4-pentanediol, vac: in vacuo, s: small, tr: trace.

Scheme 1

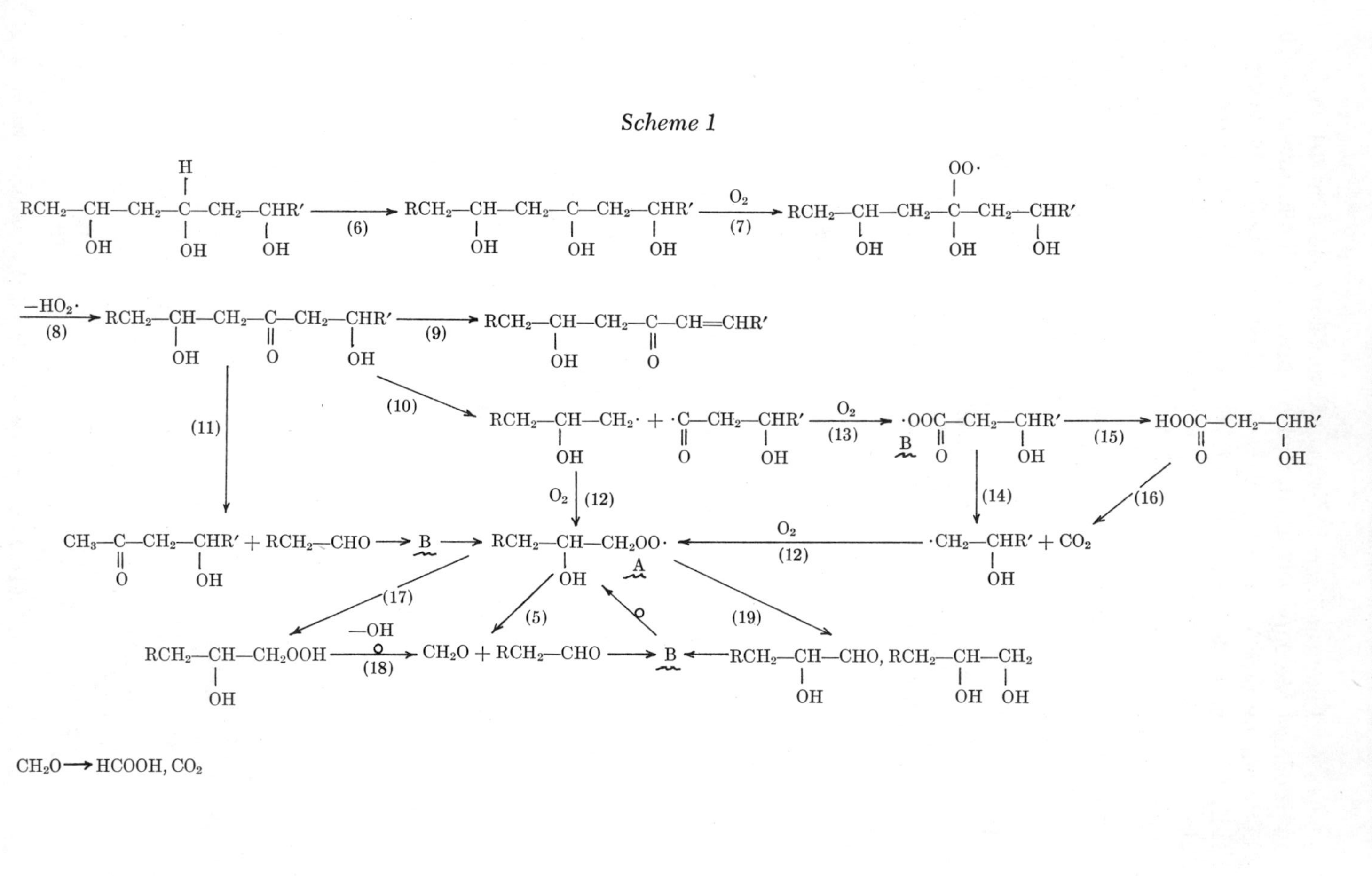

duced into the vessel. Possible initiation reactions can be induced by trace amounts of 185-nm light transmitted into the reaction vessel and/or impurities in the substrate. However, these initiation reactions must be important only at the onset of the reaction since the primary products must be more reactive toward 254-nm light and responsible for the production of radicals.

Reaction 6 is the hydrogen atom abstraction from PVA, an important propagation reaction. Since it was shown by the spin-trapping technique using 2-methyl-2-nitrosopropane that only the tertiary hydrogen of polypropylene was abstracted by the alkoxy radical (*28*), it was assumed that the first attack was also exclusively on the tertiary hydrogen of PVA. The α-hydroxyperoxy radical formed by the addition of oxygen (Reaction 7) may either decompose to ketone and a hydroperoxy radical (Reaction 8), abstract hydrogen atoms to give α-hydroxyhydroperoxide, or react with other peroxy radicals.

Several studies have been done to establish the chain-carrying species in the oxidation of alcohol (*29*). The absence of peroxide in the photooxidation of 2-propanol suggests either that the chain carrier is a hydroperoxy radical and not the 2-hydroxy-2-propylperoxy radical or that 2-hydroperoxy-2-propanol is too unstable to accumulate, if formed under these conditions. Only the hydroperoxy radical could be trapped by the spin-trapping reagent phenyl-*N*-*t*-butylnitrone in the oxidation of 2-propanol at 25°C (*28*). The reported high rate constant (*30*) for the decomposition of the 2-hydroxy-2-propylperoxy radical to acetone and hydroperoxy radical also implies that the hydroperoxy radical is the predominant chain carrier. On the other hand, Schenck and his co-workers (*31*) isolated 2-hydroperoxy-2-propanol in the photooxidation of 2-propanol with 313-nm light.

It has been observed that the intramolecular propagation plays an important role in the oxidations of 2,4-dimethylpentane (*32, 33*), 2,4,6-trimethylheptane (*34*), and polypropylene (*35, 36*). However, this intramolecular propagation must be unimportant in PVA oxidation since 2,4-pentanediol did not give 2,4-pentanedione under similar conditions (Run 18). Rust (*37*) also found no 2,4-pentanedione in the thermal oxidation of 2,4-pentanediol. Table III shows that 4-hydroxy-2-pentanone is the major primary product from 2,4-pentanediol. Therefore, the primary products from PVA must be β-hydroxyketone and hydrogen peroxide.

β-Hydroxyketone may undergo either dehydration (Reaction 9) or photolysis (Reactions 10 and 11). Dehydration yields carbonyl groups with conjugated ethylenic unsaturation, which has characteristic ultraviolet absorptions depending on the number of double bonds (*23*). Table IV shows the ketone distribution and total number of statistical chain scissions. Dehydration is favored over photolysis under the experimental conditions used.

Reactions 10 and 11 are also important competing reactions. It is known that the Norrish Type I reaction becomes more important as the wavelength of the incident light becomes shorter (*26*). As shown in Table III, the formation of acetone was smaller than other low boiling products in the photooxidation of 2,4-pentanediol, which suggests that Norrish I is more important than Norrish II photolysis under the present reaction conditions. Reaction 10 gives an α-hydroxy primary alkyl radical and an acyl radical, which must react with oxygen immediately to give the corresponding peroxy radicals, **A** and **B** (Reactions 12 and 13). Most **B** radicals must give carbon dioxide and radical **A** (Reactions 14, 15, 16, and 12), while radical **A** gives formaldehyde and radical **B** and continues the cycles. Thus, Scheme 1 shows that several molecules of carbon dioxide and acids are formed per statistical chain scission. Further, Scheme 1 explains all the products observed in the photooxidation of 2,4-pentanediol.

Table IV. Ketone Distribution and Total Number of Statistical Chain Scissions in the Photooxidation of Poly(vinyl alcohol)—Run *3*[a]

S_0		10.4
	$n = 0$	3.0
–C–C(=O)–(C=C)$_n$–C–	$n = 1$	16.6
	$n = 2$	2.8

[a] Concentrations in μmol.

Figure 1 shows that peroxides accumulate as oxidation proceeds. As already mentioned, no peroxides were observed in the photooxidation of 2-propanol. Therefore, α-hydroxyhydroperoxide cannot account for the peroxide observed. Three kinds of peroxides may be considered. One is the allylic hydroperoxide. As shown in Table I, more double bond and peroxides are formed as the conversion and temperature increase, which implies that allylic hydroperoxide is a good possibility. The other possible peroxide may be per-acids since acyl radicals are important transient radicals. Another possible peroxide may be the secondary hydroperoxide formed after the remaining acetoxy group was cleaved homolitically from the main chain to yield secondary alkyl radical. In fact, PVA irradiation under vacuum gave approximately equal amounts of carbon dioxide and methane, which suggests the formation of acetoxy radical. However, considering the small amount of acetoxy group remaining, this hydroperoxide cannot account for much of the peroxides observed.

Polymer chain ends are assumed to be carboxylic acids. They may be formed either through acyl radicals or from the β-scission of α-hydroxy-

tert-alkoxy radicals which are formed by the non-terminating bimolecular interactions of α-hydroxyperoxy radicals. This cleavage reaction may be important only when the kinetic chain length is short.

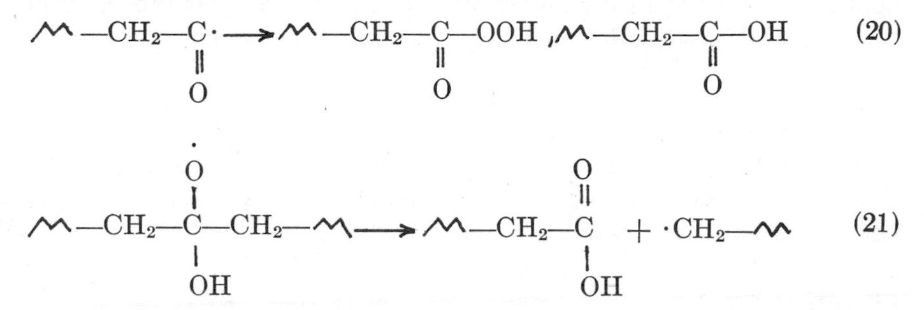

The role of the remaining acetoxy group is not clear. They may cleave homolitically as mentioned above or give acetic acid and double bond by a cyclo elimination reaction (*38*).

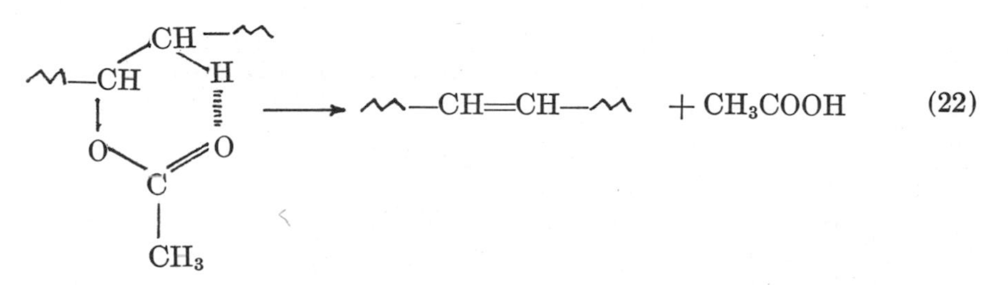

Intramolecular and intermolecular hydrogen atom abstractions by excited acetoxy groups are also proposed (*38, 39*).

The rate of initiation was not measured, and the kinetic chain length is not known. Since the rate of initiated oxidation of PVA in water in the dark is small, the kinetic chain length may be short and the bimolecular interactions of peroxy radicals may contribute considerably in determining products as well as the rate of PVA photooxidation. Among the important peroxy radicals are the hydroperoxy radical, primary alkylperoxy radical **A**, acylperoxy radical **B** and α-hydroxy-peroxy radical. The interactions of radical **A** with peroxy radicals must be terminating and proceed by Reactions 17 and 19. On the other hand, most of the interactions of the last two peroxy radicals are probably non-terminating and give acyloxy and α-hydroxy-alkoxy radicals. Acyloxy radicals should give carbon dioxide and alkyl radicals which eventually give radical **A**. The α-hydroxy-alkoxy radical may give β-hydroxyketone by hydrogen abstraction and subsequent dehydration or give acid and the alkyl radical by Reaction 21.

The hydroxy radical, which can be formed either by the photolysis of hydrogen peroxide and per-acid, by Reaction 5, or by non-terminating bimolecular interactions of hydroperoxy radicals, is another important chain-carrying species and must attack PVA rapidly and unselectively.

In conclusion, PVA is photooxidized considerably faster by 253.7-nm light in water at relatively low temperature and gives β-hydroxyketone, carbon dioxide, and acids as major products as shown in Scheme 1.

Acknowledgments

The authors thank F. R. Mayo at Stanford Research Institute for his valuable comments. We also thank Shimadzu Seisakusho for granting us permission to use the isotachophoretic analyzer.

Literature Cited

1. Sakurada, I., Matsuzawa, S., *Kobunshi Kagaku* (1963) **20**, 349.
2. Staudinger, H., Frey, K., Stark, W., *Ber. Dtsch. Chem. Ges.* (1927) **60**, 1782.
3. Sakurada, I., Okamura, S., *Kogyo Kagaku Zasshi* (1942) **45**, 1101.
4. Sakurada, I., Matsuzawa, S., *Kobunshi Kagaku* (1959) **16**, 633.
5. Marvel, C. S., Denoon, C. E., *J. Am. Chem. Soc.* (1938) **60**, 1047.
6. Sakurada, I., Matsuzawa, S., *Kobunshi Kagaku* (1961) **18**, 252.
7. Sakurada, I., Matsuzawa, S., *Kobunshi Kagaku* (1959) **16**, 565.
8. Shiraishi, M., Matsumoto, M., *Kobunshi Kagaku* (1963) **20**, 35.
9. Sakurada, I., Matsuzawa, S., *Kobunshi Kagaku* (1961) **18**, 257.
10. Shiraishi, M., Matsumoto, M., *Kobunshi Kagaku* (1962) **19**, 722.
11. Dulog, V., Kern, R., Kern, W., *Makromol. Chem.* (1968) **120**, 123.
12. Yasunaga, T., Kubota, H., Ogiwara, Y., *J. Polym. Sci., Polym. Chem. Ed.* (1976) **14**, 1617.
13. Duncalf, B., Dunn, A. S., *J. Appl. Polym. Sci.* (1964) **8**, 1763.
14. Bravar, M., Rek, V., Kostelac-Biffl, R., *J. Polym. Sci., Polym. Symp.* (1973) **40**, 19.
15. Auerbach, A., Jochsberger, T., Indictor, N., *Macromolecules* (1973) **6**, 143.
16. Auerbach, A., Indictor, N., Kruger, A., *Macromolecules* (1975) **8**, 262.
17. Auerbach, A., Indictor, N., Jochsberger, T., *Macromolecules* (1975) **8**, 632.
18. Jochsberger, T., Auerbach, A., Indictor, N., *J. Polym. Sci., Polym. Chem. Ed.* (1976) **14**, 1083.
19. Marvel, C. S., Inskeep, G. E., *J. Am. Chem. Soc.* (1943) **65**, 1710.
20. Shiraishi, M., Matsumoto, M., *Kobunshi Kagaku* (1959) **16**, 81.
21. Shiraishi, M., Matsumoto, M., *Kobunshi Kagaku* (1959) **16**, 344.
22. Imai, K., Kazusa, Y., *Kobunshi Kagaku* (1958) **15**, 249.
23. Matsumoto, M., Imai, K., Kazusa, Y., *J. Polym. Sci.* (1958) **28**, 426.
24. Clinton, N. A., Kenley, R. A., Traylor, T. G., *J. Am. Chem. Soc.* (1975) **97**, 3746, 3752, 3757, and references cited therein.
25. Kenley, R. A., Traylor, T. G., *J. Am. Chem. Soc.* (1975) **97**, 4700.
26. Calvert, J. G., Pitts, J. N., Jr., "Photochemistry," Interscience, 1966.
27. Van Sickel, D. E., Mill, T., Mayo, F. R., Richardson, H., Gould C. W., *J. Org. Chem.* (1973) **38**, 4435.
28. Ohto, N., Niki, E., Kamiya, Y., *J. Chem. Soc. Perkin Trans. 2*, in press.
29. For example: Howard, J. A., Korcek, S., *Can. J. Chem.* (1970) **48**, 2165.
30. Ilan, Y., Rabani, J., Henglein, A., *J. Phys. Chem.* (1976) **80**, 1558.
31. Schenck, G. O., Becker, H. D., Schulte-Elte, K. H., Krauch, C. H., *Chem. Ber.* (1963) **96**, 506.
32. Rust, F. F., *J. Am. Chem. Soc.* (1957) **79**, 4000.

33. Mill, T., Montorsi, G., *Int. J. Chem. Kinet.* (1973) **5**, 119.
34. Van Sickle, D. E., *J. Org. Chem.* (1972) **37**, 755.
35. Chien, J. C. W., Vandenberg, E. J., Jabloner, H., *J. Polym. Sci., Polym. Chem. Ed.* (1972) **6**, 381.
36. Niki, E., Kamiya, Y., *Bull. Chem. Soc. Jpn.* (1975) **48**, 3226.
37. Rust, F. F., Youngman, E. A., *J. Org. Chem.* (1962) **27**, 3778.
38. Geuskens, G., "Comprehensive Chemical Kinetics," Vol. 14, "Degradation of Polymers," C. H. Bamford and C. F. H. Tipper, Eds., p. 376, Elsevier, Amsterdam, 1975.
39. Ranby, B., Rabek, J. F., "Photodegradation, Photooxidation, and Photostabilization of Polymers," p. 196, Wiley-Interscience, London, 1975.
40. Matheson, M. S., Mamou, A., Silverman, J., Rabani, J., *J. Phys. Chem.* (1973) **77**, 2420.

RECEIVED May 12, 1977.

8

Photoyellowing of Polystyrene and Poly(styrene-*Alt*-Methyl Methacrylate)

ROBERT B. FOX and THOMAS R. PRICE

Chemistry Division, Naval Research Laboratory, Washington, DC 20375

*Yellowing in films of polystyrene and poly(styrene-*alt*-methyl methacrylate) induced by 254-nm radiation in vacuum and in air has been investigated by absorption and emission spectroscopy. In polystyrene, yellowing is more rapid in air than in vacuum; species related to 1,3-diphenyl-1,3-butadiene are major emissive contributors to yellowing. Photooxidation produces diene and species related to 1-phenyl-1,3-butanedione which quenches excimer in the polymer. Poly(styrene-*alt*-methyl methacrylate) initially yellows more rapidly than polystyrene. Rates of yellowing are greater in vacuum than in air and decrease with time. Yellowing is caused by species related to 1,5-diphenyl-1,3,5-hexatriene. Photolysis and photooxidation of the triene reduce the rate of yellowing. Fluorescence quenching occurs at the same rate in vacuum and air and may result from ketones formed by inter-unit interaction within the copolymer.*

Although polystyrene photochemistry has been investigated extensively, the details of the products and processes are still not understood completely (*1*). Among the physical manifestations of the photodegradation of styrene-containing polymer films, embrittlement resulting from cross-linking and yellowing resulting from photolysis products are the most obvious. Molar mass changes are difficult to measure and interpret in cross-linking systems. Yellowing, conveniently followed by changes in absorption spectra of polymer films, is a measure of the extent of photodegradation. Identification of the products that produce the yellowing is a first step toward understanding the degradation processes themselves. This first step has been undertaken by many investigators. The results have often been contradictory, (they have been reviewed by Ranby and

Rabek (*1*)). In the absence of air, photoyellowing of polystyrene has been attributed to the build up of conjugated polyene sequences and to the formation of species related to fulvene. Photooxidative yellowing has been ascribed to the formation of species related to acetophenone, benzalacetophenone, and 1,2-diones derived from the main chain of the polymer, as well as to dialdehydes or quinomethane structures resulting from photooxidation of the pendant phenyl moieties.

Of the many experimental techniques available for the identification of degradation products, optical emission spectroscopy is one of the most sensitive in the detection of emitting nonvolatile materials. This technique has been used to show that the absorption of light by polystyrene is followed by rapid energy migration to excimer-forming sites (*2*) or to impurity traps in the polymer chain (*3*). In poly(styrene-*alt*-methyl methacrylate), energy migration also takes place but excimers do not form. A comparison of the photodegradation products from polystyrene and the alternating copolymer are clearly of interest. Recently we reported (*4*) on the absorption and delayed emission spectra obtained from solutions of styrene and styrene-methyl methacrylate polymers and copolymers. Aromatic diketones formed during photooxidation under 254-nm radiation. In this chapter, we discuss the spectra obtained from films of polystyrene and poly(styrene-*alt*-methyl methacrylate) subjected to 254-nm radiation in air and in vacuum at or near room temperature.

Experimental Section

The polystyrene used in this work was prepared in tetrahydrofuran with α-methylstyrene tetramer as the initiator (*4*). The alternating polymer was described previously (*3*). Fluorescence-grade solvents were used throughout.

Films were formed by the evaporation of methylene chloride solutions on one inside surface of a quartz fluorescence cell or on quartz plates. Irradiations were carried out in a Rayonet Photochemical Reactor fitted with low-pressure mercury lamps. The incident radiation intensity on the films was 10^{-6} Einstein cm^{-2} min^{+1}. Absorbed intensity is based on energy absorbed by the styrene units of the polymers; for the alternating copolymer this was 0.95 of the total energy absorbed at 254 nm. Film thicknesses were of the order of 25 μm; a thickness effect was not observed in the 10–40 μm range, but a small correction factor of $1 - T$ (254 nm), where T is transmittance, was applied where T was greater than zero.

Following a series of exposures, the films were extracted successively with methanol and methylene chloride. Portions of these extracts were evaporated to dryness and the residues were taken up in 1:1 tetrahydrofuran–diethyl ether for delayed emission determinations at 77K. Emission spectra were measured in an Aminco–Bowman Spectrophosphorimeter fitted with a -P28 photomultiplier; the spectra are uncorrected for instrument response.

Results and Discussion

Irradiation of films of polystyrene or alternating styrene–methyl methacrylate copolymers in either air or vacuum resulted in a general increase in absorbance throughout the near-UV spectrum and extending into the visible region (5). The absorption tail in the visible region is responsible for the yellow appearance of the degraded films. Strong band formation was not observed in the absorption spectra. In irradiated films the increase in absorbance at 400 nm paralleled the increase in absorbance at any wavelength in the 300–400-nm region, indicating that absorbance changes in this spectral region can be used as an index of yellowing.

Weak absorption bands, superimposed on the general absorption, became evident as the irradiation progressed. For polystyrene in air, these bands occur at ca. 265, 275, and 340 nm. With the alternating copolymer in air or in vacuum bands appeared at ca. 278, 295, 325, and 370 nm. For reasons that will become apparent, absorbance changes at 332 and 350 nm were selected as an index of yellowing for polystyrene and the alternating copolymer, respectively. In Figure 1 absorbance increases are plotted against the 254-nm radiation absorbed by the phenyl chromophores. It is evident that there are marked differences in behavior between the two polymers.

Polystyrene. The characteristic changes that take place in the emission spectra of polystyrene subjected to 254-nm irradiation are strongly

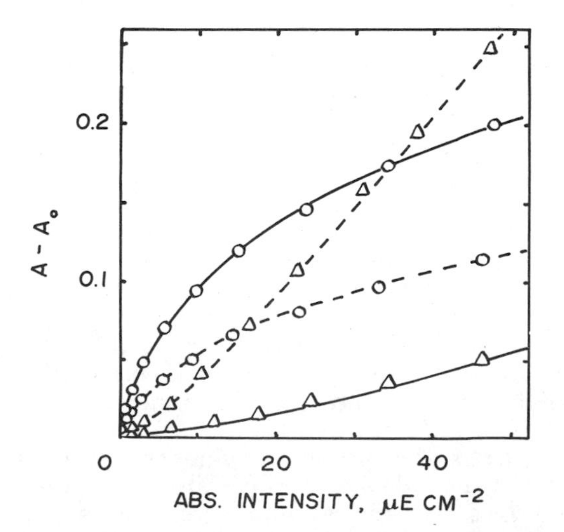

*Figure 1. Photoyellowing in films: (△), polystyrene, A at 332 nm; (○) poly(styrene-*alt-*methyl methacrylate), A at 350 nm; —— irradiation in vacuum; - - - irradiation in air*

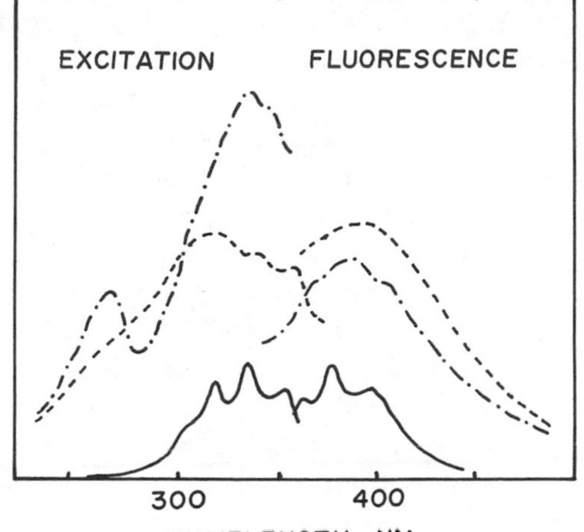

Optica and Spectroscopy (USSR)

Figure 2. Product I prompt-emission spectra: —·— polystyrene film, 37 μE cm^{-2} absorbed in vacuum, EX 300 nm; - - - methanol extract of irradiated film, EX 316 nm; —— 1,4-diphenyl-1,3-butadiene in toluene (11)

dependent on the presence or absence of oxygen and are far more striking than the changes in the absorption spectra. The most prominent features include the generation of new fluorescence with a maximum at 390 nm, the quenching of emission at 332 nm that has been identified (*6, 7*) as excimer fluorescence, and the formation of structured phosphorescence similar to that shown by phenyl alkyl ketones in the 400–500-nm region. Both the excimer quenching (*8*) and the phosphorescence (*4, 8, 9, 10*) have been reported previously.

Fluorescence spectra from a polystyrene film photolyzed in vacuum are shown in Figure 2. Similar but less intense spectra were observed in films irradiated in air. The Product I responsible for this spectrum was partially extractable from the film with methanol; the fluorescence spectrum of the extract is shown also in Figure 2. Comparison of these spectra with those of a wide variety of reasonable model compounds suggests that Product I is related to 1,3-diphenyl-1,3-butadiene since the spectral match with 1,4-diphenyl-1,3-butadiene, shown in Figure 2, is quite close. Product I spectra were obtained also from the residual films after extraction, indicating that the diene moiety may form part of a photolyzed chain as well as exist as a short-chain fragment. Fluorescence spectra that could be related to higher polyenes were not detected in the vacuum exposures. In air exposures, however, the prompt emission spectra from films did exhibit a weak shoulder superimposed on the tail of the emission in the 400–450-nm region. The excitation maximum for this shoulder was at 355 nm, suggestive of a diphenylhexatriene moiety.

For vacuum exposures plots of fluorescence intensity at 390 nm against absorbance at either 332 nm or 400 nm were linear. The excitation maximum at 332 nm corresponds roughly to the broad absorption acquired by a polystyrene film during photolysis. From the excitation spectrum of Product I it is reasonable to assume that an absorption tail extends into the visible region of the spectrum, although nonfluorescent products also may contribute to this tail. Grassie and Weir (*12*) established that hydrogen was the only volatile product formed under similar conditions; olefinic unsaturation in the residue is a reasonable concomitant product. Diphenylbutadienes have high fluorescence yields and do not phosphoresce generally. We observed no delayed emission that could be related to Product I. Thus we conclude that one or more diphenylbutadienes are major emissive species responsible for the yellowing of polystyrene that occurs during irradiation in vacuum.

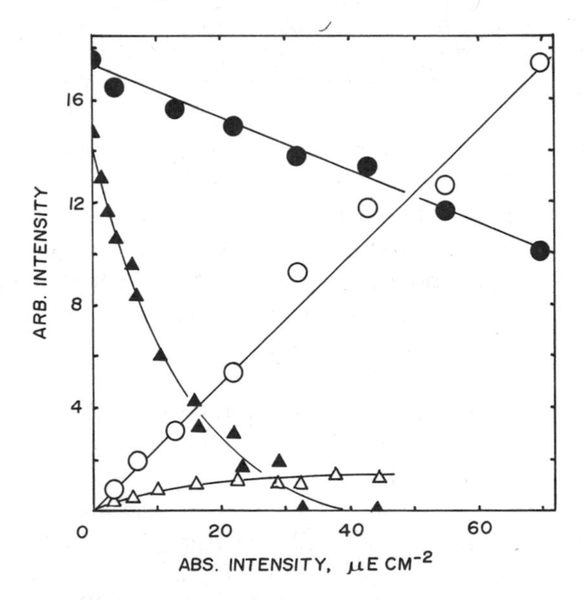

Figure 3. Prompt-emission changes during irradiation of polystyrene: (○) and (●), in vacuum; (△), (▲), in air; ○, F 390 (EX 300); ●, EF 332 (EX 260)

Excimer quenching accompanies the formation of Product I in irradiated polystyrene films, but the rates of change of excimer fluorescence intensity in vacuum and in air are quite different. They are contrasted in Figure 3. Photoconversion of excimer sites, which act as traps for migrating singlet energy (*2*), might be considered to be a cause of the reduction in emission intensity. If this were the case, an increase in polystyrene molecular fluorescence near 280 nm (*6,7*) would be expected. Such increases were not observed, but still open is the possibility of energy transfer from migrating singlets to a photolysis product. Absorbance increases were too small to attribute the reduction in excimer fluorescence to self-quenching. However, tetraphenylbutadiene is an efficient quencher of polystyrene excimer emission (*13*), and the overlap between the excimer emission spectrum and the Product I absorption spectrum is excellent. In vacuum exposures at least, a reasonable explanation for the excimer quenching lies in singlet energy transfer from either a migrating singlet or an excited excimer site to Product I, and this in turn may account for part of the observed fluorescence of Product I.

Both yellowing and excimer quenching take place more rapidly during exposure in air than in vacuum. At the same time, the generation of Product I fluorescence is slower in air (Figure 3), which suggests that Product I might not be responsible for excimer quenching in air exposures. Quenching probably is not caused by selective photooxidation of excimer sites, since Geuskens and David (*8*) showed that regeneration of excimer sites did not occur when a photooxidized film was heated above the glass transition temperature. They concluded that the quenching resulted from energy transfer to non-emitting oxidation products such as polymer peroxides and supported this view by showing that cumyl peroxide is an efficient excimer quencher.

We contend that the steady-state concentration of peroxides during air irradiation is not likely to be high enough to account for the observed quenching, although peroxides may very well be intermediates in the formation of quenching products. At no time did we observe unequivocal peroxide band formation in the ir absorption spectra of photooxidized films. On the other hand, photooxidation products that are ketonic in nature would be expected to have high triplet yields and to be readily detectable through their delayed-emission spectra. Such spectra have been reported (*4, 8, 9, 10*), and they have been ascribed reasonably to phenyl alkyl ketone moieties (*8, 9, 10*). Geuskens and David (*8*) observed, as we do also, that as photooxidation proceeded the intensities initially increased and then decreased. In an earlier report (*4*) we showed that if secondary photooxidation product spectra are considered, the delayed-emission spectra from photooxidized polystyrene correspond more closely to those obtained from 1-phenyl-1,3-butanedione. The spectra are shown in Figure 4, in which it is evident that the emissions of secondary products excited at 380 nm are quite dissimilar for the dione and acetophenone.

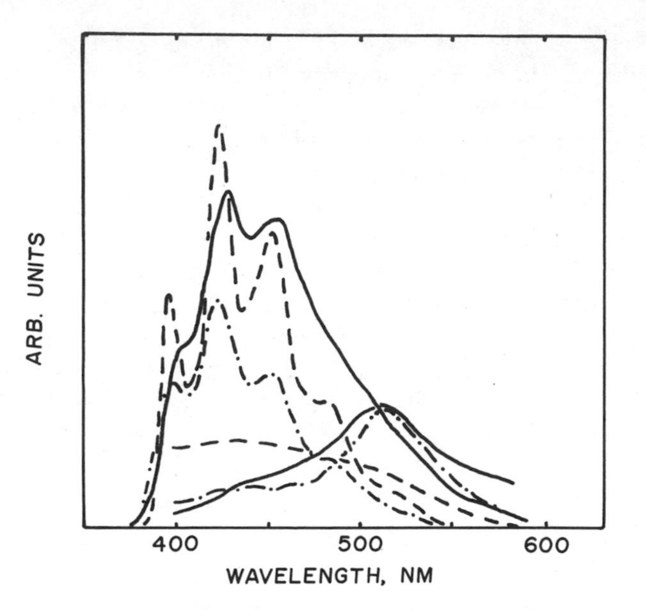

Figure 4. Delayed-emission spectra of ketonic products: —— photooxidized polystyrene film; –·– 1-phenyl-1,3-butanedione in PMMA; – – – acetophenone in PMMA; higher intensity spectra show model compounds unirradiated, EX 260 nm; lower intensity spectra show model compounds after irradiation EX 380 nm

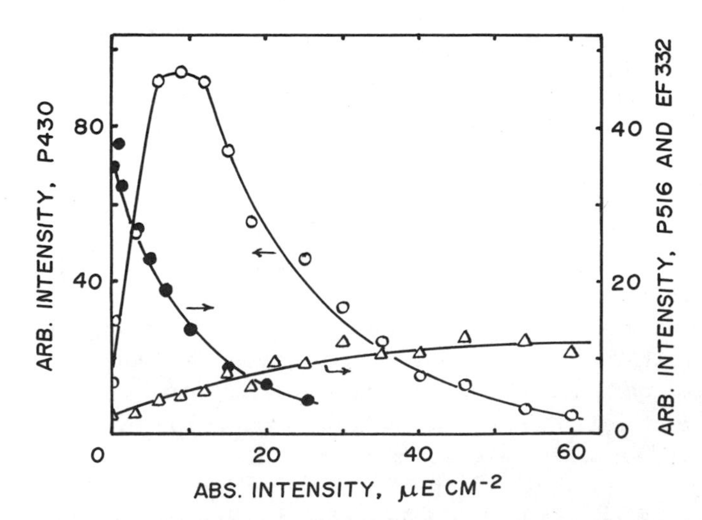

Figure 5. Emission changes during irradiation of polystyrene in air: (●), EF 332 (EX 260); (○), P430 (EX 260); (△), P515 (EX 380)

In Figure 5 the changes in the polystyrene phosphorescence intensities at 430 nm and 516 nm are compared with the excimer fluorescence intensity changes that occur during photooxidation. The initial rise in dione phosphorescence clearly parallels the decrease in excimer intensity. Secondary product formation, shown by the emission at 516 nm, may account in part for the subsequent loss in dione phosphorescence intensity. However, Geuskens and David (*8*) noted that ketone absorption in the ir spectra showed no corresponding decrease during photooxidation. Thus the reduction in dione emission is likely to be caused by a quenching process rather than a concentration loss.

Since the formation of peroxides and phenyl alkyl ketones are not ruled out by these findings, we believe that the products and the relative changes in their emission spectra are best accounted for on the basis of the photooxidation scission process proposed by Grassie and Weir (*12*):

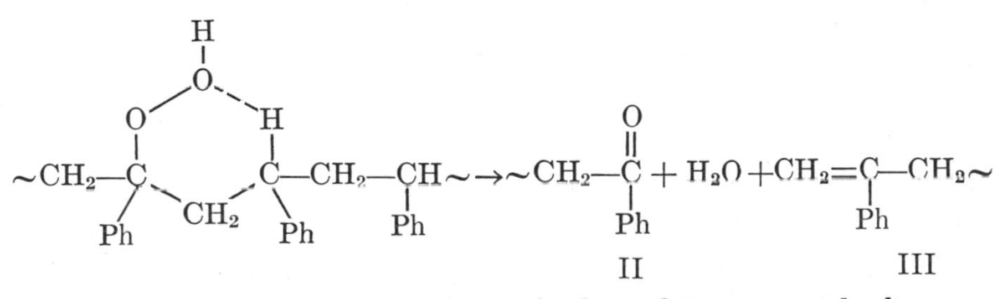

The chain fragments may undergo further photoreaction leading to dione and diene moieties, respectively.

Both dione and diene have absorption spectra that overlap very well with the excimer fluorescence spectrum, and either or both may be responsible for the observed excimer quenching through singlet energy transfer. Therefore, both dione phosphorescence and diene fluorescence would be expected to increase by this process. While the former may be the case, sensitization of diene fluorescence is not apparent in Figure 3. Photophysical interactions taking place between the dione (or other ketone) and diene may be responsible for these findings. The first of these interactions is a diene singlet quenching process

$$\text{Diene } (S_1) + \text{Dione } (S_o) \rightarrow \text{Diene } (S_0) + \text{Dione } (S_1)$$

$$\text{Dione } (S_1) \rightarrow \text{Dione } (T_1)$$

that may occur within the cage created by the scission process. Through quenching, the diene fluorescence should be far less than that expected from the true concentration of diene. The second interaction is a ketone triplet quenching process

$$\text{Dione } (T_1) + \text{Diene } (S_o) \rightarrow \text{Dione } (S_o) + \text{Diene } (T_1)$$

with the diene triplet returning to the ground state through a nonradiative transition. This quenching process, analogous to aromatic ketone triplet quenching by aliphatic dienes (*14*), may account for the ultimate reduction in ketone phosphorescence without interfering with the continuing formation of ketones by photooxidation.

Since 1-phenyl-1,3-butanedione is not yellow itself, the primary ketonic products are unlikely to be responsible for photooxidative yellowing in polystyrene, although secondary products may well contribute. A major contribution to photooxidative yellowing can be made by the diene, however, just as it did in the photolysis in vacuum. Through the scission process, diene will be expected to form far more rapidly than it did in the hydrogen abstraction process occurring in the absence of air. Thus yellowing takes place at a greater rate in photooxidation than in simple photolysis.

Poly(styrene-*alt*-methyl methacrylate). In contrast to polystyrene, films of the alternating copolymer undergo photoyellowing more rapidly during irradiation in vacuum than in air. Furthermore, as seen in Figure 1, the rate of change in absorbance at 350 nm decreased with increasing time of irradiation; corresponding changes in absorbance occurred at 400 nm. In both vacuum and air irradiations weak absorption bands with maxima ca. 278, 295, 325, 350, and 370 nm were superimposed on the generally increased absorption that appeared in the 250–400-nm spectral region. The only ir spectral changes that occurred during irradiation in air were absorption decreases in weak bands at 740 and 1275 cm^{-1}, both of which appear in the spectrum of poly(methyl methacrylate), but not in the spectrum of polystyrene.

Prompt emission spectra of films irradiated in either vacuum or air indicated the formation of at least two fluorescing species. These species are denoted as Products II and II. The most prominent emission, that of Product II, was highly structured with maxima at about 405, 435, and 455 nm. Product II can be associated with both the polymer chain and low molar mass fragments; it was extractable from the photolyzed film with methanol but not with cyclohexane, and it was present in methylene chloride extracts of cross-linked residues of films that had been extracted previously with methanol. The fluorescence spectra are shown in Figure 6; excitation maxima correspond very well with the weak absorption bands noted in the spectra of the irradiated films. Plots of absorbance increases at 350 nm against fluorescence intensity at 435 nm were linear for both vacuum and air irradiations. Therefore, it is concluded that Product II is a major emissive contributor to the yellowing of poly(styrene-*alt*-methyl methacrylate) films subjected to 254-nm irradiation.

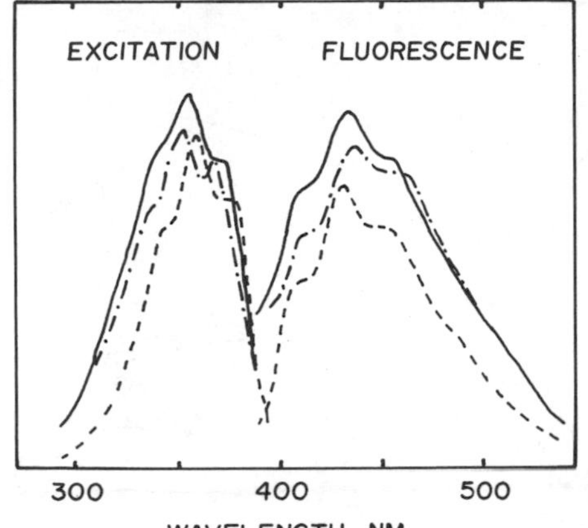

Figure 6. Product II prompt-emission spectra: —— alternating copolymer film, 16 μE cm⁻² absorbed in vacuum; –·– methylene chloride extract of irradiated film; – – – 1,6-diphenyl-1,3,5-hexatriene in methylene chloride; fluorescence is excited at 350 nm and excitation spectra are for F 435 nm

Product II can be identified empirically as a species related closely to 1,6-diphenyl-1,3,5-hexatriene, the spectra of which are shown also in Figure 6. Fluorescence and excitation spectra from an alternating copolymer film doped with this triene are identical to those for the triene in methylene chloride and very close to the spectra from photolyzed copolymer film. A six-carbon chain fragment from the alternating copolymer would have phenyl substitution in the 1,5-positions, but neither this nor methyl or carbomethoxy substitution on the chain would be expected to produce spectra greatly different from those of 1,6-diphenyl-1,3,5-hexatriene.

Product III has not been identified. It has a weak, unstructured fluorescence maximum ca. 390 nm, with an excitation maximum at 332 nm. This product was fully extractable from a photolyzed film with methanol, suggesting that it may be a secondary photolysis product arising from Product II. If nonalternating sequences exist within the copolymer, Product III might be thought to correspond to the diene observed in the photolysis of polystyrene. However, Product III formed more rapidly in vacuum than in air, the reverse of the manner in which diene appeared to form in polystyrene.

Peak fluorescence intensity changes during the irradiation of alternating copolymer films are shown in Figure 7. While the increases in fluorescence intensity of Products II and III were greater in vacuum than

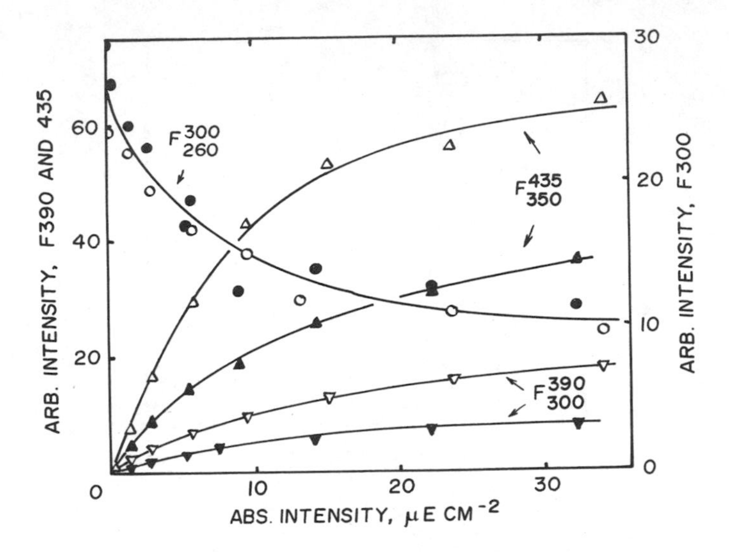

*Figure 7. Prompt-emission changes during irradiation of poly(styrene-*alt*-methyl methacrylate) films: ○, vacuum; ●, air; superscripts and subscripts refer to emission and excitation wavelengths (nm), respectively*

in air, quenching of the copolymer fluorescence occurred at the same rate in both vacuum and in air. Therefore, this quenching cannot be attributed to Product II or III. The 1,6-diphenyl-1,3,5-hexatriene itself did not quench the copolymer fluorescence in methylene chloride solution; triene absorption overlap with copolymer fluorescence is quite small.

The answer to the quenching problem may lie in an intrachain or interchain interaction in the copolymer that is independent of atmosphere. One possibility is an internal photocyclization to a 3,4-dihydro-1(2*H*)-naphthaleneone structure:

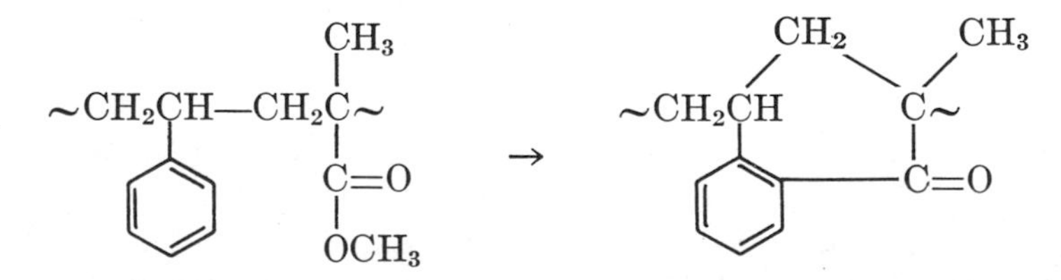

Weak delayed-emission spectra of vacuum- or air-irradiated copolymer films were similar in intensity and showed a phosphorescence maximum at 432 nm with shoulders ca. 390 nm and 450 nm on excitation at 260 nm in addition to a very weak maximum at 505 nm excited at 380 nm. These spectra are close to those shown in Figure 4 for ketone phosphorescence in photooxidized polystyrene and agree reasonably well with phosphorescence spectra for model napthaleneones (*15*). Energetically, the quenching of copolymer fluorescence by such ketones may occur through the process

$$\text{Copolymer } (S_1) + \text{Ketone } (S_o) \rightarrow \text{Copolymer } (S_o) + \text{Ketone } (S_1)$$
$$\text{Ketone } (S_1) \rightarrow \text{Ketone } (T_1)$$

Photoreactions of Product II are likely to provide an explanation for the generally decreasing rate of product formation and yellowing with increasing irradiation time and for the higher rates in vacuum relative to air. Irradiation of copolymer films doped with 1,6-diphenyl-1,3,5-hexatriene produced an increase in absorption in the 275–300-nm region and a decrease in absorption caused by triene in the 300–400-nm region. The triene fluorescence intensity decreased correspondingly without an alteration in the shape of the directly excited (at 350 nm) fluorescence band, although the fluorescence band broadened considerably with 260-nm excitation. Thus triene photolysis may compete with triene formation in copolymer photolysis without the formation of species that contribute strongly to yellowing; Product III well may be such a species. In any event, triene photolysis is likely to decrease the rate at which yellowing increases. Photooxidation reactions will also reduce the concentration of triene and thereby reduce the triene contribution to yellowing in air relative to that in vacuum. An example of such a reaction is the oxidative photocyclization of aryl polyenes to form polycyclic aromatic hydrocarbons (*16*) that generally would have lower extinction coefficients than the precursor triene in the 300–400-nm region of the spectrum. These kinds of photoreactions do not occur in the absence of oxygen or other oxidizing agents and are therefore less likely to affect the triene formed in vacuum exposures of the coplymer films.

Additional experiments are indicated to confirm many of the arguments presented in this work. Quenching experiments with low molar mass model compounds having appropriate substituents in positions corresponding to those expected from the polymer and copolymer structures would be particularly useful. An identification of naphthaleneone among the products of the photolysis of methyl 4-phenylbutyrate would support the photocyclization suggested for the alternating copolymer. Finally, a rigorous search for the non-emissive products of photodegradation is surely necessary for both polystyrene and the alternating copolymer before it can be said that the products producing photoyellowing in these materials have been identified fully.

Conclusions

Photoproducts related to 1,3-diphenyl-1,3-butadiene have been identified as major emissive contributors to the yellowing of polystyrene films subjected to 254-nm irradiation in both air and in vacuum. In photooxidative yellowing, which occurs more rapidly than yellowing in the absence of air, species related to 1-phenyl-1,3-butanedione are produced also. The dione and, to a lesser extent, the diene act as quenchers of excimer fluorescence in photodegraded polystyrene.

Poly(styrene-*alt*-methyl methacrylate) undergoes yellowing more rapidly in vacuum than in air. The major yellowing moiety is a species related to 1,5-diphenyl-1,3,5-hexatriene. Simultaneous photolysis of the triene results in a reduction in the rate of yellowing with increasing time of irradiation; oxidative reactions of the triene appear to reduce the rate of yellowing in air. Fluorescence quenching takes place by processes that are the same in air and vacuum.

Literature Cited

1. Ranby, B., Rabek, J. K., "Photodegradation, Photooxidation and Photostabilization of Polymers," pp. 165–184, Wiley, New York, 1975.
2. Hirayama, F., Basile, L. J., Kikuchi, C., *Mol. Cryst.* (1968) **4**, 83.
3. Fox, R. B., Price, T. R., Cozzens, R. F., Echols, W. H., *Macromolecules* (1974) **7**, 937.
4. Fox, R. B., Price, T. R., Cozzens, R. F., "Ultraviolet Light-Induced Reactions in polymers," *ACS Symp. Ser.* (1976) **25**, 242.
5. Reiney, M. J., Tryon, M., Achhammer, B. G., *J. Res. Nat. Bur. Stand.* (1953) **51**, 155.
6. Vala, M. T., Jr., Haebig, J., Rice, S. A., *J. Chem. Phys.* (1965) **43**, 886.
7. Hirayama, F., *J. Chem. Phys.* (1965) **42**, 3163.
8. Geuskens, G., David, C., *Proc. Int. Symp. Degrad. Stabil. Polym., Brussels* (1975) 113.
9. George, G. A., *J. Appl. Polym. Sci.* (1974) **18**, 419.
10. Burchill, P. J., George, G. A., *J. Polym. Sci., Polym. Lett. Ed.* (1974) **12**, 497.
11. Nikitina, A. N., Fedyunia, G. M., Yanovskaya, L. A., Dombrovskii, V. A., Kucherov, V. F., *Opt. Spectrosc. (USSR)* (1971) **30**, 343.
12. Grassie, N., Weir, N. A., *J. Appl. Polym. Sci.* (1965) **9**, 975, 999.
13. David C., Piens, M., Geuskens, G., *Eur. Polym. J.* (1973) **9**, 533.
14. Wagner, P. J., Hammond, G. S., *J. Am. Chem. Soc.* (1966) **88**, 1245.
15. Kanda, Y., Stanislaus, J., Lim, E. C., *J. Am. Chem. Soc.* (1963) **91**, 5058.
16. Fonken, G. J., *Chem. Ind.* (1962) 1327.

RECEIVED May 12, 1977.

9

Thermooxidative and Photooxidative Aging of Polypropylene; Separation of Heptane-Soluble and -Insoluble Fractions

H. P. FRANK and H. LEHNER

Chemie Linz AG, A–4021 Linz, Austria

Unstabilized PP was exposed to thermooxidative (150°C) and photooxidative (Xenotest 70°C) degradation. At various stages of oxidative degradation the material was separated into heptane-insoluble and -soluble fractions; in the undegraded sample (before oxidation) the insoluble fraction is essentially isotactic whereas the soluble fraction is mainly atactic. The progress of oxidation in the fractions was followed by intrinsic viscosity measurements, GPC, ir, and DSC. The heptane-soluble fraction increases gradually during oxidation (from 7.0 to 46.4 wt % under extreme conditions) because of increasing formation of very low molecular weight heptane-soluble isotactic material. In the early stages of oxidative degradation there appear to be 5 to 10 times more C–C bond scissions in the soluble fraction than in the insoluble fraction.

Thermooxidative and photooxidative degradation of polypropylene (PP) has been investigated intensively in recent years (*1, 2*). However, structural and morphological factors (tacticity, crystallinity) were taken into account comparatively rarely (*3, 4*). In this study unstabilized PP was degraded oxidatively, and at each stage the total PP sample was separated into heptane-soluble and -insoluble fractions to determine the progress of oxidative degradation in each fraction.

Experimental

Material. PP powder without additives (Petrochemie Schwechat GmbH) was compression molded under mild conditions into thin sheets

0-8412-0381-4/78/33-169-109$05.00/1

Table I. Properties of Total Samples[a]

Sample	*Oven T- Xenotest X- (hr)*	*(100 ml/g) [η]*	$\overline{M}_w \times 10^{-3}$	$\overline{M}_n \times 10^{-3}$	$\overline{M}_w/\overline{M}_n$	*C_7-Soluble Fraction (wt %)*
0	0	3.15	484	68	7.1	7.0
T1	1	2.69	—	—	—	7.7
T2	2	2.59	—	—	—	8.0
T3	3	1.39	—	—	—	13.5
T4	10	0.64	58	8.5	6.9	46.4
X1	50	2.32	—	—	—	8.3
X2	73	2.09	—	—	—	8.5
X3	109	1.89	—	—	—	9.4
X4	142	1.35	—	—	—	13.5
X5	242	0.70	70	11	6.3	24.2

[a] T- and X-degradation series.

(0.3–0.4 mm). For properties of reference sheet 0 *see* Table I.

Thermooxidation. Forced air circulation oven (150°C).

Photooxidation. Xenotest 450 LF, original Hanau (black panel temp. 70°C).

Heptane extraction. PP material was dissolved in boiling xylene, was precipitated by adding excess methanol, and was extracted with *n*-heptane (5).

Intrinsic viscosity [η]. Decalin 135°C.

Gel permeation chromatography (GPC). Waters Associates Inc., model 200, 1,2,4-trichlorobenzene 135°C.

Infrared spectrometry (ir). Perkin–Elmer 125; 30–80 μm molded PP films.

Table II. Properties

	E Fractions[b]				
Sample	*[η]*	$\overline{M}_v \times 10^{-3}$	$\overline{M}_w \times 10^{-3}$	$\overline{M}_n \times 10^{-3}$	$\overline{M}_w/\overline{M}_n$
0	0.58	50	65	9	7.2
T1	0.45	37	(42)	(7.5)	(5.6)
T2	0.43	35	(39)	(7.5)	(5.3)
T3	0.32	25	(26)	(6.5)	(4.0)
T4	0.22	16	16	6	2.7
X1	0.43	35	(39)	(7.5)	(5.3)
X2	0.40	32	(35)	(7)	(5.0)
X3	0.34	26	(27)	(6.5)	(4.2)
X4	0.26	20	(20)	(6)	(3.3)
X5	0.17	12	11	5	2.2

[a] Data in brackets obtained by linear interpolation.
[b] E- C_7-soluble extract.

Differential scanning calorimetry (DSC). Dupont 900 thermal analyzer; heating to 200°C, cooling to 70°C; second heating curve recorded (heating rate 10°C/min).

Results

The results of thermooxidative (oven, T-series) and photooxidative (Xenotest, X-series) degradation are summarized in Tables I and II ($\overline{M}_v$ was calculated from [η] data (*6*)).

The CO-bands (1700–1800 cm^{-1}) of the ir spectra, using the 974 cm^{-1} absorption as a reference for film thickness, are shown in Figure 1 for some highly oxidized fractions; absorption maxima and shoulders are observed in all cases in the range of 1700–1710 cm^{-1} (acids), 1710–1730 cm^{-1} (ketones), 1730–1740 cm^{-1} (aldehydes), 1740–1750 cm^{-1} (esters), ca. 1760 cm^{-1} (peresters), ca. 1780 cm^{-1} (γ-lactones) (*7, 8*).

In view of the CO-band's complexity, no evaluation was attempted. (The E fractions are very difficult to manipulate, i.e., the intensity of the CO-band is very sensitive to remolding of the films, etc.) However, the degree of oxidation of the E fractions is much higher than that of the corresponding R fractions.

The DSC tracings of R fractions show a single melting peak (157°–165°C) whereas the E fractions display double peaks (148°–154°C, 135°–144°C). The results are summarized in Table III.

To clarify the nature of the double melting peak at least in the undegraded E fraction (sample 0), PP was separated further on the basis of solubility in heptane at 20°C: Fraction O/E/I soluble C_7/20°C: 63 wt %, [η] = 0.76; Fraction O/E/II insoluble C_7/20°C: 37 wt %, [η] = 0.30, DSC tracings for the original O/E fraction and for O/E/I and O/E/II

of Fractions[a]

R Fractions[c]				
[η]	$\overline{M}_v \times 10^{-3}$	$\overline{M}_w \times 10^{-3}$	$\overline{M}_n \times 10^{-3}$	$\overline{M}_w/\overline{M}_n$
3.31	455	508	115	4.4
3.87	380	(407)	(92)	(4.4)
2.78	360	(385)	(87)	(4.4)
1.56	180	(193)	(44)	(4.4)
1.00	100	102	26	3.9
2.49	320	(343)	(78)	(4.4)
2.24	280	(300)	(68)	(4.4)
2.05	250	(268)	(61)	(4.4)
1.52	170	(182)	(41)	(4.4)
0.87	84	90	19	4.8

[c] R- and C_7-insoluble residue (T- and X-degradation series).

are shown in Figure 2 (ΔH_F for O/E/I = 2.7 cal/g, for O/E/II = 19.1 cal/g). From this data it is evident that only Fraction O/E/I is essentially amorphous and atactic, but not O/E/II, even though it does contain a very small and broad peak at ca. 105°C with unknown significance. (O/E/II was thought to represent stereoblock PP, crystallizing in triclinic γ-modification (*9*); an x-ray powder diffraction pattern produced reflections corresponding to the following *d* spacings (Å): 6.32 vs, 5.21 s,

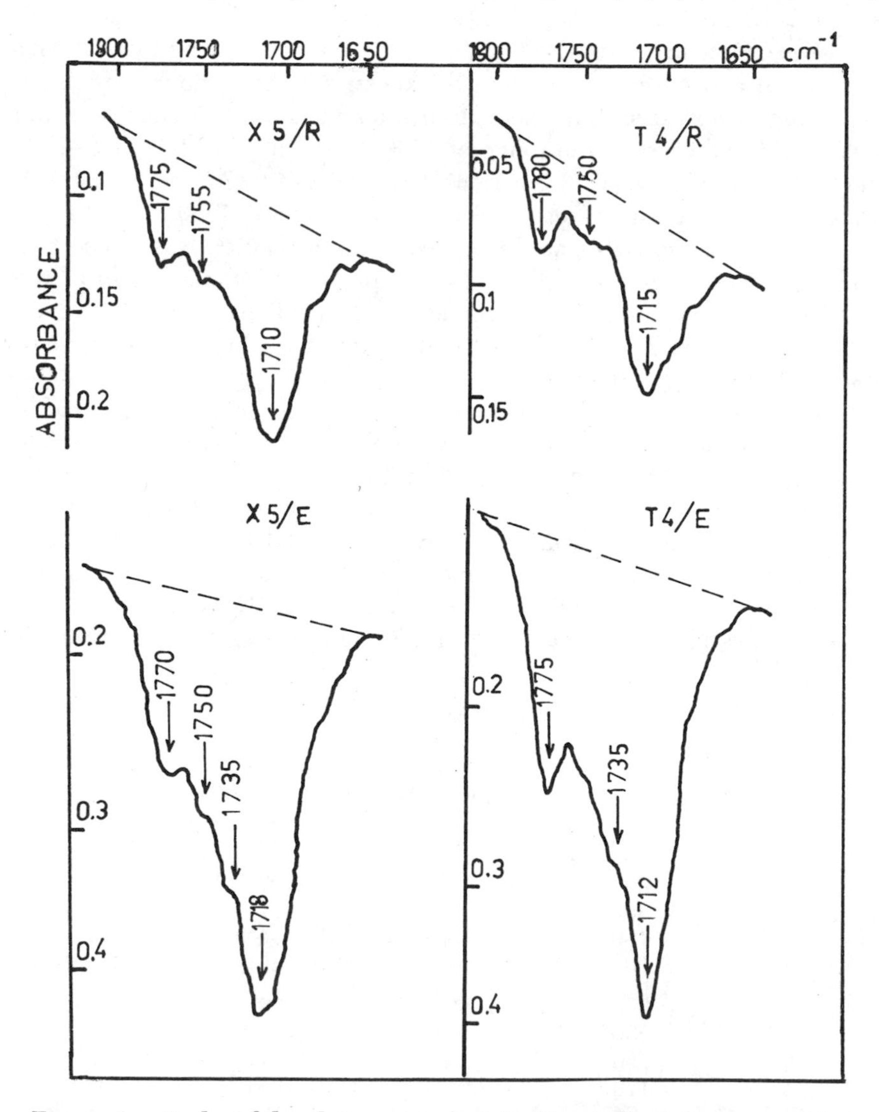

Figure 1. Carbonyl band (ir spectra) of highly oxidized fractions. Film thickness factor f = *1.0 (for X5/R and T4/R), 0.4 (for X5/E), 0.7 (for T4/E).*

Table III. DSC Data[a]

	E Fractions			*R Fractions*	
Sample	*Mainpeak (°C)*	*Sidepeak (°C)*	ΔH_F *(cal/g)*	*Peak (°C)*	ΔH_F *(cal/g)*
0	148	135	9.8	165	21.0
T3	153	140	15.2	165	22.3
T4	154	144	24.0	158	23.6
X4	153	142	17.0	165	22.7
X5	148	136	17.2	157	24.6

[a] T- and X-degradation series; E and R fractions.

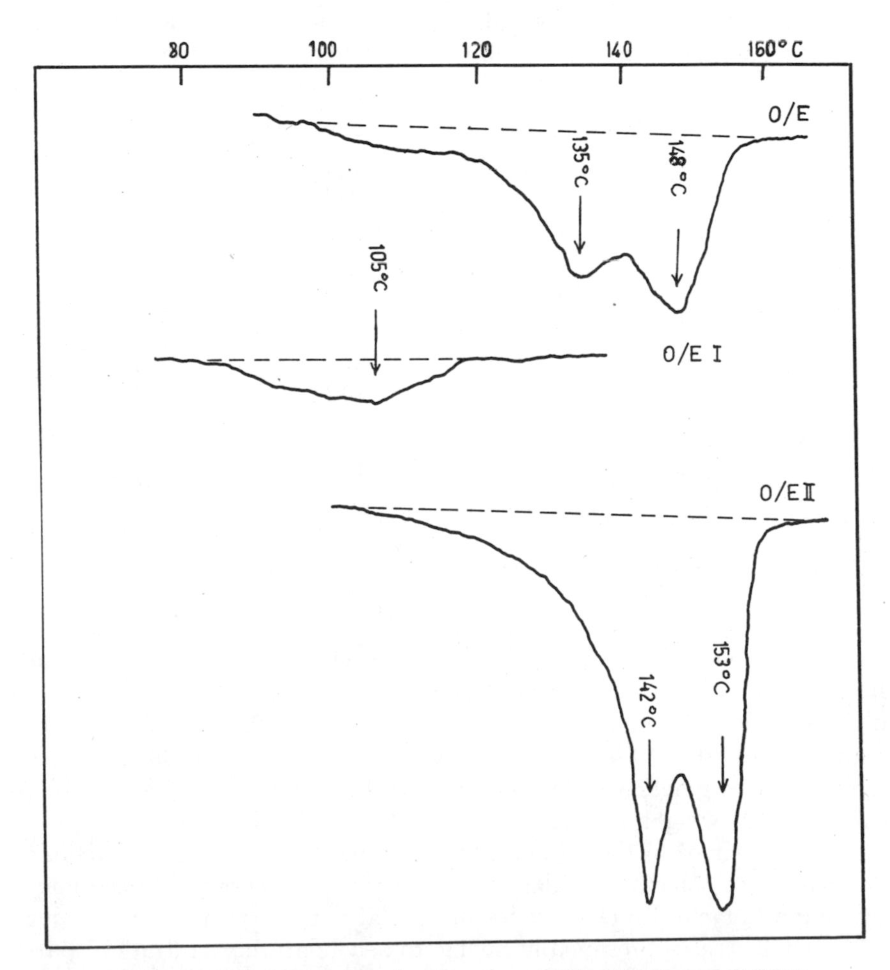

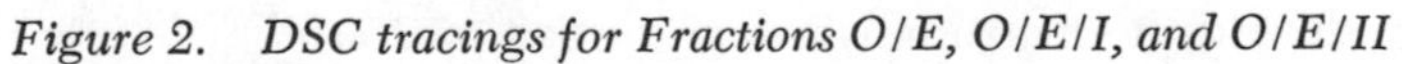
Figure 2. DSC tracings for Fractions O/E, O/E/I, and O/E/II

4.79 s, 4.14 s, 4.02 vs; this agrees well with the normal monoclinic α-modification, and any indication of γ-modification (4.42 vs) is conspicuously absent (*9, 10, 11*)).

Discussion

The C_7-soluble PP fraction obtained by the conventional extraction method contains various components which have only the solubility in heptane in common—noncrystalline atactic PP, very low molecular weight crystalline isotactic PP (species with molecular weight below about 10,000), and perhaps low molecular weight stereoblock PP (although its presence could not be demonstrated here).

At increasing degrees of degradation the C_7-soluble fraction rises slightly at first (up to 13.5% at T3, T4), but then very dramatically (T4, X5) owing to accumulation of soluble, low molecular weight, isotactic

Table IV. C–C Bond Scissions[a]

Sample	*E Fractions* ζ_E[b]	*R Fractions* ζ_R[b]	ζ_E/ζ_R
T1	0.5	0.05	10
T2	0.5	0.06	8.3
T3	0.9	0.29	3.1
T4	1.2	0.63	1.9
X1	0.5	0.09	5.6
X2	0.7	0.13	5.4
X3	1.0	0.16	6.2
X4	1.1	0.33	3.3
X5	1.9	0.92	2.1

[a] T- and X-degradation series; E and R fractions.

[a] $\zeta = \frac{C \not- C}{1000\ C-C}$ (number of scissions per 1000 C–C bonds)

PP. Evaluation of the number of C–C bond scissions (from $\overline{M}_n$ data) in the C_7-soluble (E) and C_7-insoluble (R) fractions shows a predominance in the soluble E fractions (Table IV: T1, T2; X1, X2, X3) that contain mainly atactic PP. With increasing degradation, the data no longer comply with the original conditions because of the greatly increasing soluble, low molecular weight, isotactic species (T3, T4; X4, X5).

The extent of oxidative bond scissions in the insoluble R fractions (in analogy to polyethylene data (*12*)) may be concentrated in the noncrystalline isotactic PP chain segments (total initial crystallinity is about 50%, but increases presumably owing to secondary crystallization, particularly during thermooxidative degradation at 150°C).

A continuation of this study will include stabilized PP which presents additional difficulties in view of the nonuniform distribution of stabilizing additives (*13, 14*).

Literature Cited

1. Hawkins, W. L., "Polymer Stablization," Wiley-Interscience, New York, 1972.
2. Ranby, B., Rabek, J. F., "Photodegradation, Photooxidation, and Photostabilization of Polymers," Wiley-Interscience, New York, 1975.
3. Dulog, L., Radlmann, E., Kern, W., *Makromol. Chem.* (1963) **60**, 1.
4. Kato, Y., Carlsson, D. J., Wiles, D. M., *J. Appl. Polym. Sci.* (1969) **13**, 1447.
5. Fuchs, O., *Makromol. Chem.* (1962) **58**, 247.
6. Chiang, R., *J. Polym. Sci.* (1958) **28**, 235.
7. Carlsson, D. J., Wiles, D. M., *Macromolecules* (1969) **2**, 587.
8. Adams, J. H., *J. Polym. Sci., A1* (1970) **8**, 1077.
9. Jones, A. T., Aizlewood, J. M., Beckett, D. R., *Makromol. Chem.* (1964) **75**, 134.
10. Kardos, J. L., Christiansen, A. W., Baer, E., *J. Polym. A2* (1966) **4**, 777.
11. Samuels, R. J., *J. Polym. Sci., A2* (1975) **13**, 1417.
12. Hawkins, W. L., Matreyek, W., Winslow, F. H., *J. Polym. Sci.* (1959) **41**, 1.
13. Frank, H. P., Lehner, H., *J. Polym. Sci., C* (1970) **31**, 193.
14. Billingham, N. C., Calvert, P. D., Prentice, P., Ryan, T. G., *Am. Chem. Soc., Div. Polym. Chem., Preprint* **18** (1), 476 (New Orleans, March, 1977).

RECEIVED May 12, 1977.

10

Ultraviolet Stabilization of Polymers: Development with Hindered-Amine Light Stabilizers

A. R. PATEL and J. J. USILTON

CIBA-GEIGY Corp., Ardsley, NY 10502

Significantly improved UV light stability is found with polymeric substrates stabilized with bis(2,2,6,6-tetramethyl-piperidinyl-4) sebacate, an example of hindered-amine class light stabilizers, compared with those stabilized with conventional light stabilizers. Application areas covered include polyolefin fibers, films and molded sections, poly-urethane, and styrenics. Synergistic performance with o-*hydroxyphenyl benzotriazoles in these polymers is apparent.*

Various aspects of the mechanistic efficiency of the hindered amines are discussed. Quenching of carbonyl excited states and UV absorption are not considered important modes of activity. Some effectiveness in singlet oxygen quenching and radical scavenging has been shown. Speculations on mechanistic efficiency include hydroperoxide decomposition and activity during processing to reduce sensitizer concentration. A comprehensive mechanism is not yet available.

This paper presents a review of the performance and the mechanism of stabilization of a new class of light stabilizers based on substituted piperidine derivatives. These light stabilizers have been found to have outstanding effectiveness in many polymeric substrates. In most instances, the light stability achieved with these compounds was far superior to that attained with conventional light stabilizers. Results of studies illustrating the performance of a representative member of this light stabilizer class, bis(2,2,6,6-tetramethyl-piperidinyl-4) sebacate (LS–1) in polyolefins, styrenics, and polyurethanes are presented in comparison with

0-8412-0381-4/78/33-169-116$05.00/1

conventional light stabilizers, which are specified in Appendix 1. In all cases, the level of light stabilizer specified is a total amount for the system, and 1% of a 1:1 mixture would be 0.5% of each element of the mixture.

Performance of Hindered-Amine Stabilizer LS–1

Polypropylene. HALS (LS–1) provided outstanding light stabilizing effectiveness in polypropylene fibers (16 denier per filament) (Table I):

Table I. 45° South Direct Exposure—Florida

	Kilolangley to 50% Retention of Tenacity		
		Pigmented	
550/34 – 16 dpf Base Stabilizer 0.05% AO–1	*Natural*	*1% Red BR*	*1% Phthalocyanine Blue*
0.5% LS–1 (HALS)	100	70	100
0.5% LS–6 (Nickel)	50	30	60

In .001-in. oriented polypropylene film, HALS LS–1 provided outstanding retention of tenacity as measured by kilolangleys to 50% retention of tenacity (*see* Figure 1). In relatively thick section polypropylene (.08-in., injection-molded sections, HALS LS–1 was able to provide significant retention of impact strength and gloss for outdoor direct exposure of polypropylene; in various pigment systems the HALS was far superior to conventional light stabilizers (*see* Tables II and III).

Table II. Kilolangley to 50% Retention of Dart Impact Strength

		Pigmented (0.4%)			
0.3% LS	*Natural*	*TiO_2*	*Red BR*	*Ultramarine Blue*	*Cadmium Pure Yellow 15GN*
Blank	50	50	50	50	50
HALS LS–1	300	300	300	300	300
Benzotriazole LS–4	130	50	55	50	100
Nickel LS–5	50	NT[a]	50	50	50
Nickel LS–6	80	NT[a]	NT[a]	NT[a]	NT[a]
Benzophenone LS–7	75	50	50	50	80
Benzoate LS–8	80	NT[a]	150	80	90
Nickel/benzophenone 1:1 LS–6/LS–7	75	NT[a]	NT[a]	80	95

[a] NT = Not tested.

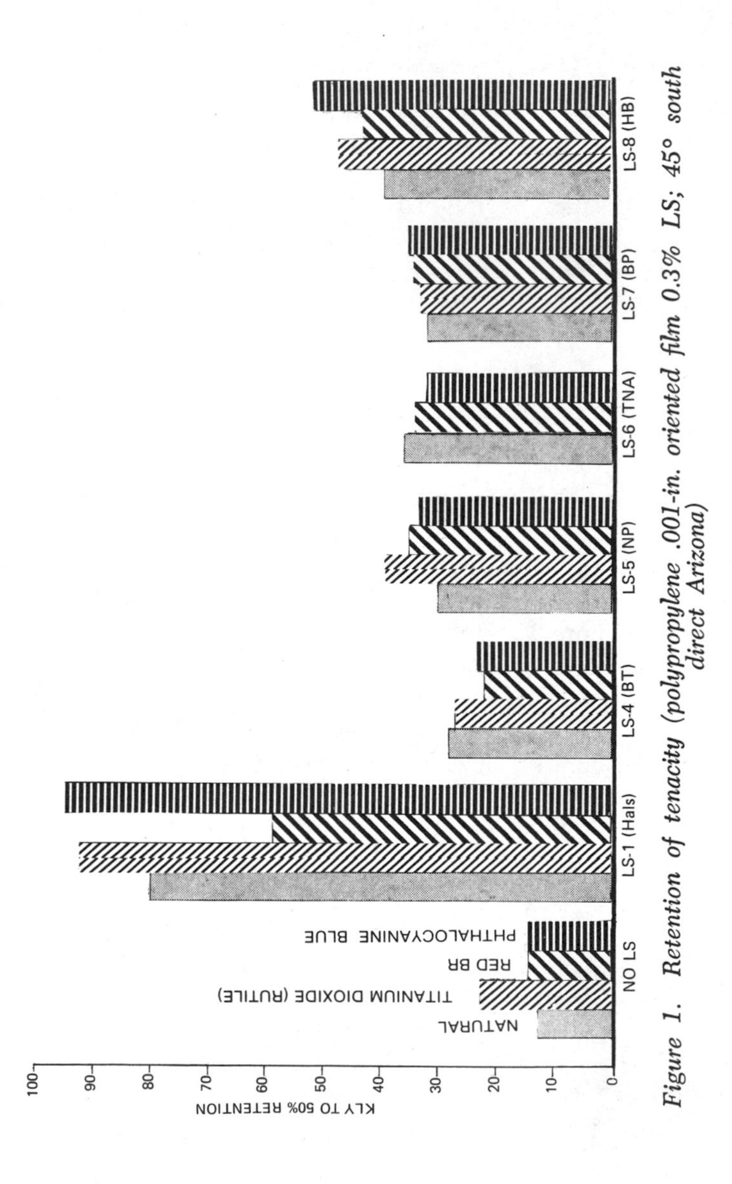

Figure 1. Retention of tenacity (polypropylene .001-in. oriented film 0.3% LS; 45° south direct Arizona)

Table III. Surface Gloss at 50% Retention of Dart Impact Strength[a]

0.3% LS	*Natural*	*TiO_2*	*Red BR*	*Ultra-marine Blue*	*Cadmium Pure Yellow 15GN*
Blank	1	3	1	1	3
HALS LS–1	5	5	4	5	3
Benzotriazole LS–4	3	3	2	3	2
Nickel LS–5	3	NT	NT	NT	3
Nickel LS–6	3	NT	NT	NT	NT
Benzophenone LS–7	3	4	3	3	3
Benzoate LS–8	3	NT	3	3	3
Nickel/benzophenone 1:1 LS–6/LS–7	3	NT	NT	3	2

[a] Rating system based on visual rating; NT = Not tested; 1 = poor; 5 = excellent.

High Density Polyethylene. In outdoor weathering in Arizona, the formulation containing 0.3% of the hindered-amine light stabilizer (LS–1) outperformed the most effective conventional light stabilizer (benzophenone LS–7) in retaining tensile impact strength and retarding the development of visual surface crazing in the presence of various pigments (*see* Figures 2 and 3). In commercially blow-molded, high-density polyethylene containers, retention of elongation on light exposure (Carbon Arc Weatherometer with Spray) was excellent for a 0.3% blend of HALS LS–1 with benzotriazole LS–4, far surpassing the performance of 0.25% benzophenone LS–7 and 0.25% blended benzophenone LS–7/nickel LS–6 (*see* Figure 4). In pigmented high-density polyethylene-molded specimens exposed to direct weathering in Florida, LS–1-stabilized samples exhibited superior retention of elongation compared with samples stabilized with other conventional light stabilizers (Figures 5, 6, 7, and 8).

Aromatic Polyester Polyurethane. The hindered-amine light stabilizer (LS–1) or a 1:1 blend of LS–1 and LS–2 provided about five to six times better performance than the best UV absorber (LS–2) alone in

Table IV. .002-In. Film—Exposure under Glass in Florida

Additive	*Hr to 50% Retention of* Tensile Strength	*Hr to 50% Retention of* Elongation	*Hr to a Hunter L–b Color of 70 Units*
None	590	250	170
2% Benzotriazole LS–2	590	925	250
0.5% HALS LS–1	1850	> 2350	> 2350
1.0% HALS LS–1	> 2350	> 2350	> 2350
1.0% HALS LS–1/BT LS–2 (1:1)	> 2350	> 2350	> 2350

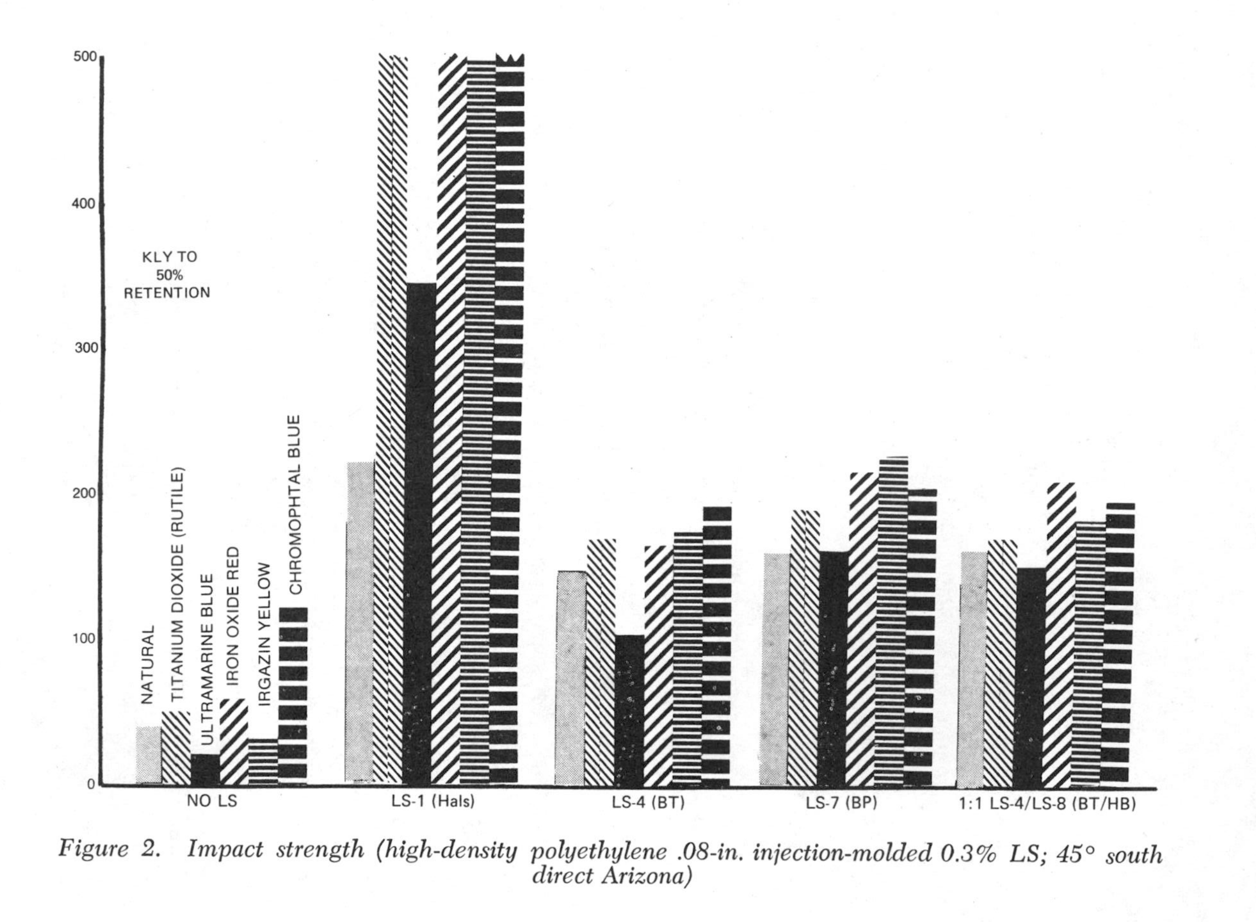

Figure 2. Impact strength (high-density polyethylene .08-in. injection-molded 0.3% LS; 45° south direct Arizona)

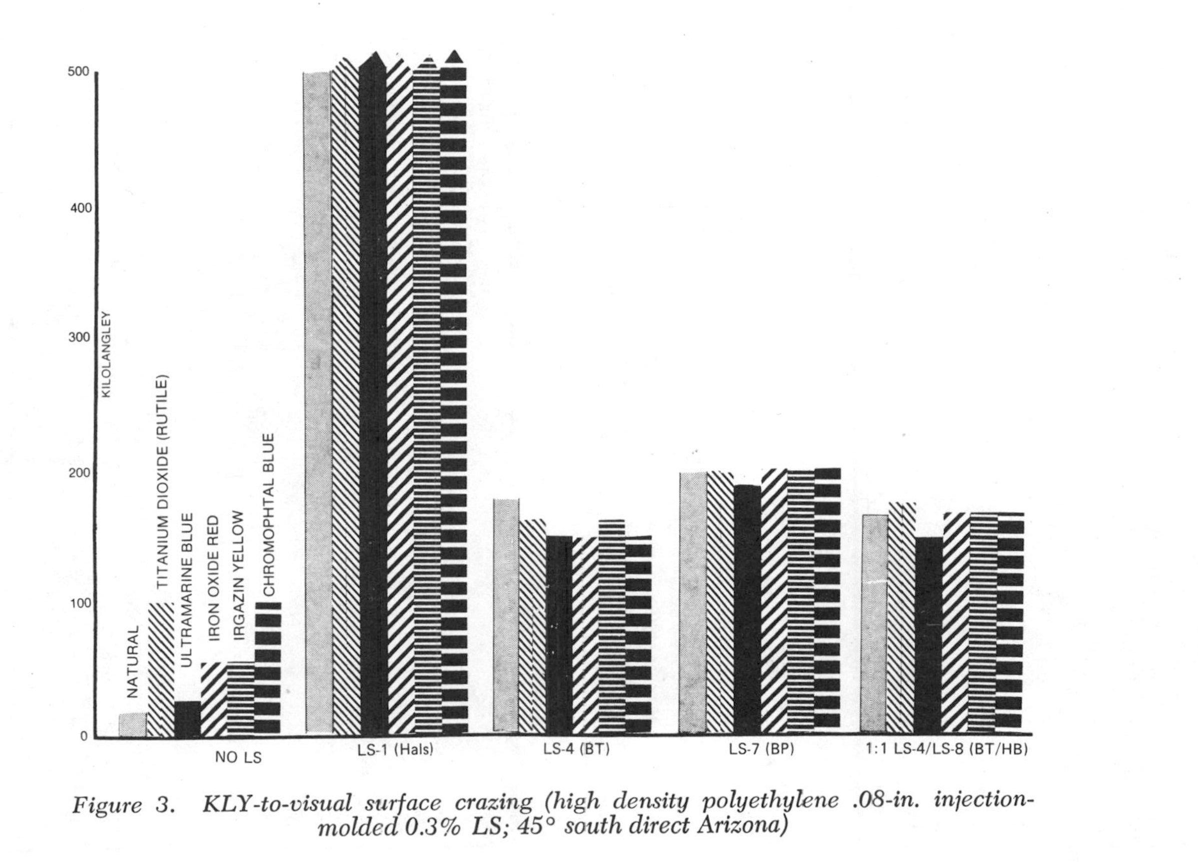

Figure 3. KLY-to-visual surface crazing (high density polyethylene .08-in. injection-molded 0.3% LS; 45° south direct Arizona)

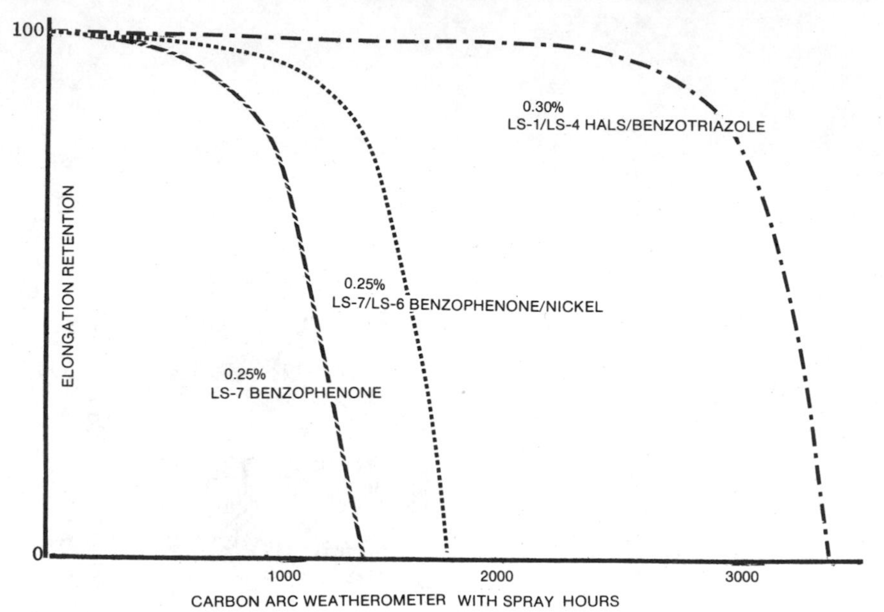

Figure 4. Commercially blow-molded HDPE containers. Elongation retention of blow-molded, high density polyethylene; .05-in. wall container sections pigmented with phthalocyanine green, chrome yellow, and TiO_2 (rutile) as a function of UV exposure.

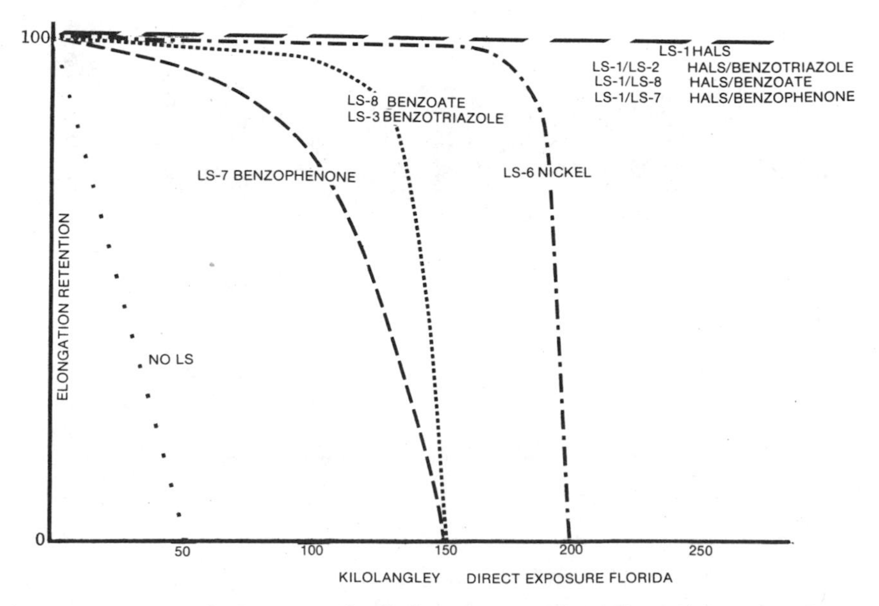

Figure 5. High-density polyethylene (2% rutile TiO_2, 0.5% LS). Elongation retention as a function of 45° south direct Florida exposure for high density polyethylene stabilized with 0.50% of various light stabilizers. Pigmented rutile TiO_2.

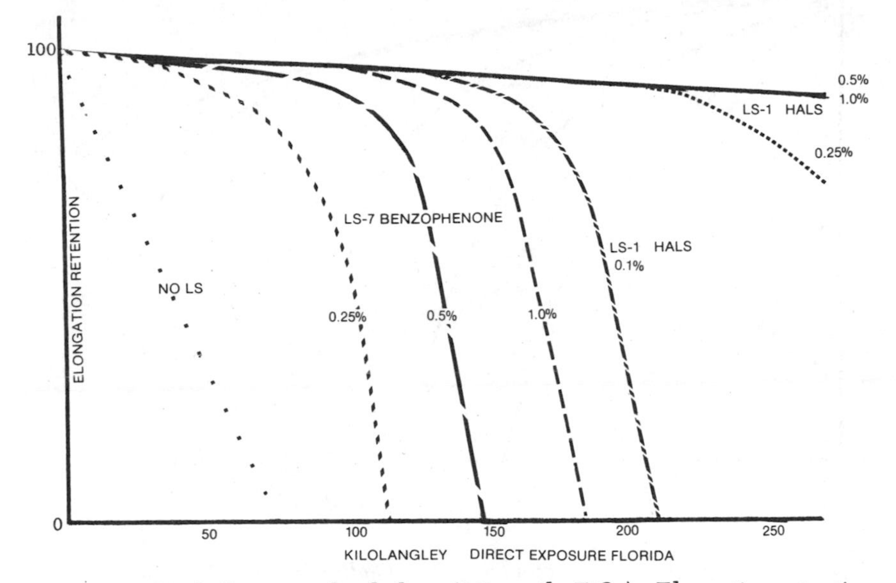

Figure 6. High density polyethylene (2% rutile TiO_2). Elongation retention as a function of 45° south direct Florida exposure for HALS and benzophenone light stabilizers at various concentrations in high density polyethylene pigmented rutile TiO_2.

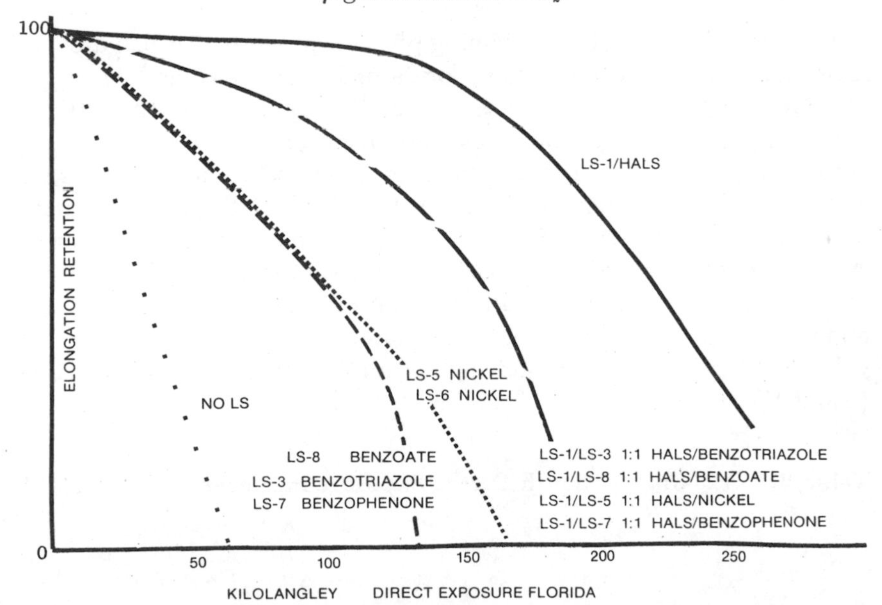

Figure 7. High-density polyethylene (2% Cadmium Yellow 0.5% LS). Elongation retention as a function of 45° south direct Florida exposure for high density polyethylene stabilized with 0.5% of various light stabilizers. Pigmented cadmium yellow.

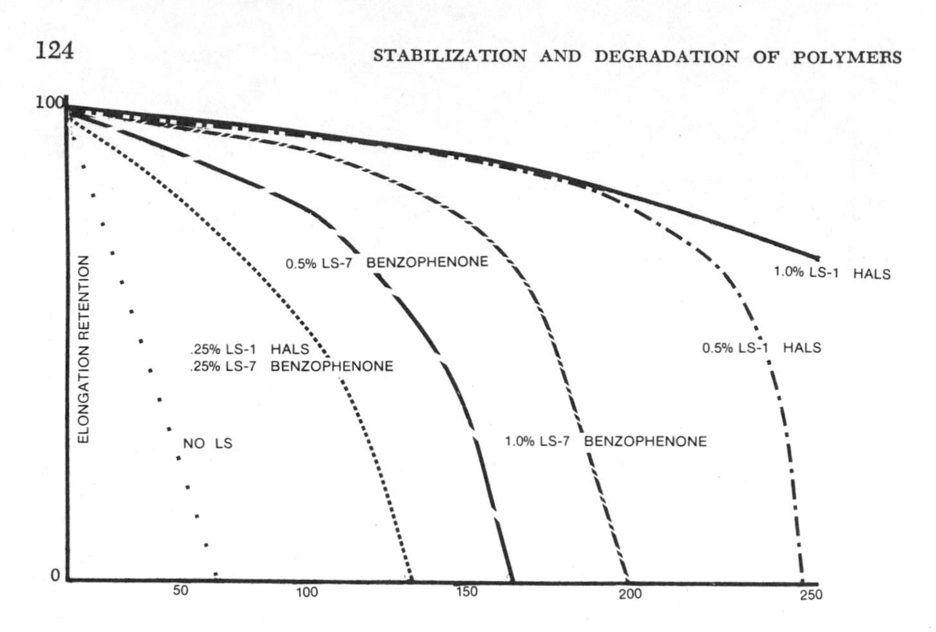

Figure 8. High-density polyethylene (2% Cadmium Yellow). Elongation retention as a function of 45° south direct Florida exposure for HALS and benzophenone light stabilizers at various concentrations in high density polyethylene pigmented cadmium yellow.

preventing yellowing and maintaining physical properties of polyurethane films after outdoor weathering in Florida under glass (Table IV).

Styrenic Polymers. The weathering stability of crystal polystyrene, ABS, and impact polystyrene is relatively poor and thus the potential for outdoor applications is limited. However, hindered-amine-based systems can improve weathering resistance. A 1:1 blend of the hindered-amine light stabilizer LS–1 with LS–2 provided a six- to sevenfold improvement in color and gloss retention compared with unstabilized samples of crystal polystyrene after outdoor weathering in Florida. It is estimated that minimal changes in color could be obtained when exposed up to two years in Florida using optimum light stabilizer systems based on LS–1 (Table V).

Table V. .125-In. 45° South Direct Exposure in Florida (0.1% AO–2)

Additive	*Kilolangleys to a 10-Unit Increase in Δ Lba Color (Hunter)*
None	23
0.2% LS–1	85
0.2% LS–2	100
0.2% 1:1 LS–1/LS–2	150

Table VI. ABS .125-In., 45° South Direct Exposure in Florida

Additive	*Kilolangleys to 50% Retention of Impact Strength (Izod)*
None	6
1.5% LS–1	11
1.5% LS–2	8
1.5% 1:1 LS–1/LS–2	27
1.5% LS–7	9

In ABS a 1:1 blend of LS–1 with LS–2 provided a four- to fivefold improvement in color and impact strength retention over unstabilized samples and it appeared to be about three times better in performance than the most effective conventional light stabilizers, benzotriazole LS–2 and benzophenone LS–7 (Table VI).

In impact polystyrene this 1:1 blend of LS–1 with LS–2 provided five to six times better performance than unstabilized formulations, and it is estimated that this blend could provide up to one year or more of protection in Florida to minimum color change and 50% retention of impact strength of impact polystyrene (*see* Figure 9).

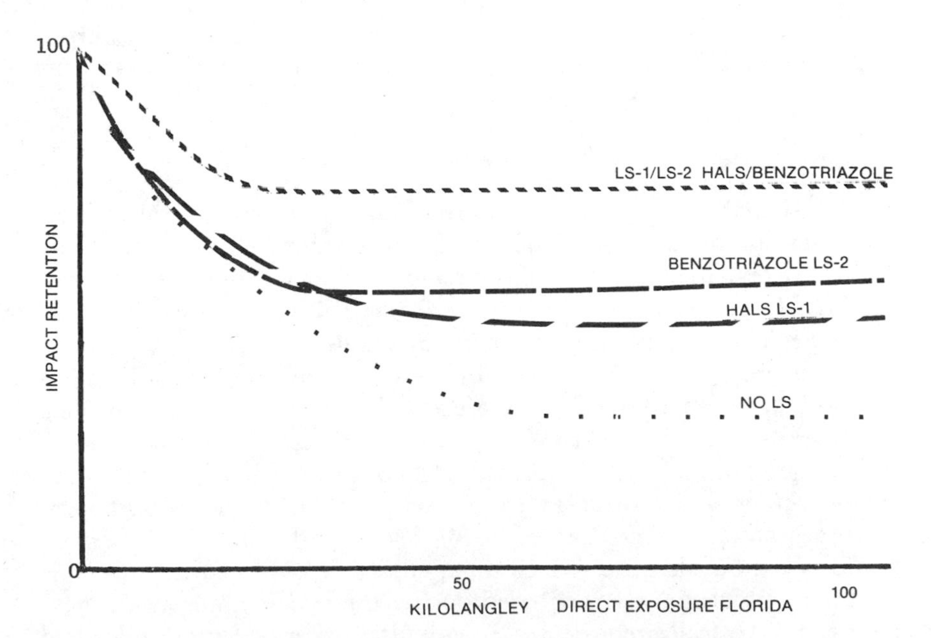

Figure 9. Impact polystyrene. Impact retention as a function of 45° south direct Florida exposure for impact polystyrene stabilized with 0.50% of various light stabilizer(s). Izod impact measurement.

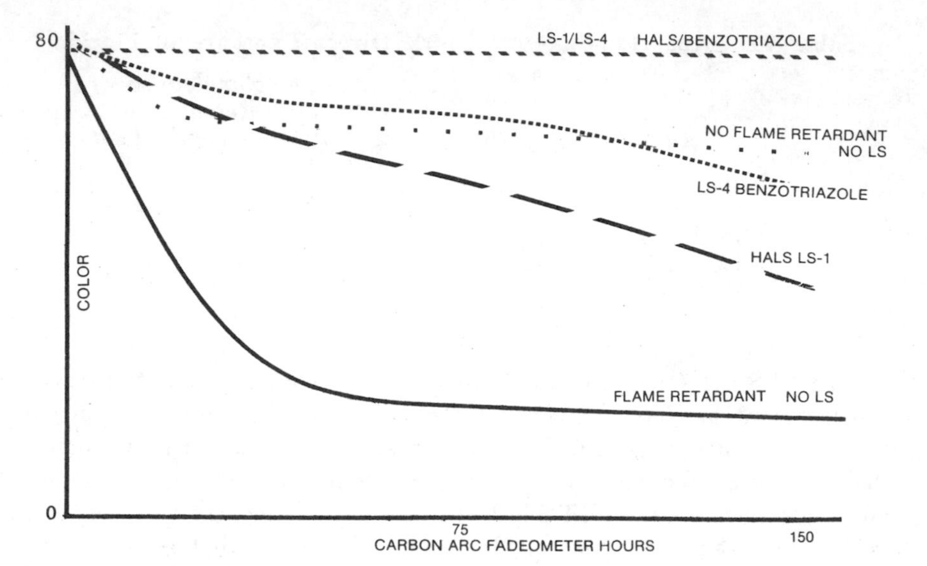

Figure 10. Impact polystyrene (12% Aliphatic flame retardant, 4% antimony oxide). Color (Hunter L–b) as a function of Carbon Arc Fadeometer exposure for flame retardant impact polystyrene stabilized with 0.50% of various light stabilizers. Flame retardant: Cities Service BC–267.

In flame-retardant impact polystyrene, a 1:1 blend of LS–1 with LS–3 or LS–4 outperformed the most effective conventional light stabilizers presently available on the basis of preventing discoloration upon exposure in the Carbon Arc Fadeometer (Figure 10).

UV Degradation

UV radiation originating from sunlight or artificial light sources results in substantial degradation of many commercial polymeric substrates. Degradation manifested as discoloration, embrittlement, cracking, and loss of other physical properties seriously reduces the useful lifetime and potential applications of polymeric materials.

Although UV radiation generally results in polymer degradation, it has been shown that typical commercial polymers have small absorption characteristics in the near-UV region. Photoactive species such as carbonyl groups and hydroperoxides introduced into the polymer during polymerization and processing are the principal points of absorption of UV radiation and the starting point for UV degradation of the polymer. A widely used package to reduce the quantity of UV radiation available to initiate UV degradation is based on the use of UV absorbers. These compounds, typically represented by *o*-hydroxyphenylbenzotriazoles and *o*-hydroxybenzophenones, preferentially absorb UV radiation which then is dissipated harmlessly. Other postulations concerning their effectiveness suggest that they can operate by the chain breaking mechanism and by triplet-state quenching as suggested in the review by Scott (*17*).

Another important class of light stabilizers are the nickel chelates. These compounds do not absorb significantly in the 300–400-nm region and may improve light stability by quenching excited state triplets (2) and excited state singlets by dipole–dipole interaction (*1*), or by activity during processing (as a UV stable antioxidant) (*18*).

In addition to the above light stabilizers, improved light stability is obtained also with antioxidants, which are efficient free radical scavengers, and with phosphites which decompose hydroperoxides.

Mechanism of Hindered-Amine Stabilizers

To date, evaluations by Heller and Blattman (*1*), Felder and Schumacher (*3*), and others have demonstrated that the hindered-amine light stabilizer LS–1 does not absorb UV and hence does not function as a typical UV absorber, nor is it believed to be a quencher of carbonyl

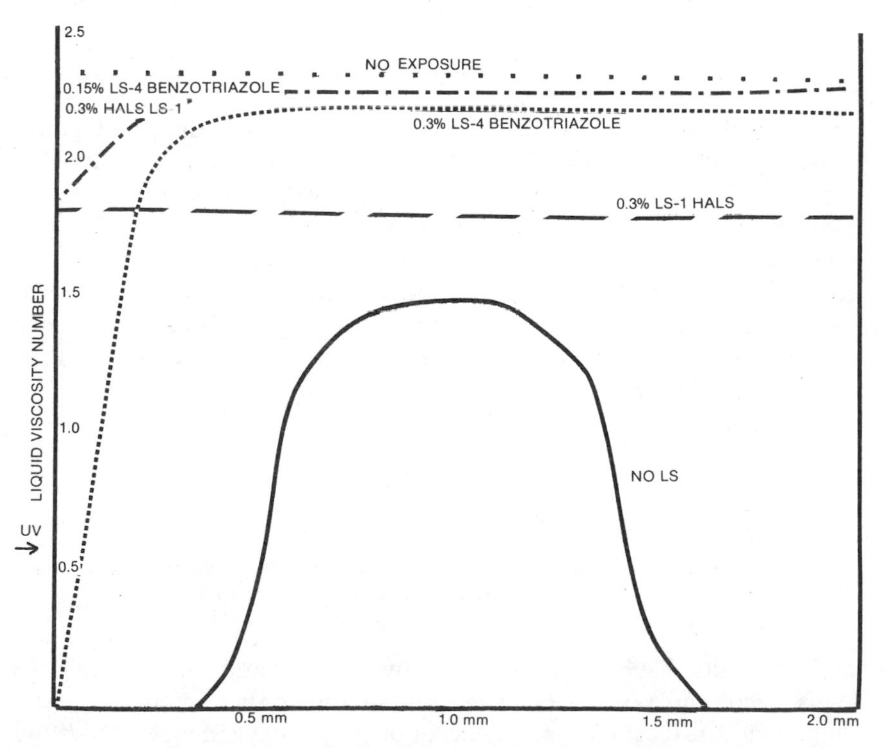

Figure 11. Cross-section profile of exposed polypropylene (0.1% AO–1). Change in viscosity (Decalin, 135°) as a function of position in a 2-mm polypropylene thick section exposed to direct weathering for 150 kly in South Africa.

singlet or triplet. There are some mechanisms of action that have been studied in some detail and are believed to make significant contribution to the light stabilizing effectiveness of HALS.

An experiment reported by Leu (*4*) shows the light stabilizing effectiveness of several types of stabilizers and stabilizer blends. The cross section profile has been determined after direct exposure of 2-mm polypropylene sections. Changes in polymer molecular weight were determined by measuring viscosity of thin sections microtomed from the plaque. With no light stabilizer, degradation is severe. With a UV absorber, degradation at the surface is severe but the inner parts of the plaque are protected; the HALS sample shows a slight uniform change with good surface stabilization; the excellent performance of the blended HALS/benzotriazole system is evident (Figure 11).

Singlet-Oxygen Quenching

Singlet oxygen formed by the quenching of excited carbonyl triplets by molecularly dissolved oxygen exhibits a higher reactivity compared

$$\left[\geq C{=}O\right]^3 \quad + \quad {}^3O_2 \quad \rightarrow \quad \left[\geq C{=}O\right]^1 \quad + \quad {}^1O_2$$

with normal triplet oxygen and it is able to oxidize the vinyl groups present in polymers to allyl hydroperoxides (*3, 5, 6, 7*). It also has been shown by Kaplan and Kelleher (*8*) that saturated hydrocarbons also are oxidized to hydroperoxides by singlet oxygen. The hydroperoxides thus formed are unstable and, by decomposition to alkoxy and hydroxy radicals, propagate decompositions. Hindered amines interfere with this mechanism by quenching the singlet oxygen to ground-state triplet oxygen. Heller and Blattman (*1*) have ranked the following stabilizers according to singlet oxygen quenching efficiency:

Singlet-oxygen quenching efficiency	
2.0	1,2,2,6,6-pentamethylpiperidines
	nickel chelates
1.5	2,2,6,6-tetramethylpiperidines-*N*-oxyls
	2,2,6,6-tetramethylpiperidines
1.0	blank

Similar results have been reported by Felder and Schumacher (*3*, Bellus, Lind, and Wyatt (*9*). The results indicate that among the tested compounds, the most efficient quencher of singlet oxygen is the *N*-methyl-substituted piperidine derivative, followed by nickel chelates and nitroxyl compounds of the hindered amines. The 2,2,6,6-tetramethylpiperidines themselves were of relatively low efficiency.

Formation of N-Oxyl from Free Amine

The hindered-amine stabilizers readily form nitroxyl compounds. Denisov (*10, 11*) has found facile formation in polypropylene. Rozantsev (*12, 13*) indicates that the mechanisms by which these sterically hindered amines act is via the generation of stable piperidinoxyl radicals, and he proposes their formation.

From the standpoint of steric hindrance, an *N*-oxyl radical should be less stable than the corresponding *N*-free radical. Murayama (*14*), however, has shown that an *N*-free radical has a strong affinity to oxygen, and forms the *N*-oxyl readily. The *N*-oxyl is a stable radical owing to the lack of hydrogen on the carbon alpha to the *N*-atom and the following resonance:

$$>N-O\cdot \longleftrightarrow >\overset{+}{\dot{N}}-O^{-} \longleftrightarrow >\dot{N}=O$$

Any significant singlet-oxygen quenching efficiency of LS–1 will require the conversion of LS–1 to the nitroxyl. Other amine derivatives are themselves efficient quenchers of singlet oxygen but can also be oxidized to the nitroxyl. Conversion to the *N*-oxyl must, however, be at a low equilibrium level—in both the present work and in Rozantsev's experiments (*13*), LS–1 -stabilized polymer is colorless throughout its lifetime even though the corresponding N-oxyl could, at higher levels, impart a weak rose color to the polymer (*13*).

Radical Scavenging

In addition to singlet-oxygen quenching, the nitroxyls formed from the amine may act as efficient radical scavengers (*10, 13, 15*). Experiments of Denisov (*10*) and Felder and Schumacher (*3*) demonstrated the efficiency of the nitroxyls in scavenging some radicals. Denisov and Rozantsev have shown that regeneration of the nitroxyl can occur by the sequential scavenging as follows:

$$>N-O\cdot + R\cdot \rightarrow >N-OR$$

$$>N-OR + R'OO\cdot \rightarrow >N-O\cdot + ROOR'$$

Other Mechanisms

Many other mechanisms such as hydroperoxide decomposition, charge transfer, etc., have been speculated on by various authors. Scott (*16*) has proposed that an important function of the nickel complex nickel dibutyldithiocarbamate in providing improved UV stability of polyolefins is by conventional chain-breaking and preventive antioxidant processes during thermal processing. It could be speculated that the presence of HALS during thermal processing may reduce the concentration of sensitizers present after processing, thus improving light stability by reducing the sites from which degradation begins. Extensive work has been done by Rozantsev, Denisov, and others investigating the use of hindered amines as antioxidants (*11, 13, 15*).

Although each of the above mechanisms may contribute to polymer stabilization, a comprehensive perspective on the mechanism of the light stabilizing effectiveness of the hindered amine is not yet available.

Appendix: Structures

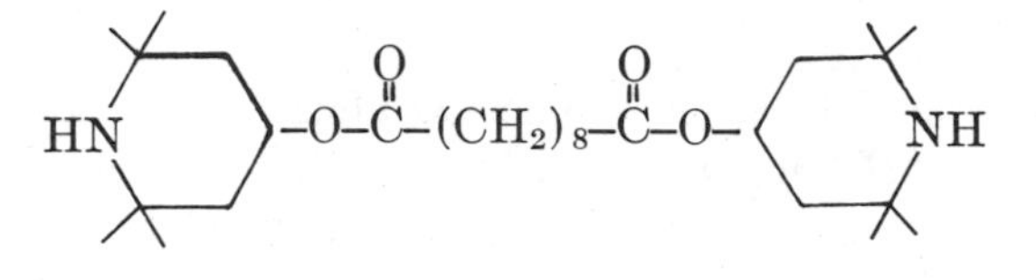

LS–1
Bis(2,2,6,6-tetramethyl-piperidinyl-4) sebacate

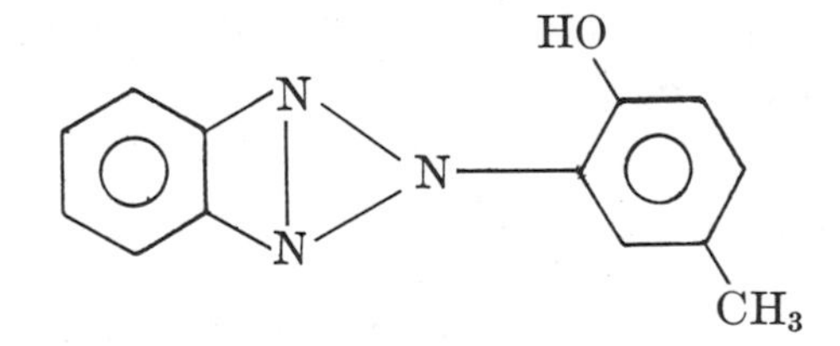

LS–2
2-(2′-Hydroxy-5′-methyl phenyl) benzotriazole

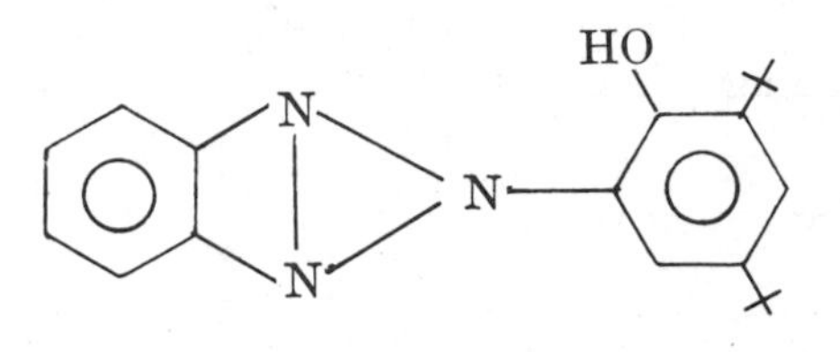

LS–3
2-(3′,5′-Di-*tert*-amyl-2′-hydroxyphenyl) benzotriazole

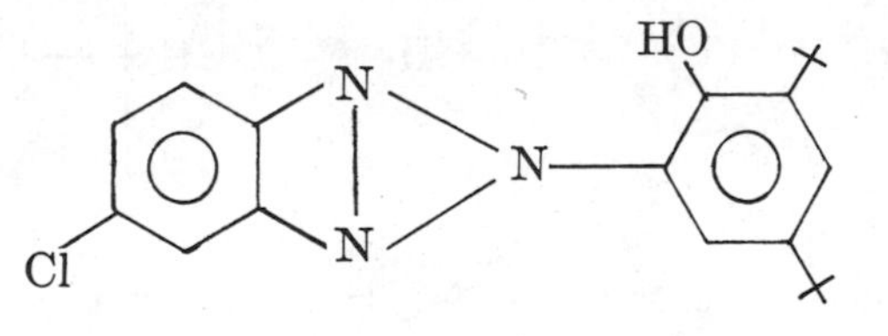

LS–4
2-(3′,5′-Di-*tert*-butyl-2′-hydroxyphenyl)-5-chlorobenzotriazole

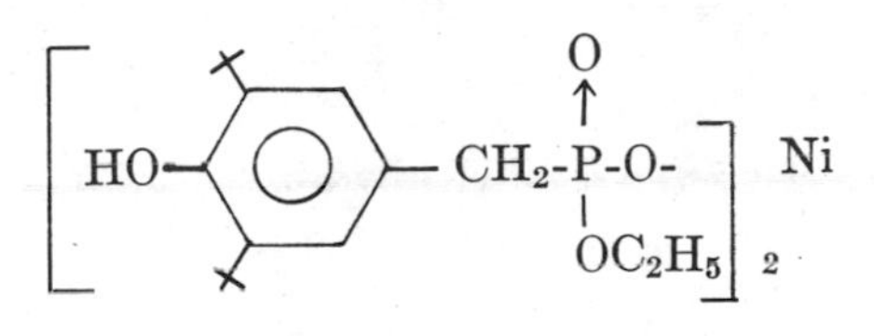

LS–5
Nickel bis[*o*-ethyl(3,5-di-*tert*-butyl-4-hydroxybenzyl)] phosphonate

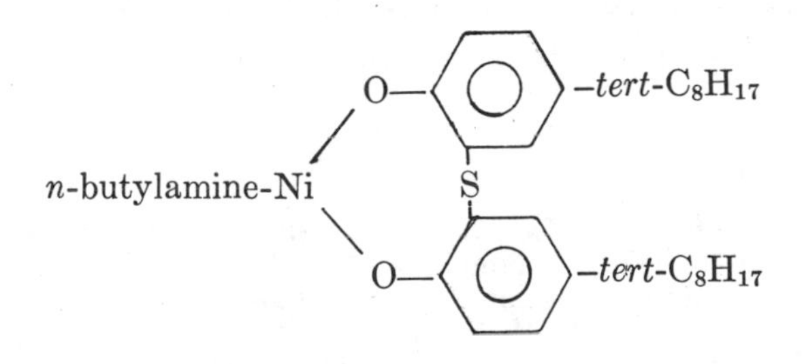

LS–6
2-2′-Thiobis(4-*tert*-octylphenylato-*n*-butylamine) nickel III

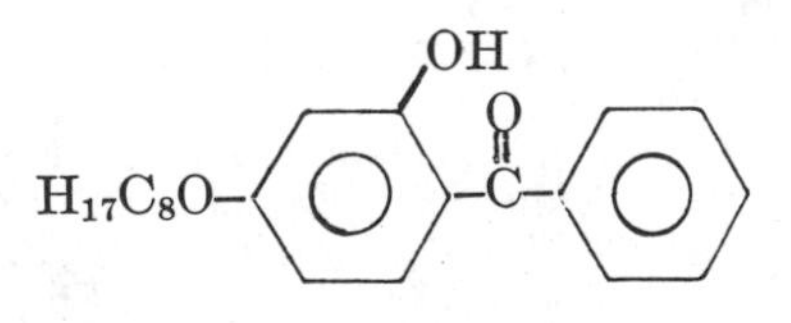

LS–7
2-Hydroxy-4-*n*-octoxybenzophenone

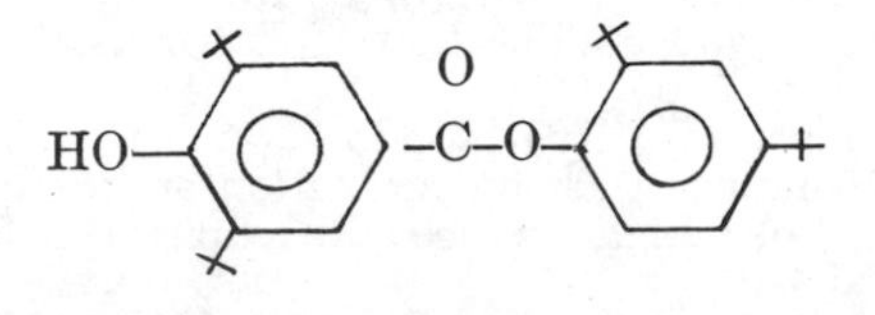

LS–8
2,4-Di-*tert*-butylphenyl-3,5-di-*tert*-butyl-4-hydroxybenzoate

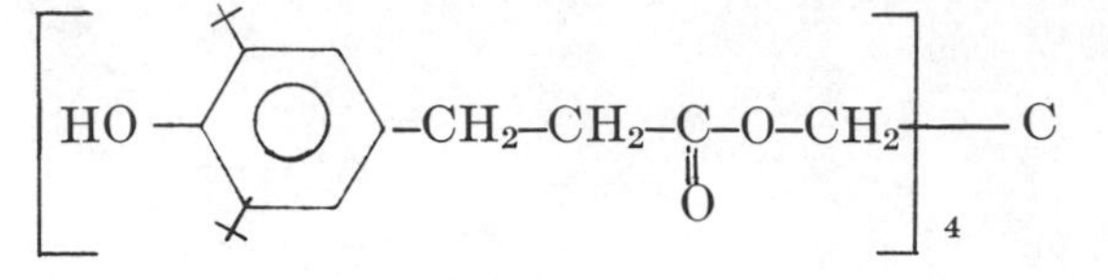

AO–1
Tetrakis[methylene(3,5-di-*tert*-butyl-4-hydroxyhydrocinnamate)]methane

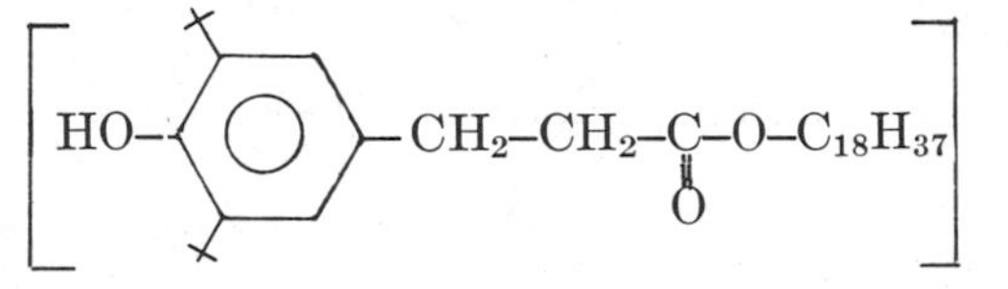

AO–2
Octadecyl-3,5-di-*tert*-butyl-4-hydroxyhydrocinnamate

Acknowledgment

The authors wish to thank the management of CIBA–GEIGY for permission to publish this work. Special thanks are in order to P. P. Klemchuk and J. Farber for their useful comments and suggestions. The work is the collective activity of the authors, the CIBA–GEIGY U.S.A. Additives Laboratory, and the CIBA–GEIGY AG, Basle Laboratories of H. Müller.

Literature Cited

1. Heller, H. J., Blattman, H. R., *Pure Appl. Chem.* (1973) **36**, 141.
2. Ranby, B., Rabek, J. F., "Photodegradation, Photo-oxidation and Photostabilization of Polymers," Wiley, New York, 1975.
3. Felder, B., Schumacher, R., *Makromol. Chem.* (1973) **31**, 35.
4. Leu, K. W., Paper presented at Plastics Institute of Australia Technical Seminar, Terrigal, N.S.W., October, 1974.
5. Trozzolo, A. M., Winslow, F. H., *Macromolecules* (1968) **1**, 98.
6. Carlsson, D. J., Wiles, D. M., *Polym. Lett.* (1973) **11**, 795.
7. Heller, H. J., *Eur. Polym. J.* (1963) **105.**
8. Kaplan, M. L., Kelleher, P. G., *Polym. Prepr., ACS, Div. Polym. Chem.* (1971) **12**, 319.
9. Bellus, D., Lind, H., Wyatt, J. F., *Chem. Commun.* (1972) 1199.
10. Denisov, Y. T., *Izv. Akad. Nauk. SSR (Eng. Transl.)* (1974) 137.
11. Denisov, Y. T., *Polym. Sci. U.S.S.R. (Eng. Transl.)* (1974) **16**, 2628.
12. Rozantsev, E. G., et al., *Polym. Sci. U.S.S.R. (Eng. Transl.)* (1973) **15**, 1165.
13. Rozantsev, E. G., et al., *Polym. Sci. U.S.S.R. (Eng. Transl.)* (1973) **15**, 3034.
14. Murayama, K., *Farumashia* (1974) **10**(8) 573.
15. Nieman, B., "Aging and Stabilization of Polymers," p. 33 ff, Consultants Bureau, New York, 1965.
16. Scott, G., *Eur. Polym. J.* (1976) **12**, 825.
17. Scott, G., "Mechanisms of Photodegradation and Stabilization of Polyolefins," *in* "Ultraviolet Light Induced Reactions in Polymers," *ACS Symp. Ser.* (1976) **25**, 340.
18. Ranaweera, R. P. R., Scott, G., *Eur. Polym. J.* (1976) **12**, 591.

RECEIVED May 12, 1977.

11

^{13}C NMR Observation of the Effects of High Energy Radiation and Oxidation on Polyethylene and Model Paraffins

F. A. BOVEY and F. C. SCHILLING

Bell Laboratories, Murray Hill, NJ 07974

H. N. CHENG

G.A.F. Corp., 1361 Alps Road, Wayne, NJ 07470

Carbon-13 NMR enables one to identify the products of the γ-irradiation of a model hydrocarbon, n-$C_{44}H_{90}$, *and of polyethylene thermal oxidation. Irradiation of* n-$C_{44}H_{90}$ *in the molten state produces cross-links (H-shaped molecules), long branches (T-shaped molecules), and* trans-*vinylene groups. Irradiation in the crystalline state produces none of these. Instead, it end-links the molecules to produce linear dimers, detected by gel permeation chromatography (GPC). Mechanisms are proposed to account for these results. Branched polyethylene thermal oxidation gives rise to long chain ketones and carboxylic acids as the principal products, with smaller amounts of long chain secondary alcohols and hydroperoxides, esters of long chain carboxylic acids with long chain secondary alcohols, and long chain γ-lactones. Branch points are about 10-fold more reactive than linear chains. The results generally agree with accepted mechanisms.*

Carbon-13 NMR is powerful for determining paraffinic polymer structures because of the sensitivity of carbon chemical shifts to branches and chain ends. Carbon shieldings are also strongly dependent on oxygen-containing groups. These ^{13}C NMR features suggest its use for the qualitative detection and quantitative estimation of the chemical effects of polyethylene and model paraffin exposure to high energy radiation and thermal oxidation. This chapter deals with its application to these problems.

0-8412-0381-4/78/33-169-133$05.00/1

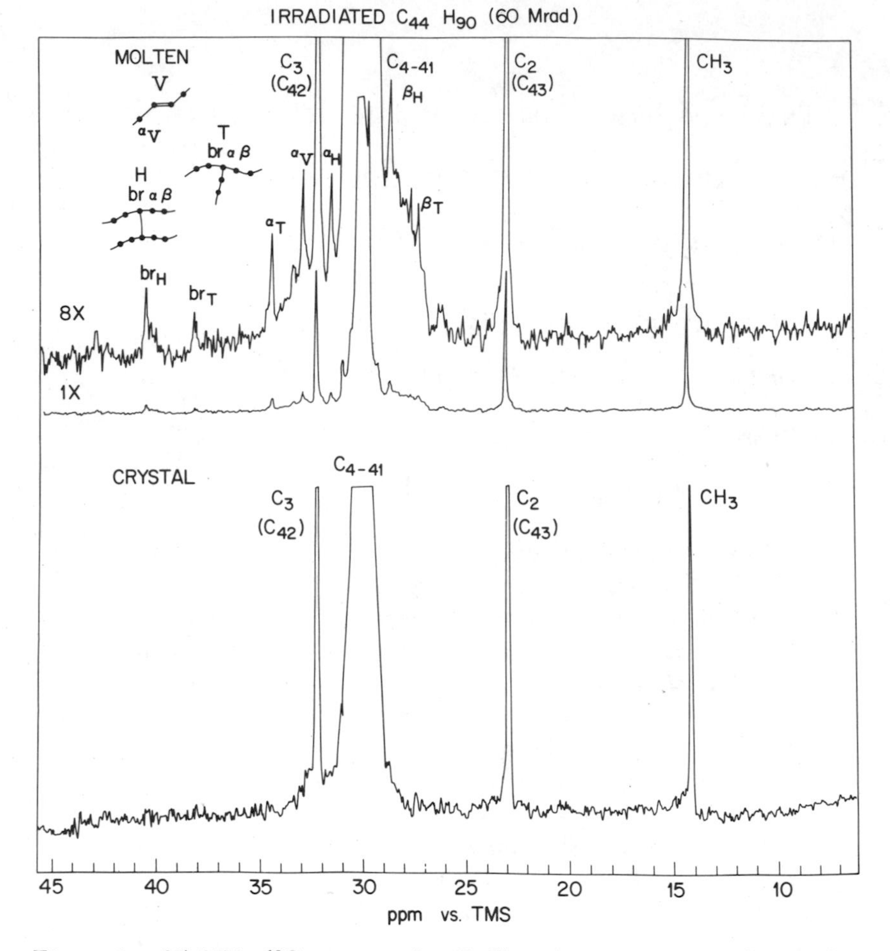

Figure 1. 25 MHz ^{13}C spectra of n-$C_{44}H_{90}$ *after exposure to 53 Mrad of γ-radiation in the molten state (top) and in the crystalline state (bottom). Peaks CH_3, C_2, and C_3 correspond to the first, second, and third carbons from the end of the chain; the fourth and successive carbons give the large peak at ca. 30 ppm.*

Gamma Irradiation

The relative susceptibilities of the crystalline and amorphous portions of polyethylene to high energy radiation has been a moot question. Present evidence indicates that for linear polyethylenes without vinyl end groups the G-values for the production of hydrogen, cross-links, and *trans*-vinylene groups are identical for the melt at 133°C and the crystalline state at 130°C (*1*). (Scission is now believed to be essentially nil; at least, its occurrence is not demonstrated).

Chemical and physical evidence indicates that cross-linking in the solid polymer occurs in the amorphous phase; this is suggested by the recent studies of Patel and Keller (*2, 3*). To reconcile these findings, one may conclude that the crystalline phase, while resisting the conspicuous effects of radiation, can transmit the absorbed energy to the amorphous phase.

^{13}C NMR has been applied to this problem by examining the spectra of n-$C_{44}H_{90}$ after exposure to γ-irradiation. This work is an extension of earlier studies by Salovey, Falconer, and Hellman (*4, 5, 6, 7*) in which gas chromatography and GPC were used to identify the cross-linked products from model paraffins. Carbon-NMR can provide more specific and conclusive information concerning the structure of such products. Model paraffins have no amorphous phase in the crystalline state and are able to accept large doses without gel formation. Figure 1 compares the spectra of $C_{44}H_{90}$ after exposure to 53 Mrad of ^{60}Co irradiation in the molten state at 115°C (top) and in the crystalline state at 25°C (bottom). (The spectra were run at 25 MHz on neat melts at 96°C; 20,000 scans were accumulated for each). In the molten state at least three new structures were formed. These can be identified by calculation and reference to model compounds (*8*) as corresponding to the expected one-bond cross-links (H structures) and *trans*-vinylene groups (V); there are also long branches (T) in numbers about equal to the cross-linked units. These structures may arise in the following way:

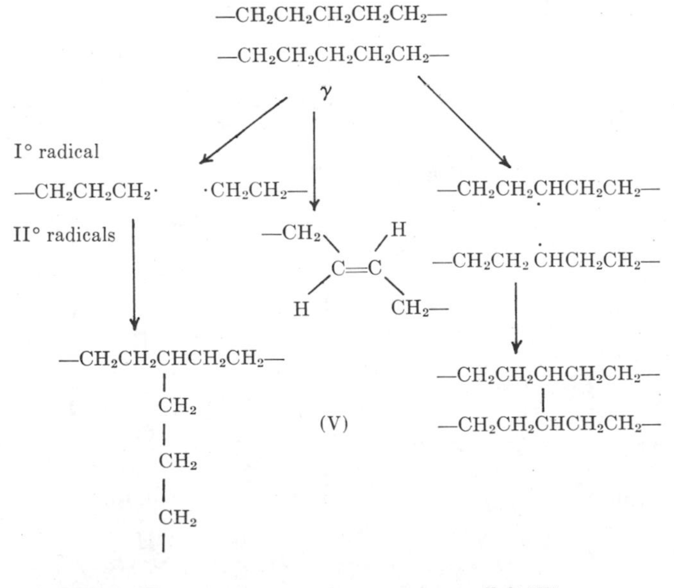

The crystal appears to undergo none of these reactions. GPC observation of the product shows, however, that about 1% of the molecules have doubled in molecular weight. This is the result of end-linking through methyls either in adjacent rows (as shown below) or adjacent layers in the crystal structure: it implies a migration of energy to the chain ends. A possible mechanism follows:

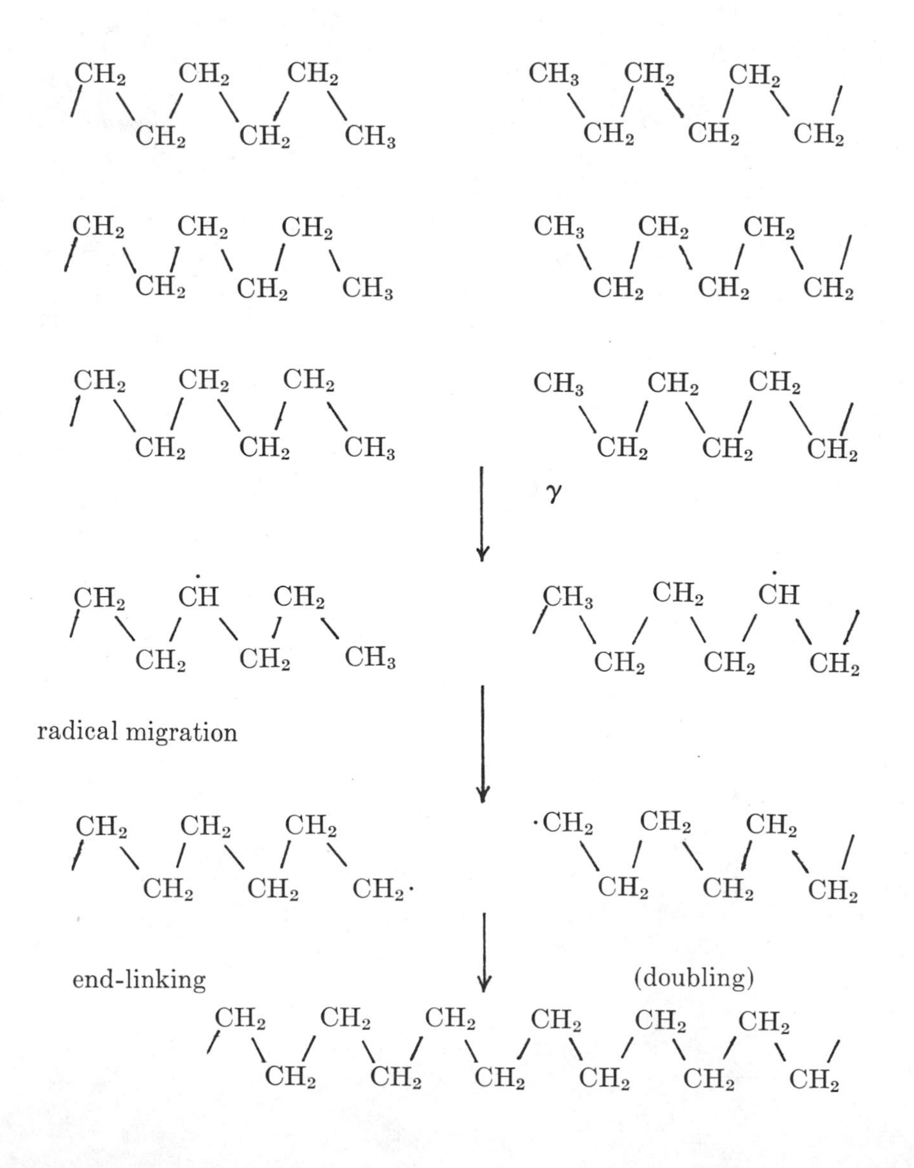

The absence of cross-linking confirms conclusions of Keller and Patel.

Recently Bennett, Keller, and Stejny (*9*) have prepared the hydrocarbon 1,1,2,2-tetra(tridecyl)ethane as a polyethylene cross-link model:

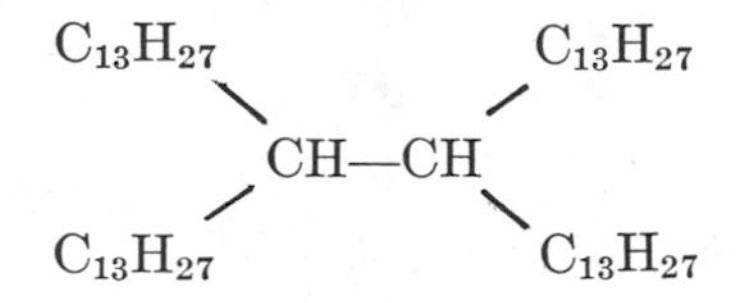

In this compound's ^{13}C spectrum, the resonances for the cross-link carbons, and for those α and β to them, appear in the same positions as those shown in Figure 1, confirming the assignments made. They also reported the spectra of eicosane (C_{20}) and hexacosane (C_{28}) which had been exposed to 500 Mrad. The principal resonances of the irradiated material corresponded to those reported here, but the very large dose caused the appearance of some additional, unidentified small peaks.

Thermal Oxidation

Although the general outlines of polyethylene thermal oxidation with molecular oxygen are understood generally, traditional investigative methods have left some questions unanswered. The exact nature of the oxidation products is not completely clear (*10, 11, 12, 13*). This is caused by heavy reliance on ir spectroscopy. IR spectroscopy suffers from band overlap, particularly in the important carbonyl stretch region near 1725 cm^{-1}, and from the necessity of establishing reliable extinction coefficients.

Because of the large range of ^{13}C chemical shifts, peak overlap usually does not present a severe problem. Since ^{13}C chemical shifts are sensitive to local structure, a more positive and detailed identification of oxygen-containing groups is possible. If spin-lattice relaxation is considered, i.e., by selecting pulse intervals equal to $3T_1$ or greater, oxidation products can be estimated quantitatively from peak intensity measurements, assuming that at the observing temperature of 110°C all nuclear Overhauser enhancements have reached their maximal value of 3.

We have studied the oxidation of branched polyethylene at 140°C using circular films ca. 5 mil thick exposed to oxygen in a conical flask held in an oil bath and connected to a mercury manometer (*14*). Figure 2 shows a branched polyethylene's ^{13}C spectrum before (top) and after (middle and bottom) thermal oxidation to the extent of 108 ml O_2 per gram of polymer, corresponding to about 4.5% toward complete oxidation. (Some H_2O is lost but probably little CO_2.) The resonances associated with branches are known (*8*) and are indicated on the spectrum. In spectra b and c the new peaks resulting from oxidation are shaded. Most of these can be unequivocally assigned by comparison with appro-

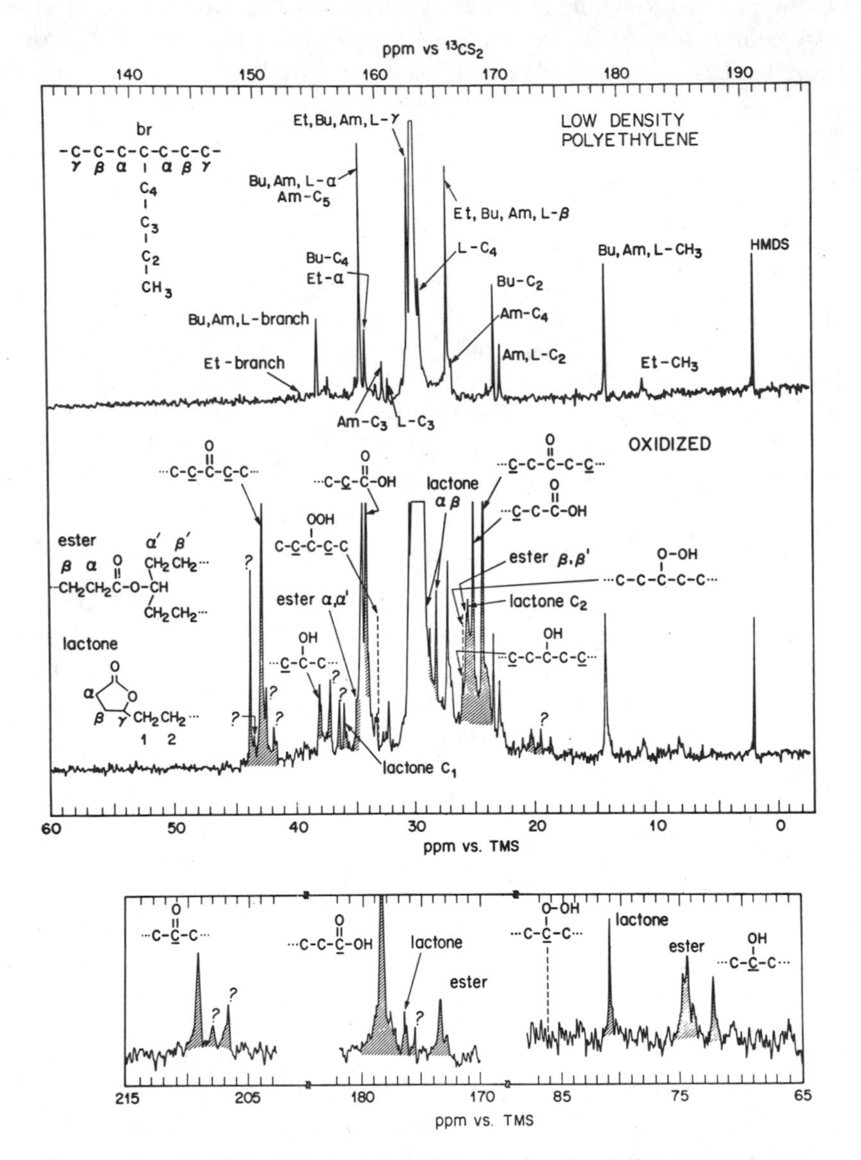

Figure 2. 25 MHz ^{13}C spectra of branched polyethylene: top, before oxidation. The peaks are labeled according to the designations of carbons on and near a branch shown at the upper left. The middle spectrum shows the new peaks (shaded) resulting from oxidation to the extent of 108 ml O_2 per gram of polymer. The bottom spectrum shows peaks of oxidation products appearing below 65 ppm. Peaks not yet identified are indicated with question marks.

priate long chain model compounds. The groups thus observed are: long chain ketone, carboxylic acid, secondary alcohol, and secondary hydroperoxides; esters of long chain carboxylic acids with long chain secondary alcohols, and long chain γ-lactones. Not observed at the present detection level (ca. 0.3%) were: aldehydes, conjugated ketones, olefins, peresters, primary and tertiary hydroperoxides, and primary and tertiary alcohols and their esters.

Figure 3 shows the distribution of established oxidation products as a function of time and extent of oxidation, expressed as mL O_2 per gram of polymer. The secondary hydroperoxide reaches a maximum at ca. 40 mL–g^{+1} and then decreases as other products accumulate. The results appear to agree with those of model reaction calculations (*15*) and support the generally accepted oxidation scheme:

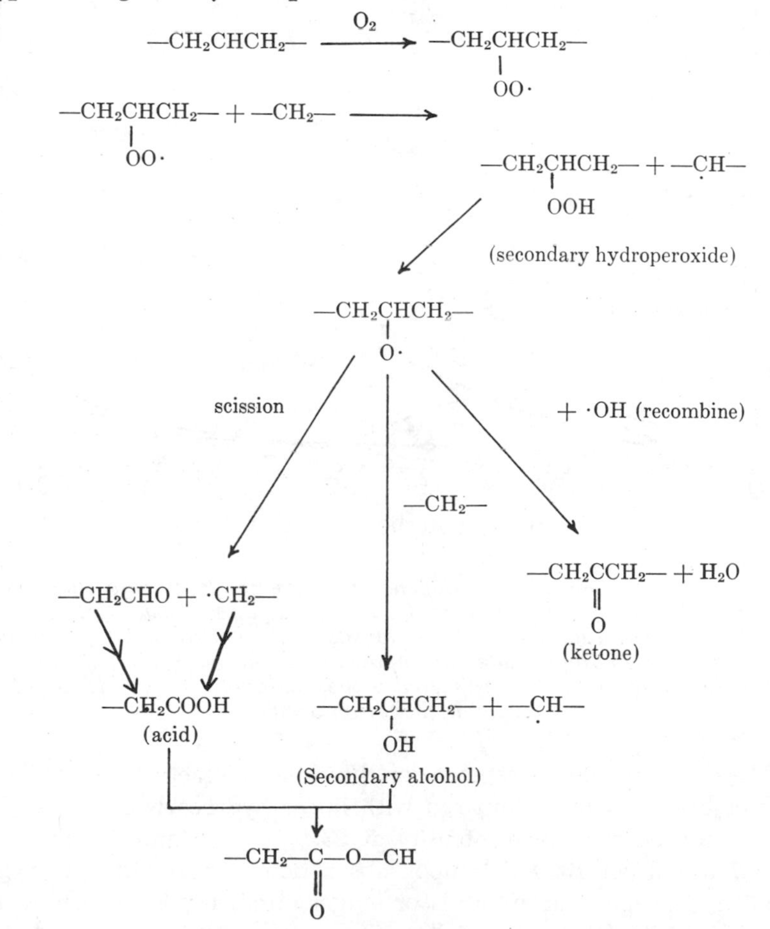

Over 80% of the oxygen consumed can be accounted for by the known products observed However, the scheme is not complete since there are products that have not been considered. One of these products is γ-lactone, also reported by Adams (*11*) in polyolefin oxidation.

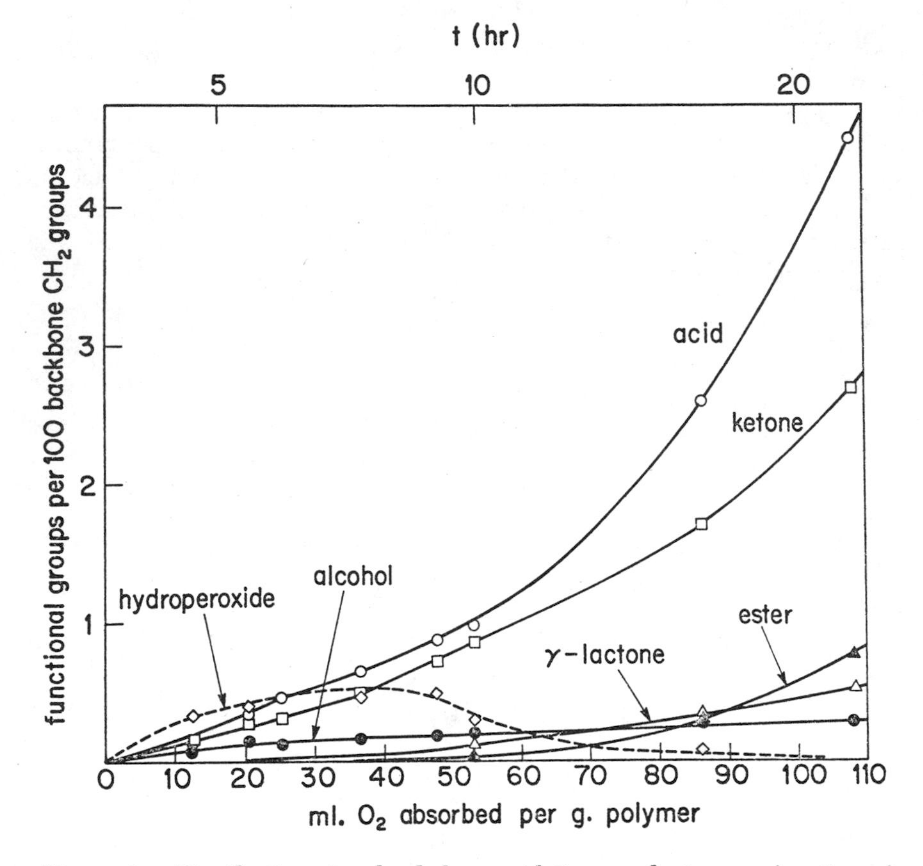

Figure 3. Distribution of polyethylene oxidation products as a function of the extent of oxidation (lower scale: O_2 absorbed; upper scale: time of reaction). The vertical scale refers to the intensity of the backbone CH_2 resonance at 30 ppm. The hydroperoxide decomposes gradually at 110°C; the concentrations shown above are the estimated values extrapolated back to zero time assuming first-order decay.

The data provide an estimate of the ratio of reactivity to oxidative attack of branch points compared with linear hydrocarbon chains. The butyl C_2 carbon resonance intensity at 23.4 ppm (Figure 2) decreases from 9.7 to 6.6 per 1000 CH_2 upon absorption of 53 ml–g^{-1} of oxygen. Oxidative cleavage at an *n*-butyl (or longer) branch point occurs as follows (*15, 16, 17*) (the tertiary alkoxy radical having been generated by steps parallel to those shown above):

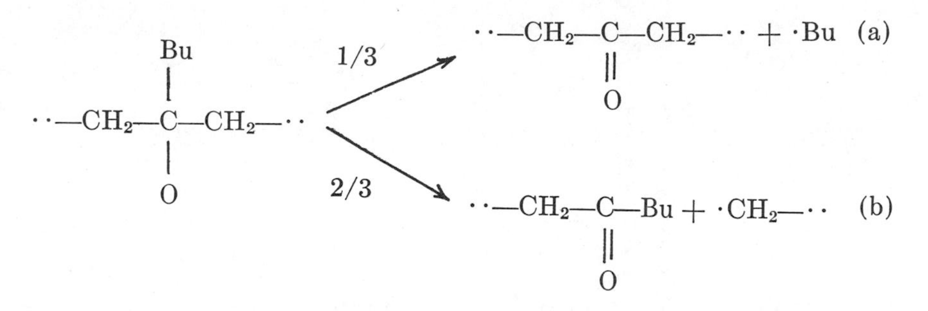

Reactions a and b occur with statistical probability giving long chain ketone and butyl ketone, which are observable in these experiments. Model compound measurements show that in the *n*-butyl ketone group, the butyl C_2 carbon resonance moves from 23.4 to 22.7 ppm, coinciding with the C_2 resonance of "amyl + long" branches, the intensity of which increases upon oxidation. By comparing these results quantitatively with the overall production of oxidized structures, the reactivity ratio of branch points to linear chains is calculated to be 9.8 ± 1.0. This result is in agreement with the value of 8 derived from model hydrocarbon oxidation studies (*15*).

Literature Cited

1. Mandelkern, L., *in* "The Radiation Chemistry of Macromolecules," Vol. I, M. Dole, Ed., p. 329, Academic, New York, 1972.
2. Patel, G. N., Keller, A., *J. Polym. Sci., Polym. Phys. Ed.* (1975) **13,** 303, 323, 333.
3. Patel, G. N., *J. Polym. Sci., Polym. Phys. Ed.* (1975) **13,** 339, 351, 361.
4. Salovey, A. R., Falconer, W. E., *J. Phys. Chem.* (1965) **69,** 2345.
5. Salovey, A. R., Falconer, W. E., *J .Phys. Chem.* (1966) **70,** 3203.
6. Falconer, W. E., Salovey, R., *J. Chem. Phys.* (1966) **44,** 3151.
7. Salovey, R., Hellman, M. Y., *Macromolecules* (1968) **1,** 456.
8. Bovey, F. A., Schilling, F. C., McCrackin, F. L., Wagner, H. L., *Macromolecules* (1976) **9,** 76.
9. Bennett, R. L., Keller, A., Stejny, J., *J. Polym. Sci., Polym. Chem. Ed.* (1976) **14,** 3027.
10. Luongo, J. P., *J. Polym. Sci.* (1960) **42,** 139.
11. Adams, J. H., *J. Polym. Sci., Polym. Chem. Ed.* (1970) **8,** 1077, 1279.
12. Émanuél, M. M., *Izv. Akad. Nauk. SSSR., Ser. Khim.* (1974) **5,** 1056.
13. Tabb, D. L., Sevick, J. J., Koenig, J. L., *J. Polym. Sci., Polym. Phys. Ed.* (1975) **13,** 815, and references therein.
14. Cheng, H. N., Schilling, F. C., Bovey, F. A., *Macromolecules* (1976) **9,** 363.
15. Allara, D. L., Edelson, D., *Rubber Chem. Technol.* (1972) **45,** 437.
16. Chien, J. C. W., Vandenberg, E. J., Jabloner, H., *J. Polym. Sci., Polym. Chem. Ed.* (1968) **6,** 381.
17. Mill, T., Richardson, H., Mayo, F. R., *J. Polym. Sci., Polym. Chem. Ed.* (1973) **11,** 2899.

RECEIVED May 26, 1977.

12

Stability of γ-Irradiated Polypropylene

I. Mechanical Properties

J. L. WILLIAMS, T. S. DUNN, and H. SUGG

Becton, Dickinson and Company Research Center,
Research Triangle Park, NC 27709

V. T. STANNETT

Department of Chemical Engineering, North Carolina State University,
Raleigh, NC 27607

Disposable medical products formed from polypropylene are readily degraded by high energy irradiation during the radiation sterilization cycle. Examination of irradiated polypropylene indicates that the degradative process is mainly oxidative. Results show that physical properties decrease rapidly with increasing absorbed dose and the physical properties continue to deteriorate with time following irradiation. This post-degradative reaction is oxidative and is initiated mainly by residual radicals following the sterilization cycle. Physical property results indicate that certain polypropylenes that may be acceptable immediately following irradiation are unacceptable after six months because of post-degradation. The degradative reaction can be accelerated by increased temperature which aids oxygen diffusion to the radical sites.

Subjecting a polypropylene item to high energy irradiation sterilization generally results in severe degradation of the article (*1, 2*). The radiation degradation problem associated with polypropylene has two aspects: discoloration and embrittlement. The latter is best characterized physically by saying that polypropylene lacks the required elongation characteristics needed to accommodate applied strains in practice.

This degradation problem manifests itself in the syringe barrel fabrication where the yellow discoloration and embrittlement cannot be

0-8412-0381-4/78/33-169-142$05.00/1

tolerated. In this study attention is given to the mechanical characterization of the property changes that occur during and after irradiation, and to the formation of radicals generated during irradiation (*3*). Our two parallel studies provide insight into the embrittlement problem of polypropylene.

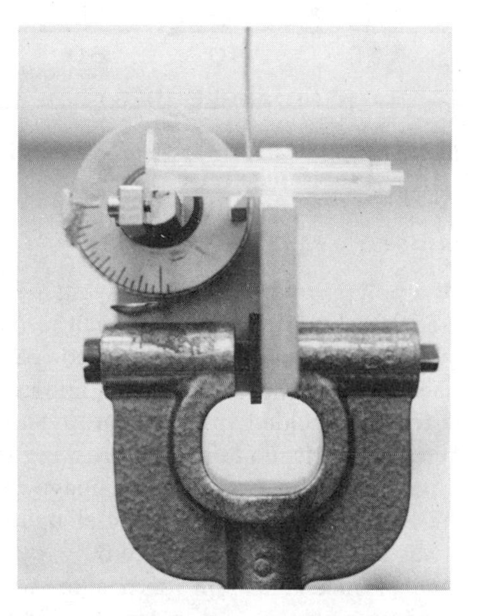

Figure 1. Instron syringe flange bending device

Experimental

Physical property measurements aided the characterization of flange strength as a function of sterilizing dose. To measure flange strength a device (Figure 1) was constructed and adapted to an Instron tester. The force required to bend the flange away from the barrel through a given angle is measured and recorded to the point of breakage or 90°C. When breakage occurs, the crack is always at the flange–barrel interface. A typical bending curve for a non-irradiated syringe flange is shown in Figure 2.

All test specimens were injection molded under controlled conditions, and a 6000 ci cobalt-60 facility was used for sample irradiation. The isotactic polypropylene was 60% crystalline with a density of 0.903 g/cm^3.

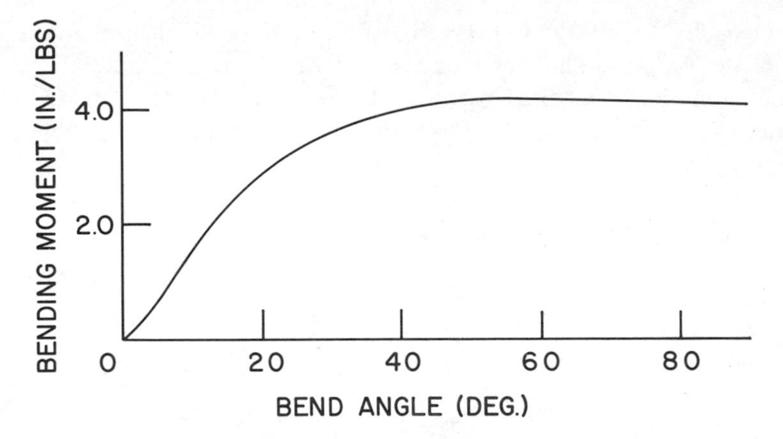

Figure 2. Typical flange bending test curve using Instron bending device

Results and Discussion

Examination of a polypropylene item following a 2.5-Mrad sterilizing dose of irradiation at 0.3 Mrad/hr indicates that the mechanical properties were severely degraded. The most dramatic change in physical properties takes place in the sample's ability to undergo extension (Figure 3) for a tensile test bar molded from polypropylene. Although there are significant changes in elastic modulus and breaking stress with irradiation, the principal parameter for irradiation damage is the reduction in breaking elongation. For instance, the breaking elongation for the sample shown in Figure 3 was reduced from 400% to 60% by a dose of 7 Mrad.

The dependence of mechanical properties on irradiation dose for elastic modulus, percent extension, and work are summarized in Figures 4, 5, and 6, respectively.

An important parameter is the strain rate at which a polypropylene sample is broken (Figure 7). It is evident from this graph that a sevenfold difference in breaking elongation can be obtained by varying the strain rate from 100%/min to 800%/min. Severe strain rate is a better indicator of radiation damage (Figure 7), i.e., at low strain rates the sample has adequate time to elongate. Therefore, a high strain rate should be used for screening various polypropylene formulations.

Polymer samples under a tensile load will break at varying elongations from sample to sample (*4*). This variation in breaking elongation will increase with gage length because of the increased probability of weak points along the sample length. Because of the variation with tensile tests, a flexural test was adopted, as described above. A typical curve obtained by bending a test bar or flange is shown in Figure 2. In this test method breaking angles for a given sample can be reproduced

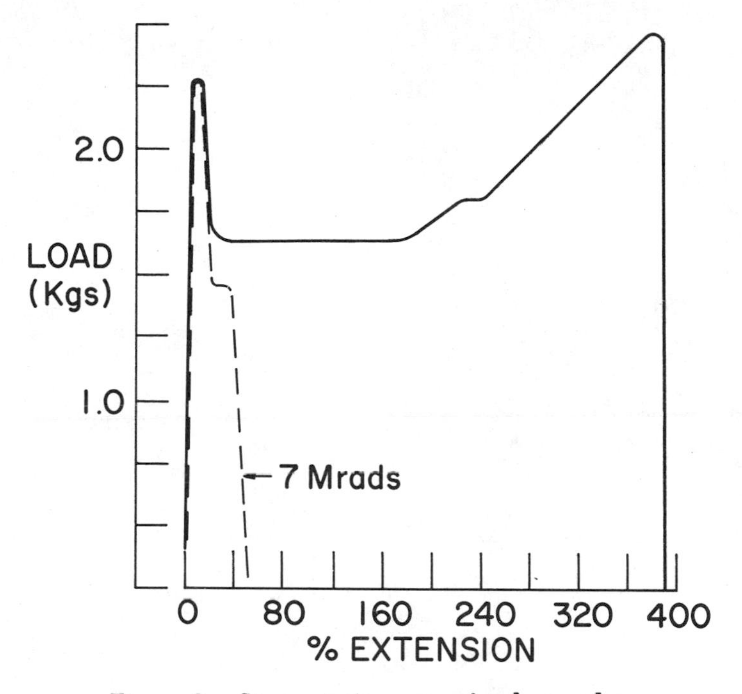

Figure 3. Stress–strain curve of polypropylene before and after 7 Mrad irradiation

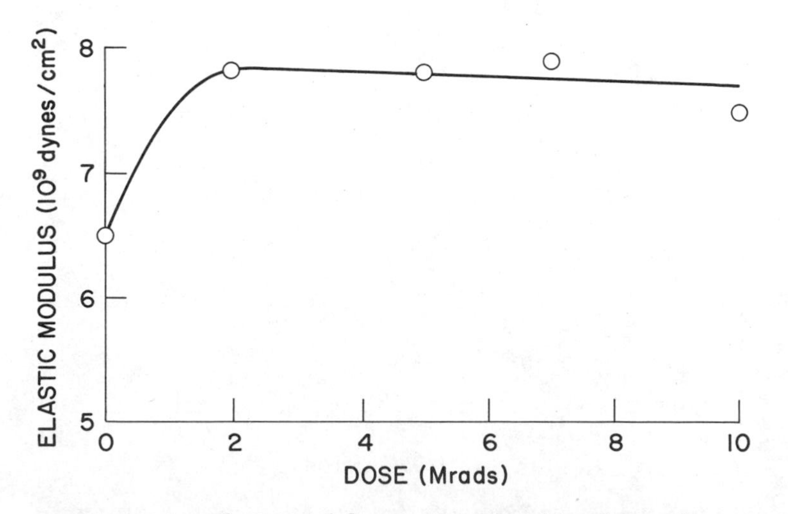

Figure 4. Elastic modulus as a function of irradiation dose

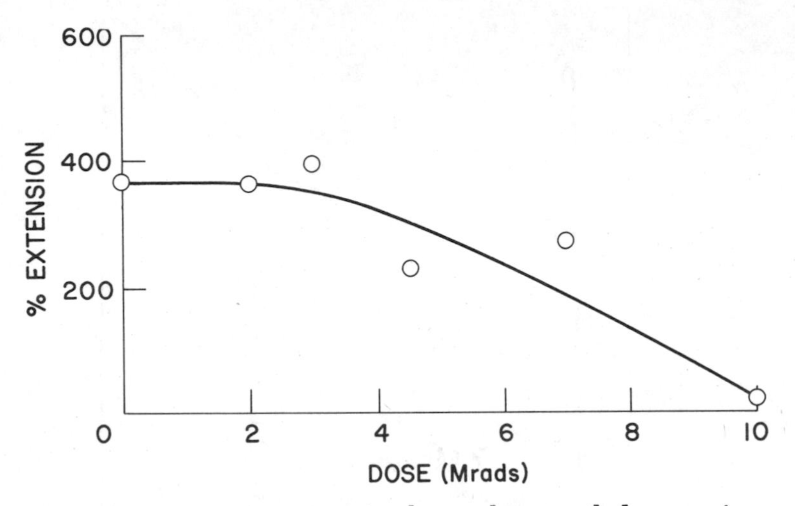

Figure 5. Percent extension of polypropylene tensile bar as a function of irradiation dose

within 10% under controlled molding conditions. The high reproducibility of breaking angles is caused by point concentration of stress when samples are broken in a bending mode.

When irradiated polypropylene syringe flanges are subjected to the bending test, their curves appear as shown in Figure 8. The main consequence of irradiation is the reduction of the angle at which the sample can be deflected before breakage occurs. As the sample is irradiated

Figure 6. Relative work as a function of irradiation for tensile bars

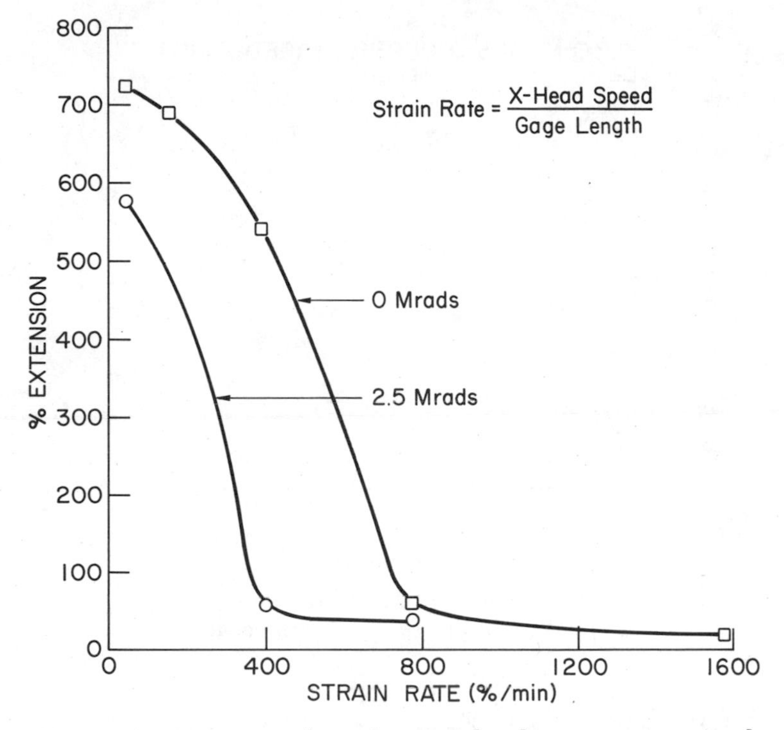

Figure 7. Strain rate dependence of breaking extension of polypropylene

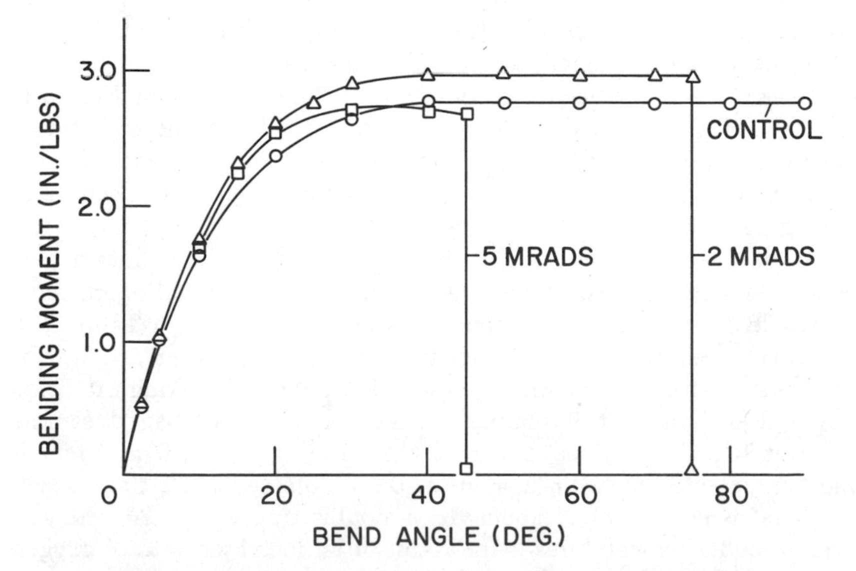

Figure 8. Irradiation enbrittlement of polypropylene syringe flange

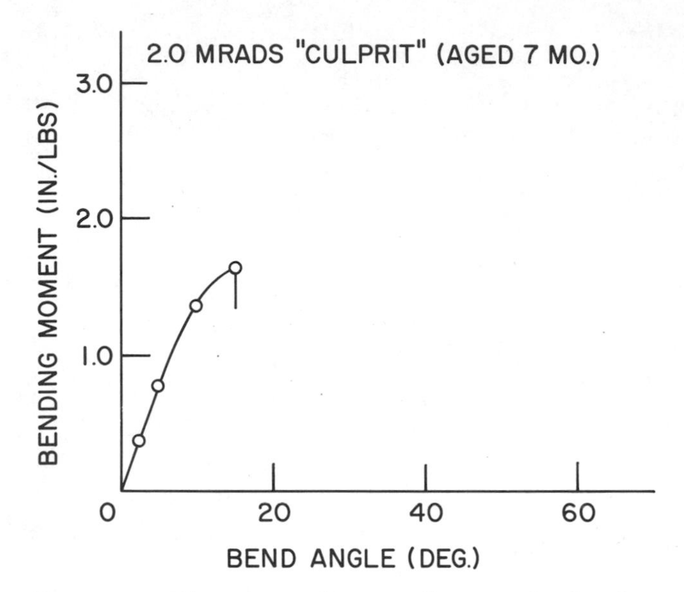

Figure 9. The post-irradiation effect on the bending angle of polypropylene syringe barrel flange

to a higher dose, the angle at the breakpoint decreases steadily. The initial flex modulus has not changed significantly and is not a good parameter for monitoring irradiation damage. The best single parameter to evaluate irradiation damage is the area under the bending curve. This area gives an excellent representation of the work which the sample can undergo before "brittle snap," a true measure of embrittlement.

Aging of an irradiated sample does not improve the mechanical properties; they continue to deteriorate with time as seen in Figure 9. Post-embrittlement is caused by post-oxidation as a result of residual radicals and is discussed later (*3*). However, the oxidative reaction initiated by the irradiation cycle does not cease with irradiation but continues for long periods of time.

An accelerated aging test has been devised to predict how a particular polypropylene would behave as a function of normal aging time. The results for several control conditions are summarized in Table I, in which each sample is the average of three test bars. An increase in temperature and oxygen concentration results in a drop in breaking extension compared with a control sample. When an accelerated-aged sample (40 hr at 100°C in O_2) is compared with true aging in air at ambient conditions, results as shown in Figure 10 are obtained. The 40-hr accelerated test is equivalent to about three-months true aging. Accelerated aging at higher temperatures is the result of an increased rate of oxygen diffusion at these elevated temperatures.

Table I. Effect of Accelerated Aging on Breaking Extension of Polypropylene

Sample Description	*% Extension*
Control	594.5
2.5 Mrad	448.8
2.5 Mrad (air @ 100°C for 40.5 hr)	70.9
2.5 Mrad (O_2 @ 100°C for 40.5 hr)	55.1
2.5 Mrad (vaco @ 100°C for 40.5 hr)	94.5
Control (O_2 @ 100°C for 40.5 hr)	393.7

Pure polypropylene does not discolor at the 2.5 Mrad required for sterilization. However, the stabilizers added to improve the polymer's thermal and irradiation stability often discolor upon irradiation. This yellowish discoloration increases with irradiation dose. This problem can be resolved by using stabilizers that do not discolor when they are irradiated.

The severe degradation of polypropylene following sterilizing doses of irradiation can be characterized mechanically by its failure to undergo the necessary work in practice. Embrittlement increases with time for an irradiated polypropylene, thus rendering an acceptable formulation totally unacceptable a few months after irradiation. Naturally, the decay of radicals can be accelerated by thermal annealing, limited by the geometrical distortion temperature.

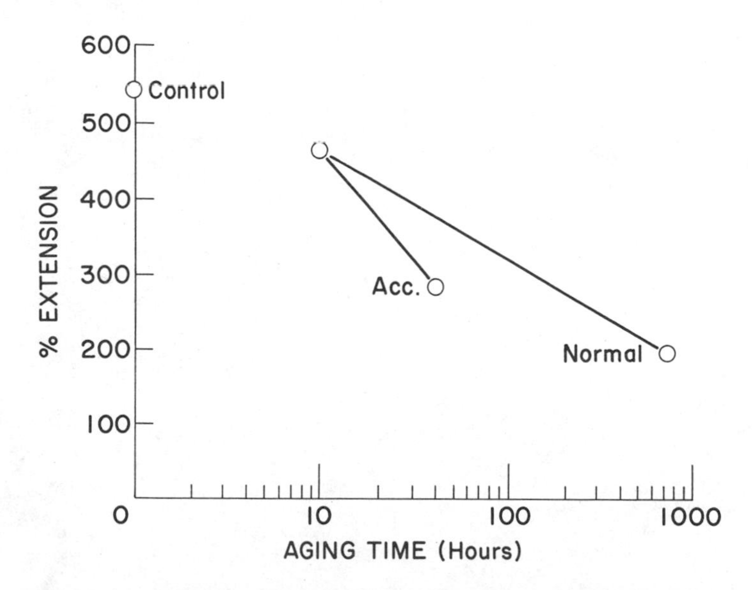

Figure 10. Aging effect of irradiated (2.5 Mrad) polypropylene in air at room conditions and in oxygen at 100°C as a function of % extension

Literature Cited

1. Geymer, D. O., *in* "The Radiation Chemistry of Macromolecules," Vol. II, p. 4, Academic Press, New York, 1973.
2. Charlesby, A., "Atomic Radiation and Polymers," Pergamon Press, London, 1960.
3. Dunn, T. S., Williams, J. L., Sugg, H., Stanett, V. T., "Stability of Gamma Irradiated Polypropylene. Part II. Electron Spin Resonance Studies," ADV. CHEM. SER. (1978) **169**, 151.
4. Billmeyer, F., "Textbook of Polymer Science," *Wiley, New York, 1971.*

RECEIVED May 12, 1977.

13

Stability of γ-Irradiated Polypropylene

II. Electron Spin Resonance Studies

T. S. DUNN, J. L. WILLIAMS, and H. SUGG

Becton, Dickinson and Company Research Center,
Research Triangle Park, NC 27709

V. T. STANNETT

Department of Chemical Engineering, North Carolina State University,
Raleigh, NC 27607

The mechanical degradation of isotactic polypropylene following γ-irradiation to sterilizing doses is significantly reduced by the inclusion of mobilizing additives to the polymer matrix. The mechanism of polypropylene stabilization to irradiation is investigated by ESR measurements. The autoxidative scheme for radical generation is substantiated. The measurements of radical concentration build-up with dose (and subsequent decay with time in the additive system) as compared with the control polypropylene, show a reduction in radical concentration by a factor of 2 for the same dose. This reduction in radical concentration reduces the overall peroxide concentration and the post-irradiation chain scission and resulting embrittlement.

In this study, attention is given to a fundamental investigation of the radicals formed during γ-irradiation sterilization of isotactic polypropylene. An autoxidative scheme for radical generation and mechanisms for radical build-up and decay were investigated by electron spin resonance (ESR). This study, in conjunction with a tandem study of mechanical properties (*1*) on identical samples, has given some insight into the reduction of polypropylene post-irradiation embrittlement by use of mobilizing additives.

0-8412-0381-4/78/33-169-151$05.00/1

Experimental

All ESR studies were performed on a 6-in. magnet JEOL model JES–ME instrument. Radical concentrations were measured at 3281 Gauss relative to a DPPH primary standard and a manganese secondary standard using a dual resonance cavity with 100-kHz matched modulation. The samples were irradiated at ambient conditions in vacuated 4-mm glass tubes at a dose rate of 1.9 Mrad/hr. After irradiation, the tubes were flame-annealed to remove color centers, and the samples were frozen in liquid nitrogen to await ESR measurements. All ESR samples were cut from test bars injection-molded under identical conditions of rapid ice-water cooling to minimize crystallinity (60%) and approximate identical crystallization conditions. The sample dimensions were 1.25 mm × 2.0 mm × 25 mm and had a density of 0.903 g/cm^3.

Results and Discussion

Polypropylene irradiated in the presence of oxygen can result in the following autoxidative reactions (*2, 3*):

$$R \rightsquigarrow 2R\cdot \tag{1}$$

$$R\cdot + O_2 \rightarrow RO_2\cdot \tag{2}$$

$$RO_2\cdot + RH \rightarrow ROOH + R\cdot \tag{3}$$

$$RO_2\cdot + R\cdot \rightarrow ROOR \tag{4}$$

$$RO_2\cdot + RO_2\cdot \rightarrow ROOR + O_2 \tag{5}$$

$$R\cdot + R\cdot \rightarrow R\text{–}R \tag{6}$$

In this scheme R represents the polypropylene chain and R· is the cleaved chain that occurs during irradiation. In this scheme, as oxygen is consumed in Reactions 2 and 3, the free radical (R·) is only consumed in Reactions 4 and 6. Therefore, the term autoxidative implies that each radical formed will consume numerous molecules of oxygen unless the oxidative reactions are prevented.

This autoxidative scheme can be verified by ESR studies. The 17-line spectrum reported in the literature to be the alkyl (*4*) radical R· formed during polypropylene irradiation at room temperature is shown in Figure 1(a) (*3, 5*). When the polypropylene irradiated in vacuum is exposed to oxygen, ESR spectra change progressively, as illustrated in Figure 1. After several days the peroxy radical (*6*) [Figure 1(d)] is the dominant species. If the sample is irradiated and then allowed to decay in vacuum, the peroxy radical does not form, as illustrated in

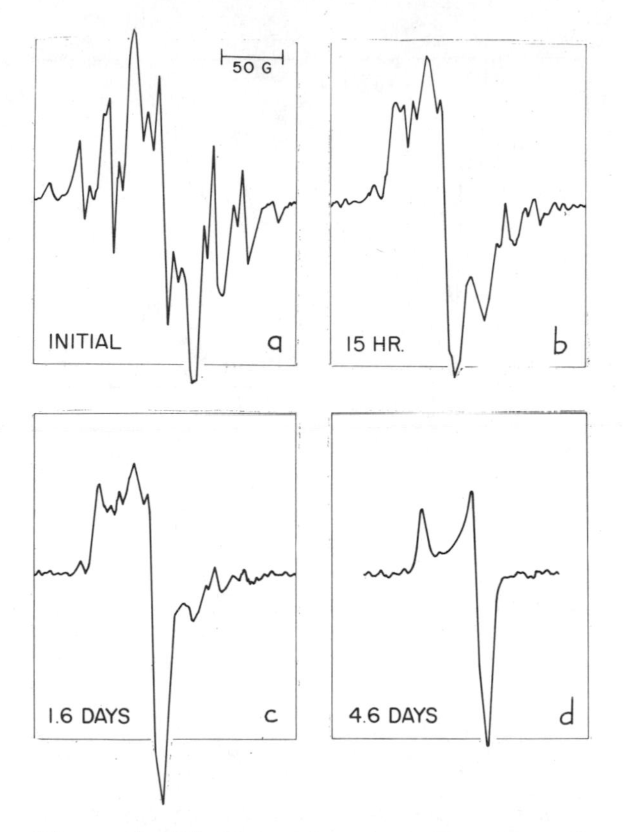

Figure 1. Electron spin resonance spectra of peroxy radical formation at room temperature (5.0 Mrads). (a) Primary radical R· under vacuum, (b) radical transformation as sample is opened to air, (c) intermediate spectrum, (d) peroxy radical RO_2·.

Figure 2. This series of spectra shows a transition from the alkyl radical to the more stable polyenyl radical (7) shown in Figure 2(d).

When the polypropylene sample is irradiated in air, as is the case during a sterilization cycle, the oxidative process is initiated at the onset of irradiation. As a consequence of oxygen diffusion during irradiation a spectrum intermediate between the alkyl and peroxy radicals is generally observed (Figure 3). Whether oxygen is present during irradiation or added later, the peroxy radical becomes the dominant radical species with time.

A study of the radical build-up with dose and subsequent decay with time illustrates that a high concentration of radicals is formed during irradiation and is long lived in the polypropylene sample. As a result of the radical longevity, oxygen has sufficient time to diffuse and

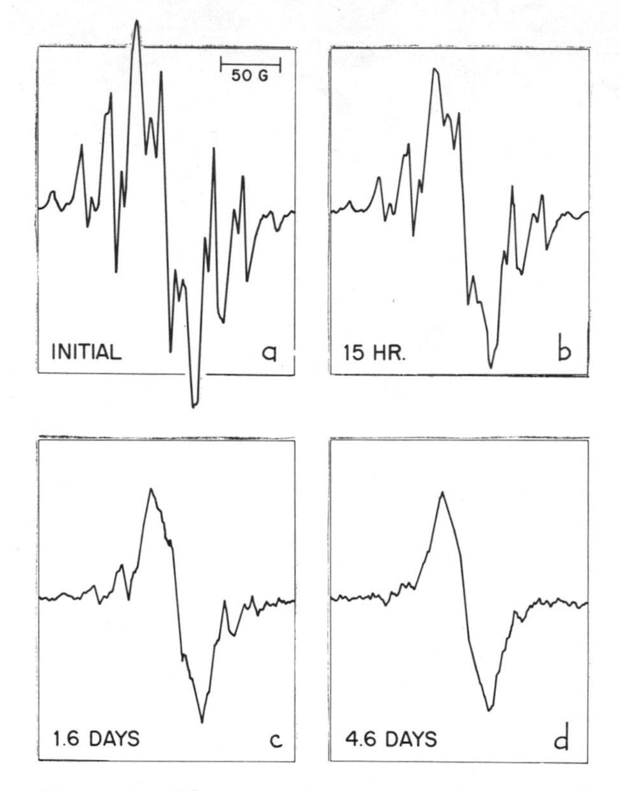

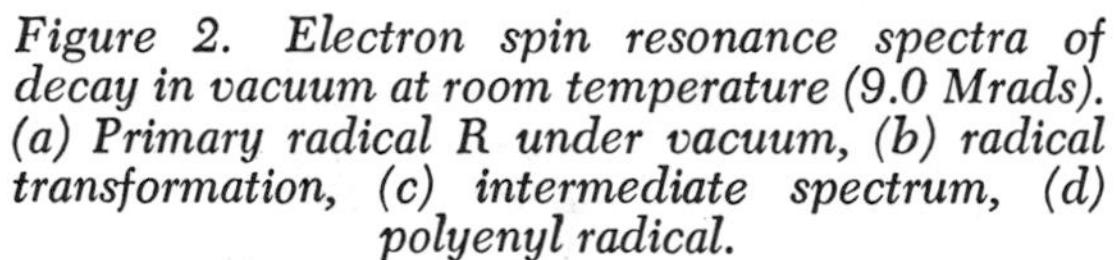

Figure 2. Electron spin resonance spectra of decay in vacuum at room temperature (9.0 Mrads). (a) Primary radical R under vacuum, (b) radical transformation, (c) intermediate spectrum, (d) polyenyl radical.

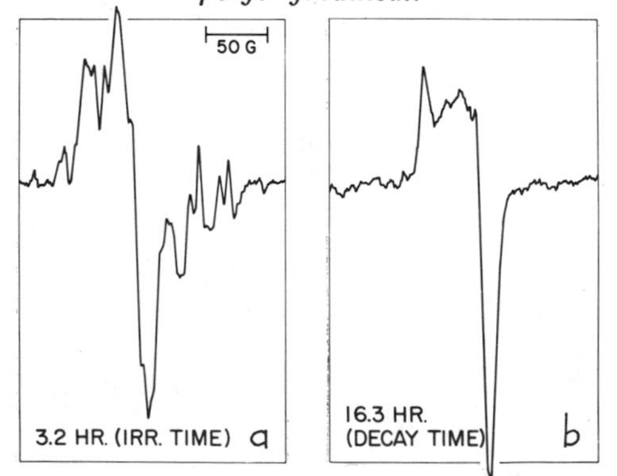

Figure 3. Electron spin resonance spectra of polypropylene irradiated in air at room temperature (5.0 Mrads). (a) Intermediate spectrum to oxygen diffusion during irradiation, (b) peroxy radical.

form peroxides, eventually leading to chain cleavage and post-irradiation substrate embrittlement. For example, after 40 days the radical population is still about 10^{16} spins/g. At this point the mechanical properties of the samples already have decreased dramatically and continue to deteriorate with time (*1*).

This study's approach to stabilizing the polypropylene to irradiation embrittlement uses an additive that effectively lowers the radical concentration during and after irradiation by enhancing main chain mobility. Experiments on polypropylene containing this additive have been completed, and the results are discussed below. The purpose of the experiments was to study the effect of the mobilizing additive on the radical concentration and to gain some insight into the mechanisms involved by rate constant determinations during and after irradiation. To simplify the system, irradiation and decay were done under vacuum; thus, the dominant radical reactions are Reactions 1 and 6.

The radical build-up and decay curves for a control polypropylene sample, and one modified by the mobilizing additive, are shown in Figure 4. Both samples were molded under the same conditions. The radical concentration is significantly lower in the modified sample.

The rate constant, k_t, in the build-up of radicals, $d[R\cdot]/dt$, can be determined by assuming second-order decay with total sample radical

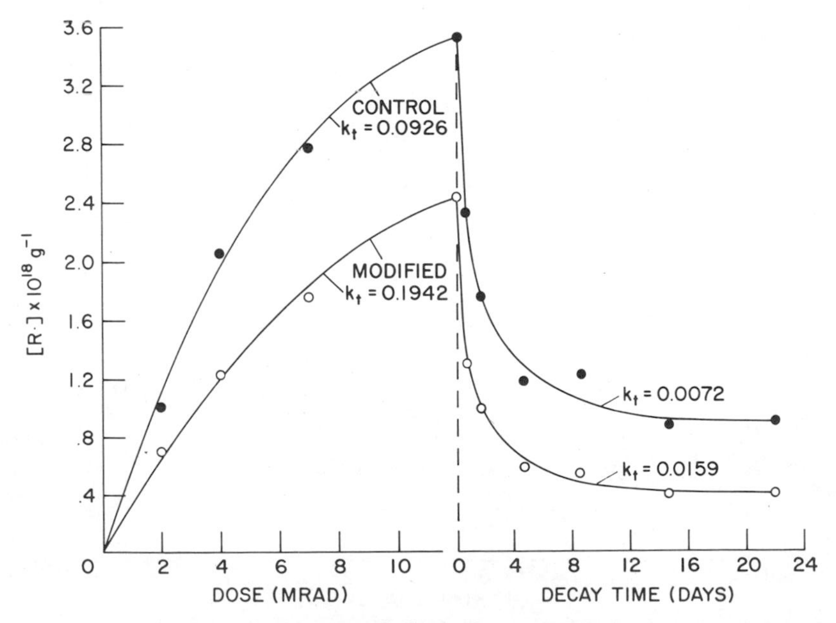

Figure 4. Radical build-up and decay in vacuum at room temperature (k_t: rate constant)

concentration, $[R\cdot]$, and linear build-up of radicals with dose rate, I, as follows:

$$\frac{d[R\cdot]}{dt} = k_1 I - k_t[R\cdot]^2 \quad (7)$$

Solving Equation 7, the radical concentration can be expressed as follows:

$$[R\cdot] = \frac{k_1 I^{\frac{1}{2}}}{k_2} \tanh\left[(k_1 k_t I)^{\frac{1}{2}} t\right] \quad (8)$$

Equation 8 can be solved transcendentally for k_t by successive iterations until $F(k_t) = 0$, where $F(k_t)$ is defined as follows:

$$F(k_t) = k_t - \frac{k_1 I}{[R\cdot]^2} \tanh^2\left[(k_1 I)^{\frac{1}{2}} t \sqrt{k_t}\right] \quad (9)$$

The values of k_t determined by this method are included in Figure 4. The rate constants for the decay region can be determined from the initial slopes of the second-order decay plots shown in Figure 5. The change in slope defining the two regions can be correlated with the transition from the primary radical to the more stable polyenyl radical illustrated in Figure 2. The k_t values of the modified sample in both regions are a factor of 2 larger than the control.

The k_t values calculated during the build-up and decay differ by a factor of 12, suggesting that the mechanism of radical recombination is

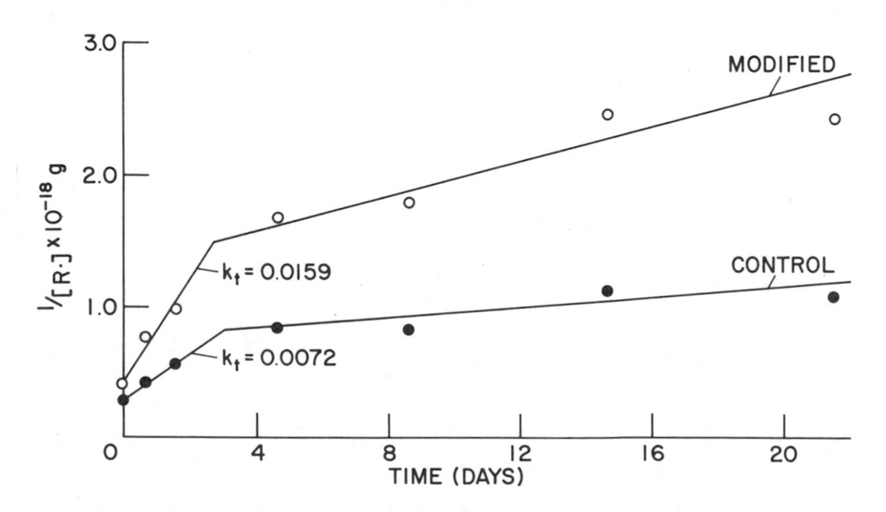

Figure 5. Radical decay in vacuum at room temperature (k_t: rate constant)

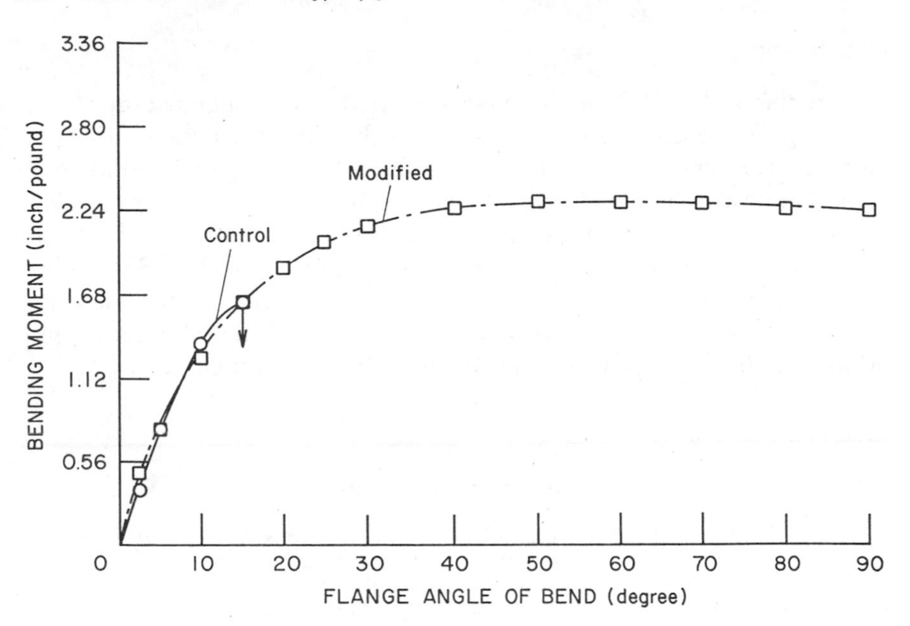

Figure 6. Mechanical properties of the modified polypropylene as compared with the control after seven months post-irradiation (2.5 Mrads) aging

different in the two regions. Two mechanisms can be proposed to explain the lower radical concentration in the modified sample. First, the incorporation of the additive increases the internal free volume in the amorphous phase of the polymer, enhancing main chain mobility and accelerating main chain recombination. This would hold true during and after irradiation. Second, during irradiation low molecular weight additive radicals will be formed and, because of their high mobility, will facilitate main chain termination. When irradiation is stopped, the additive radical concentration will quickly terminate as a result of its high mobility. Therefore, this effect can occur only during irradiation. This can explain the larger k_t value determined in the build-up region since both main chain mobility and additive radical termination could be occurring simultaneously.

Decreasing the polypropylene radical concentration lowers the concentration of peroxides that form in the presence of oxygen, thus reducing post-irradiation embrittlement. To demonstrate the mechanical properties of the modified polypropylene, two samples—one containing the mobilizing additive, the other a control—were irradiated in air to a sterilizing dose of 2.5 Mrads and aged seven months. The modified sample remains flexible after seven-months aging as shown in Figure 6, indicating that rapid radical termination minimizes the detrimental effects of oxidative embrittlement (*1*).

Conclusions

In this study ESR results have shown that the degradation of the polypropylene chain is an autoxidative mechanism that is free radical in nature. The long lifetime of the radicals formed during irradiation results in post-irradiation oxidation, leading to severe mechanical embrittlement. A mobilizing additive incorporated in the polypropylene has decreased significantly the post-irradiation embrittlement. ESR studies of radical build-up and decay suggest that the lower radical concentration in the modified sample is caused by main chain mobility and rapid additive radical main chain termination during irradiation.

Literature Cited

1. Williams, J. L., Dunn, T. S., Sugg, H., Stannett, V. T., Adv. Chem. Ser. (1978) **169**, 142.
2. Ranby, B., Rabek, J. F., "Photodegradation, Photooxidation and Photostabilization of Polymers," pp. 128–141, Wiley, New York, 1975.
3. Chien, James C. W., Boss, C. R., *J. Polym. Sci.* (1967) **5**, 3091.
4. Ascough, P. B., *J. Polym. Sci.* (1966) **4**, 503–506.
5. Fischer, H., Hellwege, K., *J. Polym. Sci.* (1962) **56**, 33–45.
6. Chien, James C. W., Boss, C. R., *J. Am. Chem. Soc.* (1967) **89**, 571.
7. Geymer, D. O., *in* "The Radiation Chemistry of Macromolecules," Vol. II, p. 10, M. Dole, Ed., Academic Press, New York, 1973.

Received May 12, 1977.

The Effect of Transition Metal Compounds on the Thermal Oxidative Degradation of Polypropylene in Solution

ZENJIRO OSAWA and TAKASHI SAITO

Department of Polymer Chemistry, Faculty of Engineering,
Gunma University, Kiryu City, Gunma, Japan 376

Effects of a series of transition metal stearates, the concentration of the copper stearate, the solvent, various additives, and other factors on the thermal oxidation of polypropylene were studied in trichlorobenzene solution. The mechanism of copper catalysis is discussed. The order of decreasing catalytic activity of the metal stearates was: Cu > Mn > Fe > Cr > Al ≈ Ni ≈ Co ≈ control ≈ Ti >> Zn >> V. The addition of propionic acid to the solvent accelerated the oxidation of the polymer. The presence of the copper leveled off oxygen uptake of the polymer after a certain time. The amount of oxygen absorbed decreased with increasing concentration of the copper, and at higher concentration (7.9×10^{-3}M) *the polymer oxidation was inhibited.*

The important role of a small amount of metals or metallic compounds in the thermal oxidative degradation of polymers has been studied extensively by many researchers (*1–19*). There are many ways that polymers encounter metals or metallic compounds, such as catalyst residues (*20–25*), pigments (*11, 12, 13*), metal particles contaminated during machine processing (*15*), insulated wire (*1–6*), etc. Therefore, understanding of the role of metals and metallic compounds in the autoxidation of polymers is of considerable practical importance.

A pioneering and comprehensive study of the effect of copper on the thermal oxidative degradation of polyolefins has been made by a Bell Laboratories research group (*1–6*). They found that surface reactions at the interface between metal and polymer are important factors in many applications including metal–polymer composites, polyolefin-insulated

0-8412-0381-4/78/33-169-159$05.00/1

copper wire, and adhesive joints. They showed further that oxamide and several of its derivatives stabilized polypropylene against both homogeneous and heterogeneous copper catalysis. Recently, Allara et al. (*26–31*) demonstrated clearly that one of the most important components of the oxidation is a surface reaction which occurs directly at the copper–polymer interface and that copper carboxylate salts (potential catalysts for the oxidation of polymers) are formed early in the reaction and are concentrated initially on the metal surface. They also stated that these carboxylates appear to decompose at higher temperatures and that their ability to persist may be responsible for the increasing role of copper catalysis as temperatures are lowered. A systematic study of metal stearate-catalyzed thermal oxidative degradations of isotactic polypropylene in bulk showed that the order of decreasing catalytic effect of the metal stearates was: Co $>$ Cr $\approx$ Mn $>$ Cu $>$ Fe $>$ V $>>$ Ni $>$ Ti $>$ Zn $>$ Al $>$ control (at 125°C) (*16*).

In this chapter, the effect of a series of transition metal stearates on the thermal oxidation of polypropylene in homogeneous solution is examined, and the results obtained are compared with that in bulk reported previously (*16*). In addition, the effects of the anion of copper compounds, the concentration of copper, the solvent, and the additives on the copper compound-catalyzed thermal oxidation of polypropylene are studied, and the mechanism of the copper catalysis in solution is discussed.

Experimental

Samples. Neat powdery isotactic polypropylene finer than 100 mesh and atactic polypropylene in white cakes (cut into ca. 1 mm^3) were made in Japan.

Reagents. Each metal (Ti, V, Cr, Mn, Fe, Co, Ni, Cu, Zn, and Al) stearate was prepared according to the literature (*16*) by double decomposition of sodium stearate with corresponding metal salts. Copper compounds (acetate, propionate, butyrate, laurate, and polyacrylate) also were prepared according to the literature (*32–34*). Commercial first-grade cupric oxide was used as a reagent. The compounds *tert*-butyl hydroperoxide and di-*tert*-butyl peroxide were products of the Nippon Oils and Fats Co. Ltd. Commercial first-grade 1,2,4-trichlorobenzene and propionic acid were used as solvents without further purification. Commercial first-grade *n*-octanol, *n*-octanoic acid, octanoic acid methyl ester, 2-octanone, and *n*-octanol were used as model compounds of oxidation products.

Thermal Oxidation. Ten mL of solvent were added to an ca. 20-mL flat-bottomed oxygen absorption cell containing 1 g of polymer and metallic compounds. The thermal oxidation was carried out under atmospheric oxygen pressure at 125°C. The mixture was stirred with a magnetic stirrer at 700 rpm, as monitored by a tachometer.

Gel Permeation Chromatography. Changes in molecular weight and distribution of the oxidized polymers were followed by gel permea-

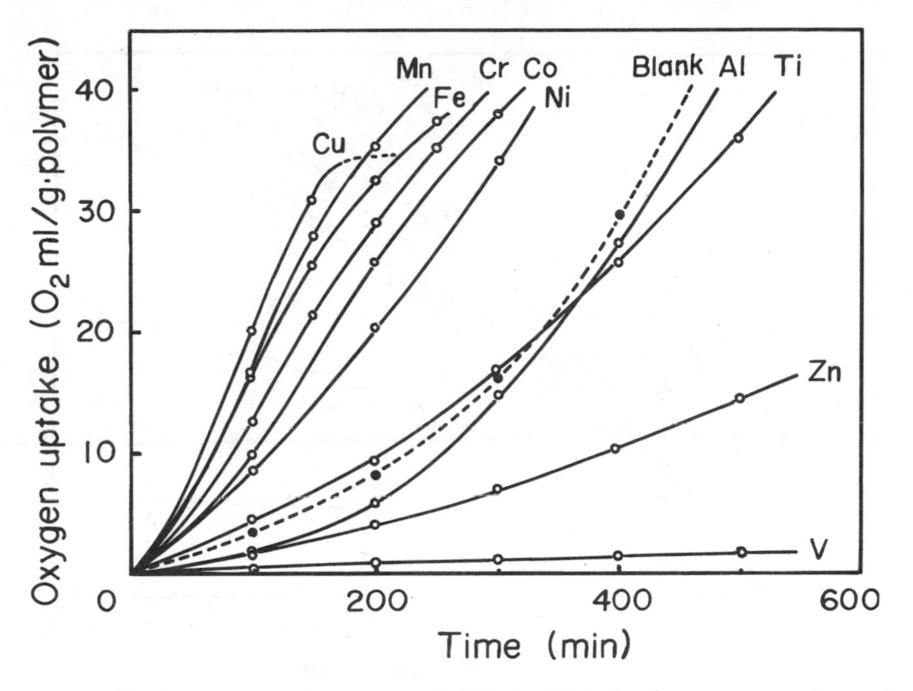

*Figure 1. Oxygen uptake curves of IPP in TCB in the presence of metal stearates at 125°C. [IPP] = 2.38*M; *[CAT.] = 7.9 × 10^{-4}*M.

tion chromatography using a Toyo Soda High Speed Liquid Chromatography HLC–801A. The instrument conditions were: elution solvent, tetrahydrofuran; temperature, 40°C; gel column, G 4000 H_6, G 2000 H_8 × 2; sample injection concentration, 0.5%; flow rate, 1.2 mL/min; pressure, 60 kg/cm^2.

Results and Discussion

Effect of Metal Stearates. The oxygen uptake curves of isotactic and atactic polypropylene in the presence of transition metal stearates in trichlorobenzene are shown in Figures 1 and 2, respectively. The order of decreasing catalytic effect of the metal stearates at the early stage of the oxidation of the polymers is: for isotactic PP, Cu > Mn > Fe > Cr > Co > Ni > Ti > control > Al >> Zn >> V and for atactic PP, Cu > Mn> Fe > Cr > Al≈ Ni ≈ Co ≈ control ≈ Ti >> Zn >> V. The order of the catalytic effect of the metals is quite different from that in bulk reported previously (*16*). In particular, V-stearate inhibits the thermal oxidation, and Co-stearate is not as effective as in bulk. In the presence of effective metal stearates such as Cu and Fe, the oxygen uptake levels off after a certain time. However, the amount of oxygen absorbed in the isotactic polypropylene is higher than that in the atactic polypropylene.

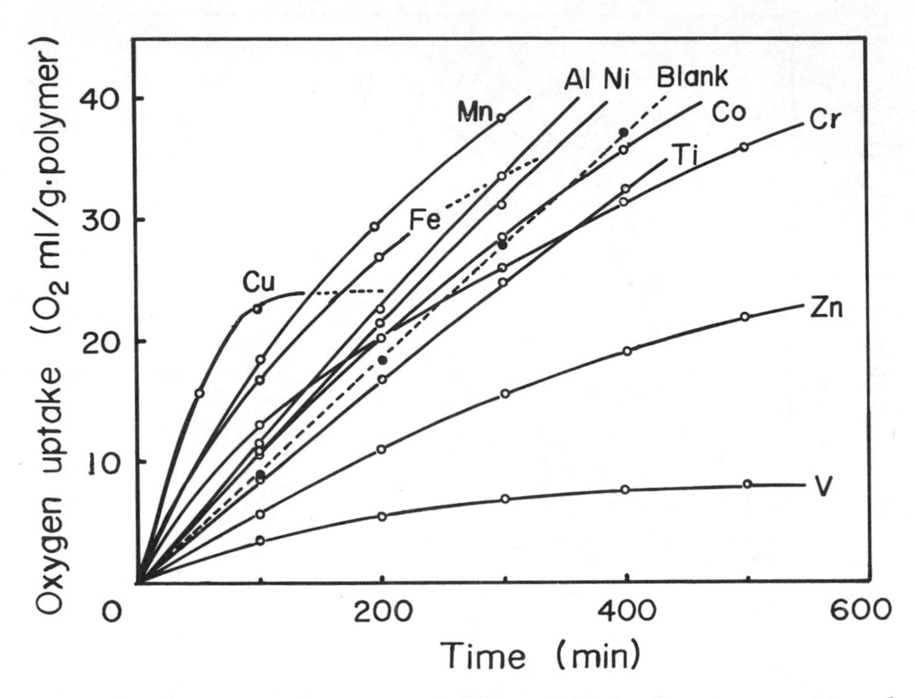

*Figure 2. Oxygen uptake curves of APP in TCB in the presence of metal stearates at 125°C. [APP] = 2.38*M*; [CAT.] = 7.9 × 10*$^{-4}$M.

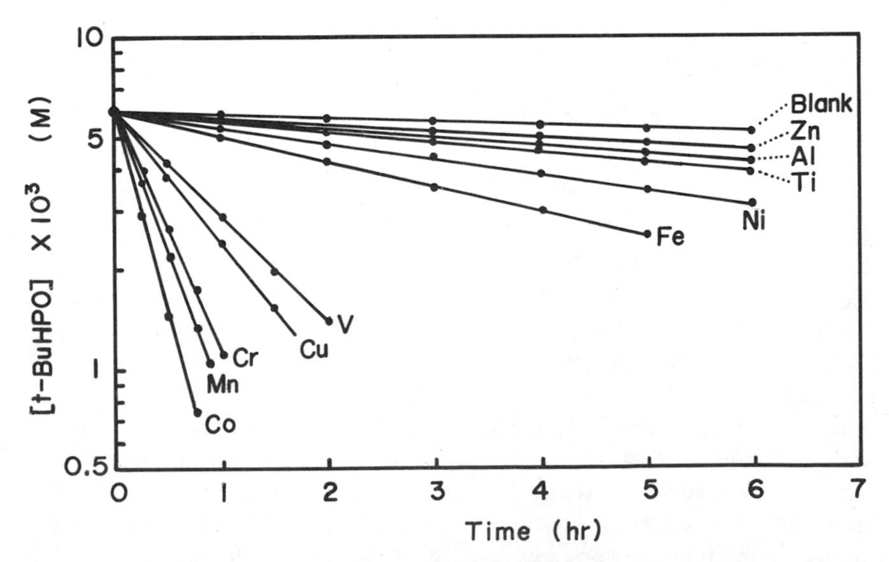

Figure 3. Decomposition of tert-*butyl hydroperoxide in TCB (at 100°C). [*tert-*BuHPO] = 6.0 ×* 10^{-3}M; *[CAT.] = 7.9 ×* 10^{-4}M.

One of the main functions of a metallic catalyst during autoxidation is believed to be promoting the breakdown of hydroperoxides to free radicals, thus continuing chain reactions (*8*). Thus, thermal decomposition of *tert*-butyl hydroperoxide was carried out in trichlorobenzene in the presence of the metal stearates. The results obtained are shown in Figure 3. It is difficult to correlate the catalytic activity of the metal stearates in the polymer oxidation with that of the decomposition of *tert*-butyl hydroperoxide in solution.

In the case of solid states, the correlation between the metal catalyst activity during the autoxidation of polypropylene and the oxidation potential of the metal has been shown (*14, 16*). However, it was difficult to observe such a correlation in the oxidation of the polymer in solution. In this context, Chalk and Smith (*32*) already reported that measurement

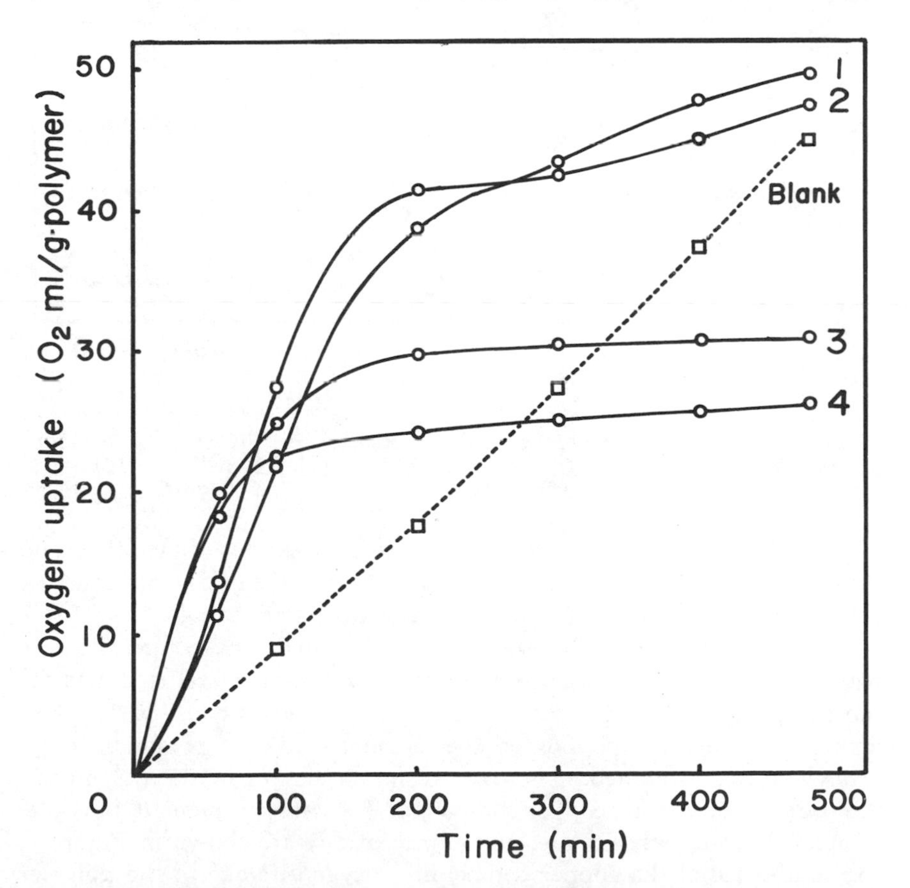

Figure 4. Effects of ligands of copper salts on the thermal oxidative degradation of APP at 125°C. [APP] = 2.38M; [CAT.] = 7.9×10^{-4}M. (1) CuO; (2) PAA-Cu^{2+}; (3) $Cu(CH_3COO)_2$; (4) $Cu(_2H_5COO)_2$, $Cu(C_3H_7COO)_2$, $Cu(C_{11}H_{23}COO)_2$, $Cu(C_{17}H_{35}COO)_2$.

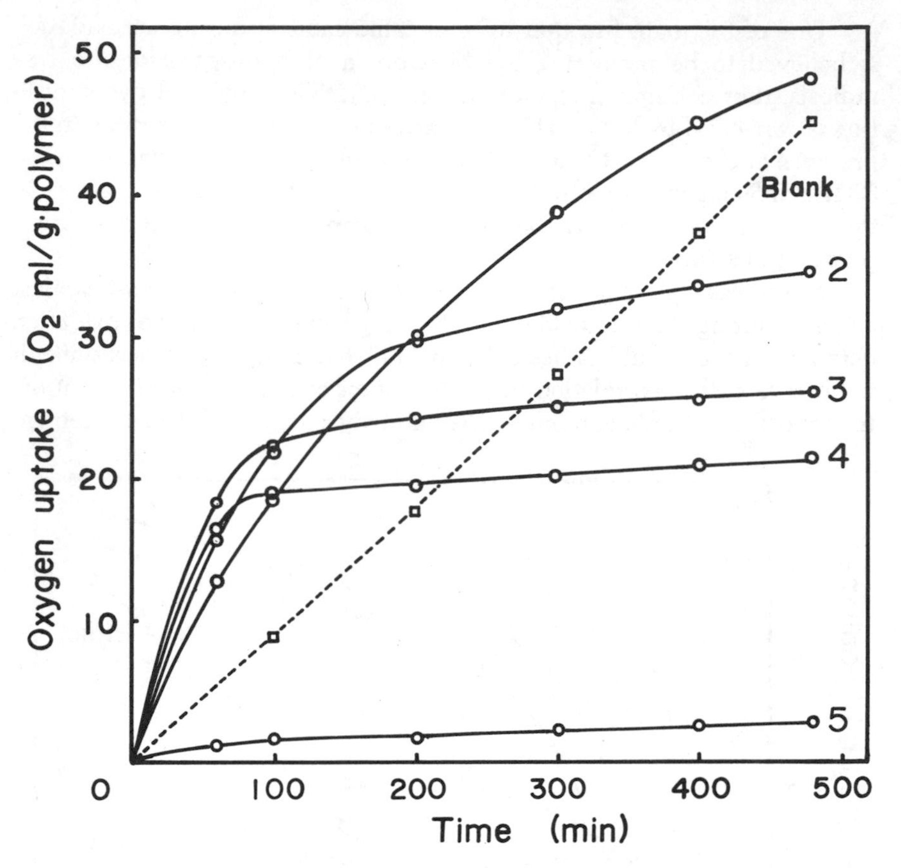

Figure 5. Effect of concentration of Cu stearate on the thermal oxidative degradation of APP at 125°C. [APP] = 2.38M. (1) 3.2×10^{-5}M; (2) 8.2×10^{-5}M; (3) 7.9×10^{-4}M; (4) 3.2×10^{-3}M; (5) 7.9×10^{-3}M.

of redox potential in nonpolar media presents serious difficulties, and little correlation exists between aqueous redox potentials and catalytic activity in the cyclohexene autoxidation in nonpolar media.

Effect of Anions in Copper Compounds. Since the copper stearate was the most effective catalyst among the transition metal stearates for the early stage of the thermal oxidation of polypropylene, the effect of anions in copper compounds on the thermal oxidation of atactic polypropylene was examined. The oxygen uptake curves of the polymer in the presence of various copper compounds (acetate, propionate, butylate, stearate, laurate, polyacrylate, and cupric oxide) are shown in Figure 4. In the absence of the copper compounds, oxygen uptake of the polymer increases linearly with time. In the presence of copper compounds of fatty acids (acetate, propionate, butyrate, laurate, and stearate), the oxygen uptake of the polymer levels off at ca. 25–30 O_2 mL/g · polymer after a certain time. In the other cases, the oxygen uptake levels off at ca. 50 O_2 mL/g · polymer. The difference is caused presumably by the lower solubility of copper polyacrylate and cupric oxide (*see* below).

Effect of Copper Stearate Concentration. As mentioned above, the thermal oxidation of the polymer catalyzed by copper compound ($7.9 \times 10^{-4}M$) proceeds very rapidly at the early stage and shows the leveling off of the oxygen uptake. Thus, the effect of the concentration of the copper stearate on the oxidation of the polymer was examined. The oxygen uptake curves of the thermal oxidation of the polymer in the presence of $3.2 \times 10^{-5}M$–$7.9 \times 10^{-3}M$ copper stearate are shown in Figure 5. In this figure, the thermal oxidation of the neat polypropylene proceeds almost linearly. However, the copper stearate-catalyzed thermal oxidation of the polymer is affected remarkably by the copper concentration. In the presence of less than $3.2 \times 10^{-3}M$ copper stearate, the

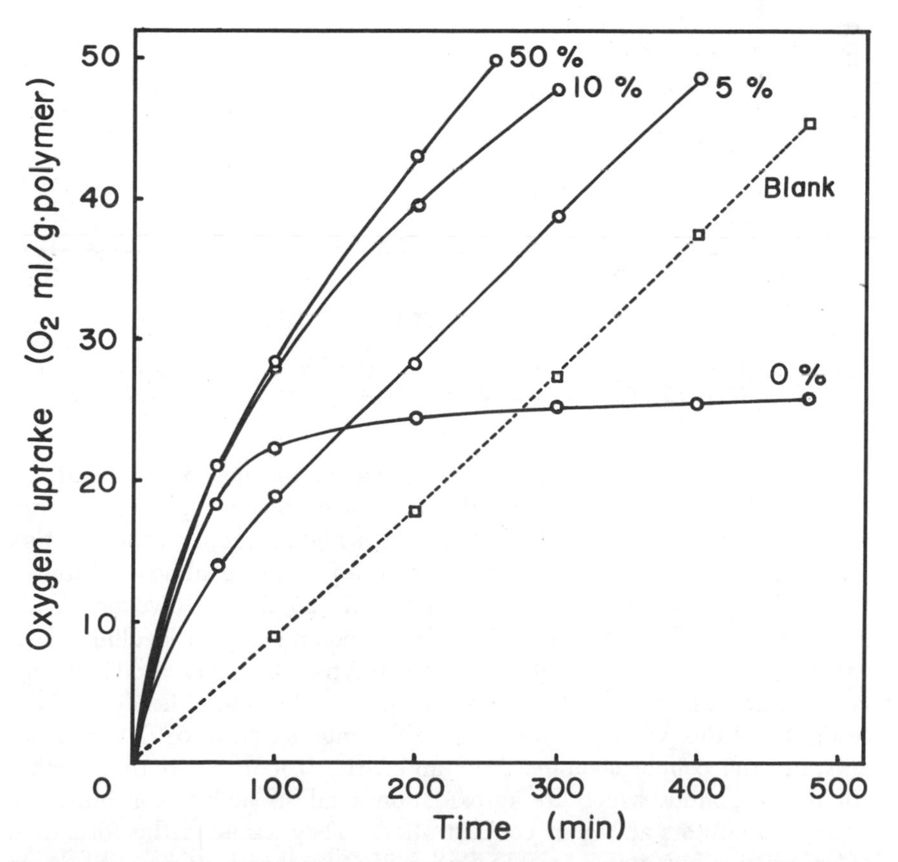

Figure 6. Effect of propionic acid on the thermal oxidative degradation of APP at 125°C. [APP] = 2.38M; [Cu stearate] = 7.9×10^{-4}M.

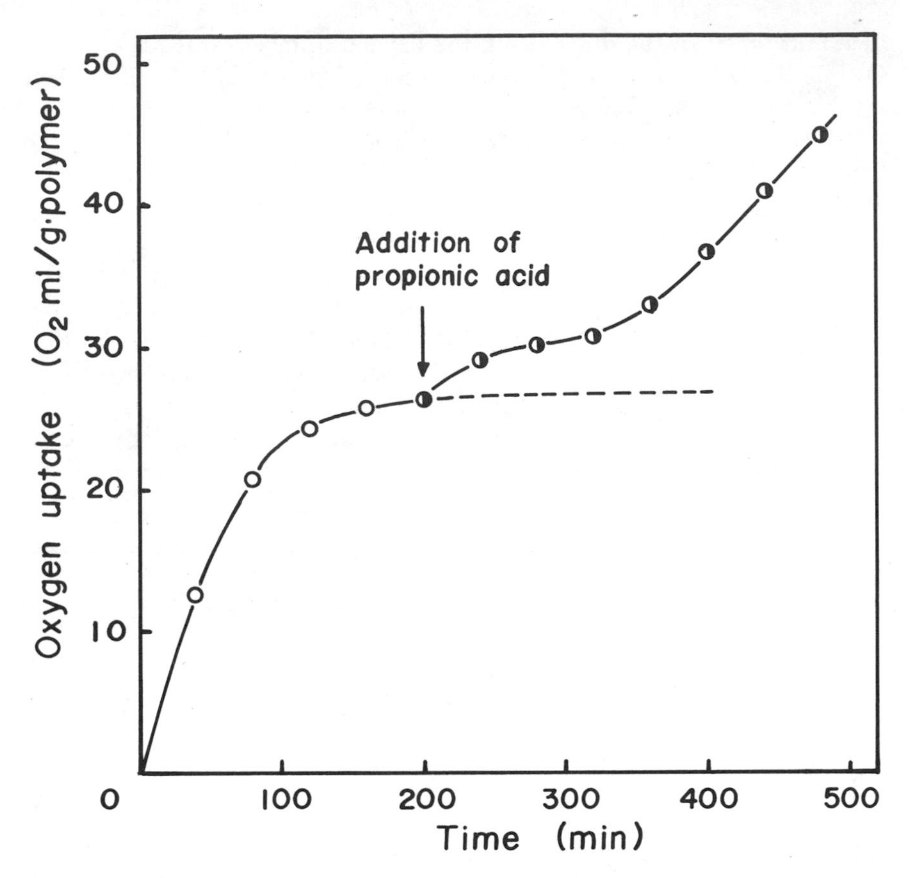

*Figure 7. Addition of propionic acid (1 mL) after leveling off of oxygen absorption; 125°C; [APP] = 2.38*M*; [Cu stearate] = 7.9 × 10^{-4}*M*.

oxidation of the polymer is faster than that of the neat polymer at the early stage of the oxidation, and the amount of oxygen uptake increases with increasing copper concentration. At a higher copper concentration ($7.9 \times 10^{-3} M$), the thermal oxidation of the polymer is almost inhibited. In this context a similar result was reported (*29, 36*). However, in the mixed solvent of trichlorobenzene and propionic acid, no leveling off of oxygen uptake was observed, and the catalytic effect increased with increasing concentration of the copper stearate. The results imply that the oxidation of the polymer is affected by the interaction of oxidation products and the copper stearate. Bets and Uri (*37*) observed that certain cobalt compounds which are autoxidation catalysts at low concentration become inhibitors at higher concentration. They assumed the formation of an inactive bidentate chelate $(CoX_2)_2 2RO_2$ in which $RO_2\cdot$ radicals act as bridges, e.g.:

$$2CoX_2 + 2RO_2\cdot \rightarrow (CoX_2)_2 2RO_2$$

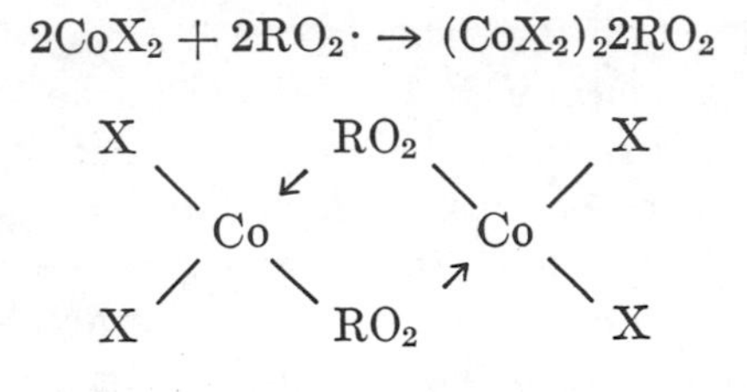

Therefore, interaction of active radicals, such as peroxy radicals, with the copper stearate at a higher concentration could be one reason for the inhibition of the thermal oxidation of the polymer.

Effect of Propionic Acid in Solvent. The thermal oxidation of the polypropylene in mixed solvent of trichlorobenzene and propionic acid was carried out, and the oxygen uptake curves obtained are shown in

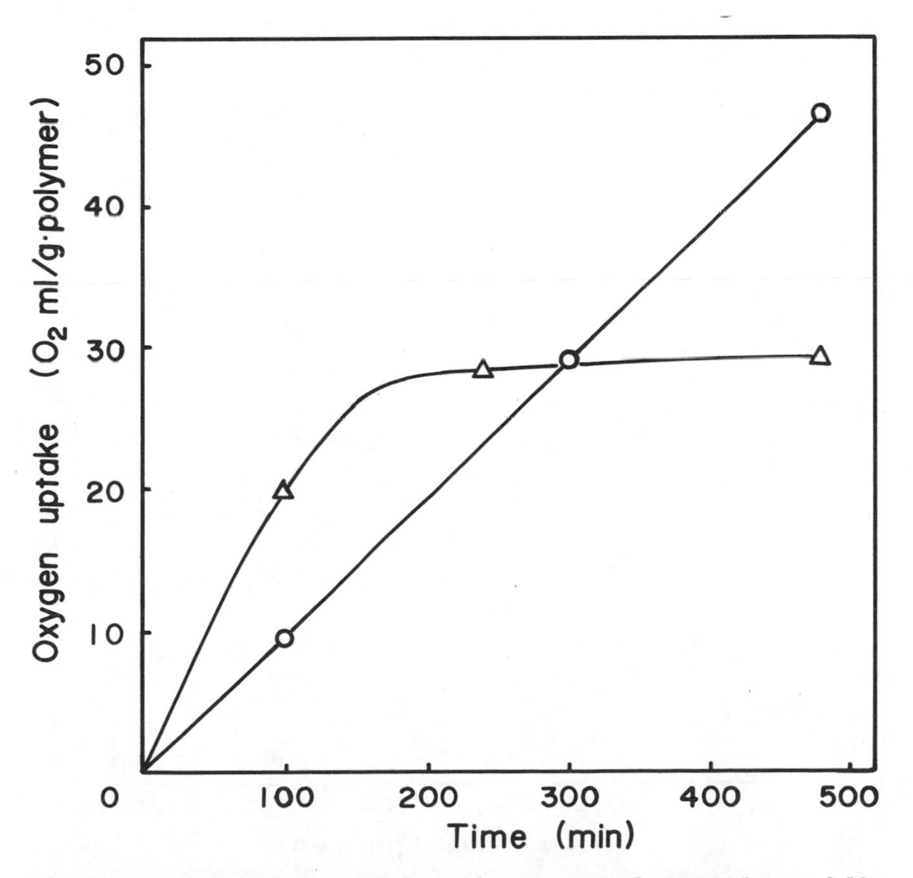

Figure 8. Preparation of oxidized samples of extracted APP (Ether insoluble–heptane soluble). Oxidation conditions: 125°C; [RH] = 2.38M. (○) = uncatalyzed; (△) = catalyzed by Cu stearate (7.9×10^{-4}M).

Figure 6. In this graph, the leveling off of oxygen uptake disappears completely in the presence of propionic acid in the solvent. The results indicate that the conversion of the copper stearate to inhibitors is prohibited by the presence of fatty acid, and the catalytic activity of the copper stearate persists through the reaction. In the absence of the copper stearate, the oxidation of the polymer was accelerated also by the addition of propionic acid in the solvent. The similar effect of organic acids was observed also and was explained by the accelerated decomposition of hydroperoxide in a polar solvent (*38, 39*). A favorable chain reaction might be another factor in the acceleration of the oxidation since the solubility of the polymer decreases by adding propionic acid.

Furthermore, when propionic acid is added to the reaction system (trichlorobenzene solution) after the leveling off of the oxygen uptake, the catalytic activity reappears, and the oxygen uptake starts (Figure 7). The results suggest that the catalytic activity of the copper stearate is inhibited by the interaction of oxidation products.

Changes in Molecular Weight and Its Distribution. The thermal oxidation of the ether–insoluble heptane–soluble fraction of the atactic

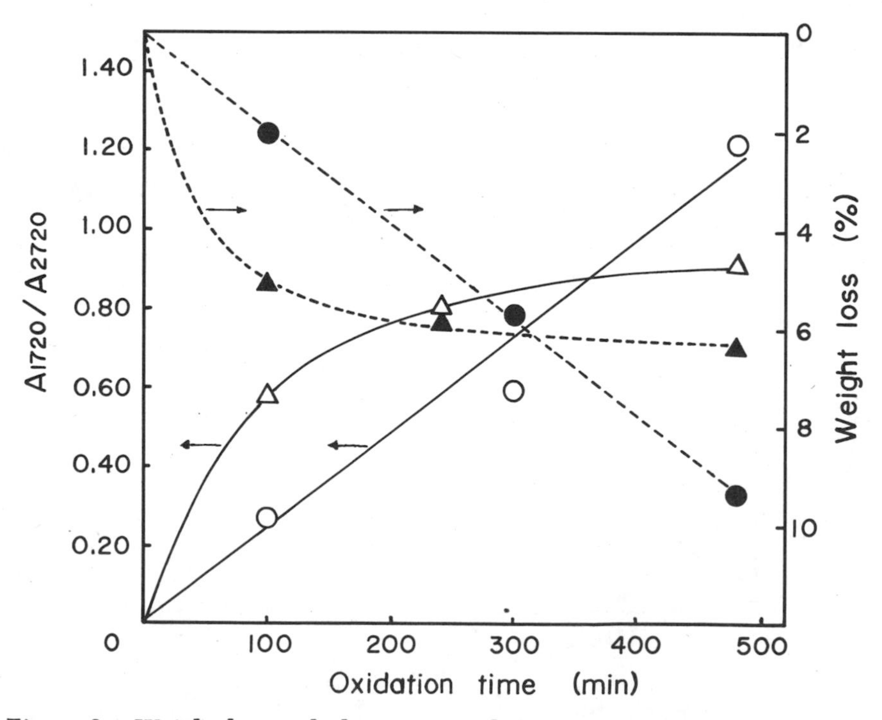

Figure 9. Weight loss and changes in carbonyl groups (A_{1720}/A_{2720}) during the thermal oxidative degradation of extracted APP. (○; ●) = uncatalyzed; (△; ▲) = catalyzed by Cu stearate (7.9×10^{-4}M).

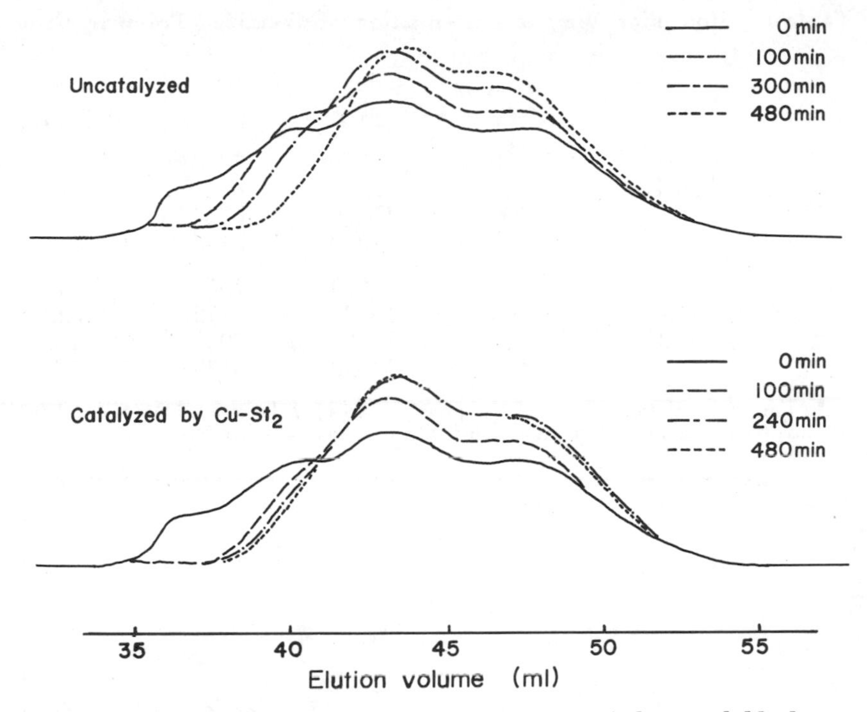

Figure 10. Gel permeation chromatograms of APP (ether insoluble–heptane soluble fraction) oxidized at 125°C in TCB

polypropylene was carried out in the presence and absence of the copper stearate, and samples with different oxygen uptake were prepared. The oxygen uptake curves and the weight loss and change in carbonyl groups of the polymer are shown in Figures 8 and 9, respectively. Good correlation is shown between the weight loss of the polymer and the amount of oxygen uptake. A similar correlation is observed also between the oxygen uptake and the change in carbonyl groups (1720 cm^{-1}) determined by ir spectrometry.

Gel permeation chromatograms of the oxidized polymers are shown in Figure 10. In Table I the weight average molecular weight ($\overline{M}_w$), the number average molecular weight ($\overline{M}_n$), and the ratio of $\overline{M}_w$ and $\overline{M}_n$ ($\overline{M}_w/\overline{M}_n$ = a measure of molecular weight distribution) are shown. In the case of the neat polymer which absorbed oxygen linearly with oxidation time, fractions of higher molecular weight decrease gradually. On the other hand, during the copper stearate-catalyzed oxidation, the molecular weight of the polymer decreases rapidly at the beginning of the oxidation and then levels off. The results are consistent with those of the oxygen uptake curves. Therefore, one can neglect the generation of oxygen ($2RO_2\cdot \rightarrow$ inert products + O_2; and the apparent oxygen uptake levels

Table I. Molecular Weight Distribution of Oxidized Polypropylene[a]

	Oxidation Time (min)	$\overline{M}_w$	$\overline{M}_n$	$\overline{M}_w/\overline{M}_n$
	0	45370	7053	6.432
Uncatalyzed	100	22290	6801	3.277
	300	15600	6354	2.455
	480	11340	5405	2.097
	0	45370	7053	6.432
Catalyzed by	100	16970	6472	2.622
Cu stearate	240	14630	6058	2.415
	480	14390	6130	2.347

[a] After the oxidation, polymer was precipitated by a large amount of methanol, and the precipitate was filtered and dried in vacuo at room temperature.

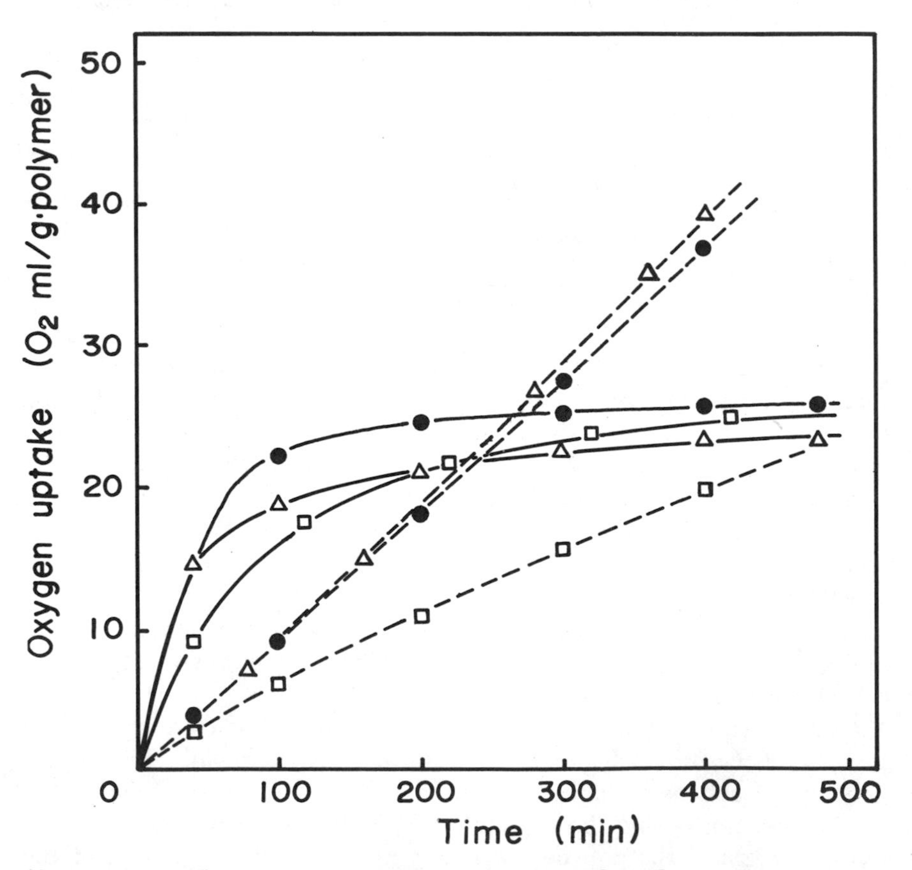

Figure 11. Effect of tert-*BuHPO on the thermal oxidative degradation of APP in the presence (——) and absence (- - -) of Cu stearate (7.9 × 10^{-4}M) at 125°C. [APP] = 2.38M; [*tert-*BuHPO] = (●) 0M; (△) 8.5 × 10^{-3}M; (□) 4.25 × 10^{-2}M.*

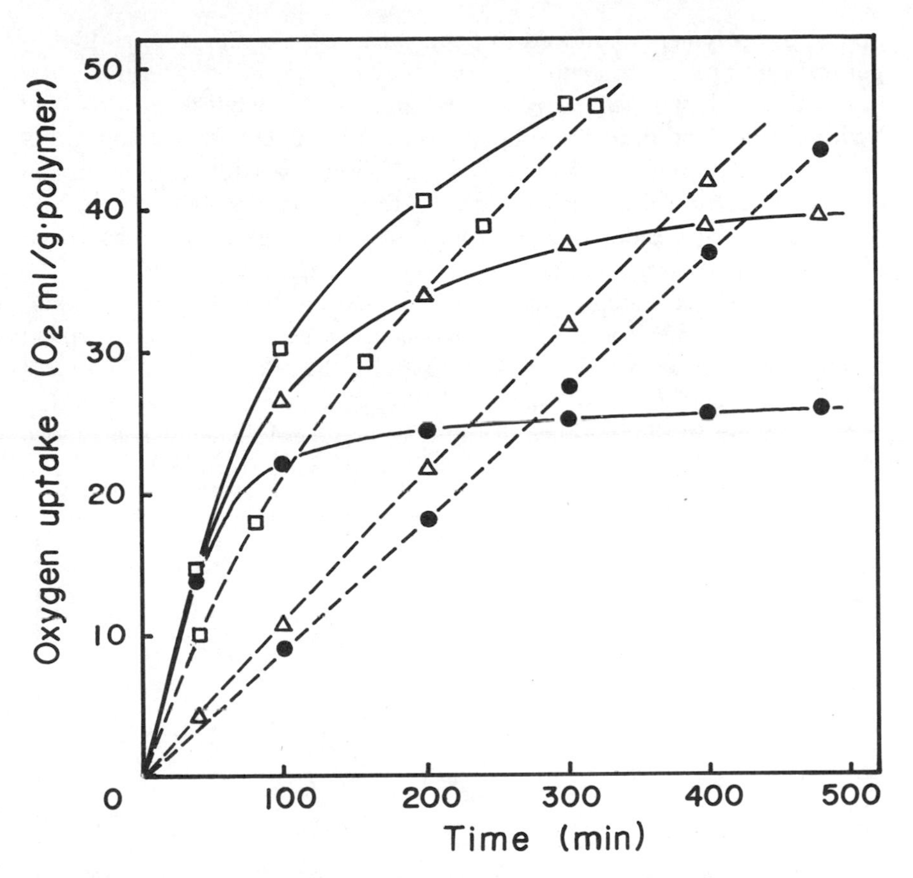

Figure 12. Effect of DTBP on the thermal oxidative degradation of APP in the presence (——) and absence (- - -) of Cu stearate (7.9 × 10^{-4}M) at 125°C. [APP] = 2.38M; [DTBP] = (●) 0M; (△) 5.34 × $^{-3}$M; (□) 2.67 × 10^{-2}M.

off) as a factor of the leveling off of oxygen uptake in the copper-catalyzed thermal oxidation of polypropylene.

Effect of Various Additives. From the results mentioned above, it is evident that the copper stearate is converted to inhibitors by the interactions of oxidation products, and the oxygen uptake appears to level off after a certain time. Thus, the effects of model compounds of oxidation products (*n*-octylaldehyde, *n*-octanoic acid, 2-octanone, *n*-octanol, and *n*-octanoic acid methyl ester), *tert*-butyl hydroperoxide, and di-*tert*-butyl peroxide on the thermal oxidation of the polypropylene in trichlorobenzene were examined.

The addition of *tert*-butyl hydroperoxide decreases the amount of the oxygen uptake both in the presence and absence of the copper stearate (Figure 11) while adding di-*tert*-butyl peroxide increases the oxidation of the polymer in both cases (Figure 12). Generally, the thermal oxidation of the polymer is believed to begin at a reactive site such as hydroperoxide groups contaminated in the polymer (*8, 43*). Therefore, the interaction of low molecular radical fragments generated from *tert*-butyl hydroperoxide and di-*tert* butyl peroxide might affect the oxidation process at the early stage. In addition, the difference between the hydrogen abstraction rates of the peroxy radical ($RO_2\cdot$ from *tert*-butyl hydroperoxide) and alkoxy radical ($RO\cdot$ from di-*tert*-butyl peroxide) also might affect the oxidation process of the polymer (*40, 41, 42*).

The effects of model compounds of oxidation products on the oxidation of the polymer in trichlorobenzene were examined. As a typical example of the oxygenated model compounds, the effect of small amounts of *n*-octanoic acid ($5 \times 10^{-3}M$, $1 \times 10^{-2}M$, $2 \times 10^{-2}M$) on the oxidation of the polymer in the presence of the copper stearate is shown in Figure 13. The effects of all model compounds of oxidation products are sum-

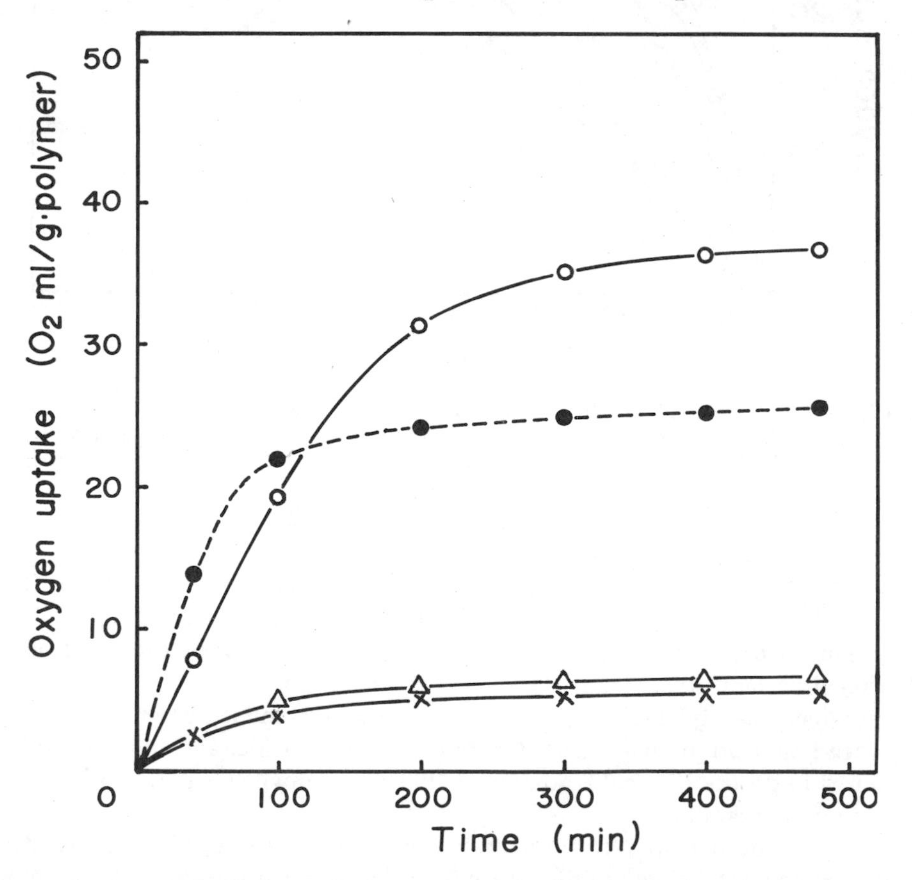

Figure 13. Effect of n-*octanoic acid on the thermal oxidative degradation of APP in the presence of Cu stearate* (7.9×10^{-4}M) *at 125°C.* [APP] = 2.38M; [n-*octanoic acid*] = (●) 0M; (×) 5×10^{-3}M; (△) 1×10^{-2}M; (○) 2×10^{-2}M.

Table II. Effects of Model Compounds of Oxidation Products on the Thermal Oxidative Degradation of APP in the Absence and Presence of Cu Stearate

Model Compounds	*Absence*	*Presence*	
	(2 × 10⁻²M)	*(2 × 10⁻²M)*	*(5 × 10⁻³M)*
n-Octylaldehyde	accel.	accel.	retard.
n-Octanoic acid	accel.	accel.	retard.
2-Octanone	accel.	retard.	retard.
n-Octanol	accel.	retard.	retard.
n-Octanoic acid methyl ester	accel.	retard.	retard.

[a] Oxidation conditions: [APP] = 2.38*M*, [Cu stearate] = $7.9 \times 10^{-4}M$; solvent, TCB; temp., 125°C.

marized in Table II. In the absence of the copper stearate all model compounds accelerate the oxidation of the polymer. On the other hand, in the presence of the copper stearate at a model compound concentration of $2 \times 10^{-2}M$, the aldehyde and acid accelerate the oxidation of the polymer, and the others retard it. At a lower concentration ($5 \times 10^{-3}M$) all model compounds retard the oxidation of the polymer. In the absence of the copper stearate, the acceleration effect of the model compounds may be explained by a polar effect (*38*). In the presence of the copper stearate, a very small amount of the oxidation products can convert the copper stearate to inhibitors, and a small amount of the aldehyde and acid ($2 \times 10^{-2}M$) can reveal the catalytic activity of the copper stearate.

Commercial polymers are believed to contain reactive groups such as hydroperoxide, carbonyl, and unsaturated groups. These groups often play important roles for the thermal oxidation processes of the polymers (*1, 2, 9, 10, 43, 44*). Therefore, one may consider also the effects of the oxidation products originating from these reactive groups at the very beginning of the oxidation, when the mechanism of the copper-catalyzed oxidation of the polymer was discussed. Further experiments on the effects of various oxidation products on the thermal oxidation of the polymer will be published.

Literature Cited

1. Hawkins, W. L., "Polymer Stabilization," Interscience, New York, 1972.
2. Hawkins, W. L., *J. Polym. Sci. C* (1976) **57**, 319.
3. Hansen, R. H., Russell, C. A., DeBenedictis, T., Martin, W. M., Pascale, J. V., *J. Polym. Sci., A* (1964) **2**, 587.
4. Hansen, R. H., DeBenedictis, T., Martin, W. M., *Polym. Eng. Sci.* (1965) **5**, 223.
5. Hansen, R. H., De Benedictis, T., Martin, W. M., *Trans. Inst. Rubber Ind.* (1963) **39**, 290.
6. Hawkins, W. L., Chan, G., Link, G. L., *Polym. Eng. Sci.* (1971) **11**, 377.

7. Mayo, F. R., Egger, K. W., Irwin, K. C., *Rubber Chem. Technol.* (1968) **41**, 271.
8. Reich, L., Stivala, S. S., "Autoxidation of Hydrocarbons and Polyolefins," Dekker, New York, 1968.
9. Osawa, Z., *Kobunshi* (1976) **25**, 406.
10. Osawa, Z., *Kobunshi* (1977) **26**, 327.
11. Takahashi, T., Suzuki, K., *Kobunshi Kagaku* (1964) **21**, 487, 494, 498.
12. Takahashi, T., Suzuki, K., *Kobunshi Kagaku* (1966) **23**, 792.
13. Kelleher, P. G., *J. Appl. Polym. Sci.* (1966) **10**, 843.
14. Reich, L., Jadrnicek, B. R., Stivala, S. S., *J. Polym. Sci., A1* (1971) **9**, 231.
15. Richeters, P., *Macromolecules* (1972) **3**, 262.
16. Osawa, Z., Shibamiyam, T., Matsuzaki, K., *Kogyo Kagaku Zasshi* (1968) **71**, 552.
17. Osawa, Z., Ishizuka, T., *J. Appl. Polym. Sci.* (1973) **17**, 2897.
18. Niki, E., Shiono, T., Ido, T., Kamiya, Y., *J. Appl. Polym. Sci.* (1975) **19**, 3341.
19. Egorenkov, N. I., Lin, D. G., Bely, V. A., *J. Polym. Sci. Chem.* (1975) **13**, 1493.
20. Fujiwara, S., "Instrumental Analysis of Polymer," p. 80, Hirokawa, 1961.
21. Voter, R. C., "Analytical Chemistry of Polymers Part I, Analysis of Monomers and Polymeric Materials, Plastics–Resins–Rubbers," p. 176, Interscience, New York, 1959.
22. Anduze, R. A., *Anal. Chem.* (1957) **29**, 90.
23. Fratkina, G. P., *Zavod. Lab.* (1958) **24**, 1373.
24. Fratkina, G. P., *Z. Anal. Chem.* (1959) **169**, 229.
25. Bolleter, W. T., *Anal. Chem.* (1959) **31**, 201.
26. Chan, M. G., Allara, D. L., *J. Colloid Interface Sci.* (1974) **47** (June), 697.
27. Tompkins, H. G., Allara, D. L., *J. Colloid Interface Sci.* (1974) **49** (Dec), 410.
28. Allara, D. L., Roberts, F. R., *J. Catal.* (1976) **45**, 54.
29. Allara, D. L., White, C. W., Meek, R. L., Briggs, T. H., *J. Polym. Sci., A* (1976) **14**, 93.
30. Allara, D. L., Chan, M. G., *J. Polym. Sci., A* (1976) **14**, 1857.
31. Allara, D. L., Preprints of the main lecture, *Amer. Chem. Soc., 173rd, New Orleans, 1977.*
32. Matsuura, R., *Nippon Kagaku Kaishi* (1960) **86**, 560.
33. Wall, F. T., Gill, S. J., *J. Phy. Schem.* (1954) **58**, 1128.
34. Nishikawa, H., Tsuchida, E., *Bull. Chem. Soc. Jpn* (1976) **49**, 1545.
35. Chalk, A. J., Smith, J. F., *Trans. Faraday Soc.* (1957) **53**, 1214.
36. Emanuel, N. M., Maizus, Z. K., Skibida, I. P., *Angew Chem* (1969) **81**, 91.
37. Betts, A. T., Uri, N., *Makromol Chem.* (1966) **95**, 22.
38. Kamiya, Y., *Org. Syn. Chem. Jpn* (1968) **26**, 957.
39. Kamiya, Y., *J. Polym. Sci., A1* (1968) **6**, 2561.
40. Niki, E., Decker, C., Mayo, F. R., *J. Polym. Sci., Pt. A1* (1973) **11**, 2813.
41. Niki, E., Kamiya, Y., *J. Am. Chem. Soc.* (1974) **96**, 2129.
42. Niki, E., Kamiya, Y., *Bull. Chem. Soc. Jpn* (1975) **48**, 3226.
43. Osawa, Z., Saito, T., Kimura, Y., *J. Appl. Polym. Sci.*, in press.
44. Sitek, F., Guillet, J. E., Heskins, M., Scott, G., *J. Polym. Sci., C* (1976) **57**, 343, 357.

RECEIVED May 12, 1977.

Pyrolysis and Oxidative Pyrolysis of Polypropylene

JAMES C. W. CHIEN and JOSEPH K. Y. KIANG

Department of Chemistry, Department of Polymer Science and Engineering, Polymer Research Institute, Materials Research Laboratories, University of Massachusetts, Amherst, MA 01003

Amorphous and semi-crystalline polypropylene samples were pyrolyzed in He from 388°–438°C and in air from 240°–289°C. A novel interfaced pyrolysis gas chromatographic peak identification system was used to analyze the products on-the-fly; the chemical structures of the products were determined also by mass spectrometry. Pyrolysis of polypropylene in He has activation energies of 5–1–56 kcal mol^{-1} and a first-order rate constant of 10^{-3} sec^{-1} at 414°C. The olefinic products observed can be rationalized by a mechanism involving intramolecular chain transfer processes of primary and secondary alkyl radicals, the latter being of greater importance. Oxidative pyrolysis of polypropylene has an activation energy of about 16 kcal mol^{-1}; the first-order rate constant is about 5×10^{-3} sec^{-1} at 264°C. The main products aside from CO_2, H_2O, acetaldehyde, and hydrocarbons are ketones. A simple mechanistic scheme has been proposed involving C–C scissions of tertiary alkoxy radical accompanied by H transfer, which can account for most of the observed products. Similar processes for secondary alkoxy radicals seem to lead mainly to formaldehyde. Differences in pyrolysis product distributions reported here and by other workers may be attributed to the rapid removal of the products by the carrier gas in our experiments.

There has been surprisingly little fundamental work reported on the chemistry of the process involved in the burning of polymers and its inhibition, despite the great technological and socio-economical importance of the subject. In our laboratories we have undertaken a funda-

0-8412-0381-4/78/33-169-175$05.75/1

mental research of this problem. The effort is divided into the study of the chemistry involved in a polymer flame and the effect of flame retardants and the study of the chemistry involved in the solid polymer phase and the effect of flame retardants. This report is concerned with the second topic.

When various polymers were burned in air in a candle-like manner (*1*), the maximum flame temperatures were 490°–740°C and the temperatures of the melt surface were 230°–540°C. Burge and Tipper (*2*) found the temperature of polyethylene to be 400°–500°C at the burning surface and decreasing to 200°–300°C at 1 cm below the surface; the temperature being greater when probed from above the sample and smaller when probed from below. Whether the degradative processes for the polymers are pure pyrolysis or oxidative pyrolysis is still largely unsettled. It has been shown that the oxygen concentration 1 mm above the surface of a burning polyethylene rod is only ca. 1% (*2*). Similar results were reported for polypropylene (*3*). From this it may be inferred that the polymer in the melt surface may be oxygen-depleted. On the other hand, the polymer further below the surface probably contains the usual amount of dissolved oxygen. It is likely that oxidative pyrolysis contributes significantly or even predominantly to the liberation of volatile and combustible fragments from the polymer. The relative importance of pure pyrolysis and oxidative pyrolysis can only be established by experimentation. For instance, if an efficient flame retardant suppresses markedly the oxidative pyrolysis but has no effect at all on the thermal pyrolysis, then further search for new flame retardants should be emphasized on compounds which could interrupt peroxy radical chain reactions. On the other hand, if nonoxidative pyrolysis dominates, then heat dissipating or intumescent additives would be effective.

In this chapter we describe a novel system for the study of pyrolysis and oxidative pyrolysis of polymers. The results of such a study on amorphous and semicrystalline polypropylene are presented and compared with other reported works. The effects of known flame retardants on the rates of pyrolysis and oxidative pyrolysis of the polymer and the product distribution have been investigated and will be published elsewhere.

Experimental

Materials. The amorphous polypropylene (APP) used is that of unstabilized Eastobond from Tennessee Eastman. The semicrystalline polypropylene (IPP) is the Profax 6501 from Hercules Incorporated. It has a crystallinity of 61% as determined from its density (*4*).

Interfaced Pyrolysis Gas Chromatographic Peak Identification System (IPGCS). The versatile IPGCS (*5*) incorporates instrumentation for thermal degradation of polymer under slow or ultra-rapid temperature rise and under inert or reactive atmosphere conditions. Evolved volatiles are transferred to a master trap manifold where precolumn procedures may be applied prior to gas chromatographic separation. Identification and analysis of individual peaks are then performed quickly by rapid scan vapor phase ir spectrophotometry, elemental analysis for carbon, hydrogen, oxygen, nitrogen, etc., functional group fingerprinting by vapor-phase thermal cracking, and molecular weight determination by differential gas density measurement (mass chromatography). Figure 1 is a block diagram of the IPGCS system. An interfaced laboratory computer provides for data acquisition, reduction, and control. In order to make clear the operation of the system, a brief discussion of individual instruments follows.

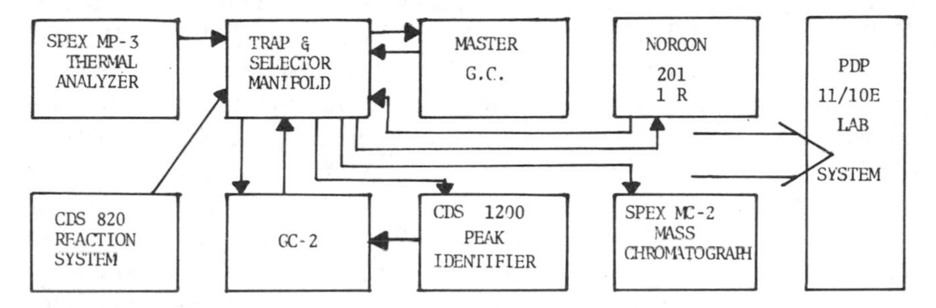

Figure 1. Block diagram of the interfaced pyrolysis gas chromatographic peak identification system

The MP–3 multipurpose thermal analyzer is manufactured by the Spex Industries. It pyrolyzes sample at programmed rates from 4°/min–40°C/min up to 1000°C. It was modified by the addition of two laminar flow controllers (HGC 187; Analabs) which provide controlled atmospheres of up to four gaseous mixtures. The CDS 820 (Chemical Data Systems) consists of a Pyroprobe 100 capable of delivering to a polymer sample a maximum heating rate of 20,000°C/sec; the CDS 820 provides a controlled atmosphere for the Pyroprobe.

The master gas chromatograph is a Varian 2760 instrument with thermal conductivity and flame ionization detection. A second Varian 2760 gas chromatograph (GC–2) serves for analysis of samples from two sources, pyrolysis products from the CDS 820 and from the structural determination function of the CDS 1200. The latter instrument (Chemical Data System) is a functional group and elemental analyzer which generates a vapor-phase thermolytic dissociation pattern for functional group analysis and also performs elemental analysis. The effluent from the master GC is split so that 10% of it is directed to the detector and 90% of it to the CDS 1200. A stop-flow valve admits one peak at a time, thus enabling the analysis of all the components of the pyrolysates without appreciable peak spreading. The use of CDS 1200 in organic analysis has been discussed by Liebman et al. (*6*).

The Norcon 201 rapid-scan, vapor-phase spectrophotometer is a double beam grating instrument which scans a spectrum from 4000 to 670 cm^{-1} in either 6 or 30 sec, with recycle times of 8 and 40 sec, respectively. Its wavelength accuracy is $\pm 0.05\mu$ with $\pm 0.025\mu$ reproducibility. The sample cell has a volume of 5 mL and is constructed of gold-coated glass (4×4-mm cross section, 30 cm in length). The path length to volume ratio is 6.0 cm^{-2}. The cell is enclosed in an oven regulated to $\pm 2.5°$, from 120°–250°C. The Norcon 201 is connected to the master GC through heated transfer line. An effluent fraction of 0.025 λ is sufficient to yield a good spectrum. The Norcon 201 is interfaced with a PDP 11/10E computer system. This system has a RT 11 foreground/background operating software, FORTRAN, and Lab Applications Program Library VO3 which contains SPARTA and THRU programs for data acquisition and a IRFLAV program for the acquisition of ir spectral data with digital filtering. Another routine, IRSPBA, performs background subtraction and spectrum print out.

The MC–2 mass chromatograph from the Spex Industries determines molecular weight of GC eluent through differential gas density measurements. A sample is split into two equal fractions; they are carried by two different gases, CO_2 and Freon 115, through two matched GC columns into density-balance detectors. The molecular weight of the unknown is obtained from

$$\text{mol wt} = \frac{(A_1/A_2)\,K \cdot \text{mol wt}_{CG1} - \text{mol wt}_{CG2}}{(A_1/A_2)\,K - 1} \qquad (1)$$

where the instrument constant K is calibrated with a substance of known molecular weight, mol wt_{st},

$$K = \left(\frac{A_1}{A_2}\right) \left(\frac{\text{mol wt}_{st} - \text{mol wt}_{CG1}}{\text{mol wt}_{st} - \text{mol wt}_{CG2}}\right) \qquad (2)$$

In these equations mol wt_{CG1} and mol wt_{CG2} are the molecular weights of the two carrier gases, and A_1/A_2 is the ratio of peak height response of a standard for the two detectors.

The accuracy for the determination of molecular weight with MC–2 depends upon the instrument constant K (Equation 2). The best results are obtained for values of K determined with known compounds having molecular weights nearly the same as the unknown. Lloyd, et al. (*7*) has developed a least-square, curve-fitting program to fit the calibration K values to a polynomial which was then used in an iterative procedure to calculate the molecular weights of the unknown. The molecular weights thus obtained are accurate to ±1 mass unit as judged by comparison with mass spectrometric results. The advantages of the simple MC–2 mass chromatography are offset by the loss of information about geometric isomerism. Therefore, the technique must be used together with mass spectrometry.

In addition to the above IPGCS system, other thermoanalytical techniques were used also in this work. This includes a conventional pyrolysis-GC-mass spectrometry system which consists of a CDS Pyroprobe 100, a Perkin–Elmer 990 GC, and a Hitachi–Perkin–Elmer RMV 66 mass spectrometer, as well as a duPont 900 thermal analyzer.

Procedures for Pyrolysis. In the pyrolysis-GC-mass spectrometry experiments, about 1 mg of polymer was weighed into a quartz tube which was inserted into the heating coil of the Pyroprobe. The latter fitted directly into the injection port of the Perkin–Elmer 990 GC. The GC was operated at a manifold temperature of 220°C, injector temperature of 210°C, interface temperature of 255°C, He flow rate of 83 mL min^{-1}, and FID detection. Samples were pyrolyzed at 600°, 650°, 700°, 750°, 800°, 850°, 900°, and 950°C at a heating rate of 20,000°C sec^{-1}. All samples were held for 20 sec at the final temperatures.

Two columns were used to separate the pyrolysis products programmed from 1°C min^{-1} to 6°C min^{-1}. A Porapak Q column (4 ft × ⅛ in. OD) was used to separate low boiling compounds and a 4% SE30 on ABS column (6 ft × ⅛ in. OD) resolves higher boiling products. The gas chromatogram was obtained first. This was followed by an identical run whereas a peak of interest was just emerging, the interface splitter valve was opened to admit the sample into the mass spectrometer. The latter was operated at an electron energy of 70 eV and a filament current of 3.4 A.

For pyrolysis with the IPGCS system, about 2 mg of polymer was weighed into a quartz tube. It was placed into the oven of the MP–3 apparatus. The polymer was pyrolyzed at 388°, 414°, and 438°C at a heating rate of 40°C min^{-1} under a He flow rate of 25 mL min^{-1}. The weight of the residue was obtained with a microbalance. The products, after passage through the master trap manifold, were collected in a 1 ft × ⅛ in. OD glass bead column at −195°C. When all of the products have been collected, the column was heated rapidly to 300°C and the pyrolyzates were back-flushed into the master GC using FID detection. Low boiling hydrocarbons were separated with a Chromosorb 102 column (12 ft × ⅛ in. OD); higher boiling products were resolved with an 8% Dexsil 300 GC on Chromosorb W–HP column (6 ft × ⅛ in. OD). At a He flow rate of 14 mL min^{-1}, the column was kept first at 37°C for 3 min, then programmed at 4°C min^{-1} to a final temperature of 250°C for the Chromosorb 102 column and to 300°C for the Dexsil column. The molecular weight of the product was determined with the MC–2 instrument calibrated with *n*-alkanes. Weight percentage of the products were calculated from the integrated area of the GC peaks taking into consideration the attenuation factors. Dietz (*8*) had shown previously that FID is equally sensitive to all aliphatic hydrocarbons.

Procedures for Oxidative Pyrolysis. All oxidative pyrolysis experiments were performed with the IPGCS system. About 5 mg of polymer were pyrolyzed in an air atmosphere at 240°, 264°, and 289°C. The same Chromosorb 102 column was used for the separation of low molecular weight products, but a Carbowax 20M on Chromosorb P column (12 ft × 1/8 in. OD) replaces the Dexsil column for the separation of high molecular weight products. The products were characterized primarily by the Norcon 201 spectrophotometer and mass spectrometry.

Results

Pyrolysis of Polypropylene. For the measurement of rates of pyrolysis, the temperature range is limited to that of conveniently measurable rates. Figure 2 shows the thermograms of APP and IPP which provides the choice of temperature for pyrolysis.

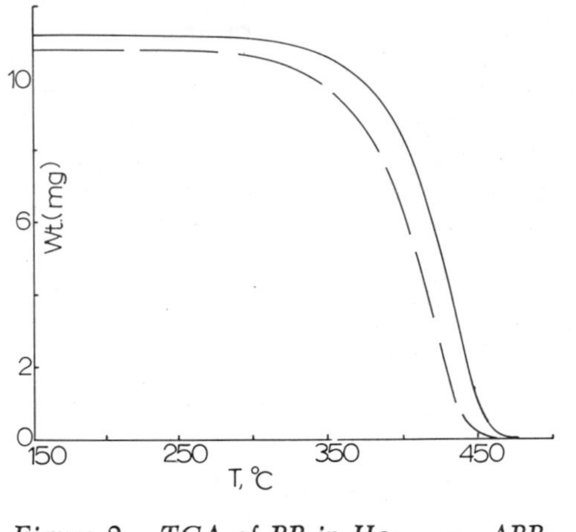

Figure 2. TGA of PP in He: ——, APP; – – –, IPP

The rate of pyrolysis was obtained by weighing the polymer sample before and after a preset hold time in MP–3. The results are given in Figure 3. The process follows first-order kinetics as shown in Figure 4 and are semilog plots of wt % of polymer residue vs. *t*. The rate constants and activation energies for pyrolysis of polypropylene are given in Table I. By comparison, IPP decomposes faster than APP at 388°C but more slowly at 438°C. This is reflected in a lower activation energy of 51 kcal mol^{-1} for the semicrystalline polypropylene, the accuracy for the activation energy is about 10%.

The chromatograms of the pyrolyzates of IPP at 438°C are shown in Figure 5. The products and their distributions at three temperatures are summarized in Tables II and III. The chemical structures of the products were obtained from mass spectrometry; Table IV presents the

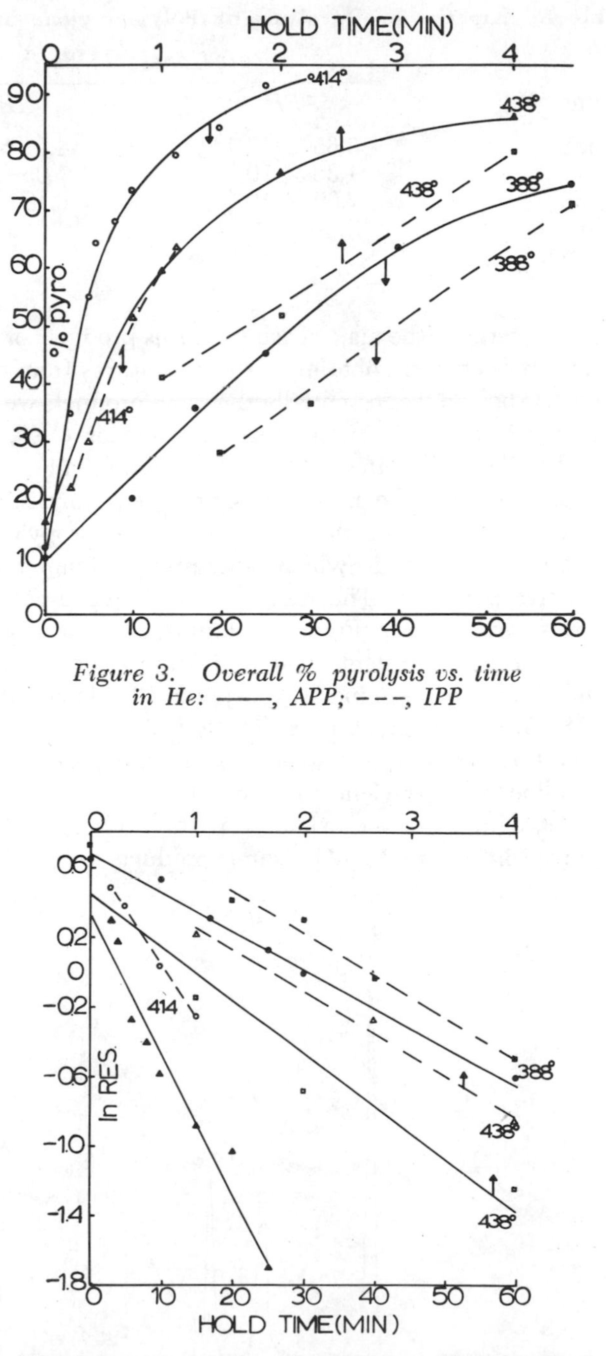

Figure 3. Overall % pyrolysis vs. time in He: ——, APP; – – –, IPP

Figure 4. First-order plot of PP pyrolysis in He: ——, APP; – – –, IPP

Table I. Kinetics of Pyrolysis of Polypropylene in He

Temp., °C	*Rate Constants, Sec⁻¹* APP	IPP
388	3.65×10^{-4}	4.0×10^{-4}
414	1.38×10^{-3}	1.05×10^{-3}
438	7.60×10^{-3}	6.18×10^{-3}
Activation energy, kcal mol^{-1}	56	51

fragmentation patterns of the major flash pyrolysis products of polypropylene. Identical products were obtained at temperatures from 600°–950°C. Tables II and III showed that essentially the same products were produced by pyrolysis at much lower temperatures except for C_2H_6 and C_4H_{10}. Figures 6 and 7 show the rates of formation of individual products at 414°C, each normalized to the amount at complete pyrolysis.

Oxidative Pyrolysis. The thermograms of polypropylene obtained in air are shown in Figure 8, which suggests the temperatures to be used for oxidative pyrolysis. The results of oxidative pyrolysis at 240°, 264°, and 289°C are given in Figure 9; Figure 10 shows the first-order kinetic plots. The rate constants and activation energies are given in Table V. Like the pyrolysis in inert atmosphere (Table II), the oxidative pyrolysis of IPP at low temperature is slightly faster than APP and slower at high temperature resulting in a somewhat smaller activation energy for the semicrystalline polypropylene (Figure 11).

The chromatograms for the oxidative pyrolyzates obtained at 363°C for IPP are given in Figure 12. The same products were formed at the

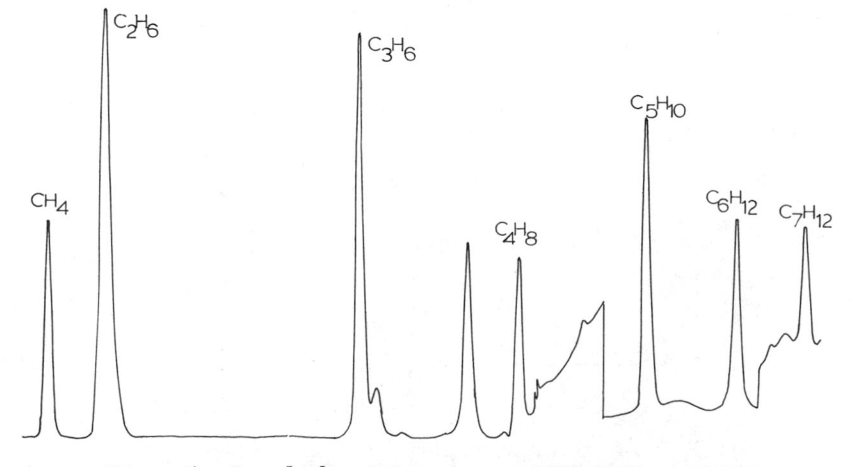

Figure 5. Low-boiling GC pyrogram of IPP in He at 438°C

Table II. Pyrolysis Products of Polypropylene at 438°C

Peak	*Mol Wt*	*Hydrocarbons*	*wt %*[c] *APP*	*IPP*
1[a]		CH_4 (methane)[b]	0.1	0.09
2[a]		C_2H_6 (ethane)	1.2	1.4
3[a]		C_3H_6 (propylene)	12.2	10.7
4	55.7	C_4H_8 (isobutylene)	3.2	2.5
5	70.0	C_5H_{10} (2-pentene)	15.8	15.5
6	81.2	C_6H_{12} (3-methyl-1-pentene)	11.4	10.2
7	96.8	C_7H_{12} (3-methyl-3,5-hexadiene)	2.0	1.7
8	110	C_8H_{16} (4-methyl-3-heptene)	[d]	[d]
9	123.5	C_9H_{16} (2,4-dimethyl-heptadiene)	[d]	[d]
10	125.9	C_9H_{18} (2,4-dimethyl-heptenc)	39.3	42.6
11	139.5	$C_{10}H_{20}$	2.4	2.1
12	157.5	$C_{11}H_{22}$ (4,6-dimethyl-3-nonene)	1.3	1.6
13	168.2	$C_{12}H_{24}$ (2,4,6-trimethyl-8-nonene)	9.5	10.2
14	180.6	$C_{13}H_{24}$	1.7	1.6

[a] Identified by retention time; molecular weight determination by MC–2 was thwarted by instability in base line during first several minutes of operation of this instrument.
[b] Structure determined by mass spectrometry.
[c] Average of two determinations.
[d] Incompletely resolved.

other temperatures, the main difference being that the product distributions are not the same (Figure 13).

Most of the products were identified by the Norcon 201 ir spectrophotometer. The spectra were compared with the atlas compiled by

Table III. Pyrolysis Products of Polypropylene at 388° and 414°C*

	Pyrolysis Temperature			
	414°C wt % of Products		*388°C wt % of Products*	
Hydrocarbons	*APP*	*IPP*	*APP*	*IPP*
CH_4	0.08	0.12	0.09	0.08
C_2H_6	1.0	1.6	0.9	0.07
C_3H_6	10.6	11.7	9.0	6.9
C_4H_8	2.7	2.1	2.7	1.8
C_5H_{10}	16.5	18.9	15.0	14.0
C_6H_{12}	10.3	11.9	12.0	10.5
C_7H_{12}	1.6	1.4	1.1	1.0
C_9H_{18}	41.3	36.0	43.0	45.2
$C_{10}H_{20}$	2.2	1.5	1.7	1.5
$C_{11}H_{22}$	1.7	2.0	2.4	3.1
$C_{12}H_{24}$	11.0	12.3	11.3	13.5
$C_{13}H_{24}$	1.3	1.2	1.4	1.6

* Same footnotes as in Table II.

Table IV. Identification of Flash Pyrolysis Products of Polypropylene by Mass Spectrometry

Product	*Fragmentation Patterns in Mass Numbers*
Methane	16, 15
Ethylene	28, 27, 26
Propylene	42, 41, 27
Isobutane	58, 43
2-Pentene	70, 55, 42, 41, 39
3-Methyl-1-pentene	84, 69 56, 41
3-Methyl-3,5-hexadiene	96, 81, 67, 55, 41, 39
4-Methyl-3-heptene	112, 69, 55, 41, 39, 27
2,4-Dimethyl-heptadiene	124, 123, 109, 95, 82, 67, 55, 41, 39
4,6-Dimethyl-3-nonene	154, 111, 85, 69, 55, 43, 41, 39
2,4,6-Trimethyl-8-nonene	168, 153, 125, 111, 97, 83, 69, 57, 43
2,4,6,8,10-Pentamethyl-3,9-undecadiene	222, 207, 179, 166, 151, 137, 123, 109, 95, 83, 69, 55, 41
2,4,6,8,10-Pentamethyl-1-undecene	224, 210, 168, 153, 141, 125, 111, 97, 83, 69, 57, 43

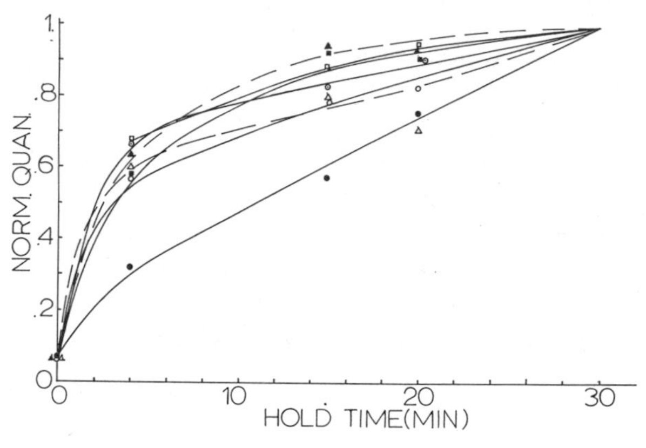

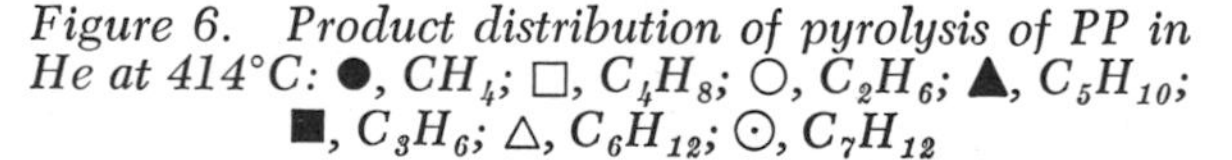

Figure 6. Product distribution of pyrolysis of PP in He at 414°C: ●, CH_4; □, C_4H_8; ○, C_2H_6; ▲, C_5H_{10}; ■, C_3H_6; △, C_6H_{12}; ⊙, C_7H_{12}

Welti (9). The quality of the spectra is uniformly good as exemplified retention times are: CO_2, H_2O, formaldehyde, acetaldehyde, methanol, propylene, isobutylene, acetone, hexene, and other unsaturated aldehydes and ketones. The actual chemical structures of these ketones and of several other products are being determined by mass spectrometry.

Discussion of Results

Pyrolysis of Polypropylene. There have been several published studies of pyrolysis of polypropylene. They are more in discord than in

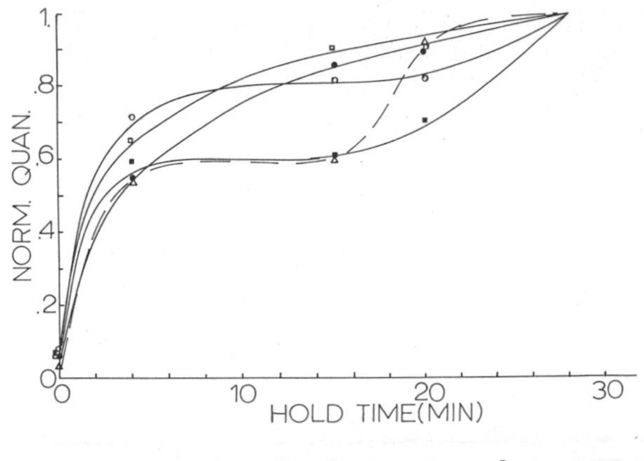

Figure 7. Product distribution of pyrolysis of PP in He at 414°C: ●, C_9H_{18}; □, $C_{12}H_{24}$; ○, $C_{10}H_{20}$; △, $C_{13}H_{24}$; ■, $C_{11}H_{22}$

agreement. Different kinetics, activation energies, and products were reported.

Wall and Straus (*10*) found the rate of volatilization of polypropylene at 375°C to increase rapidly to a maximum at 40% conversion and then decrease rapidly with further heating. Similar behaviors were observed also for polyethylene (*10*). However, branches longer than a methyl group were found to eliminate the maxima in the rate curves even when present in quite low concentration (*11*). On the other hand, Madorsky and Straus (*12*) and this work found the kinetics of pyrolysis to be first order.

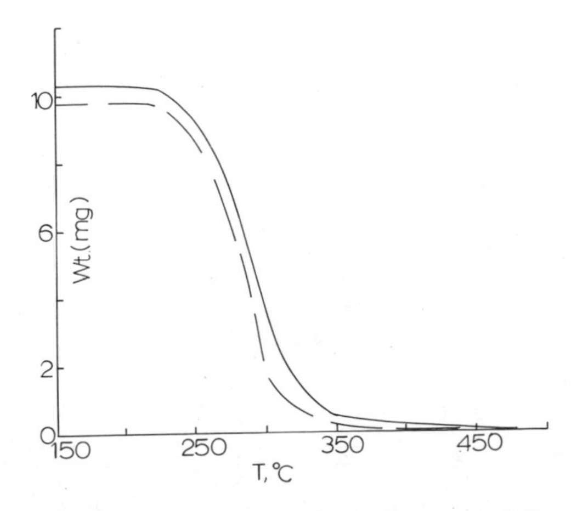

Figure 8. TGA of PP in air: ——, APP; – – –, IPP

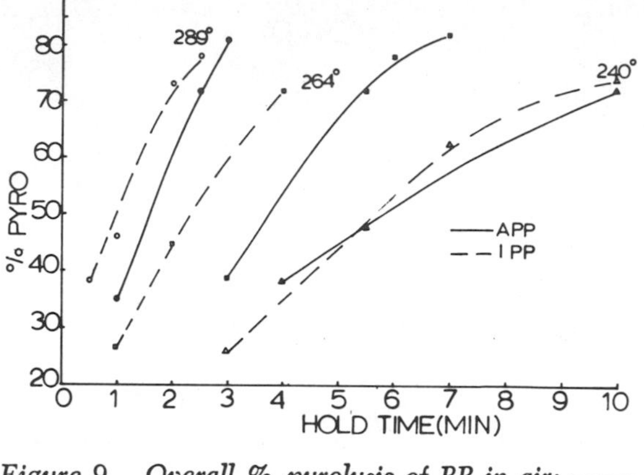

Figure 9. Overall % pyrolysis of PP in air: ——, APP; – – –, IPP

Even though the pyrolysis of polypropylene is mechanistically complicated (vide infra), the kinetics is first order because the rate determining step is the homolysis of the C–C bond describable by a well-defined rate constant. We found the activation energy for pyrolysis to be 51 and 56 kcal mol^{-1}, respectively, for IPP and APP. Other literature values are 60 kcal mol^{-1} measured at 350°–400°C by Wall and Straus (*10*), 58 kcal mol^{-1} in the range 336°–366°C reported by Madorsky and Straus (*12*) and 55 kcal mol^{-1} found by Moissev et al. (*13*) in the temperature range 320°–420°C.

We found the pyrolysis products in our work to be predominantly olefins. A small amount of methane was found at all temperatures. Minor

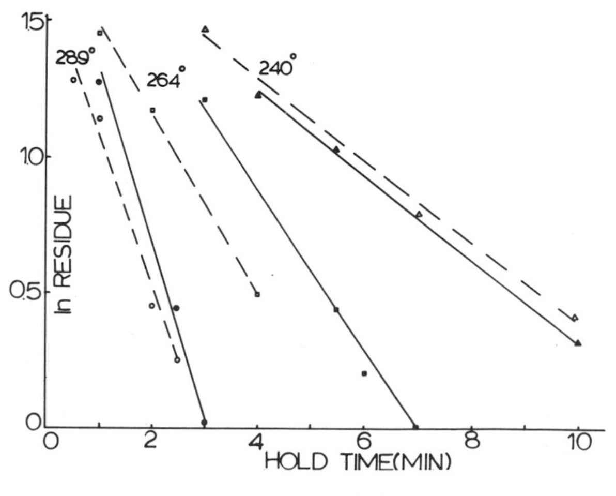

Figure 10. First-order plot of PP pyrolysis in air: ——, APP; – – –, IPP

Table V. Kinetic Results for Oxidative Pyrolysis of Polypropylene

	Rate Constants, Sec^{-1}	
Temp., °C	*APP*	*IPP*
240	2.27×10^{-3}	2.52×10^{-3}
264	5.20×10^{-3}	5.35×10^{-3}
289	1.02×10^{-2}	9.2×10^{-3}
Energy of activation, kcal mol^{-1}	17	15

quantities of ethane and $C_{11}H_{24}$ were observed at low temperatures and isobutane was found in flash pyrolysis. The major products shown in Tables II, III, and IV peaks were: 3, propylene; 4, isobutylene; 5, 2-pentene; 6, methylpentene; and 10, dimethylheptene. Significant amounts of trimethylnonene (13) were also identified. These results are most similar to those of Tsuchiya and Sumi (*14*) execpt for certain products. They found more methane than we did. Their major C_6 product was assigned to be 2-methyl-1-pentene. However, the most important discrepancy is that Tsuchiya and Sumi found pentane to be the major pyrolysis product. The main products were found by Moiseev et al. (*13*) to be propylene, isobutylene, and pentene. At the other end of the spectrum, Bailey and Liotta (*15, 16*) found that at 340°C pyrolysis of polypropylene yielded 80–90% propane and at 380°C, 65% was *n*-pentane.

Madorsky (*17*) studied the pyrolysis product over the widest temperature range from 380°–1200°C. At the low temperature regions propylene, isobutylenc, butane, pentene, pentane, and hexene are the major

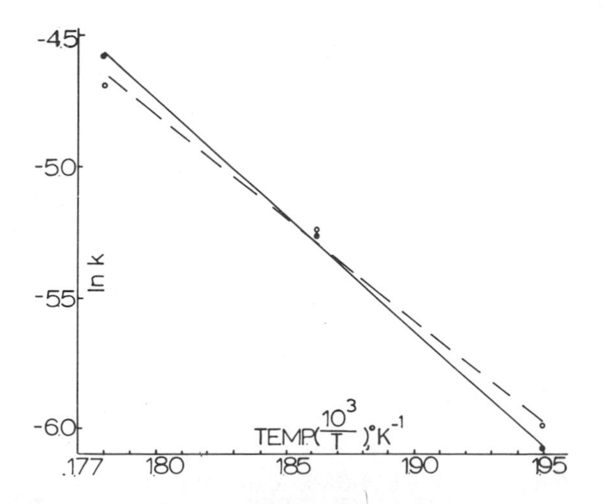

Figure 11. Arrhenius plot of PP pyrolysis in air: ——, APP; – – –, IPP

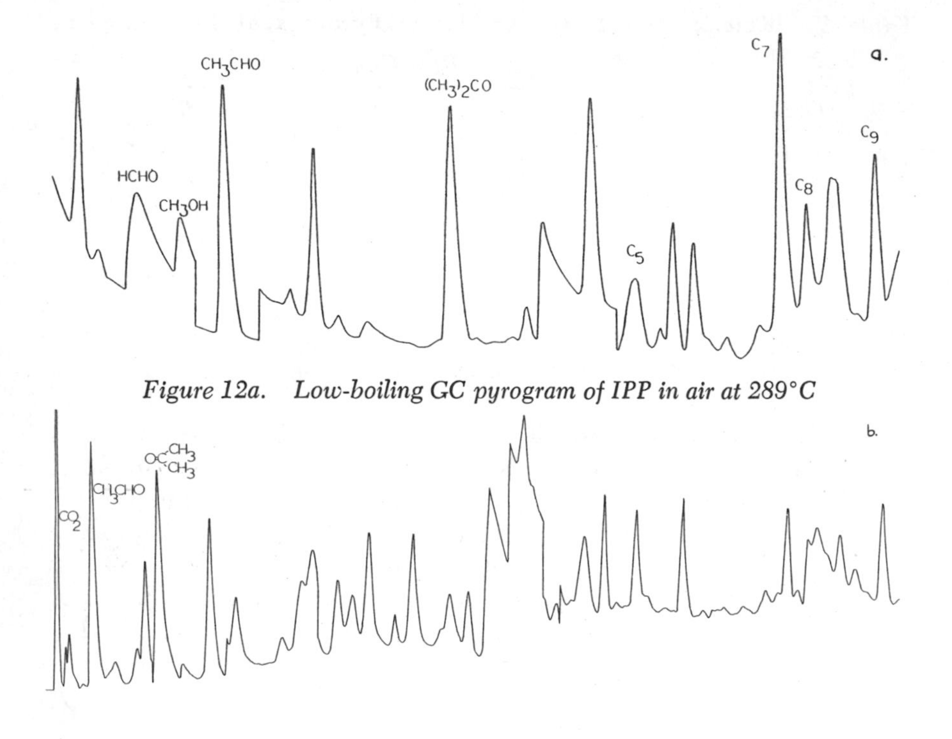

Figure 12a. Low-boiling GC pyrogram of IPP in air at 289°C

Figure 12b. High-boiling GC pyrogram of IPP in air at 289°C

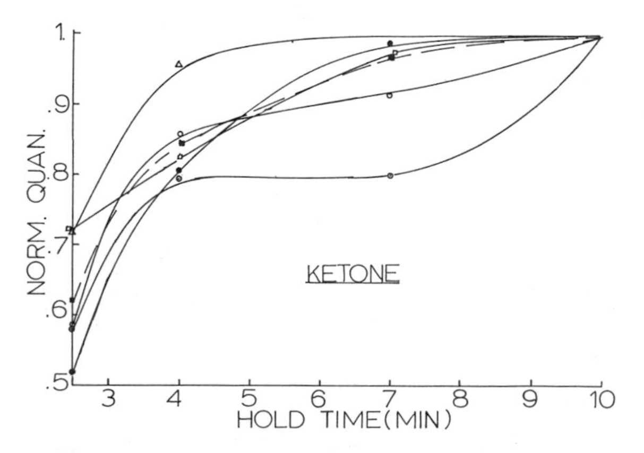

***Figure 13.** Product distribution of pyrolysis of PP in air at 289°C: ○, CH_3CHO; ●, acetone; △, C_5; ⊙, C_7; ■, C_8; □, C_9*

products. However, the amount of pentane found was only 11 mol % as compared with 65% and 31% reported by Bailey (*16*) and Tsuchiya and Sumi (*14*), respectively. Madorsky (*17*) observed that between 400° and 800°C, the major products are produced in nearly the same amounts except for propylene, whose yield increases suddenly suggesting unzipping of the polymer. Finally, APP and IPP seem to be slightly different in their pyrolysis behavior as had been noted also by Bresler et al. (*18*). However, the differences are not too great and their significance needs further verification.

In order to reconcile to some degree the results cited here, the mechanism for pyrolysis of polypropylene needs to be considered. The mechanism commonly accepted is based on those proposed for the gas-phase degradation of simple hydrocarbons (*19, 20*).

CHAIN INITIATION.

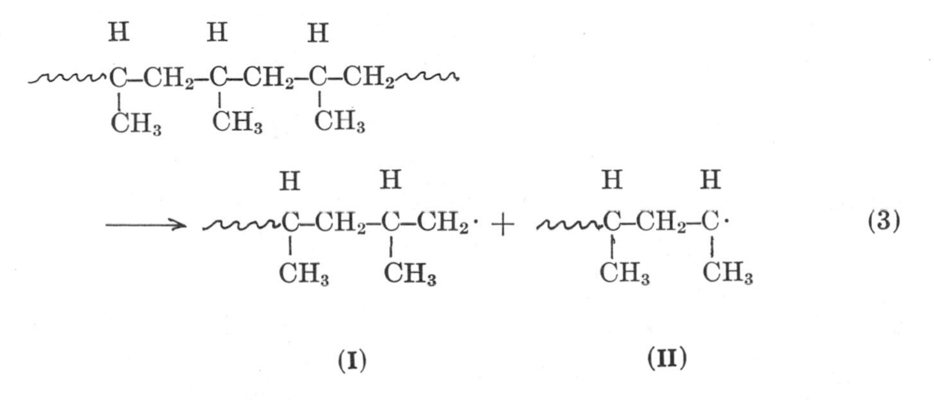

where **I** is a primary aliphatic radical and **II** is a secondary radical. In highly stereoregular and pure polypropylene initiation is the result of thermal homolysis. However, most commercial polymers contain small percentages of impurities and structural irregularities. Chain initiation via scission of weak links cannot be ruled out as a possibility.

UNZIPPING (DEPOLYMERIZATION).

$$\mathbf{I} \text{ or } \mathbf{II} \rightarrow \mathbf{I'} \text{ or } \mathbf{II'} + CH_3\text{–}\overset{H}{C}\text{=}CH_2 \text{ (propylene)} \qquad (4)$$

For polymers with low ceiling temperature, such as poly(α-methylstyrene) and poly(methylmethacrylate), unzipping is the predominant degradative process. However, this is unimportant for pyrolysis of polypropylene at low temperature. Unzipping becomes more important above 800°C but still is not the dominant reaction.

CHAIN TRANSFER.

Intramolecular Chain Transfer. Intramolecular hydrogen transfer (*20*) leads to many of the observed products. In the equations below,

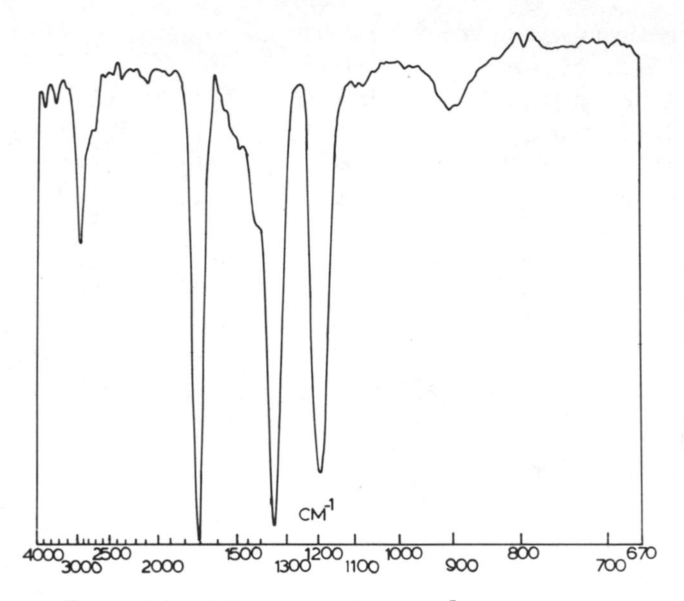

Figure 14. A 6-sec scan of vapor phase ir of acetone

the numbers above the arrow indicate the modes of hydrogen transfer. Not all of the possible products were found in our experiments; the italicized products correspond to those found in significant amounts experimentally.

From the primary radical:

$$\left.\begin{array}{l} \mathbf{I} \xrightarrow{1,2} \mathbf{II} + \textit{isobutylene} \\ \xrightarrow{1,3} \mathbf{I} + \text{4-methyl-2-pentene} \\ \xrightarrow{1,4} \mathbf{II} + \text{2,4-dimethylpentene} \\ \xrightarrow{1,5} \mathbf{I} + \textit{4,6-dimethyl-heptene} \\ \xrightarrow{1,6} \mathbf{II} + \text{2,4,6-trimethylheptene} \\ \xrightarrow{1,7} \mathbf{I} + \textit{4,6,8-trimethyl-nonene} \\ \xrightarrow{1,8} \mathbf{II} + \text{2,4,6,8-tetramethylnonene} \end{array}\right\} \quad (5)$$

From the secondary radical:

$$\left.\begin{aligned} \mathbf{II} &\xrightarrow{1,2} \mathbf{I} + \textit{2-pentene} \\ &\xrightarrow{1,3} \mathbf{II} + \textit{2-methylpentene} \\ &\xrightarrow{1,4} \mathbf{I} + \textit{4-methyl-heptene} \\ &\xrightarrow{1,5} \mathbf{II} + \textit{2,4-dimethylheptene} \\ &\xrightarrow{1,6} \mathbf{I} + \textit{4,6-dimethyl-nonene} \\ &\xrightarrow{1,7} \mathbf{II} + \textit{2,4,6-trimethylnonene} \end{aligned}\right\} \quad (6)$$

Intermolecular Transfer.

$$\sim\!\!\sim C(H)(CH_3)-CH_2\cdot + \sim\!\!\sim C(H)(CH_3)-CH_2-C(H)(CH_3)-CH_2-C(H)(CH_3)\sim\!\!\sim \longrightarrow$$

$$\sim\!\!\sim CH(CH_3)-CH_3 + \sim\!\!\sim C(H)(CH_3)-CH_2-C(CH_3){=}CH_2 + \cdot C(H)(CH_3)-CH_2\sim\!\!\sim \quad (7)$$

This process does not produce volatile products directly.

CHAIN TERMINATION.

Termination is thought to occur either by disproportionation of radi-

$$\sim\!\!\sim C(H)(CH_3)-CH_2\cdot + \sim\!\!\sim C(H)(CH_3)-CH_2-C(H)(CH_3)\cdot \longrightarrow$$

$$\sim\!\!\sim C(CH_3){=}CH_2 + \sim\!\!\sim C(H)(CH_3)-CH_2-CH_2CH_3 \quad (8)$$

cals, or others between **I** and **II** in different combinatory paths, or by combination of radicals.

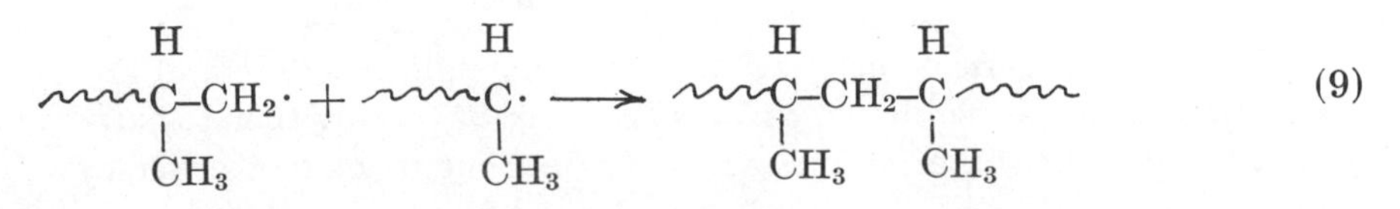

$$\sim\!\!\sim C(H)(CH_3)-CH_2\cdot + \sim\!\!\sim C(H)(CH_3)\cdot \longrightarrow \sim\!\!\sim C(H)(CH_3)-CH_2-C(H)(CH_3)\sim\!\!\sim \quad (9)$$

Our results are consistent with pyrolysis of polypropylene proceeding mainlyby intramolecular Chain Transfers 5 and 6. It accounts for the fact that most of the volatile products are olefins (Table III) and that olefins from C_3–C_{13} are all formed. Furthermore, C_9H_{18} is formed by 1,5-hydrogen transfers of both Radicals **I** and **II**; this should make this the most abundant product (this is, in fact, observed). Also, $C_{12}H_{24}$ are produced by 1,7-hydrogen transfers of both Radicals **I** and **II**, making it a significant pyrolysis product as well.

The obvious way saturated hydrocarbons such as propane, butane, and pentane can be formed in significant quantity would be if chain initiation were primarily from the polymer chain ends followed by hydrogen abstraction. For instance,

$$\sim\!\!\sim\!\!\sim \mathrm{CH(CH_3)}\text{–}\mathrm{CH_2}\text{–}\mathrm{CH(CH_3)}\text{–}\mathrm{CH_2}\text{–}\mathrm{CH(CH_3)}\text{–}\mathrm{CH_3} \rightarrow \mathbf{I} + \mathbf{II} + \mathrm{CH_3}\text{–}\dot{\mathrm{C}}\mathrm{H}\text{–}\mathrm{CH_3} + \cdot\mathrm{CH_2}\text{–}\mathrm{CH(CH_3)}\text{–}\mathrm{CH_3} + \cdot\mathrm{CH(CH_3)}\text{–}\mathrm{CH_2}\text{–}\mathrm{CH(CH_3)}\text{–}\mathrm{CH_3} \quad (10)$$

$$\sim\!\!\sim\!\!\sim \mathrm{CH(CH_3)}\text{–}\mathrm{CH_2}\text{–}\mathrm{CH(CH_3)} \rightarrow \mathbf{I} + \mathbf{II} + \mathrm{CH_3}\text{–}\mathrm{CH_2}\cdot + \cdot\mathrm{CH_2}\text{–}\mathrm{CH_2}\text{–}\mathrm{CH_3} + \cdot\mathrm{CH(CH_3)}\text{–}\mathrm{CH_2}\text{–}\mathrm{CH_2}\text{–}\mathrm{CH_3} \quad (11)$$

Termination involving these radicals could also lead to alkanes and some olefins as well. The reason why initiation at polymer chain ends dominates in some works is not clear, and may be enlightened by pyrolysis studies of polypropylene prepared by initiation with different alkyls such as $Al(Me)_3$, $Al(Et)_3$, or $Al(iBu)_3$.

One major difference between our experimental procedure and those of others is that the pyrolysis products are swept away by the He-carrier gas, whereas most other studies were carried out in closed, evacuated vessels. It seems that secondary reactions are much less likely to occur under the former conditions. Dienes are secondary pyrolysis products—so are some of the alkanes.

Oxidative Pyrolysis. Autoxidation of polypropylene below its melting point has been investigated quite thoroughly with our own laboratory among the most active ones (*21, 22, 23, 24*). This is not the proper place to discuss that topic; a chapter written by one of us dealing with the subject can be found in this volume. It suffices to say that the main chain propagating species is the peroxy radical and the chain branching species is the hydroperoxide.

We are not aware of any significant study of autoxidation of polypropylene at elevated temperatures, i.e. oxidative pyrolysis. On the other hand, much is known about gas-phase oxidation of hydrocarbons at high temperatures and the cool-flame limit (*25, 26, 27, 28*). The reactions are recognized as free radical chain reactions propagated by peroxy radicals and hydroperoxides which was essentially a development of Bäckstrom's scheme for the oxidation of aldehydes (*29*). These mechanisms may be adapted to the oxidative pyrolysis of polypropylene.

CHAIN INITIATION.

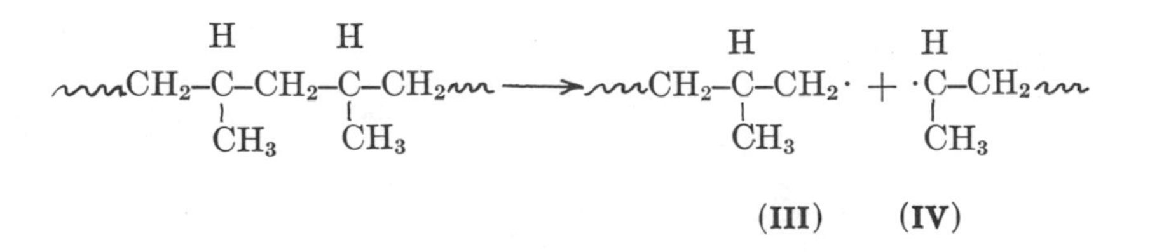

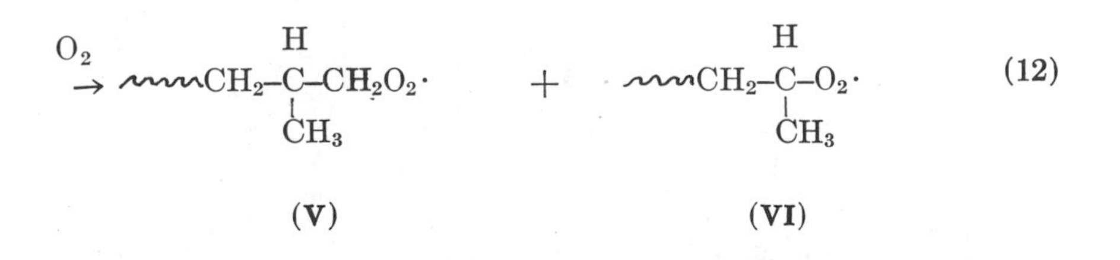

CHAIN PROPAGATION. Radicals V and VI can extract a hydrogen atom intermolecularly to give the corresponding hydroperoxides of transient stability. Loss of ·OH leads to the respective Alkoxy Radicals VII and VIII. Scission of C–C bonds accompanied by H or CH_3 transfer yields the products shown below where the numbers above the arrow indicate the particular C–C bond severed. As before, the italicized products correspond to those found in significant amounts experimentally.

$$\sim C(H)(CH_3)-\left(CH_2-C(H)(CH_3)\right)_2-CH_2O\cdot \quad (VII)$$

$\xrightarrow{1,2}$	IV	+	*formaldehyde*	(13)
$\xrightarrow{2,3}$	III	+	propan-1-al	
$\xrightarrow{3,4}$	IV	+	2-methyl-propan-1-al	
$\xrightarrow{4,5}$	III	+	2-methyl-pentan-1-al	
$\xrightarrow{5,6}$	IV	+	2,4-dimethyl-pentan-1-al	

$$\sim C(H)(CH_3)-\left(CH_2-C(H)(CH_3)\right)_2-CH_2-C(H)(CH_3)-O\cdot \quad (VIII)$$

$\xrightarrow{1,2}$	III	+	*acetaldehyde*	(14)
$\xrightarrow{2,3}$	IV	+	*acetone* + propan-1-al	
$\xrightarrow{3,4}$	III	+	*pentan-2-one* + 3 methyl-butan-1-al	
$\xrightarrow{4,5}$	IV	+	4-methyl-pentan-2-one + 3-methyl-pentan-1-al	
$\xrightarrow{5,6}$	III	+	4-methyl-hexan-2-one + 3,5-dimethylhexan-1-al	

In addition, intramolecular hydrogen abstraction, followed by loss of OH, C–C scission, H or CH_3 transfer, and H abstraction (*30–36*) can lead to the same products of Reactions 13 and 14. This is illustrated for abstraction of γ–H by Radical **VI.**

$$\text{(VI)} \rightarrow \sim CH_2-\dot{C}(CH_3)-CH_2-C(H)(CH_3)-OOH \xrightarrow{-OH}$$

$$\sim CH_2-\dot{C}(CH_3)-CH_2-C(H)(CH_3)-O\cdot \rightarrow \text{III} + \cdot C(H)(CH_3)-CH_2-C(=O)CH_3 \qquad (15)$$

$$\xrightarrow{+H} \text{pentan-2-one}$$

The fact that only formaldehyde, but not its higher analogs, was observed is understandable because of the energetically favorable process of forming the stable formaldehyde and a tertiary alkyl radical from **VII.** In Reaction 14, CH_3 migration must be much less facile than H transfer since the former produces aldehydes which were not found. Finally, CO_2 is probably the oxidation product of formaldehyde.

It should be pointed out that whereas the terminal radicals are responsible for the volatile products observed, similar reactions occur for the backbone radicals. The backbone tertiary or secondary peroxy radicals can isomerize by H transfer from β, γ, or σ carbons intramolecularly. Subsequent reactions are: (a) loss of HO· produces O-heterocycles such as oxinan, furan and pyran (*30, 31, 32*); (b) loss of HO· with C–C bond scission, group migration, and H abstraction forming carbonyl compounds with rearranged polymer backbone (*33, 34, 35*); (c) loss of HO· with C–C bond scission and H transfer leading to terminal carbonyl and olefinic functionalities; and (d) loss of HO_2· producing internal double bonds (*36*). The backbone peroxy radicals can isomerize also via group transfer (*37, 38*) followed by O–O bond scission to give internal and terminal carbonyl compounds as well as low molecular weight and high molecular weight alkoxy radicals.

CHAIN TERMINATION. This occurs undoubtedly by a bimolecular process such as

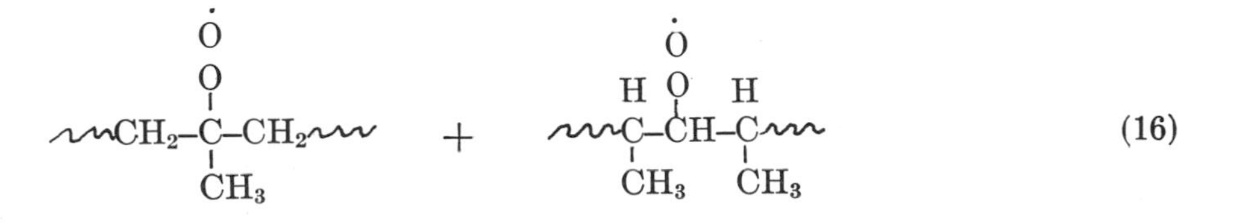

$$\rightarrow \sim CH_2\text{–}C(OH)(CH_3)\text{–}CH_2\sim \; + \; \sim C(H)(CH_3)\text{–}C(=O)\text{–}C(H)(CH_3)\sim \; + \; O_2$$

Under our experimental conditions of oxidative pyrolysis (small sample under a rapidly flowing stream of air) all the alkyl radicals react immediately with O_2 before any other chemical processes, such as isomerization and dissociation, can occur. For bulk polymers this would not be true. Then Reactions 5, 6, 13, and 14 of Radicals **I, II, VII,** and **VIII** are all important, their relative importance being governed by the local oxygen concentrations. The volatile products would be expected to be unsaturated carbonyl compounds.

Conclusions

A novel interfaced pyrolysis gas chromatographic peak identification system has been used to study the pyrolysis and oxidative pyrolysis of polypropylene. The major pyrolysis products are olefins: polypropylene, isobutylene, methylpentene, pentene, dimethylheptene, and trimethylnonene. A mechanism of intramolecular chain transfer processes can account for the formation of these compounds very well. The major oxidative pyrolysis products are ketones which are formed by rearrangement of peroxy radicals.

The pyrolysis products identified in this work are quite different from those observed by other investigators. The principal difference is that in our experiments the volatile products are flushed away rapidly by the carrier gas. This condition perhaps better simulates the state at the surface of a burning polymer. In any event, it is important in any study of the effect of flame retardants that identical techniques and experimental conditions be used in the study of polymer with and without the additives.

Acknowledgment

This work is supported partly by a grant from the National Bureau of Standards.

Literature Cited

1. Volans, P., Conference on Flame Resistance in Polymers, London, 1966.
2. Burge, S. J., Tipper, C. F. H., *Combust. Flame* (1969) **13**, 495.
3. Stuetz, D. E., Barnes, B. P., DiEdwardo, A. H., Zitomer, F., Polymer Conference Series, University of Utah, June 15–20 (1970).
4. Wang, D. S. T., Ph.D. Thesis, University of Massachusetts, 1974.
5. Uden, P. C., Henderson, D. E., Lloyd, R. J., *J. Chromatogr.* (1976) **126**, 225.
6. Liebman, S. A., Ahlstrom, D. H., Nauman, C. D., Averitt, R., Walker, J. L., Levy, E. J., *Anal. Chem.* (1973) **45**, 1360.
7. Lloyd, R. J., Henderson, D. E., Uden, P. C., *Anal. Chem.* (1976) **48**, 1645.
8. Dietz, W. A., *J. Gas Chromatogr.* (1961) 68.
9. Welti, D., Infrared Vapor Spectra, Heyden, Sadtler (1970).
10. Wall, L. A., Straus, S., *J. Polym. Sci.* (1960) **44**, 313.
11. Wall, L. A., "High Tempearature Resistance and Thermal Degradation of Polymers," S.C.I. Monograph No. 13, p. 145, Macmillan Co., N.Y., 1961.
12. Madorsky, S. L., Straus, S., *J. Res. Nat. Bur. Stand.* (1954) **53**, 361.
13. Moiseev, V. D., Neiman, M. B., Krilkova, A. I., *Vysokomol. Soedin.* (1959) **1**, 1552.
14. Tsuchiya, Y., Sumi, K., *J. Polym. Sci., Part A-1* (1969) **7**, 1599.
15. Bailey, W. J., *Proc. Symp. Polypropylene Fibers,* Birmingham, Alabama, 1964, p. 121.
16. Bailey, W. J., Liotta, C., *Polym. Prepr., Am. Chem. Soc., Div. Polym. Chem.* (1964) **5**, 333.
17. Madorsky, S. L., "Thermal Degradation of Organic Polymers," Interscience, N.Y., 1964.
18. Bresler, S. E., Os'minskaya, A. T., Popov, A. G., *Vysokomolekul. Soedin.* (1960) **2**, 930.
19. Rice, F. O., Rice, K. K., "The Aliphatic Free Radicals," Johns Hopkins Press, Baltimore, 1935.

20. Kossiakoff, A., Rice, F. O., *J. Am. Chem. Soc.* (1943) **65**, 590.
21. Chien, J. C. W., Vandenberg, E. J., Jabloner, H., *J. Polymer Sci., A-1* (1968) **6**, 381.
22. Chien, J. C. W., Jabloner, H., *J. Polym. Sci., Part A-1* (1968) **6**, 393.
23. Chien, J. C. W., Boss, C. R., *J. Polym. Sci., Part A-1* (1967) **5**, 3071.
24. Chien, J. C. W., Boss, C. R., *J. Polym. Sci., Part A-1* (1967) **5**, 1683.
25. Ubbelohde, A. R., *Proc. Roy. Soc., Part A* (1935) **152**, 354.
26. Hinshelwood, C. N., *Disc. Faraday Soc.* (1947) **2**, 117.
27. Walsh, A. D., *Trans. Faraday Soc.* (1946) **42**, 269.
28. Norrish, R. G. W., *Disc. Faraday Soc.* (1951) **10**, 269.
29. Backstrom, H. L. J., *J. Am. Chem. Soc.* (1927) **49**, 1460.
30. Bailey, H. C., Norrish, R. G. W., *Proc. Roy. Soc., Part A* (1952) **212**, 311.
31. Kyryacos, G., Menapace, H. R., Boord, C. E., *Anal. Chem.* (1959) **31**, 222.
32. Chung, Y. H., Sandler, S., *Combustion and Flame* (1962) **6**, 295.
33. Trimm, D. L., Cullis, C. F., *J. Chem. Soc.* (1963) 1430.
34. Cullis, C. F., Hardy, F. R. F., Turner, D. W., *Proc. Roy. Soc., Part A* (1959) **251**, 265.
35. Raley, J. H., Rust, F. F., Vaughan, W. E., *J. Am. Chem. Soc.* (1948) **70**, 88.
36. Zeelenberg, A. P., Bickel, A. F., *J. Chem. Soc.* (1961) 4014.
37. Zeelenberg, A. P., *Rec. Trav. Chim.* (1962) **81**, 720.
38. Semenov, N. N., "Some Problems of Chemical Kinetics and Reactivity," London, Vol. 1, p. 99, Pergamon Press, 1958.

RECEIVED May 12, 1977.

16

Kinetic Studies on Degradation in Polyimide Precursor Resins

DAVID E. KRANBUEHL, JEAN TAKEUCHI, DEBORAH GIBBS, and GEORGE TSAHAKIS

Department of Chemistry, College of William and Mary, Williamsburg, VA 23185

The stability of a series of BTDA–DABP and BTDA–MDA polyimide precursor resins in dimethylacetamide is investigated by measuring the number average molecular weight ($\overline{M}_n$) as a function of time. The dependence of the degradation rate on the chemical nature of the monomer unit and the geometric structure of the amide linkage is discussed in terms of a previously proposed degradation mechanism. The effect of the presence of water, the concentration of the resin, and the temperature are investigated also. The degradation rate is doubled by adding 0.5 vol % H_2O to the anhydrous resin solution. Increasing the concentration of the resin solution, the presence of an electron donating group between the phthalic acid groups and a para diamine amide linkage increase the stability.

Polyimides with aromatic groups in their backbone have generated considerable interest in recent years because of their thermal and thermooxidative stability. They can be used in air between 200–300°C and for short periods at 400–500°C (*1, 2*). Therefore, they are being considered as matrix resins for high performance composites in the National Aeronautics and Space Administration space shuttle program. However, the high glass transition temperatures of polyimides make processing difficult. Thus, the acid and amine monomers often are polymerized in solution to form a poly(amic-acid) resin. This polyimide precursor resin solution may be applied then to a fiber backing and cured at 200–300°C to form a thermally stable polyimide composite.

The effect of chemical structure on the thermooxidative and thermomechanical properties of the cured polyimide has been studied (*1, 2, 3, 4, 5*). However, limited kinetic information is available on the storage

0-8412-0381-4/78/33-169-198$05.00/1

stability of the poly(amic-acid) resin. Viscosity measurements show that poly(amic-acid) solutions are sensitive to moisture, temperature, and concentration (*1, 7, 8, 9*). In this work, the degradation rate of several poly(amic-acid) solutions is determined by measuring the rate of change of the number average molecular weight ($\overline{M}_n$) as a function of the chemical and geometric structure of the diamine as well as of temperature, moisture content, and polymer concentration.

Experimental

A series of 15 wt % isomeric polyimide precursor resin solutions was prepared by the National Aeronautics and Space Administration Langley Research Center. The polymerization procedure has been described previously (*2, 3*). The poly(amic-acid) resin solutions were made from 3,3′,4,4′-benzophenonetetracarboxylic acid dianhydride (BTDA), 4,4′-oxydiphthalic anhydride (ODPA), diaminobenzophenone (DABP), and methylenedianiline (MDA) monomers.

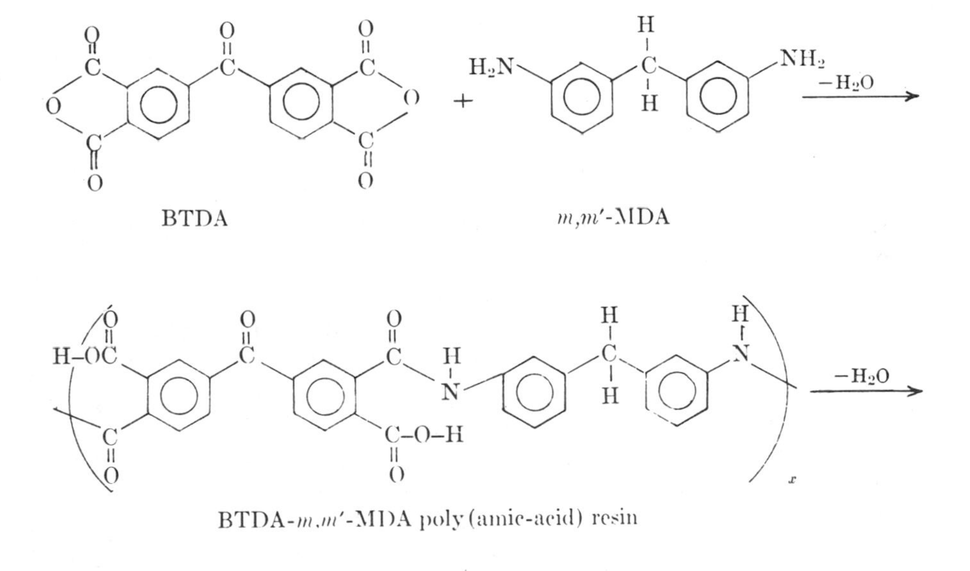

BTDA-*m,m′*-MDA poly(amic-acid) resin

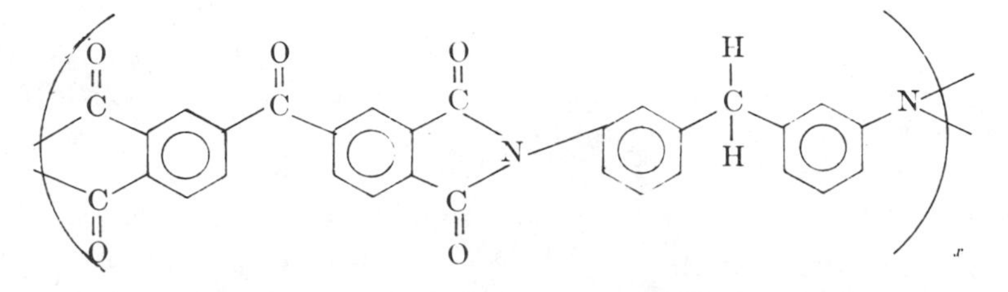

cured resin—BTDA-*m,m′*-MDA-polyimide

The initial number average molecular weights ($\overline{M}_n$) for the poly(amic-acids) were between 10,000 and 20,000, and no attempt was made to optimize them. Dimethylacetamide solvent was distilled from calcium hydride under vacuum and stored under nitrogen. The resin solutions were stored also under nitrogen. The water content of the solvent was measured with a Porapak Q column in a Hewlett Packard 5750 gas chromatograph. Unless otherwise stated, the water concentration in the solvent remained below 0.05 vol %. $\overline{M}_n$ was determined from osmotic pressure measurements on a Hewlett Packard–Mechrolab 502 osmometer using ArRo Lab gel cellophane 600D membranes.

Results and Discussion

The change in $\overline{M}_n$ caused by degradation of the resin solution was observed as a function of time for 6- to 52-day periods or until M_n fell below 5000. Assuming degradation occurs by random scission and that the initial fraction of functional groups reacted, p_o, remains near unity:

$$\frac{1}{\overline{M}_n(t)} = \frac{1}{\overline{M}_n(o)} + \frac{kp_ot}{M_o}$$

where M_o is the molecular weight of the monomer unit. The rate of degradation was characterized by the constant kp_o/M_o which was determined from a least squares fit of $1/\overline{M}_n(t)$ vs. time. The maximum deviation from the least squares fit was 6% for $\overline{M}_n > 10{,}000$ and 8% for $10{,}000 > \overline{M}_n > 5{,}000$. The uncertainty in k, based on measurements of several different polymerization batches, is estimated to be 25%.

Table I lists values of the degradation rate expressed as kp_o/M_o. The first part of Table I consists of values of kp_o/M_o for a series of BTDA poly(amic-acid) solutions that differ in their chemical and geometric structure. The smaller values of kp_o/M_o for BTDA poly(amic-acids) made from DABP amines show that the BTDA–DABP poly(amic-acids) are more stable than their MDA counterparts. For both the DABP and MDA polymers, the para amide linkage is more stable than the meta or ortho structures. Methyl-substituted anilines are two pK_a units greater than methyl carbonyl-substituted anilines (*10*). However, based on the basicity of substituted anilines (*11*) and electron charge density calculations (*2*), a small increase in methylene dianiline basicity is expected in going from the ortho to the para isomer. Thus, the small variation in the BTDA–DABP and the BTDA–MDA degradation rates suggests that the rate of degradation does not depend on the basicity of the diamine. The data do suggest, however, that the degradation rate depends on the geometric linkage at the amide bond. Steric effects are an important rate determining factor in hydrolysis of the carboxamide group (*10*).

The ODPA-*m,m'*-MDA was one of the most stable poly(amic-acids) studied. It is four times more stable than BTDA-*m,m'*-MDA. Because of the electron donating character of the methoxy group that destabilizes the carboxylic anion, methoxy-substituted benzoic acids are weaker acids than acetyl-substituted benzoic acids. The ODPA data suggest that the degradation rate is related to the acidity of the phthalamic acid group and that the presence of an electron-donating bridge group in the diphthalic monomer increases the stability of the poly(amic-acid) resin solution.

Data on variations in degradation rates with temperature are reported in Table I. The values of the activation energy for degradation all fall within 60–125 kJ/mol. These are values normally observed in amide hydrolysis (*12*). For the three compounds studied, the activation energy increases as the room temperature stability of the resin solution increases.

Table I shows also that increasing the weight percent of polymer in the resin solution increases the stability—a factor that was observed previously in viscosity measurements (*8*).

The effect of the presence of water in the solvent on the stability of the resin solution is listed in Table I. The degradation rate for BTDA-*p,p'*-DABP increases by a factor of three in the presence of 1.0 vol % water. The BTDA-*m,m'*-DABP data indicate that the degradation rate kp_o/M_o depends linearly on the water concentration. The slope of the three-point line is 7.5×10^{-6}. The intercept kp_o/M_o(anhydrous) is $.031 \times 10^{-4}$ days^{-1} in agreement with the nearly anhydrous $kp_o/M_o(H_2O < .05\%)$ value of $.032 \times 10^{-4}$ days^{-1}.

The hydrolysis rate of phthalamic acid is 10^5 times faster than that of benzamide. Formation of a phthalic anhydride intermediate has been demonstrated in C^{13}- and O^{18}-labeling experiments (*13*). These results and the degradation data are consistent with the previously proposed mechanism of intramolecular displacement of the protonated amide moiety by the carboxylate anion (*7, 13, 14*).

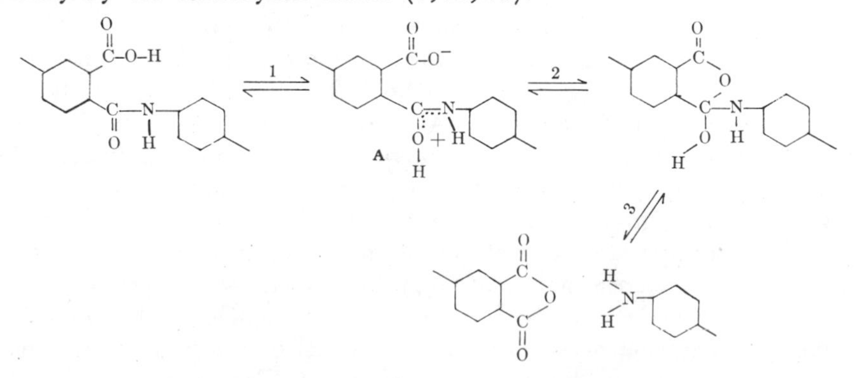

Table I.

Effect of	*Polymer*
Structure	BTDA-*m,m'*-DABP
	BTDA-*p,p'*-DABP
	BTDA-*o,p'*-MDA
	BTDA-*m,m'*-MDA
	BTDA-*p,p'*-MDA
	ODPA-*m,m'*-MDA
Temperature	BTDA-*p,p'*-MDA
	BTDA-*p,p'*-MDA
	BTDA-*p,p'*DABP
	BTDA-*p,p'*DABP
	BTDA-*p,p'*DABP
	BTDA-*m,m'*-MDA
	BTDA-*m,m'*-MDA
	BTDA-*m,m'*-MDA
Concentration	BTDA-*m,m'*-MDA
	BTDA-*m,m'*-MDA
	BTDA-*m,m'*-MDA
Water	BTDA-*m,m'*-DABP
	BTDA-*m,m'*-DABP
	BTDA-*m,m'*-DABP
	BTDA-*m,m'*-DABP
	BTDA-*p,p'*-DABP
	BTDA-*p,p'*-DABP

This mechanism is supported by the dependence of the degradation rate on the chemical and geometric nature of the monomer unit. Electron-donating bridge groups in the diphthalic acid monomer would decrease the tendency of the free carboxylic acid group to ionize. Thus, equilibrium (Step 1) is shifted to the left and the stability of the poly(maic-acid) increased, as is observed in ODPA-*m,m'*-MDA. The formation of a tetrahedral intermediate (**A**) is consistent with a dependence of kp_o/M_o on the steric–geometric structure of the amide linkage. The basicity of the diamine would have little effect on stability of the protonated carboxylic group in intermediate (**A**). The basis for the polymer and water concentration dependence is not as clear. The polymer concentration dependence of kp_o/M_o may be caused by a decrease in the probability of the ortho carboxylic group in intermediate (**A**) to ionize at higher polymer concentrations. Attack by water on the protonated amide

Degradation Rate

Vol % H_2O	*Temp*	*Wt % Polymer*	$\frac{kp_o}{M_o} \times 10^4$ *Days*
$< .05$	27	0.5	.032
$< .05$	27	0.5	.010
$< .05$	27	0.5	.057
$< .05$	27	0.5	.044
$< .05$	27	0.5	.021
$< .05$	27	0.5	.011
$< .05$	27	0.5	.021
$< .05$	40	0.5	.102
$E_A = 94$ kJ/mol			
$< .05$	27	0.5	.010
$< .05$	37	0.5	.085
$< .05$	50	0.5	.20
$E_A = 107$ kJ/mol			
$< .05$	27	0.5	.044
$< .05$	35	0.5	.092
$< .05$	50	0.5	.24
$E_A = 60$ kJ/mol			
$< .05$	27	0.5	.044
$< .05$	27	7.5	.028
$< .05$	27	15.0	.021
< 0.05	27	0.5	.032
0.10	27	0.5	.038
0.30	27	0.5	.053
0.50	27	0.5	.068
0.05	27	0.5	.010
1.00	27	0.5	.032

(**A**) may compete with the proposed intramolecular degradation process. The water and polymer concentration dependence is also partially caused by hydrolysis of phthalic anhydride, making it less likely to recombine in Step 3, and by transamidation reactions involving trace amounts of secondary amines released by dimethylacetamide (*1, 7*).

Conclusions

When stored under anhydrous conditions at room temperature, the time for $\overline{M}_n$ to decrease to one-half the initial value varied from 14 to 90 days. In the presence of a small amount of water, the degradation rate depended linearly on the water concentration. The presence of 1.0 vol % H_2O in dimethylacetamide increased the rate of degradation threefold. Increasing the concentration of the resin increased the room temperature stability of the resin solution. The poly(amic-acids) were most stable when the geometric structure at the amide linkage was para and when the diphthalic acid contained an electron-donating bridge group.

Literature Cited

1. Scroog, C. E., *Macromol. Rev.* (1976) **11,** 161–208.
2. Bell, V., Stump, B., Gager, H., *J. Polym. Sci., A1* (1976) **14,** 2275–2292.
3. Bell, V. L., *Org. Coatings Plastics Chem. Prepr.* (1973) **33,** 153.
4. Dine–Hart, R. A., Wright, W. W., *Makromol. Chem.* (1972) **153,** 237.
5. Critchley, J., White, M., *J. Polym. Sci., A1* (1972) **10,** 1809.
6. Gillham, J. K., Gillham, H. C., *J. Polym. Sci., B* (1968) 821–825.
7. Frost, L. W., Kesse, I., *J. Appl. Polym. Sci.* (1964) **8,** 1039.
8. Sroog, C. E., Endrey, A. L., Abramo, S. V., Berr, C. E., Edwards, W. M., Oliver, K. L., *J. Polym. Sci., A3* (1965) 1373–1390.
9. Bower, G. M., Frost, L. W., *J. Polym. Sci., A1* (1963) 3135–3150.
10. Patai, S., "The Chemistry of the Amino Group," Interscience, London, 1968.
11. Young, P., McairN, H., *Anal. Chem.* (1975) **47,** 756–759.
12. Meloche, I., Laidler, K., *J. Am. Chem. Soc.* (1951) **73,** 1712–1714.
13. Bender, M. L., Chow, Y. L., Chloupek, J. F., *J. Am. Chem. Soc.* (1958) **80,** 5380.
14. Zabicky, J., "The Chemistry of Amides," Interscience ,London, 1970.

RECEIVED May 12, 1977. This work was supported by grant NASA–NSG 1095 from the National Aeronautics and Space Administration.

17

The Effects of Some Structural Variations on the Biodegradability of Step-Growth Polymers

S. J. HUANG, M. BITRITTO, K. W. LEONG, J. PAVLISKO, M. ROBY, and J. R. KNOX

Department of Chemistry, Biological Science Group, and Institute of Materials Science, University of Connecticut, Storrs, CT 06268

Many new step-growth polymers containing one or more types of linkages such as amide, ester, urea, and urethanes were biodegradable by fungi and enzymes. Suitable substituents, combinations of hydrophilic and hydrophobic segments and long repeating units, contribute to the increase of biodegradability of the polymers. The presence of hydrolyzable linkages and conformational flexibility are some of the important requirements for biodegradability.

Activity in the study of biodegradable polymers has been increasing in recent years (*1–4*). Since most of the currently available polymers have not been found to be biodegradable, efforts have been directed toward the syntheses of new polymers that are biodegradable. In this chapter, our recent results and those of related reports by other researchers are discussed in terms of the effects of structural variations on the biodegradabilities of step-growth polymers.

Concerns with how to prevent or retard attack on polymer products by bacteria, fungi, insects, rodents, and other animals provided the early incentive for the study of the biodegradations of synthetic polymers. In recent years, the disposal of the mostly bioresistant polymer products now in use has become increasingly difficult. The disposal of biodegradable polymers, on the other hand, is less difficult. This provides the current incentive for the study of biodegradable polymers. Moreover, biodegradable polymers are useful for the preparation of surgical implants, sutures, controlled-release drugs, fertilizers, fungicides, and agricultural mulch. Successful uses of biodegradable polymers in these newly developing areas will result in more effective utilization of resources.

0-8412-0381-4/78/33-169-205$05.00/1

Attempts to prepare biodegradable addition polymers have not been very successful. Incorporation of photodegradable units into polymer chains and mixing photosensitizers into polymer composites to prepare photodegradable materials so the products of photolyses might become biodegradable were the alternative approaches for making addition polymers biodegradable (*2*).

Most of the recently discovered biodegradable polymers contain hydrolyzable linkages such as amide, ester, urea, and urethane along the polymer chains. These polymers are prepared generally by step-growth polymerizations. Increasing the hydrophilicity of the polymer, introducing substituents for better interactions between the polymer chains and enzymes, and increasing the polymer chain flexibility by copolymerization of different monomers have been some of the approaches taken by researchers with various degrees of success.

Testing Methods

Although soil burial testing affords a way to test samples for breakdown close to actual conditions of waste disposal, it lacks reproducibility because of the difficulties in controlling climatic factors and the populations of various biological systems that are involved. For more reproducible results, degradations by cultured fungi and bacteria as well as degradations by purified enzymes are used.

A large part of the biodegradations of polymers is studied by using the polymers as the carbon and/or the nitrogen sources for the growth of microorganisms. Fungi are used more frequently than bacteria. The degrees of degradation of the polymer samples are determined by studying (*a*) the evidence of colony growth (the ASTM method) (*3–6*), (*b*) the production of carbon dioxide (*7, 8*), (*c*) oxygen consumption (*9*), (*d*) the increase in cell count or cell mass (*9*), (*e*) product formation analysis (*9*), (*f*) the changes in the polymers' physical properties such as molecular weight, solution viscosity, tensile strength, etc. (*10*), (*g*) the weight loss of solid samples (*10, 11*), and (*h*) visible destruction of the samples (*3, 4*).

Degradation by enzymes is carried out generally by incubation of the samples with buffered enzyme solutions with suitable blanks to correct for buffer degradation and contamination proteins and/or degradation of the enzymes (*11*). The degrees of degradation of samples can be followed by methods (*e–h*) mentioned above.

Since degradation by microorganisms might be the result of multistep reactions catalyzed by enzymes and reactions not involving enzymes, they are very complex and the interpretation of data in a quantitative manner is often very difficult. On the other hand, microorganisms can often utilize wide ranges of nutrients of different structures so positive results for biodegradation by microorganisms might be obtained without too much difficulty. Degradations by enzymes are comparatively simpler and there is a better chance for quantitative treatment of data. However, the high degrees of substrate specificity associated generally with enzymes make the selection of the right enzyme for degradation rather difficult. We used both methods in our study (*9*) and we found that fungi growth was a good screening test, whereas enzymes are found to be more effective degradation agents since relatively high concentrations of enzyme can be used.

The ultimate but very time consuming approach of carrying out degradation by microorganisms, isolating the enzyme that is responsible for the degradation, and then studying the details of degradation by purified enzyme has been reported by the research groups of Suzuki (*12, 13*) and Okada (*14*).

Materials

Our step-growth polymers were synthesized and characterized by standard methods. Details can be found in the references cited in the following sections.

Results and Discussion

Polyesters. In 1968, Darby and Kaplan reported that polyethylene adipate, mol wt 2,390, poly(trimethylene adipate), mol wt 5,240, and poly(tetramethylene adipate), mol wt 1,950, supported intense growth of fungi when tested with seven microorganisms: *Aspergillus niger, A. flavus, A. versicolor, Chaetomium globosum, Penicillium funiculosum, Pullaria pullulans,* and *Trichoderma* (*15*). Potts, et al. studied several polyesters and found that aliphatic polyesters derived from ε-caprolactone, succinic acid, and adipic acid supported heavy growth of fungi when tested with *A. niger, A. flavus ,P. funiculosum,* and *C. globosum* according to the ASTM method (*3, 4*). The unsaturated poly(hexamethylene fumarate) supported only light growth and the polyesters derived from aromatic diacids did not support any growth. In a study using a fungus of *Penicillium* sp. strain 26–1 capable of utilizing high molecular weight polycaprolactone, Tokiwa, Ando, and Suzuki found that among a series of polyethylene esters prepared from ethylene glycol and alkane diacids, polyethylene sebacate has the highest biodegradability. They found also that the polyesters derived from aromatic diacids were inert (*9*). These results suggest that replacements of the relatively flexible alkylene chain with more rigid olefin or aromatic rings prevent or retard degradation.

Fields, Rodriguez (*10*), and Potts et al. (*34*) reported that polyesters of low molecular weight supported better fungal than polymers of the same structures but having higher molecular weight.

To study the effect of changing the hydrophilicity of the polymer on its biodegradability we studied the poly(alkene D-tartrates) prepared from C_2–C_{12} diols and D-tartaric acid (*16*). The relative abilities of these polyesters to support the growth of *A. niger* are shown in Table I. The polyesters derived from medium size diols (C_6 and C_6) are more degradable than those derived from smaller and larger diols. It is interesting to note that although polyethylene tartrate is water-soluble and noncrystalline, its ability to support *A. niger*'s growth is less than that of the partially crystalline, more hydrophobic and water-insoluble polyesters derived from 1,6-hexanediol and 1,8-octanediol. All polyesters tested were low molecular weight materials with $\overline{M}_n$ between 1,200–1,500. A balance of hydrophilicity and hydrophobicity appears to give the best result. Changing the hydrophilicity of poly(dodecamethylene tartrate) by acetylation or phosphoration decreased its ability to support *A. niger* growth.

Table I. Extent of Growth of *A. Niger* on Poly(alkylene D-tartrates) after 14 Days at 37°C

$$-\!\!\left[OC\underset{|}{\overset{}{C}}H\text{—}\underset{|}{C}HCOO(CH_2)_nO \right]\!\!-$$
$$\quad OH \quad OH$$

n	*Extent of Growth*[a]
2	1
4	1
6	4
8	4
10	3
12	3
12 acetate[b]	1
12 phosphate[b]	1

[a] ASTM rating: 4–60 to 100% surface covered; 3–30 to 60% surface covered; 2–10 to 30% surface covered; 1 < 10% surface covered; and 0 = no visible growth.
[b] 50% of available hydroxy groups reacted.

Studies on enzyme degradations of polyesters provided some insight into the mechanism of the biodegradation. From the *Penicillium* sp. strain fungus that degraded polycaprolactone and other polyesters, Tokiwa and Suzuki isolated the enzyme and found that it splits endogenous ester bonds of the polyesters (*12, 13*). They found also that the enzyme degraded polycaprolactone and polypropiolactone but not poly-D,L-hydroxybutyrate (the D-isomer of which is known as one of the bacterial and algal storage materials). We found that the viscosity molecular weight of a sample of polycaprolactone was reduced from 13,000 to 10,000 by exposure to a buffered solution of acid protease from *Rhizopus chinensis* for 10 days (*17*). Copolymers derived from phenyllactic acid and lactic acid were hydrolyzed by chymotrypsin with the rate of hydrolysis changing with the phenyllactic acid residue content in the polymer (*18*). It seems that some of the fungal degradations of polyesters proceed by enzyme-catalyzed hydrolysis of the ester linkages with certain degrees of substrate specificity.

Polyamides. Although high molecular weight polyamides such as nylon-6, nylon-6,6, and nylon-12 resisted microbial (*3, 4, 19*) and enzyme attack (*17*), low molecular weight cyclic and linear oligomers of ϵ-aminocaproic acid were utilized by certain bacteria isolated from the effluent water of a nylon-6 plant. These include *Corynebacterium aurantiacum B-2* reported by Fukumura (*20, 21*) and *Achromobacter guttatus KI 72* reported by Okada et al. (*22*).

Bailey and co-workers reported that an alternating copolymer of glycine and ϵ-aminocaproic acid was degradable by soil microorganisms (*7*). The corresponding copolymer derived from serine was water soluble and could be degraded more rapidly.

Since many proteolytic enzymes are specific in cleaving peptide linkages adjacent to substituent groups, we decided to prepare substituted polyamides, anticipating that the introduction of the substituents would make the polyamides more degradable. Oligomeric benzylated nylon-6,3, prepared from the achiral benzylmalonic acid, was hydrolyzed by chymotrypsin (7% hydrolysis owing to enzyme in addition to 40% hydrolysis owing to buffer in 10 days) (*11*). In the case of nylons-*n*,6 derived from D,L-α-benzyladipic acid, very little degradation was observed. However, all of the enzymes tested (chymotrypsin, subtilisin, thermolysin, pepsin, and elastase) were absorbed on the polymer surface after a 10-day testing period. It is possible that the D-isomers of the D,L-polymers acted as enzyme inhibitors.

$$-\!\!\left[-NH(CH_2)_6NHCO\underset{\displaystyle CH_2Ph}{\underset{|}{C}}HCO-\right]\!\!-$$

benzylated nylon 6,3
mp 140–145°C $\overline{M}_n$ 2,000

$$-\!\!\left[-NH(CH_2)_nNHCO(CH_2)_3\underset{\displaystyle CH_2Ph}{\underset{|}{C}}HCO-\right]\!\!-$$

benzylated nylons *n*,6 $n = 2, 4, 6, 8$

Table II. Degradation of Methyl- and Hydroxy-Substituted Polyamides by Fungi and Enzymes

$$-\!\!\left[\mathrm{NHCH(CH_3)CH_2NHCO(CH_2)_nCO}\right]_x\!\!-\!\!\left[\mathrm{NHCH_2CH(OH)CH_2NHCO(CH_2)CO}\right]_y\!\!-$$

				Extent of Fungal Growth[a]		*% Hydrolysis by Enzymes*[b]	
x/y[c]	n	*mp, °C*	$\overline{M}_n$	A. niger	A. flavus	*Chymo-trypsin*	*Elas-tase*
1/0	4	240–245	13,400	2	2		
1/0	7	180–190	10,400	2	3		
1/0	8	223–228	9,470	1	2	16.0	4.5
1/1	8	200–210	13,000	4	4	9.3	13.5
0/1	8	225 dec.	20,000	2	2	0	32.0

[a] ASTM rating: 4–60 to 100% surface covered; 3–30 to 60% surface covered; 2–10 to 30% surface covered; 1 < 10% surface covered; and 0 = no visible growth. 14-day exposure at 35–37°C.

[b] Calculated on the % of total amount of susceptible amide linkages. Detected by ninhydrin analysis of the amino groups produced after a 5-day exposure at 30°C. Chymotrypsin was buffered (imidazole) at pH 7.5 and elastase was buffered (borax-boric acid) at pH 8.8 ± 0.4%.

[c] Ratio of diamines in the polyamides.

A series of polyamides containing methyl and/or hydroxy substituents were prepared from the polymerizations of 1,2-diaminopropane and/or 1,3-diamino-2-propanol with diacid chlorides. All of the substituted polymers supported the growth of *A. niger* and *A. flavus* whereas the unsubstiuted nylons prepared from ethylenediamine and 1,3-diaminopropane were resistant to fungal attack (Table II) (*23, 24*). It is interesting to note that the copolymer containing both the methyl and the hydroxy groups was the most degradable. When the polyamides were tested with elastase the polyamide degradability increased wtih increas-

Table III. Extent of Fungal Growth on Polyureas after 14 Days at 37°C

$$-\!\!\left[\mathrm{CONH(CH_2)_6NHCONH(CH_2)_4CH(COOR)NH}\right]\!\!-$$

Sample	*Extent of Growth*[a] A. niger	A. flavus
L-lysine methyl ester · HCl	3	0
L-lysine ethyl ester · HCl	2	3
Polyurea H = Me Mn 5,900	2	0
Polyurea R = Et Mn 17,000	1	3

[a] Same as Table I, note a.

ing hydroxy content in the polymer. The reverse pattern was found for chymotrypsin (*25*).

Polyureas. Substituted polyureas were prepared from L-lysine methyl and ethyl esters (*23*). The methyl polyurea ester and the ethyl polyurea ester supported the growth of *A. niger* and only the ethyl polyurea ester supported the growth of *A. flavus* (Table III).

After exposure to buffered chymotrypsin and subtilisin for seven days, the methyl polyurea became completely water-soluble. Ninhydrin analysis of the amino groups produced revealed 5% and 9% hydrolyses of the susceptible urea linkages by chymotrypsin and subtilisin, respectively. Since the polyurea chains contain 18 repeating units on the average, it is reasoned that in addition to the hydrolysis of urea linkages, ester-linkage cleavage must have occurred in order to give water-soluble products.

A poly(ester–urea) of $\overline{M}_n$ 1,930 prepared from L-phenylalanine glycol ester was hydrolyzed by chymotrypsin up to 22.9% in 10 days (*26*). Most of the hydrolysis was found to be cleavage of the ester linkages in keeping of the known specificity of chymotrypsin (acyl cleavage adjacent to the phenylalanine benzyl group). The corresponding poly(ester–urea) prepared from the glycine glycol ester was not degraded by chymotrypsin.

$$-\!\!\left[NHCH(R)COOCH_2CH_2OOCCH(R)NHCONH(CH_2)_6NHCO \right]\!\!-$$

Poly(ester–ureas)

R = H, not degraded

R = CH_2Ph, degraded

Polyurethanes. Polyester base polyurethanes were reported to support fungal growth better than polyether base polyurethanes (*15*). Polyurethanes derived from cellulose hydrolysates were degraded by cellusin (*27*). We found that the polyurethane obtained by reacting poly(dodecamethylene D-tartrate) with 1,6-di-isocyanatohexane supported the growth of *A. niger* with an ASTM rating of 4 (*16*). A similar polyurethane derived from poly(hexamethylene tartrate) was degraded in vivo (*28*). These reports suggest that biodegradation of polyurethanes does not necessarily involve the cleavage of urethane linkages. Although several polyurethanes have been used as surgical implants (*29*), there is very little known about their degradation in vivo.

In order to find out if the urethane linkages are biodegradable, we studied the degradation of two simple polyurethanes containing only urethane linkages. Although we have not found a purified enzyme that

Table IV. Degradation of Polyurethanes by Axion in 10 Days at 30°C

$-\!\!\!+ OCH_2CH_2OOCNH—R—NHCO +\!\!\!-$

R	*[η]*	*% Hydrolysis* Buffer[a] c, d	Axion[b] c, d
1,6-hexamethylene	0.20	10.2 (1.0)	15.1 (6.2)
2,4-tolylene	0.21	14.4 (1.3)	24.0 (7.1)

[a] Phosphate buffer pH 8.0 + sodium dodecylsulfate.
[b] Axion solution pH 8.0.
[c] Weight loss of solid samples, ±1%.
[d] Ninhydrin analysis of amino groups increase in solution based on the total amount of susceptible urethane linkages, ±0.4%.

will effectively degrade the polyurethanes the enzyme containing detergent Axion degraded the polyurethanes, Table IV. In addition to the hydrolysis caused by the basic buffer, degradations were caused also by the enzyme and/or other additives in Axion.

One of the major differences between natural proteins and synthetic polymers is that proteins generally do not have repeating units along the polypeptide chains. This irregularity provides the protein chains with conformational flexibility which allows them to fit into the enzyme active sites. It is very likely that this contributes to the biodegradabilities of proteins. The synthetic polymers, on the other hand, generally have short repeating units and this regularity results in conformational rigidity which inhibits a close fit between the polymer chains and the enzyme active sites. Thus no effective enzyme catalysis will occur. We reasoned that synthetic polymers with long repeating units might be conformationally flexible and thus biodegradable. We prepared several poly(amide urethanes) with rather long repeating units to test this hypothesis (*30*). The degradation by subtilisin results are encouraging (Table V).

Table V. Degradation of Poly(amide urethanes) by Subtilisin in 10 Days at 30°C

$-\!\!\!+ NHCH_2CH_2OOCHN(CH_2)_6NHCOOCH_2CH_2NHCO(CH_2)_xCO +\!\!\!-$

x	$\overline{M}_n$	*mp, °C*	*% Wt Loss* Buffer[a]	Buffered Enzyme[a]
2	6,200	190–213	8	25
4	9,100	195–205	8	22
8	8,800	172–192	2	7

[a] Phosphate buffer, pH 7.2.

Conclusion

Many new step-growth polymers were biodegradable. Most of the biodegradation of synthetic polymer systems are complex multicomponent and multiphase systems. The surface area, morphology, and the molecular weights of the polymers should have significant effects on the biodegradability of the polymer samples. Information in these areas still awaits future research. Although it is premature to draw final conclusions on all of the factors affecting the biodegradability of polymers, several points can be made on the existing information.

Although amide, ester, urea, and urethane linkages are biodegradable, the flexible ester-containing polymers are generally more degradable than polymers containing the more rigid amide, urea, and urethane groups. Replacement of the flexible alkylene segments with the more rigid olefinic and aromatic systems retards or inhibits degradation.

Changing the hydrophilic–hydrophobic characteristics of the polymer samples alters the biodegradability. In general, the presence of both hydrophilic and hydrophobic segments gives the best results of degradation. Proper stereoisomers of substituted polymers are more degradable thanthe corresponding unsubstituted analogs. Introduction of substituents increases the conformational flexibility of polymer chains and provides favorable hydrophobic or hydrophilic interaction between the polymer chains and the active sites of enzymes, thus improving the catalysis by enzymes. Increasing the repeating unit length has a similar effect.

Copolymers, both in terms of substituents and linkages, are generally more degradable than the corresponding homopolymers. Again, this is probably caused by the fact that copolymers are more flexible than the more regular homopolymers with the exception of some polyesters.

In addition to the presence of hydrolyzable linkages, it seems that conformational flexibility is one of the most important requirements.

Acknowledgment

We thank the National Science Foundation (Grant DMR 75–16912) and the University of Connecticut Research Foundation for financial support.

Literature Cited

1. Huang, S. J., Roby, M., Knox, J. R., *Enzyme Technol.* (1976) **5**, 135.
2. Guillet, J., Ed., "Polymer and Ecological Problems," Plenum, New York, 1973.
3. Potts, J. E., Clendinning, R. A., Ackart, U.S., E.P.A. Contract No. CPE–70–124, 1972, p. 22.

4. Guillet, J., Ed., "Polymer and Ecological Problems," pp. 61–79, Plenum, New York, 1973.
5. 1970 Annual of ASTM Standards, Part 27, ASTM-D-1924, p. 593.
6. 1970 Annual of ASTM Standards, Part 26, ASTM-D-2676, p. 758.
7. Bailey, W. J., Okamoto, Y., Kuo, W.-C., Narita, T., *Proc. Int. Biodegradation Symp., 3rd* (1976) 765.
8. Nykvist, N. B., *Proc. Conf. Degrad. Polym. Plast.* (1973) 1.
9. Tokiwa, Y., Suzuki, T., *J. Ferment. Technol.* (1974) **52**, 393.
10. Fields, R. D., Rodriquez, F., *Proc. Int. Biodegradation Symp., 3rd* (1976) 775–784.
11. Huang, S. J., et al., *Proc. Int. Bidegradation Symp., 3rd* (1976) 731–741.
12. Tokiwa, Y., Suzuki, T., *Agric. Biol. Chem.* (1977) **41**, 265.
13. Tokiwa, Y., Ando, T., Suzuki, T., *J. Ferment. Technol.* (1976) **54**, 603.
14. Kinoshita, S., Bisaria, V. S., Sawada, S., Okada, H., *Abst. Annu. Meet. Soc. Ferment. Technol.* (1974) 110.
15. Darby, R. T., Kaplan, A. M., *Appl. Microbiol.* (1968) **16**, 900.
16. Bitritto, M. M., Bell, J. P., Brinkle, G. M., Huang, S. J., Knox, J. R., *J. Appl. Polym. Sci.*, in press.
17. Bell, J. P., Huang, S. J., Knox, J. R., U.S. NTIS, AD-A Rep. No. 009577 (1974).
18. Tabushi, I., Yamada, H., Matsuzaki, H., Furukawa, J., *J. Polym. Sci. Polym. Lett.* (1975) **13**, 447.
19. Rodriquez, F., *Chem. Technol.* (1971) 409.
20. Fukumura, T., *Plant Cell Physiol.* (1966) **7**, 93.
21. Fukumura, T., *J. Biochem.* (1966) **59**, 537.
22. Kinoshita, S., Kageyama, S., Iba, K., Yamada, Y., Okada, H., *Agr. Biol. Chem.* (1975) **39**, 1219.
23. Huang, S. J., Leong, K. W., Knox, J. R., unpublished data.
24. Leong, K. W., Ph.D. Dissertation, University of Connecticut, 1976.
25. Huang, S. J., Pavlisko, J., unpublished results.
26. Huang, S. J., Bansleben, D. A., Knox, J. R., *J. Appl. Polym. Sci.*, in press.
27. Kim, S., Stannett, V. T., Gilbert, R. D., *J. Macromol. Sci., Chem.* (1976) **A10**, 671.
28. Wang, P. Y., Arlitt, B. B., *Polym. Sci. Technol.* (1975) **1**, 173.
29. Wilkes, G. L., "Polymers in Medicine and Surgery," R. L. Kronenthal, Z. Oser, E. Martin, Eds., pp. 45–76, Plenum, New York, 1976.
30. Huang, S. J., Roby, M., Knox, J. R., unpublished data.

RECEIVED May 12, 1977. Publication 944 from Institute of Materials Science.

18

Stabilization Fundamentals in Thermal Autoxidation of Polymers

J. REID SHELTON

Department of Chemistry, Case Western Reserve University,
Cleveland, OH 44106

There are two ways in which stabilizers can function to retard autoxidation and the resultant degradation of polymers. Preventive antioxidants reduce the rate of initiation, e.g., by converting hydroperoxide to nonradical products. Chain-breaking antioxidants terminate the kinetic chain by reacting with the chain-propagating free radicals. Both mechanisms are discussed and illustrated. Current studies on the role of certain organic sulfur compounds as preventive antioxidants are also described. Sulfenic acids, RSOH, from the decomposition of sulfoxides have been reported to exhibit both prooxidant effects and chain-breaking antioxidant activity in addition to their preventive antioxidant activity as peroxide decomposers.

Organic materials are susceptible to oxidative degradation by reaction with elemental oxygen and thus require protection against the autoxidation reaction. This protection is provided by the addition of stabilizers. The initial product of the reaction is hydroperoxide which decomposes under appropriate conditions to give free radicals capable of initiating the free-radical chain reaction (*1*). The decomposition is accelerated by heat, light, and the presence of certain metal catalysts.

Uninhibited autoxidation of hydrocarbons in the absence of added initiators or terminators involves the following reactions (*2*):

Initiation: $$\mathrm{ROOH} \rightarrow \mathrm{RO}\cdot + \mathrm{HO}\cdot$$

$$2\,\mathrm{ROOH} \xrightarrow{k_1} \mathrm{RO}\cdot + \mathrm{RO_2}\cdot + \mathrm{H_2O}$$

0-8412-0381-4/78/33-169-215$05.00/1

Propagation: $RO_2\cdot + RH \xrightarrow{k_p} ROOH + R\cdot$

$$R\cdot + O_2 \xrightarrow{\text{fast}} RO_2\cdot$$

Termination: $2\,R\cdot \rightarrow R\text{–}R$

$$R\cdot + RO_2\cdot \rightarrow RO_2R$$

$$2RO_2\cdot \xrightarrow{k_t} \text{nonradical products} + O_2$$

There are ways in which stabilizers can retard the oxidation process. They can reduce the rate of peroxide initiation or intercept the chain-propagating free radicals and thus terminate the chain mechanism. Scott (*3*) classified such stabilizers as preventive and chain-breaking. Both types of antioxidant are known and include a variety of compounds that can act in several different ways (*2*):

- Preventive antioxidants: (a) light absorbers, (b) metal deactivators, (c) peroxide decomposers (nonradical products).
- Chain-breaking antioxidants: (a) free-radical traps, (b) electron donors, (c) hydrogen donors.

Since stabilization against photooxidation and metal-catalyzed oxidation are covered elsewhere in this symposium, this discussion is restricted to protection against thermal autoxidation. I will first review the mechanism by which typical chain-breaking antioxidants function and then describe some of our current studies on the way in which certain organic sulfur compounds act as preventive antioxidants.

Chain-Breaking Antioxidants

The widely used hindered phenol and aryl amine antioxidants contain reactive O–H and N–H functional groups capable of reacting with oxy radicals by transfer of hydrogen (*4*). Electron transfer is also a possibility, and some antioxidants, or their reaction products, may function as traps for alkyl radicals. The hydrogen donation mechanism is capable of terminating two kinetic chains:

$$RO_2\cdot + AH \rightarrow ROOH + A\cdot$$

$$RO_2\cdot + A\cdot \rightarrow RO_2A$$

A kinetic deuterium isotope effect would be expected if transfer of hydrogen were the rate-controlling reaction. Initial attempts by Hammond and co-workers (*5, 6*) to observe such an isotope effect in the

AIBN-initiated oxidation of cumene and Tetralin in the presence of deuterated amines were unsuccessful. They proposed an alternative mechanism involving reversible formation of a complex of antioxidant with peroxy radical as the kinetically controlling process. We observed an isotope effect, $k_D/k_H = 1.8$, consistent with the hydrogen-donation mechanism in the retarded oxidation of SBR polymer with deuterated amines (*7, 8*). Our results were confirmed by observation of significant isotope effects in the initial stage of oxidation of purified *cis*-1,4-polyisoprene with both hindered phenols and amines (*9*). Table I shows the effect of temperature and antioxidant concentration on the rates of oxidation and the observed deuterium isotope effects.

Table I. Oxidation Polyisoprene, 2,6-di-*tert*-butyl-4-methylphenol, one atm O_2

			Rate, R [*ml* O_2*(22°)/g/hr*]		*Isotope Effect* (R_D/R_H)	
Temp.	*Conc. (mol/g)*	*Inhib.*	*First*	*Second*	*First*	*Second*
90°C	4.41×10^{-5}	IN-H	0.0584	0.0861	1.27	1.25
		IN-D	0.0750	0.1077		
	13.2	IN-H	0.0273	0.0393	0.79	0.82
		IN-D	0.0216	0.0321		
75°C	8.82	IN-H	0.00968	0.0132	1.56	1.49
		IN-D	0.0151	0.0197		
	13.2	IN-H	0.00710	0.00962	0.92	0.95
		IN-D	0.00653	0.00913		
60°C	4.41	IN-H	0.00251	—	1.76	—
		IN-D	0.00441	0.00549		
	13.2	IN-H	0.00124	—	1.16	—
		IN-D	0.00144	—		

Prooxidant effects were observed at higher antioxidant concentrations and at higher temperatures. Reversal of the direction of the isotope effects observed under these conditions showed that initiation by direct reaction of the antioxidant with oxygen is an important initiation reaction. Peroxide decomposition is quite slow at 90°C and begins to contribute significantly to initiation only at the start of a second stage of more rapid, but still retarded, autoxidation. We have suggested (*4*) that some oxidation product of polymer or antioxidant may induce hydroperoxide decomposition.

$$AH + O_2 \rightarrow HO_2\cdot + A\cdot \xrightarrow{O_2} AO_2\cdot$$

$$ROOH \rightarrow RO_2\cdot, \text{etc.}$$

$$\uparrow$$

Oxy groups

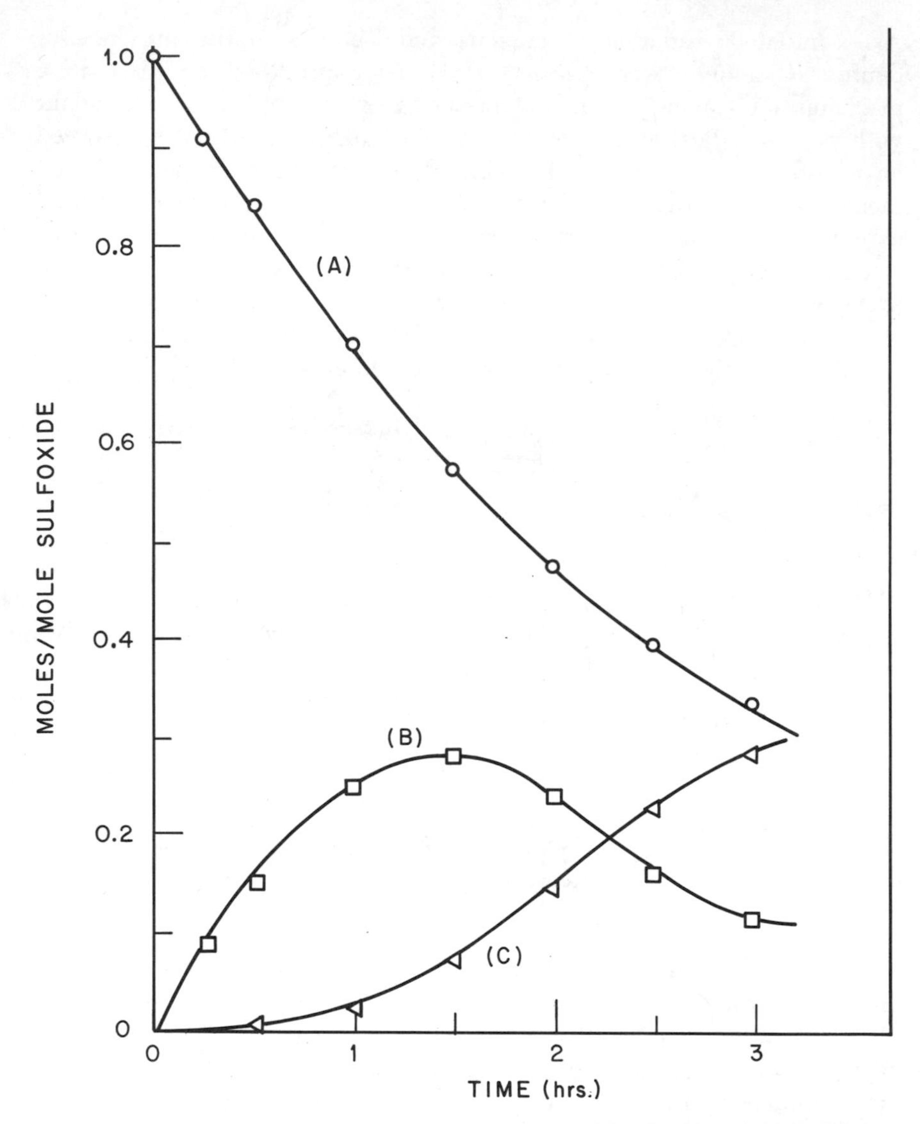

International Journal of Sulfur Chemistry

*Figure 1. Concentrations (by NMR) of sulfoxide (A), sulfenic acid (B), and thiolsulfinate (C), as a function of time of heating 0.5M di-*tert*-butyl sulfoxide in benzene at 80°C (13)*

Preventive Antioxidants

Various organic sulfur, nitrogen, and phosphorus compounds are known to accelerate the decomposition of organic hydroperoxides without production of free radicals. A polar mechanism has been suggested (3) to account for the preventive antioxidant activity that results from the decreased availability of hydroperoxide for initiation by homolytic dissociation. Many alkyl and aryl sulfides and disulfides, and the corresponding sulfoxides and thiolsulfinates formed by reaction with hydroperoxides, have been shown to have antioxidant activity (*10, 11*). In order to learn more about the chemistry and mechanisms involved, we have carried out extensive studies of the decomposition of sulfoxides and the nature of the initial and final products.

Our observation by NMR (*12, 13*) of the presence of 2-methyl-2-propanesulfenic acid, *tert*-BuSOH, in the reaction mixture from the thermal decomposition of di-*tert*-butylsulfoxide, and proof of its identity was the first demonstration that simple organic sulfenic acids could be prepared and characterized. In the absence of hydroperoxide and trapping agents the sulfenic acid is converted to the corresponding thiolsulfinate, *tert*-Bu(SO)St-Bu, as shown in Figure 1. Block (*14*) has shown that thiolsulfinates of this type decompose thermally with the formation of a thiosulfoxylic acid, RSSOH. The following reactions illustrate the formation of sulfoxide and thiolsulfinate and their initial decomposition products:

$$(CH_3)_3CSC(CH_3)_3 \xrightarrow{\text{ROOH}} (CH_3)_3C\overset{\overset{O}{\uparrow}}{S}C(CH_3)_3$$

$$(CH_3)_3C\overset{\overset{O}{\uparrow}}{S}C(CH_3)_3 \xrightarrow{65°\text{–}100°C} (CH_3)_3CSOH + (CH_3)_2C{=}CH_2$$

tert-Butanesulfenic acid
(*tert*-BuSOH)

$$2(CH_3)_3CSOH \xrightarrow{-H_2O} (CH_3)_3C\overset{\overset{O}{\uparrow}}{S}SC(CH_3)_3$$

tert-Butyl *tert*-butanethiolsulfinate

$$(CH_3)_3CSSC(CH_3)_3 \xrightarrow{\text{ROOH}} (CH_3)_3C\overset{\overset{O}{\uparrow}}{S}SC(CH_3)_3$$

$$(CH_3)_3C\overset{\overset{O}{\uparrow}}{S}SC(CH_3)_3 \xrightarrow{\Delta} (CH_3)_2C{=}CH_2 + (CH_3)_3CSSOH$$

tert-Butanethiosulfoxylic acid
(*tert*-BuSSOH)

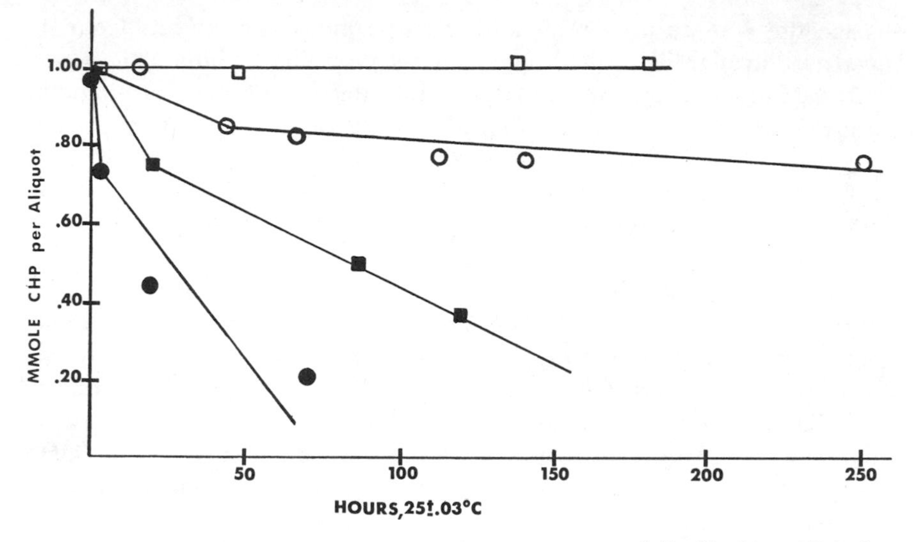

Figure 2. Effect of base on decomposition of cumene hydroperoxide in the presence of tert-*butanesulfenic acid. Concentrations, mmol/1 in benzene: CHP 2.0 ± 0.1; A, □ $CaCO_3$ 0.08; B, ● tert-BuSOH 0.2; C, ■ tert-BuSOH 0.2, $CaCO_3$ 0.08; D, ○ tert-BuSOH 0.2, $CaCO_3$ 2.5* (15).

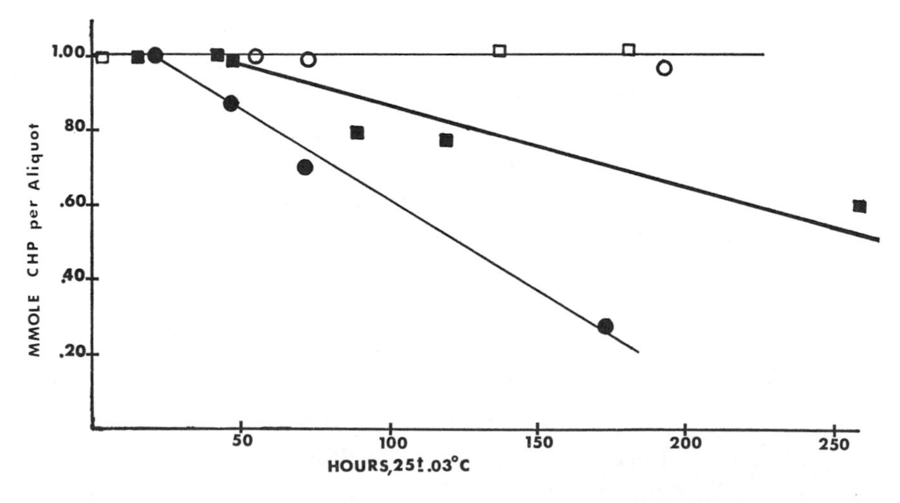

Figure 3. Effect of base on decomposition of cumene hydroperoxide in the presence of tert-*butanethiosulfoxylic acid from thiolsulfinate decomposition. Concentrations, mmol/1 in benzene: CHP 2.0 ± 0.1; A, □ $CaCO_3$ 0.08; B, ●* tert-*BuS(O)S*-tert-*Bu 0.2; C, ■* tert-*BuS(O)S*-tert-*Bu 0.2, $CaCO_3$ 0.08; D, ○* tert-*BuS(O)S*-tert-*Bu 0.2, $CaCO_3$ 2.5* (15).

Recently we reported a study of the peroxide decomposing activity of sulfoxides, sulfenic acids, thiolsulfinates, and their oxidation or decomposition products (*2, 15*). A benzene solution of the sulfenic acid reacted rapidly with both *tert*-butyl hydroperoxide and cumene hydroperoxide, consuming two moles of hydroperoxide per mole of sulfur compound. A slower catalytic process destroyed many additional moles of hydroperoxide per mole of sulfur compound. A similar solution of cumene hydroperoxide in benzene with thiolsulfinate present in a ratio of 10 moles of hydroperoxide per mole of sulfur compound showed no change for 22 hr at 25°C. At that time a catalytic decomposition started which destroyed many moles of hydroperoxide per mole of sulfur compound.

These experiments were repeated in the presence of $CaCO_3$ to see if the base would affect the activity of the sulfur compounds as peroxide decomposers. The initial reaction of sulfenic acid with the hydroperoxide was slowed, as shown in Figure 2, but ultimately consumed two moles of ROOH per mole of RSOH. The subsequent catalytic decomposition was almost completely stopped with excess of base consistent with neutralization of an acid catalyst, presumed to be the sulfonic acid formed by oxidation of the sulfenic acid:

$$\textit{tert}\text{-BuSOH} + 2\text{ROOH} \rightarrow \textit{tert}\text{-BuSO}_3\text{H} + 2\ \text{ROH}$$

$$\textit{tert}\text{-BuSO}_3\text{H} + \text{CaCO}_3 \rightarrow \textit{tert}\text{-BuSO}_3\text{CaHCO}_3$$

When only a small amount of $CaCO_3$ was added to the benzene solution of cumene hydroperoxide containing thiolsulfinate (0.04 mol per mol of sulfur compound), the time to onset of peroxide decomposition was doubled, and the subsequent catalytic reaction was slowed as shown in Figure 3. Excess solid $CaCO_3$ completely stopped the reaction showing that an acidic product from the thiolsulfinate decomposition is responsible for the hydroperoxide decomposition observed in the absence of base (*15*). The nature of the reactions occurring in this system is still being investigated, and my co-worker will describe the work in progress in the next chapter.

The products from cumene hydroperoxide decomposition induced by organic sulfur compounds were determined by quantitative NMR except for phenol by high-pressure liquid chromatography and cumene hydroperoxide by iodometric titration (*16*). Cumyl alcohol is produced in the initial oxidation of sulfenic acid to sulfonic acid, and subsequently most of it is converted to α-methylstyrene as shown in Table II. The major products (40–45%) are phenol and acetone consistent with an acid-catalyzed decomposition of cumene hydroperoxide. Considerable

Table II. Products from Cumene Hydroperoxide Decomposition Induced by Organic Sulfur Compounds (*16*)

	Mol % Products[a] from 0.2M CHP in Benzene with 0.02M Sulfur Compound, 25°C, N_2						
	CHP	*Cumyl Alcohol*	*Acetone*	*Phenol*	*Dicumyl Peroxide*	*Acetophenone*	*α-Methylstyrene*
tert-BuSOH							
25 min	77	21	tr	tr	tr	0	0
171 hr	3	4	42	37	13	0	25
tert-$BuSO_3H$							
237 hr	21	6	28	40	13	4	8
tert-BuS(O)S-*tert*-Bu							
262 hr	11	6	42	47	12	3	3

[a] Concentrations by quantitative NMR except phenol by liquid chromatography and CHP by titration.

dicumyl peroxide is formed (13%), presumably by a polar reaction since little β-scission occurred and any cumyloxy radicals would abstract hydrogen from the hydroperoxide to form alcohol which is observed. Some free-radical products are evidently produced (10–15%) by reactions induced by the sulfur compounds since no hydroperoxide decomposition was observed at 25°C in their absence. Polar processes consumed 86% of the original hydroperoxide when sulfenic acid was present in a ratio of 1 mol per 10 mol of ROOH as shown in Table III.

We have thus established that sulfides, disulfides, and their initial oxidation products are not the actual preventive antioxidants. The active peroxide decomposers are the sulfenic acid from sulfoxide decomposition, the thiosulfoxylic acid from thiolsulfinate decomposition, and the acidic products formed when they react with hydroperoxides. The catalytic

Table III. Polar and Radical Decomposition of Cumene Hydroperoxide Induced by Organic Sulfur Compounds (*16*)

	Percent CHP Consumed in Various Types of Decomposition 0.2M CHP in Benzene with 0.02M Sulfur Compound, 25°C, N_2		
	tert-*BuSOH*	tert-$BuSO_3H$	tert-*BuS(O)S*-tert-*Bu*
Initial oxidation	20	0	0
Polar decomposition	66	61	69
Radical decomposition	9	18	12
Undecomposed CHP	3	21	11
Original CHP accounted for	98	100	92

destruction of peroxides by sulfoxides and their reaction products has also been studied by Scott and co-workers (*17, 18, 19, 20*) who have arrived at similar conclusions independently.

Roles of Antioxidants in Retarded Autoxidation

The observed effects of stabilizers in autoxidation include both prooxidant and antioxidant effects. Scott (*17, 19*) has proposed that both sulfoxides and sulfenic acids may induce free-radical formation under certain conditions based on observed prooxidant effects. We have suggested (*2*) a hydrogen-bonded association of the sulfenic acid with hydroperoxide which could induce a homolytic cleavage. Sulfenic acids also function in part as chain-breaking antioxidants (*17, 19, 21*) in addition to the peroxide decomposing activity which we have reported. Under autoxidation conditions the sulfenyl free radical could react with oxygen and hydrocarbon substrate to form sulfur dioxide which is known to be an efficient peroxide decomposer (*10, 17, 19*). These suggested roles are illustrated for *tert*-butanesulfenic acid in the following reactions:

Prooxidant:

$$\mathrm{RO\overset{\displaystyle H}{\overset{|}{O}}\text{---}HOS\text{-}\mathit{tert}\text{-}Bu \rightarrow RO\cdot + H_2O + \mathit{tert}\text{-}BuSO\cdot}$$

$$\mathrm{RO\cdot + R'H \rightarrow ROH + R'\cdot \xrightarrow{O_2} R'O_2\cdot}$$

Antioxidant:

Chain stopper

$$\mathit{tert}\text{-}\mathrm{BuSOH + RO_2\cdot \rightarrow ROOH +} \mathit{tert}\text{-}\mathrm{BuSO\cdot}$$

Peroxide decomposer

$$\mathit{tert}\text{-}\mathrm{BuSO\cdot + O_2 \xrightarrow{R'H}} \mathit{tert}\text{-}\mathrm{Bu\overset{\displaystyle O}{\overset{\uparrow}{S}}O_2H \rightarrow} \mathit{tert}\text{-}\mathrm{BuOH + SO_2}$$

$$\mathrm{ROOH \xrightarrow[\mathit{tert}\text{-}BuSOH]{SO_2} Nonradical\ products}$$

The mechanism of retarded autoxidation in the presence of stabilizers of both preventive and chain-breaking type thus includes the possible participation of the antioxidant in all stages of the process (*1*):

Peroxide destruction:

$$ROOH + AH \rightarrow \text{nonradical products}$$

Initiation:

$$n\text{-}ROOH \rightarrow RO\cdot, RO_2\cdot, HO.$$

$$AH + O_2 \rightarrow A\cdot + HO_2\cdot$$

Propagation:

$$RO_2\cdot + RH \rightarrow ROOH + R\cdot \xrightarrow{O_2} RO_2\cdot$$

Chain transfer:

$$RO_2\cdot + AH \rightarrow ROOH + A\cdot \xrightarrow[O_2]{RH} AO_2H + RO_2\cdot$$

Termination:

$$RO_2\cdot + AH \rightarrow ROOH + A\cdot \qquad 2\,RO_2\cdot \rightarrow \text{nonradical products}$$

$$RO_2\cdot + A\cdot \rightarrow RO_2A \qquad RO_2\cdot + R\cdot \rightarrow RO_2R$$

$$2\,A\cdot \rightarrow A\text{—}A \qquad 2\,R\cdot \rightarrow R\text{—}R$$

Acknowledgment

Studies of the kinetic deuterium isotope effects which established the chain-breaking mechanism of antioxidant action by hydrogen donation were carried out in our laboratories by E. T. McDonel, J. C. Crano, and D. N. Vincent. Studies of sulfoxides, sulfenic acids, thiolsulfinates, and their reactions with hydroperoxides which illustrate the chemistry of the processes involved in their activity as preventive antioxidants were done by K. E. Davis, J. V. Webba, E. R. Harrington, and D. M. Kulich.

These studies were made possible by the continuing financial support of the Goodyear Tire and Rubber Co. and, in part, by grants from the National Science Foundation and the Petroleum Research Fund of the American Chemical Society.

Literature Cited

1. Shelton, J. R., "Polymer Stabilization," W. L. Hawkins, Ed., Ch. 2, Wiley, New York, 1972.
2. Shelton, J. R., *Rubber Chem. Technol.* (1974) **47**, 949.

3. Scott, G., "Atmospheric Oxidation and Antioxidants," Elsevier, Amsterdam, 1965.
4. Shelton, J. R., *Rubber Chem. Technol.* (1972) **45**, 359.
5. Boozer, C. E., Hammond, G. S., *J. Am. Chem. Soc.* (1954) **76**, 3861.
6. Hammond, G. S., Boozer, C. E., Hamilton, C. E., Sen, J. N., *J. Am. Chem. Soc.* (1955) **77**, 3238.
7. Shelton, J. R., McDonel, E. T., *J. Polym. Sci.* (1958) **32**, 75.
8. Shelton, J. R., McDonel, E. T., Crano, J. C., *J. Polym. Sci.* (1960) **42**, 289.
9. Shelton, J. R., Vincent, D. N., *J. Am. Chem. Soc.* (1963) **85**, 2433.
10. Hawkins, W. L., Sautter, H., *Chem. Ind. (London)* (1962) 1825.
11. Hawkins, W. L., Sautter, H., *J. Polym. Sci.* (1963) **1A**, 3499.
12. Shelton, J. R., Davis, K. E., *J. Am. Chem. Soc.* (1967) **89**, 718.
13. Shelton, J. R., Davis, K. E., *Int. J. Sulfur Chem.* (1973) **8**, 197, 205.
14. Block, E., *J. Am. Chem. Soc.* (1972) **94**, 642, 644.
15. Shelton, J. R., Harrington, E. R., *Rubber Chem. Technol.* (1976) **49**, 147.
16. Harrington, E. R., Ph.D. Thesis, Case Western Reserve University, 1976.
17. Scott, G., *Mech. React. Sulfur Compd.* (1969) **4**, 99.
18. Scott, G., *Br. Polym. J.* (1971) **3**, 24.
19. Scott, G., *Pure Appl. Chem.* (1972) **30**, 267.
20. Armstrong, C., Plant, M. A., Scott, G., *Eur. Polym. J.* (1975) **11**, 161.
21. Koelewijn, P., Berger, H., *Recl. Trav. Chim. Pays-Bas* (1972) **91**, 1272.

RECEIVED May 12, 1977.

19

The Role of Certain Organic Sulfur Compounds as Preventive Antioxidants

III. Reactions of *tert*-Butyl *tert*-Butanethiolsulfinate and Hydroperoxide

DONALD M. KULICH and J. REID SHELTON

Department of Chemistry, Case Western Reserve University, Cleveland, OH 44106

Thiolsulfinates and their reaction products play an important role in the preventive antioxidant activity observed with organic sulfides and disulfides. An investigation of the decomposition of cumene hydroperoxide in benzene at 25°C in the presence of tert-*butyl* tert-*butanethiolsulfinate has shown that the actual peroxide decomposer is an acidic species whose activity is affected by the basic character of the S–O group in the parent thiolsulfinate and in the sulfoxides. Alternative mechanisms for generating the acidic species are discussed. Although hydroperoxide decomposition occurs primarily by a polar process, the results also indicate the involvement of radical generating processes.*

Additives that destroy intermediate hydroperoxides provide an effective means of stabilization of polymers against oxidative degradation. A wide variety of sulfur compounds, including sulfides and disulfides, can be effective stabilizers. Pro-oxidant as well as antioxidant behavior is observed, indicating the involvement of reactions generating radicals and peroxide decomposition. Oxidation studies show that preventive antioxidant behavior by disulfides is exhibited only after the absorption of oxygen, resulting in the formation of the corresponding thiolsulfinate. Thiolsulfinates are also thermolysis products of sulfoxides. Recent work has shown that thiolsulfinate itself does not catalytically decompose hydroperoxide (*1*). Thus, the oxidation of disulfides to thiolsulfinates is only the first in a series of reactions leading to the formation of the active preventive antioxidant. A study of cumene hydroperoxide decomposition

0-8412-0381-4/78/33-169-226$05.00/1

has indicated that this induced decomposition of the hydroperoxide is primarily a polar process (*1, 2*). This chapter presents the intial results of a systematic investigation of the hydroperoxide decomposing activity exhibited by *tert*-butyl *tert*-butanethiolsulfinate. Various alternative mechanisms for generating active species from the thiolsulfinate in the presence of cumene hydroperoxide are examined. Another aspect of the reaction mechanism considered is the ability of the highly polar S–O group in the thiolsulfinate and sulfoxide to complex and deactivate the acidic peroxide decomposing agent. This proposal was tested by examining the effects of various sulfoxides on the decomposition of cumene hydroperoxide by strong acids. These results are compared with a determination of the relative hydrogen bonding ability of thiolsulfinate and sulfoxides.

Experimental

Unless otherwise indicated, hydroperoxide decomposition studies were carried out in spectral grade benzene (Mallinckrodt). Preparation and purification of the sulfoxides, *tert*-butyl *tert*-butanethiolsulfinate, and cumene hydroperoxide were discussed previously (*1, 2*). Reagent grade phenol was purified by vacuum sublimation. The carbon tetrachloride was spectroscopic grade dried over P_2O_5. Decomposition studies were carried out under nitrogen in screw-top vials stored in a constant temperature bath at $25 \pm 0.05°C$ (*1, 2*). Cumene hydroperoxide concentrations were iodometrically determined by the Hercules Method I reviewed by Mair and Graupner (*3*). Blanks were 0.001–0.003 meq of iodine.

Possible Mechanisms

The polar nature of the catalytic decomposition of cumene hydroperoxide with added *tert*-butyl *tert*-butanethiolsulfinate has been clearly established (*1*). The thiolsulfinate is converted into an active peroxide decomposer capable of destroying many moles of hydroperoxide per mole of sulfur compound. The acidic character of the active species was demonstrated by its effective neutralization with the added base calcium carbonate. Formation of the active peroxide decomposer may be envisaged as involving one or more of the following three reaction types: concerted process, ionic processes, and free-radical processes.

Concerted Process. Thiolsulfinates are thermally labile, and those with appropriate alkyl groups are known to undergo an intramolecular cycloelimination reaction on heating (*4*):

$$H_2C(H\cdots O{=})C(CH_3)_2\text{–}S(O)\text{–}S\text{–}R \rightarrow (CH_3)_2C{=}CH_2 + RSSOH \qquad (1)$$

Formation of the thiosulfoxylic acid intermediate has been demonstrated by mass spectroscopic studies and trapping experiments. The thiosulfoxylic acid formed may then react with hydroperoxide through a mechanism that destroys many moles of hydroperoxide per mole of sulfur compound (*1*).

Alternatively, the formation of thiosulfoxylic acid may be only the first in a series of consecutive reactions leading to the effective peroxide decomposing species.

Polar Processes. Kice (*5*) has presented extensive evidence that cleavage of the S–S bond can be catalyzed by electrophilic and nucleophilic assistance. Heterolytic cleavage would generate sulfenic acid as the active peroxide decomposing species. Traces of water and sulfur-containing compounds may be considered as potential nucleophiles.

WATER AS A NUCLEOPHILE. Trace sulfenic acid formation may occur by thiolsulfinate reacting with water.

$$\overset{O}{RS}SR + H_2O \rightleftharpoons 2RSOH \quad (2)$$

It has been estimated that the reverse of the above equilibrium is favored by 10^6 for R = aryl, and to an even greater extent for R = alkyl (*4*). However, in the presence of hydroperoxide sulfenic acid would be readily oxidized to sulfonic acid, a catalytic peroxide decomposer (*1*).

SULFUR NUCLEOPHILES. Thiolsulfinates are known to readily disproportionate with the generation of intermediate sulfenic and sulfinic acids as in reactions 3, 4, and 5 (*4*).

$$\overset{O}{RS}SR \xrightarrow{H^+} [\overset{\overset{H}{O}}{RS}SR]^+ \quad (3)$$

$$\overset{O}{RS}SR + [\overset{\overset{H}{O}}{RS}SR]^+ \rightarrow RSOH + [\overset{O}{RS}\underset{R}{S}SR]^+ \quad (4)$$

$$[\overset{O}{RS}\underset{R}{S}SR]^+ + H_2O \rightarrow RSO_2H + RSSR + H^+ \quad (5)$$

$$RSO_2H \xrightarrow{ROOH} RSO_3H \quad (6)$$

Initially the acid catalyst for Reaction 3 would be supplied by decomposition of the thiosulfinate to thiosulfoxylic acid as an induction period.

Reactions 3, 4, 5, and 6 would then ensue with sulfonic acid being the catalytic peroxide decomposer, and the major source of H^+ for the induced decomposition of the thiolsulfinate. The disproportionation of thiolsulfinates is accelerated markedly by the addition of substances such as alkyl sulfides that contain a more nucleophilic sulfur atom than found in thiolsulfinates.

Homolytic processes: Evidence also has been presented for the radical-induced decomposition of thiolsulfinates (*6, 7*). Homolytic cleavage is facilitated by the weak S–S bond ($\sim$ 40 kcal). The availability of sulfidic sulfur for radical attack is indicated by the observation that thiolsulfinates strongly retard the free radical polymerization of vinyl monomers (*8*).

Results and Discussion

Information regarding the importance of the above processes in the formation of the active antioxidant from thiolsulfinate in the presence of hydroperoxide can be provided by examining the effects of added water, organosulfur nucleophiles, radical trapping agents, and determining the nature of the sulfur-containing products. The following results were obtained using cumene hydroperoxide and *tert*-butyl *tert*-butanethiolsulfinate.

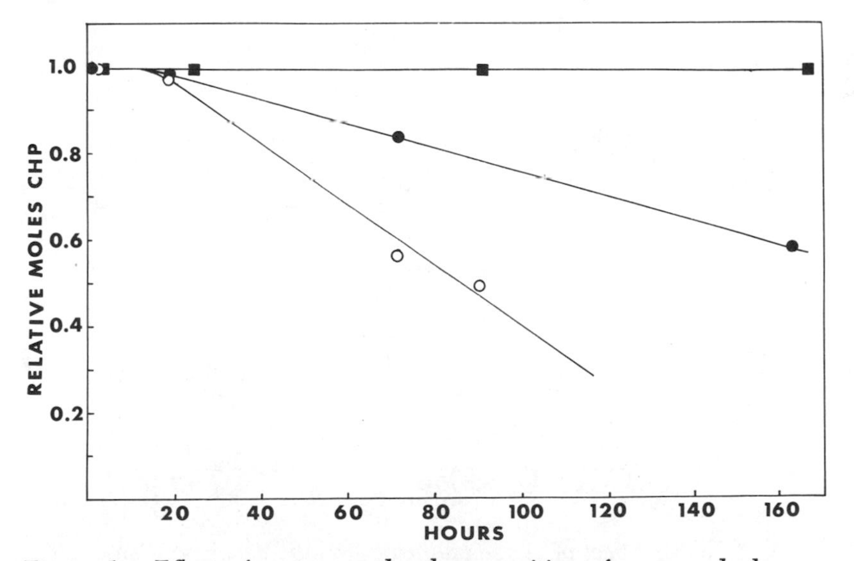

Figure 1. Effect of water on the decomposition of cumene hydroperoxide in the presence of tert-*butyl* tert-*butanethiolsulfinate at 25°C:* ● *0.20M CHP and 0.020M* tert-*BuSS(O)*tert-*Bu in spec. benzene (0.05% H_2O);* ○ *0.20M CHP and 0.021M* tert-*BuSS(O)*tert-*Bu in spec. benzene dried over CaH_2 and distilled;* ■ *0.20M CHP and 0.020M* tert-*BuSS(O)-*tert-*Bu in spec. benzene and $\sim$ 1% water added*

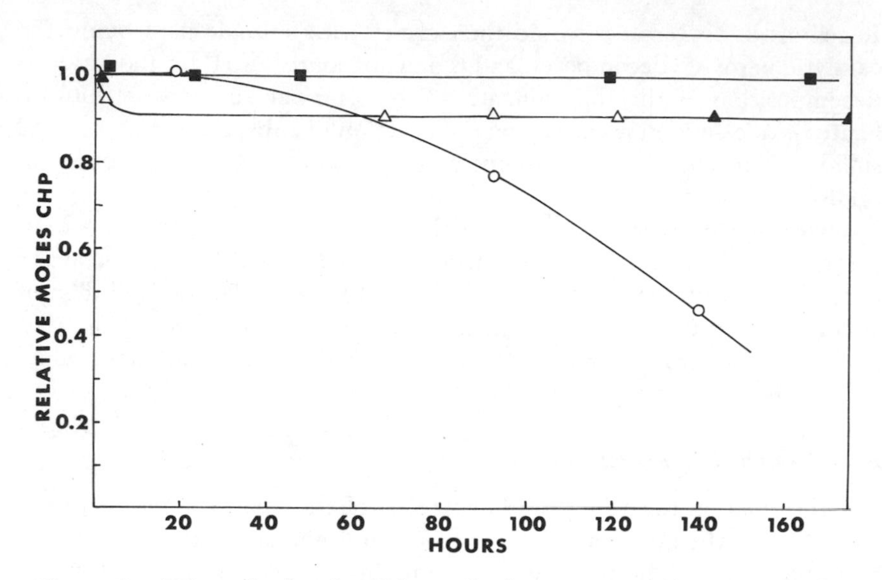

Figure 2. Effect of n-*butyl sulfide on the decomposition of cumene hydroperoxide in the presence of* tert-*butyl* tert-*butanethiolsulfinate at 25°C:* ■ *0.20*M *CHP;* ○ *0.19*M *CHP and 0.020*M tert-*BuSS(O)*tert-*Bu;* △ *0.20*M *CHP, 0.020*M tert-*BuSS(O)*tert-*Bu, and 0.018*M (n-*Bu)*$_2$*S;* ▲ *0.21*M *CHP and 0.020*M (n-*Bu)*$_2$*S*

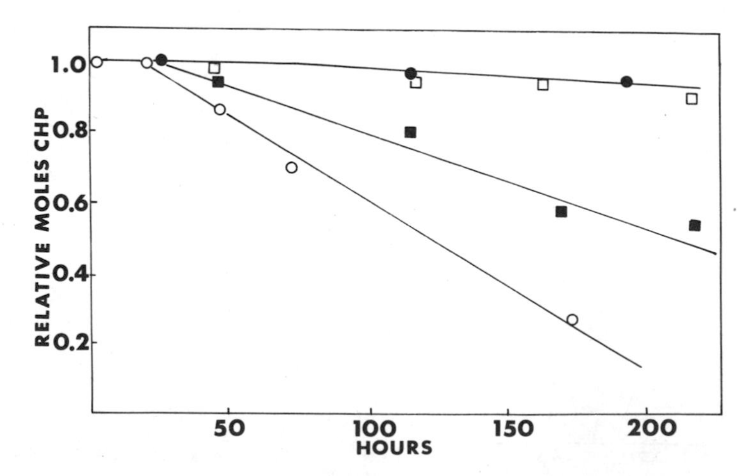

Figure 3. Effect of various sulfoxides on the decomposition of cumene hydroperoxide in the presence of tert-*butyl* tert-*butanethiolsulfinate at 25°C (from Ref. 2):* ○ *0.21*M *CHP and 0.02*M tert-*BuSS(O)*tert-*Bu;* ■ *0.20*M *CHP, 0.02*M $C_6H_5S(O)C_6H_5$*, and 0.02*M tert-*BuSS(O)*tert-*Bu;* □ *0.20*M *CHP, 0.02*M $CH_3S(O)CH_3$*, and 0.02*M tert-*BuSS(O)*tert-*Bu;* ● *0.20*M *CHP, 0.02*M tert-*BuS(O)*tert-*Bu, and 0.02*M tert-*BuSS(O)*tert-*Bu*

The effect of water on the rate of cumene hydroperoxide decomposition in the presence of thiolsulfinate is shown in Figure 1. The displacement of thiolsulfinate by water is not the primary means of generating the active peroxide decomposer. To the contrary, water inhibits peroxide decomposition. Water may be expected to hydrate the thiolsulfinate (9) and the acidic decomposing species thereby decreasing the observed hydroperoxide decomposition.

Since thiolsulfinates are very weakly nucleophilic, cleavage of the S–S bond should be subject to nucleophilic assistance if Reaction 4 operates. Figure 2 shows the effect of added *n*-butyl sulfide on the thiolsulfinate hydroperoxide reaction. From Figure 2, the addition of sulfide —whether thiolsulfinate is present or absent—results in the consumption of one mole of hydroperoxide per mole of sulfide presumably forming the sulfoxide and thereby preventing further hydroperoxide decomposition. The inhibitory effect of sulfoxides on the thiolsulfinate hydroperoxide reaction has been noted previously (*see* Figure 3) (2).

Table I. Spectral Shifts

Proton Acceptor	*Molarity*[a]	$\Delta\nu$[b] *(cm^{-1})*
tert-BuS(O)*tert*-Bu	1.5×10^{-2}	413
	6.2×10^{-3}	413
$CH_3S(O)CH_3$	1.9×10^{-2}	368[c]
tert-BuSS(O)*tert*-Bu	2.1×10^{-2}	292
	8.7×10^{-3}	289
$C_6H_5S(O)C_6H_5$	9.4×10^{-3}	285[d]

[a] Solutions are in CCl_4, $4.3 \times 10^{-3}M$ in phenol.
[b] Shift relative to ν "free" at 3613 cm^{-1} at 20°C.
[c] Value of 360 cm^{-1} reported by Engberts, J. B. R. N., Zuidema, G., *Rec. Trav. Chim. Pays–Bas* (1970) **89**, 1202.
[d] Value of 294 cm^{-1} reported by Gramstad, T., *Spectrochim. Acta* (1963) **89**, 829.

The relative ability of *tert*-butyl *tert*-butanethiolsulfinate, methyl sulfoxide, *tert*-butyl sulfoxide, and phenyl sulfoxide to participate as proton acceptors was studied by ir spectroscopy. Phenol in CCl_4 was used as the proton donor. A dilute solution of the phenol in CCl_4 displayed a single absorption band because of the O–H stretch (*see* Table I). The addition of the proton acceptor gave rise to a new, broad, and intense band at a lower frequency, but the free peak position and appearance changed little except for a decrease in intensity. Variation of the concentrations resulted in no significant changes in $\Delta\gamma$. Therefore, the $\Delta\gamma$ values should be considered reasonable approximations of 1:1 complexes at infinite dilution. The spectral shifts are indicative of ground state electron availability and basicity. Table I shows that *tert*-butyl *tert*-butanethiolsulfinate is comparable to phenyl sulfoxide and is not as efficient as alkyl sulfoxides in its ability to bind to hydrogen.

In Figure 3 phenyl sulfoxide has a small but distinct retarding effect on the decomposition of hydroperoxides by thiolsulfinate solutions. Since the thiolsulfinate is comparable in basicity to the phenyl sulfoxide, the thiolsulfinate should exert a similar effect, and no decomposition should occur until sufficient acid is generated. The addition of the more basic methyl sulfoxide or *tert*-butyl sulfoxide prevents hydroperoxide decomposition by effectively complexing the active acidic species.

The inability of complexed acid to decompose hydroperoxide was verified by adding sulfoxide to a solution of cumene hydroperoxide decomposing under acid catalysis. In Figure 4 the addition of *tert*-butyl sulfoxide to a solution of cumene hydroperoxide at 122 hr halted the ability of either sulfuric or *tert*-butyl sulfonic acid to further decompose hydroperoxide.

At elevated temperatures *tert*-butyl sulfoxide functions as a peroxide decomposer. We have found that the ability of the sulfoxide to complex the acidic species formed is significant under these conditions also. At elevated temperatures the decomposition of *tert*-butyl sulfoxide to acidic species is rapid. Thus, the concentration of the acid species soon exceeds the capacity of the remaining sulfoxide to effectively complex it. Therefore, the basic influence of the sulfoxide is indicated by a pronounced induction period.

In contrast to *n*-butyl sulfide, *tert*-butyl sulfide is not oxidized by cumene hydroperoxide at 25°C. The addition of *tert*-butyl disulfide did

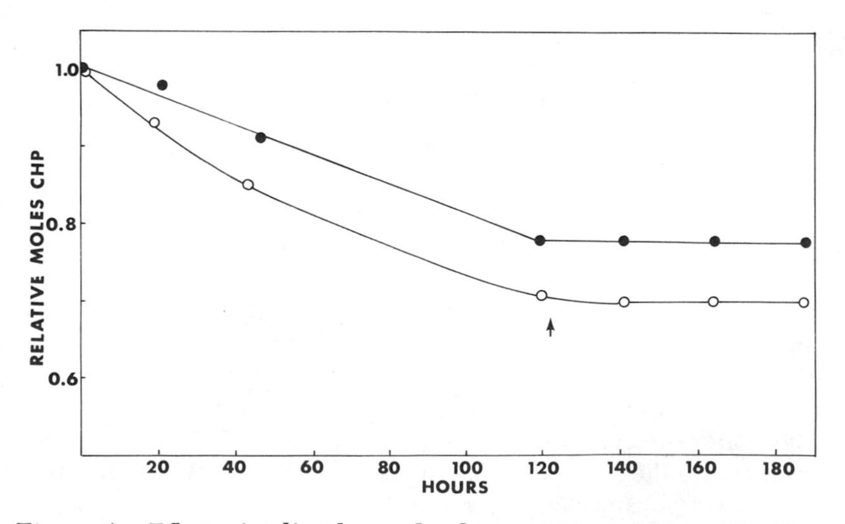

Figure 4. Effect of sulfoxide on the decomposition of cumene hydroperoxide in the presence of acids at 25°C: ○ *0.20*M *CHP and 0.0005*M *H_2SO_4 with 0.042*M tert-*BuS(O)*tert-*Bu added at 122 hr;* ● *0.20*M *CHP and 0.001*M tert-*$BuSO_3H$ with 0.041*M tert-*BuS(O)*tert-*Bu added at 122 hr*

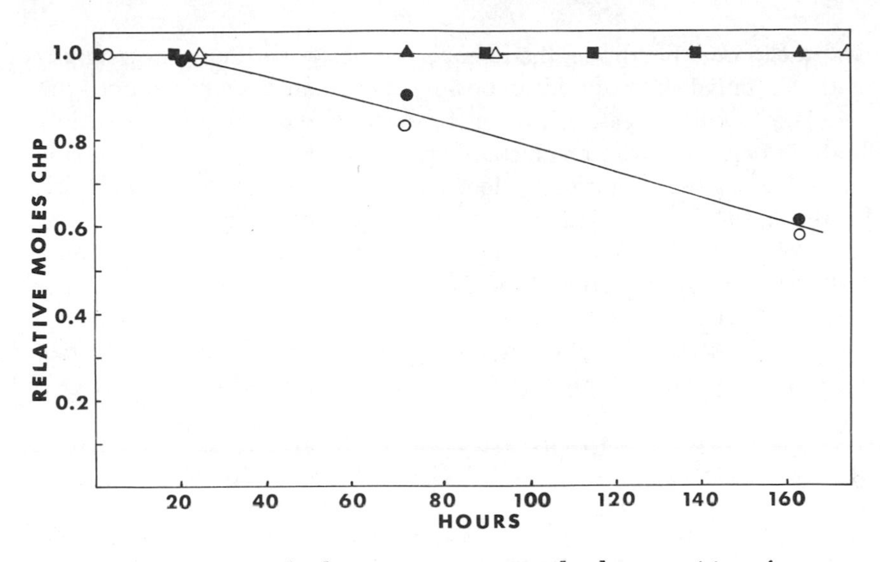

Figure 5. Effect of radical trapping agents on the decomposition of cumene hydroperoxide in the presence of tert-*butyl* tert-*butanethiolsulfinate at 25°C:* △ *0.20M CHP, 0.020M* tert-*BuSS(O)*tert-*Bu, and 0.019M β-naphthol;* ▲ *0.20M CHP, 0.020M* tert-*BuSS(O)*tert-*Bu, and 0.006M β-naphthol;* ■ *0.20M CHP, 0.022M* tert-*BuSS(O)*tert-*Bu, and 0.011M 2,6-di-*tert-*butyl-4-methylphenol;* ● *0.20M CHP, 0.022M* tert-*BuSS(O)*tert-*Bu, and 0.024M cyclohexanol;* ○ *0.20M CHP and 0.020M* tert-*BuSS(O)*tert-*Bu*

not accelerate the decomposition of cumene hydroperoxide in the presence of thiolsulfinate. Similarly, no accelerative effect was observed when the nucleophile was added along with benzoic acid. Thus, a carboxylic acid is not strong enough to catalyze the decomposition under the conditions indicated. The induction period remained at ~ 20 hr for the thiolsulfinate hydroperoxide reaction in the presence of added benzoic acid.

It has been established that the decomposition of cumene hydroperoxide in the presence of thiolsulfinate occurs primarily via a polar process (*1*). However, a homolytic process may be involved in the conversion of the thiolsulfinate to the active peroxide decomposer. This was probed by adding the radical inhibitors β-naphthol and 2,6-di-*tert*-butyl-4-methylphenol (*see* Figure 5). The inhibitors totally suppressed the decomposition of hydroperoxide by thiolsulfinate. In contrast to β-naphthol and 2,6-di-*tert*-butyl-4-methylphenol, the addition of cyclohexanol had no significant effect. Similarly, the addition of methanol only reduced the decomposition of hydroperoxide by the thiolsulfinate by ~ 1% after 186 hr.

To insure that the phenolic inhibitor was not simply interfering with the acid-catalyzed hydroperoxide decomposition, 2,6-di-*tert*-butyl-4-methylphenol was added after hydroperoxide decomposition commenced. In contrast to initial addition, the phenolic inhibitor had no detectable effect once the active peroxide decomposing agent had been generated.

The following reaction sequence is suggested. Initially the thiolsulfinate undergoes an intramolecular cycloelimination reaction as in Reaction 1. Extensive supporting evidence for this postulate has been presented by Block (*4*) from studies on the decomposition of thiolsulfinate under mild conditions ($< 100°C$). Since Reaction 1 is favored over homolysis and a reactive species, thiosulfoxylic acid, is formed, Reaction 1 is probably the initial step.

The thiosulfoxylic acid may be expected to undergo oxidation to the corresponding thiosulfurous and thiosulfuric acid (*2*).

$$RSSOH \xrightarrow{ROOH} RSSO_2H \xrightarrow{ROOH} RSSO_3H \qquad (7)$$

The analogous sulfenic acid has been shown to undergo rapid oxidation consuming 2 mol of cumene hydroperoxide at 25°C (*1*).

The thiosulfuric acids expected from Reaction 7 are unstable compounds. Acidified solutions of their salts are reported to undergo oxidation forming the corresponding disulfide (*10*).

$$2RSSO_3H + H_2O \xrightarrow{[O]} RSSR + 2H_2SO_4 \qquad (8)$$

The sulfuric acid, if formed in our reactions, would be a particularly effective peroxide decomposer.

The parent thiolsulfinate will complex with the acidic species formed:

$$R\overset{O}{S}SR + HA \rightleftharpoons [RS\underset{R}{S}O \cdots HA] \text{ complex} \qquad (9)$$

When sufficient acid (HA) is generated, hydroperoxide decomposition ensues:

$$ROOH \xrightarrow{HA} \text{nonradical products} \qquad (10)$$

Accordingly, if a more basic species (i.e., an alkyl sulfoxide) is added, the acidic species will be complexed effectively preventing the decomposition of cumene hydroperoxide by thiolsulfinate solutions at 25°C.

Radical involvement is indicated also by the inhibitory effect of radical trapping agents and product analysis, since 10–15% of the products from cumene hydroperoxide decomposition induced by the organic sulfur compounds result from free-radical processes (*2*). Acids will hydrogen bond to hydroperoxides and can influence rupture of the O–O bond in the hydroperoxide leading to formation of free radicals (*11*).

$$\underset{\text{H}}{\text{ROO}}\cdots\text{HA} \rightarrow \text{radicals (RO}\cdot\text{, etc.)} \tag{11}$$

Thus, it has been proposed that the homolytic decomposition of hydroperoxides can be induced by sulfenic acid (*12, 13*). There is evidence that various carboxylic acids can promote radical formation from hydroperoxides at elevated temperatures (*11, 14*). The intermediate thiosulfurous acid (Reaction 7) itself may function as the source of radicals, since sulfinic acid is known to initiate the radical polymerization of vinyl monomers at 20°C (*15*). Based on the AIBN-initiated oxidation of cumene, Koelewijn and Berger (*16*) proposed that pro-oxidant effects arise from catalysis of the radical decomposition of hydroperoxides by intermediate compound formation between the hydroperoxide and sulfoxide. However, under our conditions hydroperoxide was stable in the presence of sulfoxide alone.

The radicals generated may induce decomposition of the thiolsulfinate as proposed by Barnard and Percy (*6*). An alternative fate of generated radicals will be an attack on thiosulfoxylic acid:

$$\text{RSSOH} + \text{RO}_2\cdot \rightarrow \text{RSSO}\cdot + \text{RO}_2\text{H} \tag{12}$$

This reaction is facilitated by formation of the stabilized RSSO· radical that is isoelectronic with the stabilized polysulfide radical, $RS_3\cdot$. The analogous sulfenic acids are effective radical scavengers reacting with peroxy radicals with a rate constant of $10^7 M^{-1}$ sec^{-1} at 60°C (*16*). The S–S bond in the thiolsulfinate is weak, and the corresponding bond in the thiosulfoxyl radical should be considerably less stable. Thus, the thiosulfoxyl radical may function as a source of sulfur oxides:

$$\text{RSSO}\cdot \rightarrow \text{RS}\cdot + [\text{SO}] \tag{13}$$

Thermolysis of the analogous butyl sulfoxylate produces sulfur and sulfur dioxide presumably via sulfur monoxide (*17*). Further studies regarding the participation of free radicals are in progress.

In conclusion, several alternative mechanisms for the thiolsulfinate hydroperoxide reaction have been suggested. The results indicate that nucleophilic displacement on *tert*-butyl *tert*-butanethiolsulfinate by water, sulfides, and disulfide does not play a significant role under the conditions cited. The possibility of a homolytic process has been considered, and evidence indicates that radicals play an essential role in formation of the active peroxide decomposing species, that then function primarily via a polar mechanism. Since the active peroxide decomposer is an acidic species, its activity can be affected significantly by the basic character of the parent thiolsulfinate or sulfoxide.

Acknowledgment

The authors wish to thank the Goodyear Tire and Rubber Company for financial support of this research project.

Literature Cited

1. Shelton, J. R., Harrington, E. R., *Rubber Chem. Technol.* (1976) **49**, 147.
2. Harrington, E.R., PhD dissertation, Case Western Reserve University (1976).
3. Mair, R. D., Graupner, A. J., *Anal. Chem.* (1964) **36**, 1944.
4. Block, E., O'Connor, J., *J. Am. Chem. Soc.* (1974) **96**, 3929.
5. Kice, J. L., Cleveland, J.P., *J. Am. Chem. Soc.* (1973) **95**, 109.
6. Barnard, D., Percy, E. J., *Chem. Ind.* (1960) 1332.
7. Koch, P., Ciuffarin, E., Fava, A., *J. Am. Chem. Soc.* (1970) **92**, 5971.
8. Barnard, D., Bateman, L., Cole, E. R., Cunneen, J. I., *Chem. Ind.* (1958) 918.
9. Barnard, D., *J. Chem. Soc.* (1957) 4675.
10. Milligan, B., Swan, J., *Rev. Pure Appl. Chem.* (1962) **12**, 72.
11. Emanuel, N. M., Denisov, E. T., Maizus, Z. K., "Liquid-Phase Oxidation of Hydrocarbons," p. 81, Plenum, New York, 1967.
12. Shelton, J. R., *Rubber Chem. Technol.* (1974) **47**, 949.
13. Scott, G., *Mech. React. Sulfur Compd.* (1969) **4**, 99.
14. Privalova, L. G., Maizus, Z. K., *Izv. Akad. Nauk SSSR* (1964) 281; *Chem. Abstr.* (1964) **60**, 11867.
15. Overberger, C. G., Godfrey, J. J., *J. Polym. Sci.* (1959) **40**, 179.
16. Koelewijn, P., Berger, H., *Rec. Trav. Chim. Pays–Bas* (1974) **93**, 63.
17. Mathey, F., Lampin, J. P., *Tetrahedron Lett.* (1972) 3121.

RECEIVED May 12, 1977.

20

Inhibited Autoxidation of Polypropylene

DALE E. VAN SICKLE and DAVID M. POND

Research Laboratories, Tennessee Eastman Company, Division of Eastman Kodak Company, Kingsport, TN 37662

*For di-*tert*-butyl peroxide-initiated oxidations in benzene solution at 120°C, singly hindered phenols appear to have twice the chain-stopping power of the di-*tert*-butylphenols; e.g., a given molar quantity of 2-*tert*-butyl-4-methylphenol will give twice the inhibition period of 2,6-di-*tert*-butyl-4-methylphenol. The stoichiometry of the polypropylene peroxy radicals reacting with 2,6-di-*tert*-butyl-4-alkylphenols is independent of their attachment in large molecules; all of the phenolic residues in these polyfunctional molecules were utilized in stopping oxidation chains. Although inhibition periods in polypropylene oxidations are proportional to the inhibitor concentration, the overall kinetics of the inhibited oxidations are complex. Attempts to demonstrate the superiority of singly hindered phenols in oven-aging tests were only partially successful.*

Evaluation of antioxidants for polyolefin stabilization is done empirically by compounding resin with additive, subjecting the composition to elevated temperatures, and later measuring some extensive property such as flexibility or impact strength of the sample. Many properties of the inhibitor besides its reactivity toward peroxy radicals enter into its overall effectiveness; e.g., volatility, thermal stability during compounding, and compatibility. For designing a superior antioxidant, it is desirable to separate these factors and to evaluate the chemical antioxidant activity of an inhibitor independently of the ancillary properties of the material. The phenomenon of inhibited oxidation of simple, low molecular weight substrates [styrene (*1, 2, 3*), tetrahydronaphthalene (*4, 5*), and cumene (*6*)] has been investigated thoroughly, but extension of these studies to polymeric substrates has not been done analytically.

A suitable set of conditions for the solution oxidation of low molecular weight crystalline polypropylene has been described (*7, 8, 9*).

0-8412-0381-4/78/33-169-237$05.00/1

The solution oxidation technique allows the study of polyolefin autoxidation under conditions where the temperature, concentration of reactants, and rates of radical initiation can be controlled. The results should be considered as a useful prelude to any fundamental understanding of the autoxidation processes which occur in neat polymers where the effects of very high viscosity, partial crystallinity, and oxygen diffusion rates are included. The objective of our work was to determine the kinetics and stoichiometry of the inhibited autoxidation of polypropylene in solution. A relatively detailed study of the oxidation of polypropylene inhibited by 2,6-di-*tert*-butyl-4-methylphenol [butylated hydroxytoluene (BHT)] has been made for comparison with data obtained in polypropylene oxidations inhibited by a variety of other stabilizers which include commercial polyfunctional antioxidants. Singly hindered phenols appeared to be superior in the inhibited-solution oxidation of polypropylene, and the application of this finding to stabilization technology was investigated briefly.

Experimental

Materials. The polypropylene (PP) used was a low molecular weight crystalline material which has been described previously (*8*). The polymer was extracted with hexane and acetone before use and dried at 0.1 torr on a steam bath. The extracted pellets had a number average molecular weight (M_n) of ca. 29,000 and were completely soluble in benzene at the 120°C reaction temperature. The 9,10-dihydroanthracene (DHA) from Aldrich Chemical Company was recrystallized from ethanol before use; mp 106°–107°C. Benzene, used directly, was MCB Chromatoquality; di-*tert*-butyl peroxide (*tert*-Bu_2O_2) 99% from Lucidol, also was used directly. Azobis(2-methylpropionitrile) (ABN) was Eastman white label grade which was recrystallized from acetone–methanol.

2,6-Di-*tert*-butyl-4-methylphenol Eastman Tenox BHT, food grade, appeared to be very pure by GLC analysis and was used directly. The other simple alkyl derivatives of phenol were obtained from the usual fine chemical sources (Eastman Organic Chemicals, Aldrich, Columbia, and K and K). All were distilled and/or recrystallized and the assigned structures were confirmed by NMR. Other commercial stabilizers were obtained from their manufacturers, recrystallized (usually from ethanol or ethanol–water), and submitted for NMR confirmation of the assigned structure and purity.

2,4,6-Tris[3-*tert*-butyl-4-(hydroxybenzyl)]mesitylene (III) was synthesized starting from 2-*tert*-butyl-6-chlorophenol, which in turn was prepared by the method of Kolka et al. (*10*). The 2-*tert*-Butyl-6-chlorophenol was alkylated with $\alpha^1,\alpha^3,\alpha^5$-(trihydroxyhexamethyl)benzene (Aldrich) with the use of boron trifluoride etherate catalyst (*11*). A mixture of I and II was obtained, separated by fractional crystallization from methylcyclohexane. The structures of the two products were assigned on the

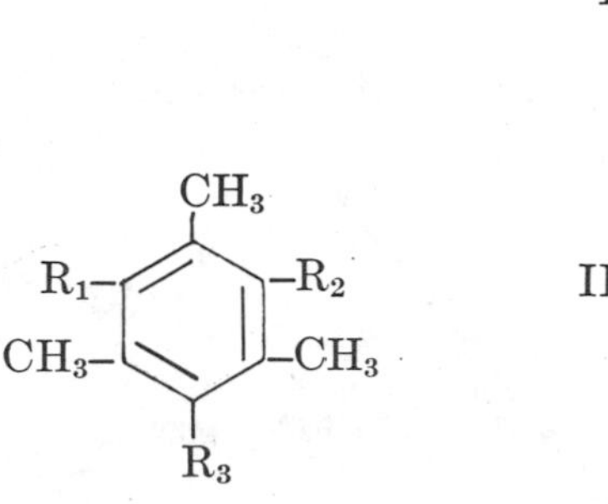

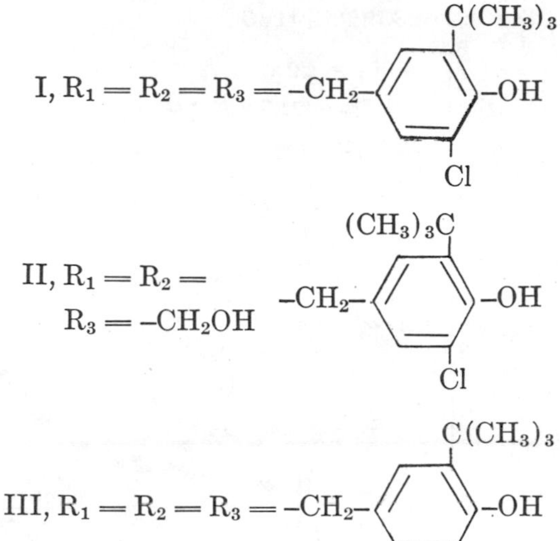

basis of NMR spectra. I was dehalogenated by a sodium–liquid ammonia technique (*12*) to yield first a product containing 5.5% chlorine and, with a repeated pass through the dehalogenation procedure, a product containing 2.76% chlorine. The NMR spectrum of this material indicated that it was 85 to 90% III, the desired product.

2,4,6-Tris(3-*tert*-butyl-4-hydroxy-6-methylbenzyl)mesitylene (IV) was obtained by alkylation of 2-*tert*-butyl-5-methylphenol with $\alpha^1,\alpha^3,\alpha^5$-(trihydroxyhexamethyl)benzene by the procedure indicated previously. From 0.020 mol of the phenol and 0.0048 mol of the tris(hydroxymethyl)-benzene, 4.27 g of crude product was obtained from which 2.35 g of white powder, mp 220°–240°C, was isolated by stirring with benzene. The NMR spectrum of the product was in agreement with the proposed structure, but it indicated that some loss of *tert*-butyl groups had occurred.

Oxidation Procedure. The constant-volume/falling pressure oxidation apparatus has been described previously (*8*). Inhibitor solutions in benzene were made up in volumetric flasks and a weighed portion was transferred to the reaction bulb which contained weighed quantities of polypropylene and initiator. Total volumes of reactant were usually 56–58 mL. The bulb was shaken in the thermostat with oxygen at a total pressure of 5–6 atm. The vapor pressure of benzene at the reaction temperature of 120°C is 3 atm (in the absence of dissolved polypropylene) so that the net partial pressure of oxygen in the bulb void was 2–3 atm during the experiment. Time, oxygen reservoir temperature, and pressure readings were recorded at appropriate intervals, and the pressure–time readings were corrected then (*8*) to a constant oxygen reservoir temperature. Oxygen consumption was plotted as a function of time, and the best lines were drawn to calculate inhibited (R_{inh}) and uninhibited (R_o) rates. A typical result for polypropylene is shown in Figure 1. The inhibition period, t_{inh}, was taken as the time axis projection of the point of intersection of the two lines representing the inhibited and uninhibited rates.

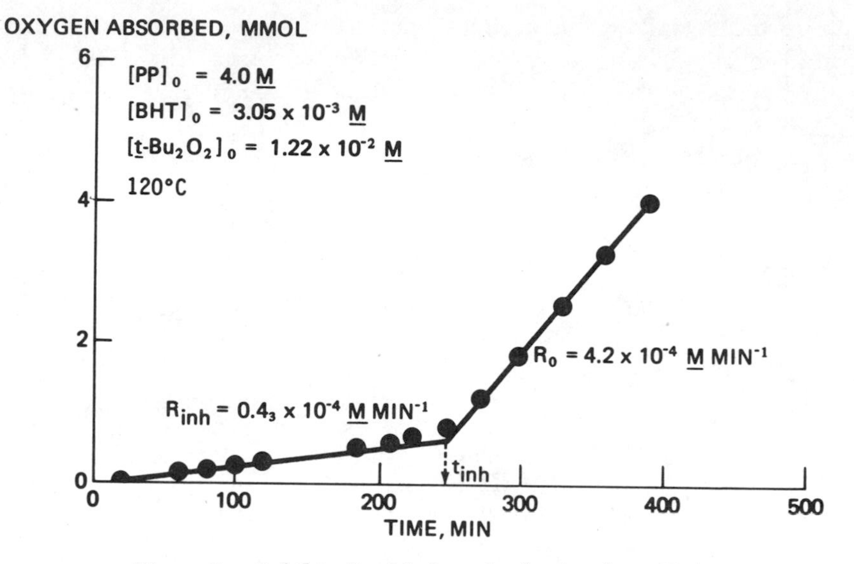

Figure 1. Inhibited oxidation of polypropylene (PP)

Results and Discussion

The results of the inhibited oxidation experiments with polypropylene are summarized in Tables I and II. A few experiments where 9,10-dihydroanthracene was used as the substrate are summarized in Table III.

Mechanism of Antioxidant Action. The effect of antioxidant additives (AH) on an oxidizing hydrocarbon can be discussed profitably in

Table I. Oxidation of Polypropylene (PP) in Benzene Solution

Run No.	$[BHT]_o$[a] $\times 10^3$, M	$[PP]_o$,[a,b] M	[tert-$Bu_2O_2]_o$[a] $\times 10^2$, M
1	0.0	2.52	1.23
2	0.0	2.52	1.27
3	0.40	2.50	1.19
4	0.90	2.52	1.22
5	1.22	2.48	1.02
6	3.01	2.55	1.24
7	6.01	2.52	1.21
8	4.07	0.0	1.27
9	3.93	0.55	1.40
10	3.49	1.24	1.21
11	3.05	4.00	1.22
12	6.57	2.52	3.58
13	6.11	2.51	7.10

[a] Initial concentrations.
[b] Concentration in monomer units, mol wt = 42.1.

terms of the following equations, where RH is the substrate molecule and I_2 is an initiator. Where the substrate is polypropylene, R is assumed to be $[\ CH_2\text{–}(CH_3)\underset{|}{C}\ ]$.

$$I_2 \xrightarrow{k_1} 2I\cdot \qquad R_i = 2ek_1[I_2] \tag{1}$$

$$I\cdot\ [\text{or } R\cdot \text{ from Equation 3}] + O_2 \xrightarrow{k_2} IO_2\cdot \quad (\text{or } RO_2\cdot) \tag{2}$$

$$A\cdot\ [\text{from Equation 6}] + O_2 \underset{k_{-2}'}{\overset{k_2'}{\rightleftarrows}} AO_2\cdot \tag{2'}$$

$$RO_2\cdot + RH \xrightarrow{k_3} RO_2H + R\cdot \tag{3}$$

$$RO_2\cdot + R\cdot \xrightarrow{k_4} \text{nonradical products} \tag{4}$$

$$2RO_2\cdot \xrightarrow{k_5} \text{nonradical products} + O_2 \tag{5}$$

$$RO_2\cdot + AH \underset{k_{-6}}{\overset{k_6}{\rightleftarrows}} RO_2H + A\cdot \tag{6}$$

at 120°C Inhibited by 2,6-Di-*tert*-butyl-4-methylphenol (BHT)

$R_{inh} \times 10^4$, M(min^{-1})	t_{inh} (min)	$R_o \times 10^4$, M(min^{-1})	Time, (min)	$[O_2]_f$[c] $\times 10$, M
—	< 10	2.6	450	1.08
—	~ 5	2.5	350	0.81
0.84	42	2.7	350	0.85
0.70	75	2.7	360	0.80
0.40	114	2.7	350	0.55
0.47	234	3.2	340	0.30
0.43	420	2.6	540	0.50
0.17	90[d]	0.37[d]	507	0.16
0.37	220	1.5	480	0.39
0.29	230	1.8	460	0.48
0.43	235	4.2	410	0.78
0.73 (1.1)	152	6.3	365	1.15
2.7	75	12.0	180	1.13

[c] Total oxygen absorbed over the time of the experiment.
[d] Continuous acceleration to final rate.

Table II. Inhibited Oxidation of Polypropylene

Run No.	*Inhibitor*
1	Pentaerythritol tetrakis[3-[3,5-di-*tert*-butyl-4-(hydroxyphenyl)]propionate]
2	Pentaerythritol tetrakis[3-[3,5-di-*tert*-butyl-4-(hydroxyphenyl)]propionate]
3	1,3,5-Tris[3,5-di-*tert*-butyl-4-(hydroxybenzyl)]-2,4,6-trihydroxytriazine
4	1,3,5-Trimethyl-2,4,6-tris[3,5-di-*tert*-butyl-4-(hydroxybenzyl)]benzene
5	"Alkylated" 2-*tert*-butyl-5-methylphenol
6	4,4′-Methylenebis(2,6-di-*tert*-butylphenol)
7	Dioctadecyl 3,5-di-*tert*-butyl-4-(hydroxybenzyl)phosphonate
8	Monoethyl 3,5-di-*tert*-butyl-4-(hydroxybenzyl)phosphonate, nickel salt
9	2,6-Bis(2-octadecyl)-4-methylphenol
10	Dilauryl 3,3′-thiodipropionate
11	2-*tert*-Butyl-4-methylphenol
12	2-*tert*-Butyl-4-methylphenol
13	2-*tert*-Butyl-4-methylphenol
14	2-*tert*-Butyl-4-methylphenol + BHT
15	2-*tert*-Butyl-4-ethylphenol
16	Methyl 3-[3-*tert*-butyl-4-(hydroxyphenyl)]propionate
17	Butylated hydroxy anisole (BHA) [d]
18	2-*tert*-Butyl-5-methylphenol
19	2-*tert*-Butyl-6-methylphenol
20	2-*tert*-Butyl-4,6-dimethylphenol
21	2-*tert*-Butylphenol
22	2,4-Di-*tert*-butylpnenol
23	2,6-Di-*tert*-butylphenol
24	Methyl 3-[3,5-di-*tert*-butyl-4-(hydroxyphenyl)]propionate
25	2,6-Diisopropylphenol
26	Phenol
27	Phenol
28	*N*-(2-hydroxyphenyl)benzamide
29	*N*-Phenyl benzamide

[a] [Polypropylene]$_o$ = 2.5M [*tert*-Bu_2O_2]$_o$ = 1.2 × $10^{-2}M$.
[b] Total phenolic group concentration.
[c] No inhibition period; steadily decreasing rate from 2.1 (initial) to 0.6 (final) M min^{-1}.
[d] A mixture of 2-*tert*-butyl-4-methoxyphenol and 2,6-*tert*-butyl-4-methoxyphenol.

Polypropylene in Benzene Solution at 120°C[a]

$[Inhibitor]_o \times 10^3$, M	t_{inh} (min)	R_{inh}, $\times 10^4$, M (min^{-1})	$R_o \times 10^4$, M (min^{-1})
0.595	200	0.55	2.7
(2.38) [b]			
0.93	300	0.49	3.1
(3.72) [b]			
0.858	225	0.61	2.4
(2.58) [b]			
0.915	210	0.79	2.2
(2.75) [b]			
1.31	310	0.43	1.8
(3.93) [b]			
1.65	90	0.33	2.0
(3.30) [b]	260	0.72	
1.06	76	0.65	2.2
1.19	140	0.57	2.1
(2.38) [b]			
1.64	170	0.51	2.4
2.28	0	2.1 [c]	0.61 [c]
1.12	180	0.33	2.6
1.71	300	0.26	2.4
3.59	660	0.26	1.9
1.79	550	0.56	2.6
1.81			
1.47	230	0.57	2.7
1.73	265	0.45	2.3
3.90	> 500 [e]	0.32	[e]
2.74	230	0.32	2.1
3.74	240	0.41	2.3
2.42	204	0.38	3.0
2.44	210	0.53	1.5
3.67	600	0.63	1.8
2.10	196	0.52	2.5
2.88	195	0.35	3.1
2.38	235	0.47	1.8
7.61	[f]	1.2 [f]	[f]
2.50	[f]	1.6 [f]	[f]
2.43	> 500 [e]	0.34	0.7 [e]
3.03	0	2.5 [g]	

[e] No indication of acceleration to an R_o of 2–2.8*M* min^{-1} within 500 min; strong orange color developed.

[f] No obvious inhibition periods; slow acceleration of oxidation rate to constant value after ~ 100 min.

[g] Initial (and constant) rate.

Table III. Inhibited Oxidations of

Run No.	*Inhibitor*
1	None
2	2,6-Di-*tert*-butyl-4-methylphenol (BHT)
3	2,6-Di-*tert*-butyl-4-methylphenol (BHT)
4	Pentaerythritol tetrakis[3-[3,5-di-*tert*-butyl-4-(hydroxyphenyl)]propionate]
5	4,4′-Methylenebis(2,6-di-*tert*-butylphenol)
6	2-*tert*-Butyl-4-methylphenol
7	2-*tert*-Butyl-4-methylphenol
8	2-*tert*-Butyl-4-methylphenol

[a] Solution 60°C, $[DHA]_o = 0.63M$, $[ABN]_o = 0.014M$.

$$RO_2\cdot + A\cdot \xrightarrow{k_7} \text{nonradical products} \quad (7)$$

$$A\cdot + RH \xrightarrow{k_8} AH + R\cdot \quad (8)$$

$$A\cdot + A\cdot \xrightarrow{k_9} \text{nonradical products} \quad (9)$$

$$AH + O_2 \xrightarrow{k_{10}} A\cdot + HO_2\cdot \quad (10)$$

Where *tert*-butyl peroxide is the initiator, Equation 1 takes two steps:

$$tert\text{-}Bu_2O_2 \xrightarrow{k_{1a}} 2\ tert\text{-}BuO\cdot \qquad R_i = 2\,k_{1a}[Bu_2O_2] \quad (1a)$$

$$tert\text{-}BuO\cdot + RH \xrightarrow{k_{1b}} tert\text{-}BuOH + R\cdot \quad (1b)$$

Use of *tert*-butyl peroxide also introduces a possible competing reaction for Equation 6:

$$tert\text{-}BuO\cdot + AH \xrightarrow{k_6'} tert\text{-}BuOH + A\cdot \quad (6')$$

For AH to be an efficient inhibitor, it is necessary that $k_6[AH] >> k_3[RH]$ and that $k_7[RO_2\cdot] >> k_8[RH]$ and $k_{-6}[RO_2H]$. Although usually not discussed, it is also necessary that equilibrium (Equation 2′) lie on the left. Another and possibly final requirement is that chains not be started by a direct oxygen–AH reaction (Equation 10).

9,10-Dihydroanthracene in Benzene[a]

[Inhibitor]$_o$ $\times 10^3$, M	t_{inh} *(min)*	R_{inh} $\times 10^4$, M *(min^{-1})*	R_o $\times 10^4$, M *(min^{-1})*
0.0	~ 0	—	26
1.90	175	1.4	13.9
3.69	~ 400	1.0	13
0.646 (2.58)	270	1.8	10.5
1.63 (3.26)	320	1.3	9.7
1.00	94	3.0	17.4
1.86	~ 250	2.2	12.5
2.97	~ 365	1.4	10.4

Rate Laws. When the conditions just given are met, the rate of consumption of oxygen (R_{inh}) during the inhibition period, t_{inh}, has been given by Howard and Ingold (*4, 5*) as

$$R_{inh} = \frac{ek_1k_3\,[RH]\,[I_2]}{k_6\,[AH]} \quad (11)$$

where e is the efficiency factor for the production of radicals from the initiator I_2 [= 0.60 for 2,2′-azobis[2-methylproprionitrile (ABN)]. This equation was developed for application to inhibited oxidations of tetrahydronaphthalene where kinetic chain lengths are significantly above unity and termination by Equations 4 and 5 is ignored. As pointed out by Howard and Ingold (*4, 5*), the kinetic form is independent of which termination (Equation 7 or 9) is important as long as transfer reaction k_{-6} is negligible. In our case, where initiation was by *tert*-butyl peroxide and Reaction 6′ was considered, the usual steady-state treatment gives the following equation for the inhibited rate of oxidation:

$$R_{inh} = \frac{R_i\,[RH]}{k_{1b}\,[RH] + k_6'\,[AH]} \left\{ \frac{k_3}{2k_6\,[AH]} (k_{1b}\,[RH] - k_6'\,[AH]) + k_{1b} \right\} \quad (12)$$

where $R_i = 2k_{1_a}[tert\text{-}Bu_2O_2]$ = the rate of initiation.
Where Reaction 6′ is omitted from the scheme ($k_6' = 0$), Equation 12 reduces to Equation 13,

$$R_{\text{inh}} = \frac{R_i \, k_3 \, [\text{RH}]}{2k_6 \, [\text{AH}]} + R_i \qquad (13)$$

which is identical to Equation 11 except for the additional term, R_i.

BHT Inhibition of Polypropylene Oxidation at 120°C. The results of applying the measured, inhibited oxidation rates listed in Table I to Equation 13 are described here. We can confirm that the inhibited oxidation rate (R_{inh}) shows a first-order dependence on initiation rate (R_i) up to a value of $6.5 \times 10^{-5} M$ min^{-1} for R_i. Figure 2 shows the results of plotting R_{inh} against [*tert*-Bu_2O_2]. The initiation rate corresponding to the *tert*-Bu_2O_2 concentrations can be computed from Equa-

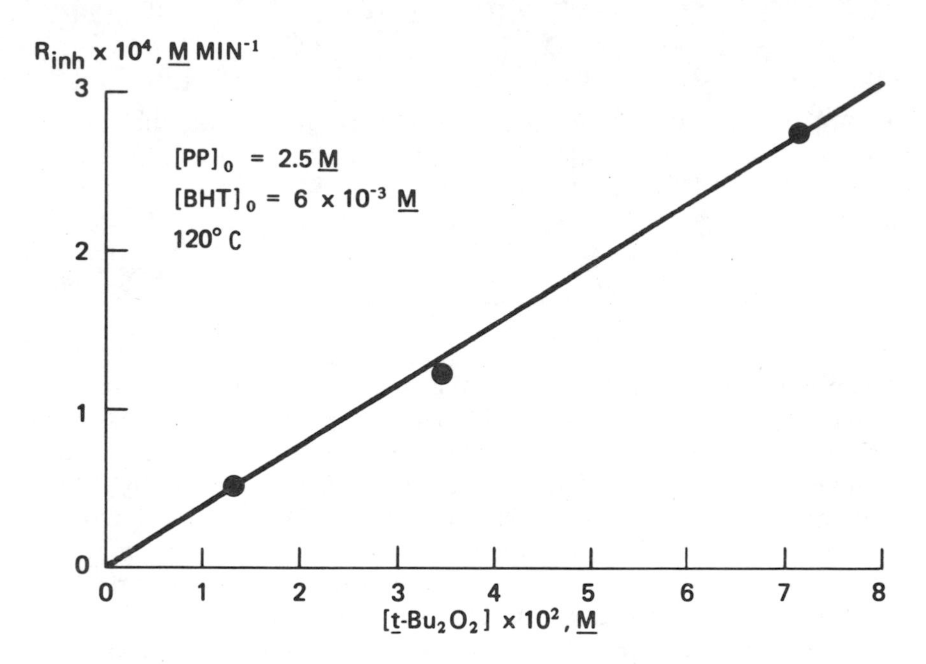

Figure 2. Inhibited-rate dependence on initiator concentration

tion 1a where k_{1_a} has the value 4.7×10^{-4} min^{-1} at 120°C. Uninhibited oxidations of polypropylene (8) are ~ 1/2-order dependent on R_i, but a plot of R_{inh} vs. [*tert*-Bu_2O_2]$^{1/2}$ shows definite curvature.

BHT-inhibited polypropylene oxidations may be inverse-order dependent on BHT concentration as suggested by Equation 13, but zero-order dependence on inhibitor concentration is not rigorously excluded. The appropriate plot is shown as the lower line of Figure 3. Experiments at sufficiently low BHT concentration to distinguish zero-order from inverse-order dependence were not practical since inhibition times would be too short to accurately estimate R_{inh}. According to Equation 13 the

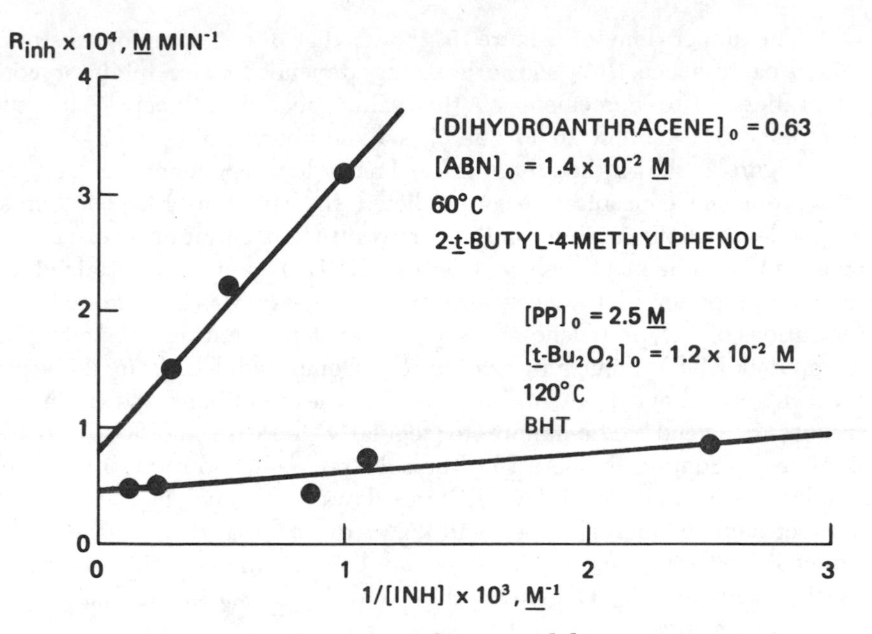

Figure 3. Inhibited-rate dependence on inhibitor concentration

intercept of the line of Figure 3 should be equal to R_i. The apparent value of the intercept is $0.4 \times 10^{-4} M$ min^{-1}, whereas the value of R_i at $1.2 \times 10^{-2} M$ *tert*-Bu_2O_2 is calculated from Equation 1a to be $0.1 \times 10^{-4} M$ min^{-1}. Uncertainty in the value of the intercept does not seem to be the cause of the discrepancy.

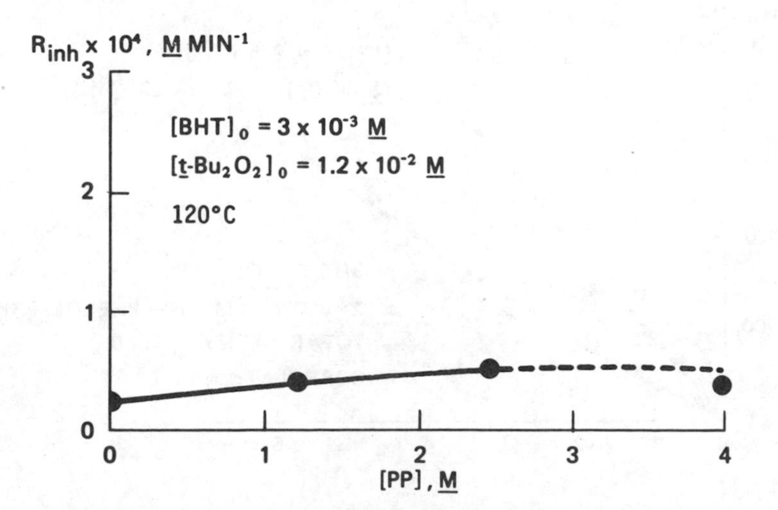

Figure 4. Inhibited-rate dependence on polymer concentration

The upper line of Figure 3 shows that for a simple substrate, dihydroanthracene, R_{inh} shows a strong dependence on inhibitor concentration. The agreement of the value of the intercept with the computed R_i ($\sim 0.1 \times 10^{-4} M$ min^{-1}) is again poor.

Figure 4 shows that there is no first-order dependence of R_{inh} on polypropylene concentration as predicted by Equation 13. The actual dependence of the inhibited rate on substrate concentration is essentially zero, and useful kinetic chain lengths (KCL's) cannot be attained by increasing polypropylene concentration up to 4.0*M*. Uninhibited (*8*) oxidations of polypropylene are $\sim$ 0.7-order dependent on substrate, and computed KCL's range upwards to 40. Computed KCL's in this case, i.e., R_{inh}/R_i, range from one to four for the conditions shown in the Figure 4 legend. The failure to regularly increase the KCL of the inhibited oxidation by increasing the substrate concentration discourages application of more complex (*13*) rate laws to the polymer system. A correct kinetic analysis (*14*) will have to make use of a theory of concentrated polymer solutions where conformational and interpenetrating network effects greatly alter the simple statistical basis upon which descriptive kinetics is based.

Inhibition Periods of Polypropylene Oxidations. Although the kinetics of BHT-inhibited polypropylene oxidations may be complex, the consumption of the inhibitor seems to follow simple stoichiometry, at least up to $3 \times 10^{-3} M$ BHT. In Figure 5, the t_{inh} values for polypropylene oxidations $[PP]_o = 2.5M$, $[tert\text{-}Bu_2O_2]_o = 1.2 \times 10^{-2} M$ are plotted against the initial BHT concentrations (triangles). The heavy line close to these

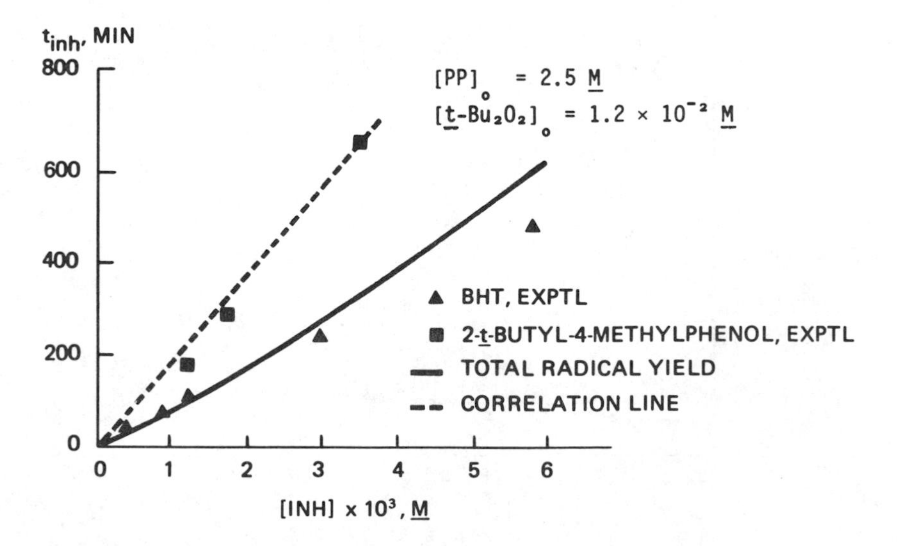

Figure 5. Inhibition of polypropylene (PP) oxidation

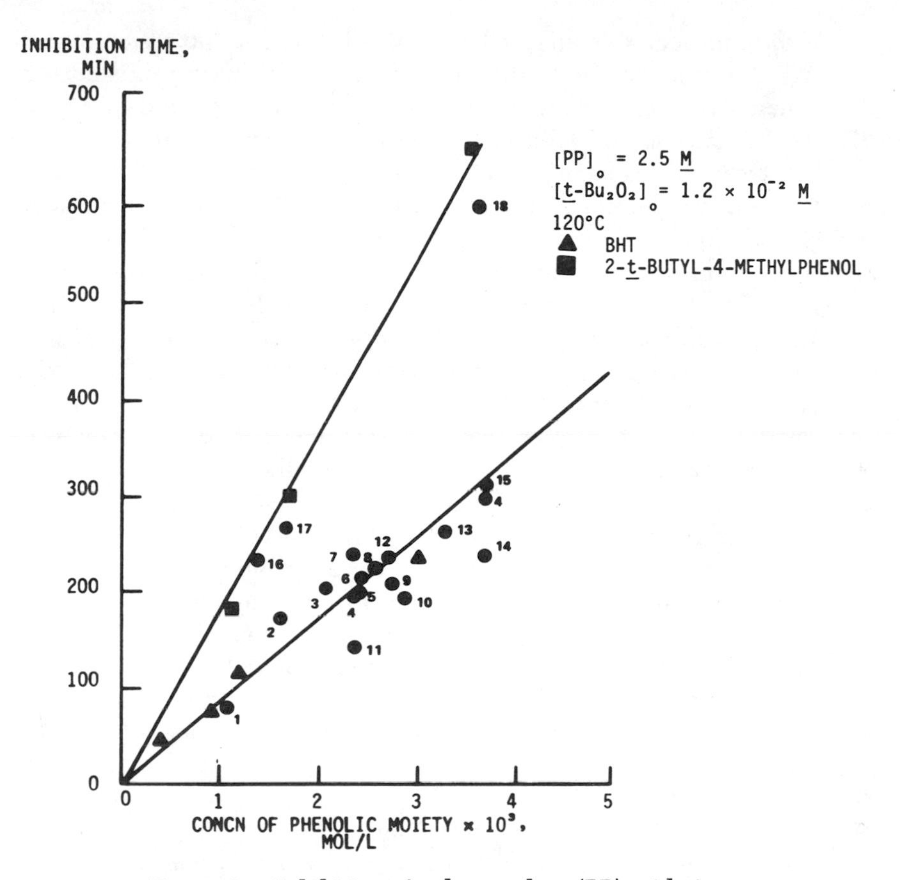

Figure 6. Inhibition of polypropylene (PP) oxidation.

*1 = Dioctadecyl 3,5-di-*tert*-butyl-4-(hydroxybenzyl)phosphonate, 2 = 2,6-Bis(2-octadecyl)-4-methylphenol, 3 = 2,6-Di-*tert*-butylphenol, 4 = Pentaerythritol tetrakis[3-[3,5-di-*tert*-butyl-4-(hydroxyphenyl)]propoinate], 5 = 2-*tert*-Butyl-4,6-dimethylphenol, 6 = 2-*tert*-Butylphenol, 7 = 2,6-Diisopropylphenol, 8 = 1,3,5-Tris[3,5-di-*tert*-butyl-4-(hydroxybenzyl)]-2,4,6-trihydroxytriazine, 9 = 1,3,5-Trimethyl-2,4,6-tris[3,5-di-*tert*-butyl-4-(hydroxybenzyl)]benzene, 10 = Methyl 3-[3,5-di-*tert*-butyl-4-(hydroxyphenyl)]propionate, 11 = Monoethyl 3,5-di-*tert*-butyl-4-(hydroxybenzyl)-phosphonate, nickel salt, 12 = 2-*tert*-Butyl-5-methylphenol, 13 = 4,4'-Methylene-bis(2,6-di-*tert*-butylphenol), 14 = 2-*tert*-Butyl-6-methylphenol, 15 = "Alkylated" 2-*tert*-butyl-5-methylphenol, 16 = 2-*tert*-Butyl-4-ethylphenol, 17 = Methyl 3-[3-*tert*-butyl-4-(hydroxyphenyl)]propionate, 18 = 2,4-Di-*tert*-butylphenol.*

points represents the total number of radicals generated from decomposition of the *tert*-butyl peroxide initiator as a function of time and was computed from the relationship

$$\text{total radicals} = 2\,[tert\text{-Bu}_2\text{O}_2]_o[1 - \exp(-k_{1a}t)]$$

$$k_{1a} = 4.7 \times 10^{-4}\ \text{min}^{-1}\ \text{at}\ 120°\text{C}$$

The coincidence of the line and the experimental points is very good although there is some indication that significant departure intrudes at the highest BHT concentrations; the last point, corresponding to 0.006*M* BHT concentration, falls well below the line. Reaction 6′ may compete significantly with Reaction 1b under these conditions.

Most other inhibitors, as long as they contain the basic 2,6-di-*tert*-butylphenol structure, can be included in this correlation. The circles plotted in Figure 6 represent data for most of the phenols tested besides BHT and 2-*tert*-butyl-4-methylphenol. The lower line appears to be a reasonable correlation for both the monofunctional phenols as well as commercial stabilizers when the actual molarity of the phenolic groups is plotted. The full utilization of all of the phenolic groups on the polyfunctional inhibitors is somewhat surprising, but is consistent with other aspects of this investigation as described later.

The most interesting data pertain to the effectiveness of singly hindered phenols as antioxidants. The top line plotted in Figure 5 correlates data for oxidations inhibited by 2-*tert*-butyl-4-methylphenol. Some other 2-*tert*-butyl-4-alkylphenols are included along the top line of Figure 6. Clearly, a given concentration of the singly hindered phenol gives a significantly greater inhibition period, actually by a factor of two. Exceptions to this classification include *o*-*tert*-butylphenol, which seems to have the same inhibiting stoichiometry as the di-*tert*-butylphenols. 2-*tert*-Butyl-5-methylphenol appears also to correlate best with doubly hindered phenols.

Figures 5 and 6 show that two polypropylene peroxy radicals are scavenged by the singly hindered phenols, whereas only one is scavenged by the di-*tert*-butylphenols. For most inhibitors acting on the low molecular weight substrate cumene, a stoichiometry of two radicals consumed per inhibitor molecule has been found (*15*). Apparently for polypropylene substrate, 2-*tert*-butyl-4-methylphenol more nearly approaches this ideal. Thus, for BHT-inhibited oxidations Reaction 9 applies, whereas with singly hindered phenols, Reaction 7 is appropriate. Reaction 7 along with 2:1 stoichiometry is commonly assumed for neat polymer systems so that the number of radicals generated during an oxidation can be determined (*16, 17*).

In the absence of analytical data for the products formed from the antioxidants, we can only speculate on the reason for the observed stoichiometry. However, a reviewer has suggested that an explanation may lie in the instability of peroxy radical–phenoxy radical products at 120°C (*18*) and that the only effective terminations are those that occur between two phenoxy radicals. For the singly hindered phenoxy radical product, there is the possibility of terminating two additional peroxy radicals as follows.

For the doubly hindered phenoxy radical products, there are no active hydrogens to be scavenged by the peroxy radicals; further, disproportionation of the dimer back to the parent phenol and a quinone methide must be assumed to be slow (*19*).

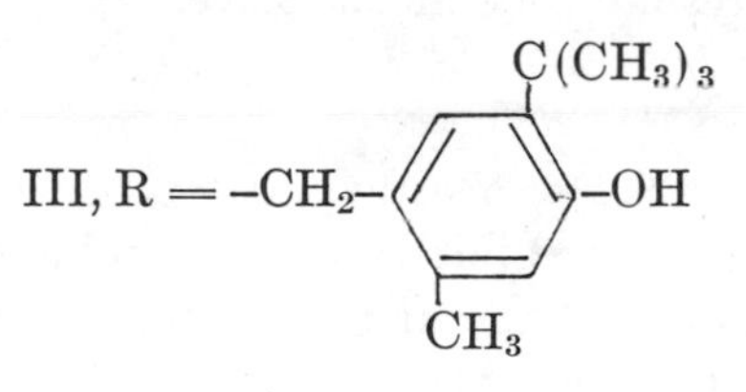

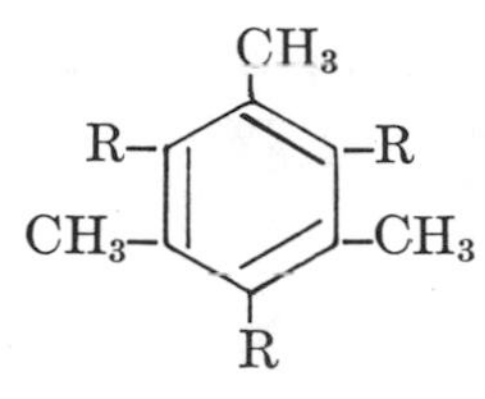

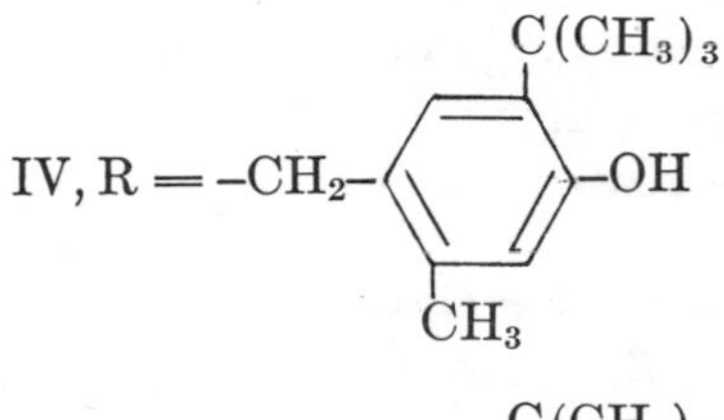

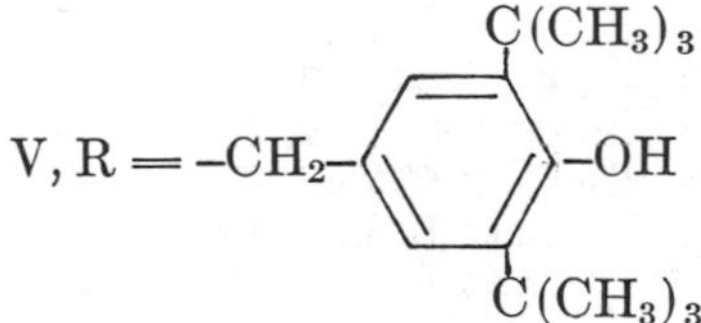

Table IV. Oven Aging of Polypropylene in the Presence of Antioxidants

Run No.	*Polypropylene Formulations: Unstabilized, Crystalline Polypropylene + the Following:*	*Oven Life, Hr for 5-Mil Film Samples*	
		160°C	*140°C*
1	Control (no stabilizer)	1	1
2	+ III, 0.1 wt %	15	340
3	+ IV (0.1%)	15	200
4	+ V (0.1%)	20	300
5	+ III (0.1%) + DLTPD[a] (0.3%)	90	600
6	+ IV (0.1%) + DLTDP (0.3%)	40	650
7	+ V (0.1%) + DLTDP (0.3%)	70	230

[a] Dilauryl 3,3′-thiodipropionate.

Application to Oven-Aging Tests. The results of the preceding section suggest that a superior polyolefin antioxidant could be obtained by keeping all phenolic moieties in the molecule singly hindered with one *tert*-butyl group. Two high molecular weight antioxidants whose structures were assigned as III and IV were synthesized and evaluated in oven-aging tests on polypropylene. These materials were compared with the similar commercial stabilizer V that contains the 2,6,-di-*tert*-butylphenol moiety. In oven-aging tests at 140°C on thin films with the synergist (*20*) dilauryl 3,3′-thiodipropionate added, stabilizers III and IV appeared to be superior to V (*see* Table IV). Under other conditions, the results were inconclusive.

Conclusion

This investigation has shown that most antioxidants based on the 2,6-di-*tert*-butyl-4-alkylphenol moiety behave similarly as inhibitors in the solution oxidation of polypropylene. Although the kinetics of oxidations are complex, the length of the inhibition period is proportional to the phenolic group concentration. With 2-*tert*-butyl-4-alkylphenols, the apparent stoichiometry for reaction with polypropylene peroxy radicals is increased from one to two.

Literature Cited

1. Howard, J. A., Ingold, K. U., *Can. J. Chem.* (1962) **40,** 1851.
2. Ibid. (1963) **41,** 1744, 2800.
3. Ibid. (1964) **42,** 1044.
4. Ibid. (1964) **42,** 2324.
5. Ibid. (1965) **43,** 2724.
6. Thomas, J. R., *J. Am. Chem. Soc.* (1963) **85,** 2166.
7. Van Sickle, D. E., *J. Polym. Sci., Part A-1* (1972) **10,** 275.
8. Ibid. (1972) **10,** 355.
9. Van Sickle, D. E., *Macromolecules* (1977) **10,** 474.
10. Kolka, A. J., Napolitano, J. P., Filbey, A. H., Ecke, G. C., *J. Org. Chem.* (1957) **22,** 642.
11. Fieser, L. F., "Reagents for Organic Synthesis," p. 171, John Wiley, New York, 1967.
12. Hoff, M. C., Greenlee, K. W., Boord, C. E., *J. Am. Chem. Soc.* (1951) **73,** 3329.
13. Mahoney, L. R., *J. Am. Chem. Soc.* (1967) **89,** 1895.
14. Morawetz, H., *Acct. Chem. Res.* (1970) **3,** 354.
15. Boozer, C. E., Hammond, G. S., Hamilton, C. E., Sen, J. N., *J. Am. Chem. Soc.* (1955) **77,** 3233.
16. Decker, C., Mayo, F. R., *J. Polym. Sci., Part A-1* (1973) **11,** 2847; correction (1975) **13,** 2415.
17. Denisova, L. N., Denisov, E. T., *Kinet. Katal.* (Eng. trans.) (1976) 17, 519.
18. Mahoney, L. R., DeRooge, M. A., *J. Am. Chem. Soc.* (1975) **97,** 4722.
19. Weiner, S. A., Mahoney, L. R., *J. Am. Chem. Soc.* (1972) **94,** 5029.
20. Shelton, J. R., "Polymer Stabilization," W. L. Hawkins, Ed., p. 110, Wiley–Interscience, New York, 1972.

RECEIVED July 1, 1977.

21

Nonmigrating Antioxidants via Sulfonyl Azide Intermediates

STEPHEN E. CANTOR

Uniroyal Chemical, Elm St., Naugatuck, CT 06770

Upon heating, sulfonyl azide compounds containing a hindered phenolic antioxidant moiety graft to the polymer matrix. This technique is applicable to saturated and unsaturated hydrocarbon polymers, copolymers, and terpolymers such as NBR, NR, EPDM, polyethylene, and polypropylene. Even after extraction with an organic solvent, the oxidative stability of the polymers is much greater than that of the polymers containing conventional antioxidants. The preparation of the sulfonyl azide antioxidants and the chemistry and application of the nitrene insertion mechanism are described.

A serious disadvantage of conventional antioxidants is their loss from the polymer matrix either by solvent extraction or volatility, resulting in loss of the useful life of the products via degradation or embrittlement (*1*). This disadvantage is surmounted by chemically attaching the antioxidant to the polymer. The recent literature describes several methods and techniques to accomplish this (*2*). For example, aromatic nitroso compounds (*p*-nitrosodiphenylamine or *N,N*-diethyl-*p*-nitrosoaniline) bind to rubber when an allylic hydrogen is present (*3*).

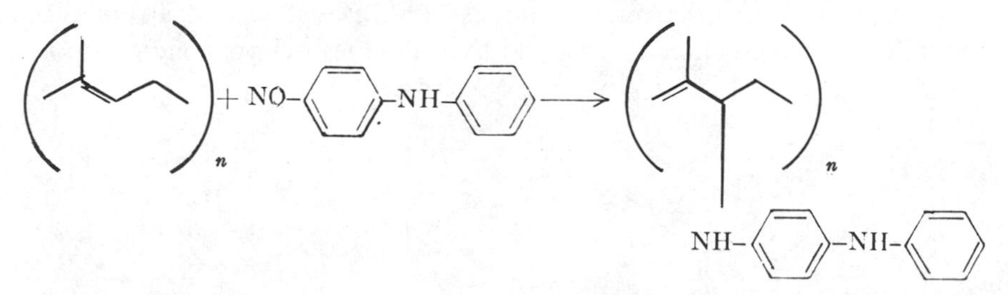

Epoxy groups on the polymer are attacked by β-naphthylamine to produce a built-on antioxidant (*4*).

0-8412-0381-4/78/33-169-253$05.00/1

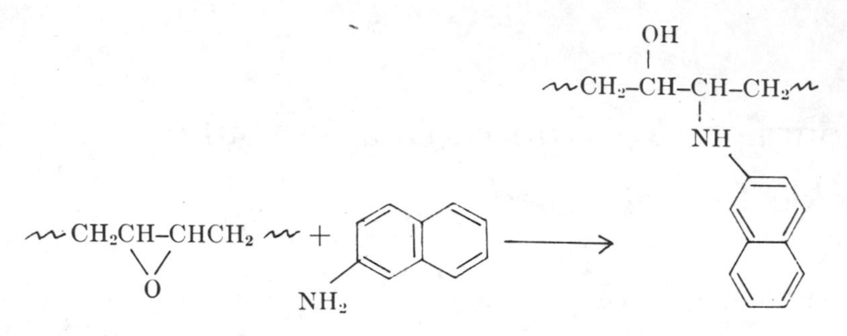

Acid chloride groups have been introduced that are then treated with amine antioxidant (5). However, all of these techniques require some type of functionality on the polymer chain and will fail with saturated hydrocarbon polymers such as polyethylene or polypropylene.

In the case of synthetic elastomers formed by free radical emulsion techniques, monomers that contain both an antioxidant moiety and a polymerizable function can be copolymerized (6). A series of esters and amides containing the 4-anilinophenyl group or the 3,5-di-*tert*-butyl-4-hydroxyphenol group and a polymerizable α, β-unsaturated acyl group such as acryloyl or methacryloyl were incorporated into the polymer.

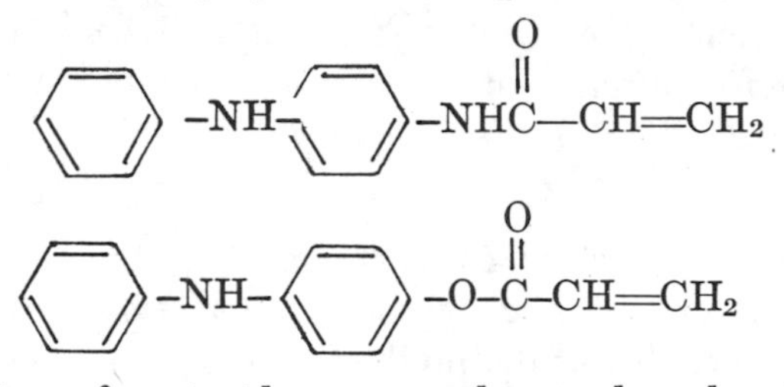

Direct grafting of antioxidants on polypropylene has been accomplished by treatment with 3,5-di-*tert*-butyl-4-hydroxylbenzyl acrylate and a photoactivator such as benzophenone (7).

Polymer stabilization has been realized by generating a highly reactive carbene species that can insert into the polymer chain. Kaplan et al. described the preparation and use of a phenolic diazooxide that generates a carbene upon heating and that attaches the phenolic group to the polymer chain (8).

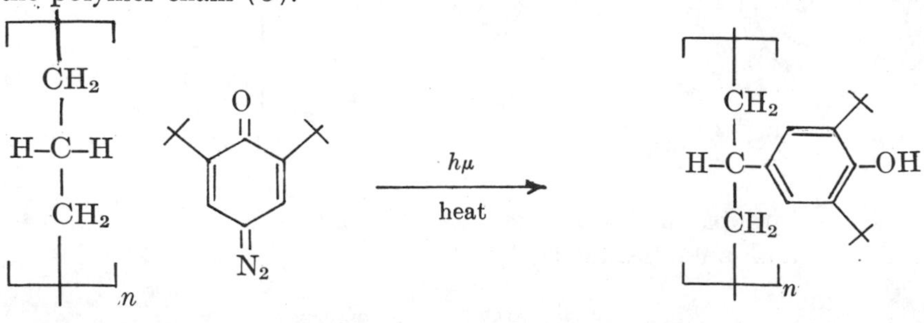

Recently, Porter and Waisbrot demonstrated that azidoformyl compounds containing a hindered phenolic antioxidant moiety decomposed thermally to form carbalkoxynitrene intermediates that insert into a natural rubber matrix (*9*). This chapter describes the preparation and use of antioxidants containing the sulfonyl azide group, that decomposes to liberate nitrogen and the sulfonyl nitrene species when heated either in a vulcanization process or in an extruder.

$$RSO_2N_3 \xrightarrow[\text{heat}]{} RSO_2\ddot{N}\!: + N_2$$

"nitrene"

Both alkane and arene sulfonyl nitrenes insert into the carbon–hydrogen bonds of saturated hydrocarbons although not as efficiently as carbenes (*10*).

$$C_5H_{11}SO_2N_3 + \text{(cyclohexane)} \xrightarrow[\text{heat}]{} C_5H_{11}SO_2NH\text{–(cyclohexyl)} + N_2$$

With the discovery that sulfonyl nitrenes insert into a carbon–hydrogen bond, a number of patents have described the use of mono- and polysulfonyl azides for the cross-linking of hydrocarbon polymers such as polypropylene and polyisobutylene (*11, 12, 13*). Pendant sulfonyl azide groups have been attached to linear macromolecules that are cross-linked via the sulfonyl nitrene insertion mechanism upon photolysis or heating (*14*). Silanesulfonyl azides of the general formula $(CH_3O)_3$–Si–R–SO_2N_3 have been prepared and found to be effective agents for coupling inorganic fillers to polyolefins, polystyrene, ABS, and other thermoplastic polymers (*15*).

Theoretical

To prepare several sulfonyl azides containing the required antioxidant moiety, an intermediate was chosen that would react readily with known hindered phenols and that would still possess the requisite precursor for sulfonyl nitrene insertion. Therefore, 4-isocyanatobenzenesulfonyl azide was prepared following the procedure described by Danhaeuser and Pelz (*16*).

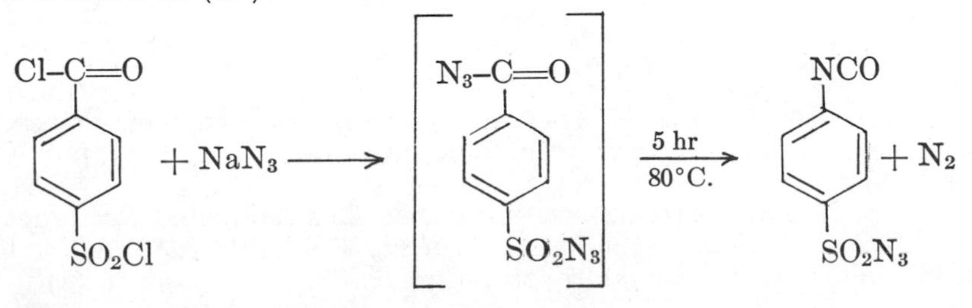

The isocyanate was allowed to react with antioxidant amines and alcohols to prepare sulfonyl azides containing either a urea or a carbamate linkage.

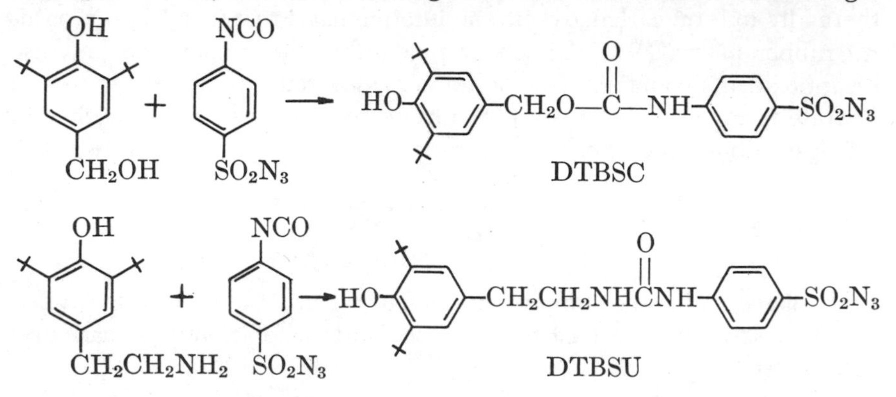

Another useful building block was 2-hydroxyethyl 4-azidosulfonyl carbanilate prepared from 4-isocyanatobenzenesulfonyl chloride and ethylene glycol, followed by treatment with sodium azide (*17*).

The hydroxyl function was esterified readily with 3-(3,5-di-*tert*-butyl-4-hydroxy phenyl) propionyl chloride to produce an antioxidant sulfonyl azide containing an ester linkage.

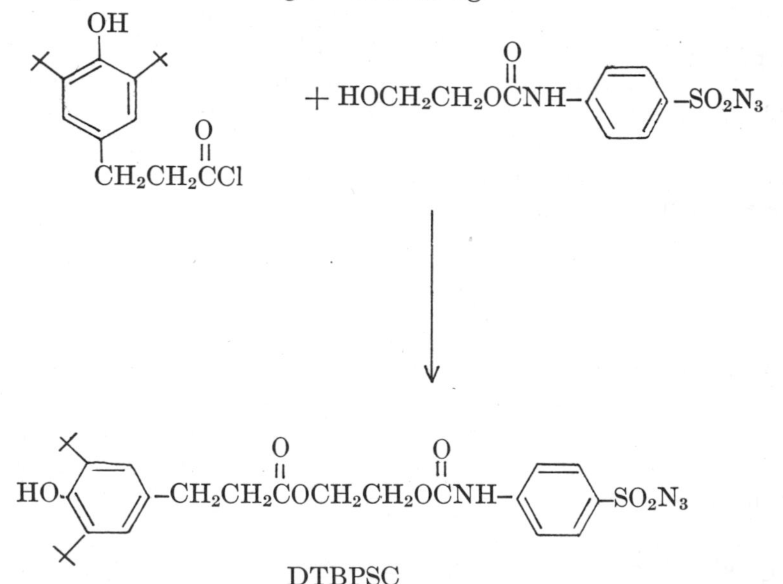

Synthetic Procedures. The amines and hydroxy compounds used as intermediates in this work were purchased or prepared by published procedures.

4-Isocyanatobenzenesulfonyl azide. This compound was prepared in a 70% yield from *p*-chlorosulfonylbenzoyl chloride (*18*) and sodium azide following the literature procedure (*16*). The product, with mp

37–39°C, was stored in a tightly sealed vessel to protect the isocyanate function from moisture.

N-[2-(3,5-DI-*tert*-BUTYL-4-HYDROXYPHENYL)ETHYL] *N'*-(4-AZIDOSULFONYLPHENYL) UREA (DTBSU). 4-Isocyanatobenzenesulfonyl azide (11.2 g, 0.05 mol) and dry acetonitrile (200 mL) were placed in a 500-mL three-necked flask fitted with a thermometer, stirrer, and drying tube. The flask was maintained at 20–25°C and 12.4 g (0.055 mol) of 2-(3.5-di-*tert*-butyl-4-hydroxyphenylethylamine (*19*) dissolved in 200 mL acetonitrile were added slowly. After 2 hr of stirring at room temperature ir spectrophotometric examination of an aliquot indicated complete loss of the isocyanate function. The acetonitrile was removed in a rotary evaporator and the remaining material poured into ice to precipitate the product in an 87% yield. The product melted at 122–127°C and evolved nitrogen at 140–150°C. The ir spectrum contains bands at 3210 cm^{-1} (NH), 2120 cm^{-1} (azide), 1675 cm^{-1} (amide carbonyl), and 1160 cm^{-1} (sulfone).

3,5-DI-*tert*-BUTYL-4-HYDROXYBENZYL 4-AZIDOSULFONYLCARBANILATE (DTBSC). 3,5-Di-*tert*-butyl-4-hydroxybenzyl alcohol (Ethyl 754) (4.28 g, 0.02 mol) was dissolved in 75 mL of acetonitrile and added to 4.48 g (0.02 mol) of 4-isocyanatobenzenesulfonyl azide dissolved in 100 mL of acetonitrile. After 2 hr of stirring at room temperature all of the isocyanate was consumed. The excess solvent was removed in vacuo and the remaining oil was poured into ice water. The solid product was obtained in 78% yield. It melted at 130–133°C with loss of nitrogen and decomposition at 175°C. The ir spectrum contained bands at 3450 cm^{-1} (NH), 2120 cm^{-1} (azide), 1725 cm^{-1} (carbamate), and 1160 cm^{-1} (sulfone).

2-[3-(3,5-DI-*tert*-BUTYL-4-HYDROXYPHENYL)PROPIONOXY] ETHYL 4-AZIDOSULFONYLCARBANILATE (DTBPSC). 3-(3,5-Di-*tert*-butyl-4-hydroxyphenyl)propionyl chloride (20 g, 0.7 mol) (*20*) and 200 mL of dry DMF containing a few drops of pyridine were placed in a 500-mL flat-bottomed flask. The solution was stirred at room temperature as 21 g (0.7 mol) of 2-hydroxyethyl-4-azidosulfonylcarbanilate (*15*) dissolved in 150 mL of DMF were added. After stirring at room temperature, the DMF solution was poured into ice water and the solid was collected and dried to yield 22 g (58%) of product (mp 70–75°C). It decomposed at 175°C with evolution of nitrogen. The ir spectrum contained absorption bands at 3420 cm^{-1} (NH), 2120 cm^{-1} (azide), 1735 cm^{-1} (ester carbonyl), 1725 cm^{-1} (carbamate), and 1165 cm^{-1} (sulfone).

Preparation of Samples. Polyethylene and polypropylene powder were treated with 0.5% by weight of the antioxidants in tetrahydrofuran. The solvent was removed at reduced pressure, and thin films were molded at 160°C for 15 min.

EPDM (Royalene 505) dissolved in benzene was treated with 1% by weight of the antioxidant, followed by removal of the solvent. The gum elastomer was molded at 160°C for 10 min.

Preparation of Vulcanizates. For filled compounds prepared from NR and NBR, a Brabender was used to mix the ingredients. A small laboratory mill was used to add the curatives and the antioxidants. The formulations are shown in Tables II and III.

Extraction. A methanol/acetone/chloroform azeotrope (28:35:29 mL, bp 57.5°C) was used as the organic solvent. Thin sheets for oxygen absorption measurements were extracted continuously with the azeotropic mixture in a Soxhlet apparatus. Tensile specimens from the vulcanizates were cut to shape by a DeMattia die and were extracted continuously also. All samples were dried in vacuo prior to testing.

Testing. Oxygen absorption was measured at 150°C under pure oxygen. The vulcanizates were conditioned in an oxygen bomb, according to ASTM method D-572 at 70°C for either 24 or 96 hr prior to tensile testing by ASTM method D-412.

Results and Discussion

O_2 Absorption Study. Two commercially available conventional antioxidants were chosen for evaluation and comparison with the network bound antioxidants. The time in minutes for the absorption of 20 cc of oxygen by 1 g of the samples is found in Table I. The sulfonyl azide

Table I. O_2 Absorption of Polymers at 150°C

Polymer	*Added Stabilizer*	*Min to 20 cc of O_2/g of Polymer* — *Unextracted*	*Extracted*[a]
EPDM	DTBSU	1212	355
	DTBSC	744	105
	DTBPSC	986	135
	Irganox 1076	900	48
Polypropylene	DTBSU	6930	850
	Ethyl 754	4050	177
Polyethylene	DTBSU	720	720
	Ethyl 754	947	157

[a] Extracted with an azeotropic mixture of methanol/chloroform/acetone for 24 hr.

antioxidants are certainly equivalent to the conventional antioxidants prior to extraction. However, only those samples treated with the sulfonyl azide antioxidants retained properties to a marked degree after the extraction. With the EPDM, DTBSU was seven times more efficient than the Irganox 1076 after extraction and testing. This effect is noted again in the protection afforded to both polyethylene and polypropylene by the sulfonyl azide antioxidants.

Oxygen Aging of Vulcanizates. Examination of Table II (NBR) and Table III (NR) demonstrates clearly that vulcanizates treated with the sulfonyl azide antioxidants retain physical properties to a marked degree after solvent extraction, followed by aging in oxygen. The conventional antioxidants used for comparison, such as PBNA in the NR or Irganox 1076 in the NBR, are removed by the extraction process and therefore afford no protection, resulting in samples too brittle to test.

Table II. Nonmigrating vs. Conventional Antioxidant in NBR Vulcanizate[a]

	DTBSU			
	Unextracted		*Extracted*	
	Original	*Aged*[b]	*Original*	*Aged*[b]
Tensile MPa	19.5	18.6	12.4	15.2
Elongation	250	210	180	130
Shore A	78	80	72	75
	Irganox 1076			
Tensile MPa	18.6	19.8	11.8	too brittle to test
Elongation	210	220	160	
Shore A	78	80	75	

[a] NBR = 100, N-744 black = 70, zinc oxide = 5, stearic acid = 1. Monex = 0.2, Delac S = 1,0, sulfur = 2.0, and antioxidant = 2.0 cured 20 min at 152°C.
[b] 96 hr at 70°C.

Table III. Nonmigrating vs. Conventional Antioxidant in NR Vulcanizate[a]

	DTBSU			
	Unextracted		*Extracted*	
	Original	*Aged*[b]	*Original*	*Aged*[b]
Tensile MPa	22.3	13.5	20.5	18.3
Elongation	400	330	270	280
Shore A	65	71	74	76
	PBNA			
Tensile MPa	24.5	14.9	16.3	too brittle to test
Elongation	470	380	310	
Shore A	61	65	67	

[a] Recipe: SMR-5 = 100, N-330 black = 45, zinc oxide = 5, stearic acid = 2, Delac NS = 1.0, DPG = 0.3, sulfur = 2.0, and antioxidant = 2.0 cured 30 min at 152°C.
[b] 24 hours at 70°C.

Glossary

Commercial antioxidants

Irganox 1076: *n*-Octadecyl 3-(3,5-di-*tert*-butyl-4-hydroxyphenyl)propionate

Ethyl 754: 3,5-di-*tert*-butyl-4-hydroxybenzyl alcohol

PBNA: *N*-phenyl-β-naphthylamine

Acknowledgment

The author is indebted to Richard Gencarelli for his assistance and synthesis of starting materials and to Uniroyal Chemical for permission to publish this work.

Literature Cited

1. Spacht, R. B., Hollingshead, W. S., Bullard, H. L., Wills, D. C., *Rubber Chem. Technol.* (1964) **37**, 210.
2. Spacht, R. B., Hollingshead, W. S., Bullard, H. L., Wills, D. C., *Rubber Chem. Technol.* (1965) **38**, 134.
3. Kline, R. H., Miller, J. P., *Rubber Chem. Technol.* (1974) **46**, 96.
4. Cain, M. E., Knight, G. T., Lewis, P. M., Saville, B., *J. Rubber Res. Inst. Malaya* (1969) **22**, 289.
5. Kirpichev, V. P., Yakubchik, A. I., Maglyph, G. M., *Rubber Chem. Technol.* (1970) **43**, 1225.
6. Blatz, P. S., Maloney, D. E., U.S. **3,441,545** (1969).
7. Meyer, G. E., Kavchok, R. W., Naples, F. J., *Rubber Chem. Technol.* (1974) **46**, 106.
8. Evans, B. W., Scott, G., *Eur. Polym. J.* (1974) **10**, 453.
9. Kaplan, M. L., Kelleher, P. G., Bebbington, G. H., Hartless, R. L., *Polym. Lett.* (1973) **11**, 357.
10. Porter, D. D., Waisbrot, S. W., U.S. **3,991,131** (1976).
11. Sloan, M. F., Breslow, D. S., Renpow, W. B., *Tetrahedron Lett.* (1964) 2905.
12. Breslow, D. S., Spenlin, H. M., U.S. **3,058,944** (1962).
13. Johnstone, P. L., U.S. **3,137,745** (1964).
14. Newburg, N. R., U.S. **3,287,376** (1966).
15. Ulrich, H., Stuber, F. A., Sayigh, A. A. R., *Polym. News* (1970) **1**, No. 8.
16. Thomson, J. B., U.S. **3,706,592** (1972).
17. Danhaeuser, J., Pelz, W., Belgium Pat. **665,429** (1965), C.A. (1966) **64**, 12901.
18. Sayigh, A. A. R., Tucker, B. W., Ulrich, H., U.S. **3,652,599** (1972).
19. Imai, Y., Okunoyama, H., *J. Polym. Sci., A2* (1972) **10**, 2257.
20. Cohen, L. A., Jones, W. M., *J. Am. Chem. Soc.* (1962) **84**, 1629.
21. Dexter, M., Steinberg, D., U.S. **3,657,309** (1972).

RECEIVED May 12, 1977.

22

The Distribution of Additives and Impurities in Isotactic Polypropylene

T. G. RYAN, P. D. CALVERT, and N. C. BILLINGHAM

School of Molecular Sciences, University of Sussex, Brighton BN1 9QJ, England

Additive and impurity rejection at the growing crystal front leads to uneven distribution in a crystalline polymer. This redistribution process has been studied by UV and fluorescence microscopy and by an electron microscope with energy dispersive x-ray analysis. In polymer samples which are quenched after rapid crystallization, the additive distribution is kinetically determined and may be modeled in a computer as a three-dimensional zone-refining process. In annealed polymer samples, low molecular weight additives are uniformly concentrated in the amorphous phase. The additive distribution reflects that of crystalline material within the polymer. Antioxidant and uv stabilzer redistribution probably does not have a major effect on polymer stability, but the redistribution of partially oxidized, impure polymer may be important.

Oxidative degradation of a crystalline polyolefin is a complex reaction involving a dissolved gas and a two-phase, impure, inhomogeneous solid. Factors affecting the reaction rate are antioxidant concentrations, crystallinity, UV illumination intensity, UV absorber concentration, and the sample's previous oxidation history. Failure of a sample is often mechanical rather than chemical, and cannot be regarded as occurring at a particular degree of oxidation.

To understand oxidative degradation it is necessary to consider the factors that influence the overall process and to study their effects separately. This knowledge is then recombined into a comprehensive understanding of degradation in real systems. One important group of problems concerns reactions between the various components in homogeneous systems such as melts and solutions. A second important group of problems with which this chapter is concerned, is the effects of crystalline

0-8412-0381-4/78/33-169-261$05.00/1

morphology on oxidation. This can be observed at three levels. On the smallest scale, oxygen and antioxidants are excluded from the crystalline regions in polyethylene and isotactic polypropylene so that oxidation is localized in the disordered regions (*1*). This is supported by a decrease in oxidation rate and an increase in limiting oxygen uptake with increasing crystallinity (*2, 3*). The reverse relations hold in poly(4-methyl-1-pentene), and it has been shown that oxygen can penetrate the crystalline regions (*4*). Keith and Padden (*5*) have shown that high molecular weight impurities are also concentrated in the intercrystalline regions, and there is good reason to believe that this rejection will be almost complete for most impurities.

On the scale of the whole spherulite it is known that nonuniform distributions of small molecule and polymeric impurities are produced during crystallization. Price (*6*) showed that a wave of noncrystallizable material is pushed ahead of the growing spherulite, and Keith and Padden (*5*) demonstrated that such impurities affect spherulite structure and growth rate. Moyer and Ochs (*7*) found nonuniform distributions of radio-labeled additives in polyethylene, polypropylene, and polystyrene by autoradiography. In this context, impurity means any species which is less likely to enter the crystal phase than are the bulk of the polymer chains. It includes additives, dissolved gases, solid particles, and short, atactic or partially oxidized polymer molecules. In this chapter, impurity redistribution during spherulite growth and its importance for oxidative degradation are discussed.

On the scale of the whole sample, loss of additives from the sample surface will lead to concentration gradients, while UV intensity within the polymer will decrease with depth, particularly if UV absorbers are present. Oxygen consumption within the polymer leads to a concentration gradient if the oxidation is sufficiently rapid. Surface transcrystallinity may cause concentration gradients of additives and impurities near the surface. Surface oxidation and flow patterns during molding will lead to gradients in the concentration of partially oxidized material. Degradation measurements such as embrittlement or induction time, can be affected by degradation rate variation within the sample.

Redistribution of Impurities by Growing Spherulites

As shown by Price (*6*), rejected noncrystallizable impurities will be pushed ahead of the growing spherulite as a wave, leaving a lower concentration within the spherulite than in the original melt. Frank and Lehner (*8*) and Curson (*9*) have used UV transmission microscopy to observe additive distributions in crystalline polyolefins. This method can be used to observe the redistribution process in action by quenching partially crystallized polypropylene to freeze the additive concentrations

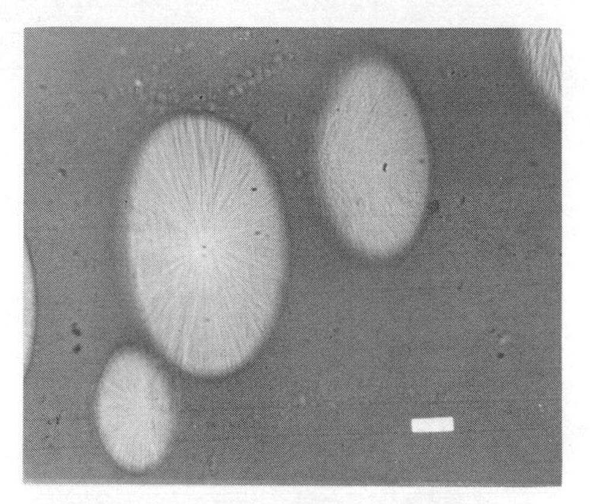

Figure 1. Polypropylene section containing 0.5% Uvitex OB partly crystallized at 125°C and quenched. Viewed in UV transmission. Bar is 20 μm.

around a growing spherulite. Figure 1 shows a microtomed section of a polypropylene sample containing 0.5% Uvitex OB [2,5-di(5-*tert*-butyl-2-benzoxazolyl)thiophene], an optical brightener, which was partially crystallized at 125°C and quenched. Features noted are a low additive concentration within the spherulite with a central dip, a uniform high concentration in the region that was molten before quenching, and a higher concentration ring around the spherulite boundary. Uvitex OB is a good subject since it has a large UV absorption coefficient and is observable in fluorescence as shown in Figure 2. The central dip in con-

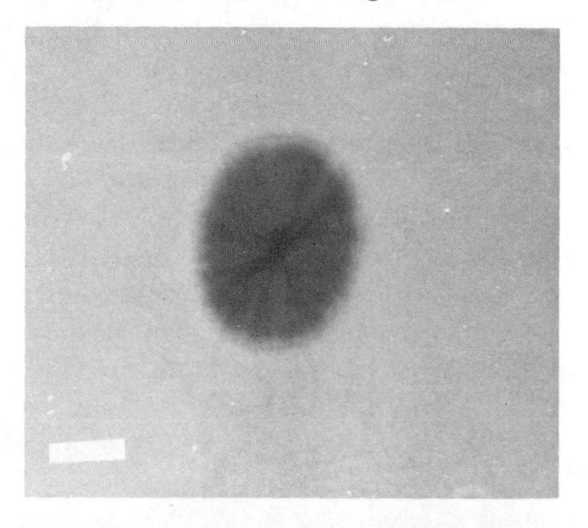

Figure 2. Polypropylene section containing 0.1% Uvitex OB partly crystallized at 130°C and quenched. Viewed by fluorescence. Bar is 20 μm.

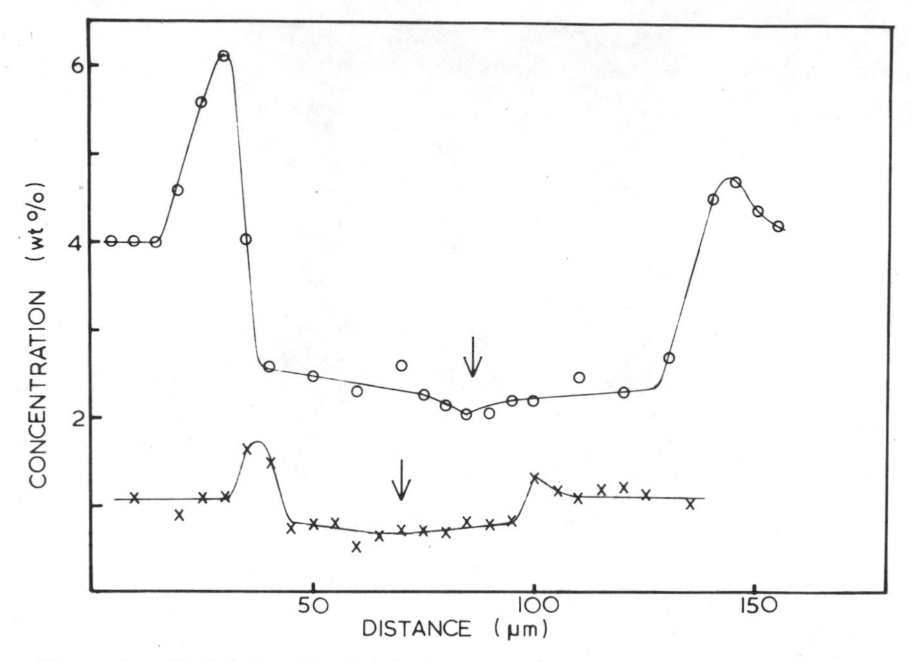

Figure 3. Nickel EDAX distribution of UV1084 in polypropylene partly crystallized and quenched. (○) 4 wt % UV1084, (×) 1 wt % UV1084, ↓ Spherulite center.

centration is more apparent since the fluorescence mode is sensitive at low concentrations while absorption is best at high concentrations. A number of other additives, including a range of phenolic antioxidants, behave similarly using UV absorption. These are harder to see since they have lower UV absorption coefficients.

Another technique we used to observe these distributions is scanning electron microscopy with energy dispersive x-ray analysis (EDAX). Concentrations of Cyasorb UV 1084, [2-2′-thiobis(4-*tert*-octylphenolato)-*n*-butylamine nickel], a nickel-containing UV absorber, were point counted to obtain nickel concentrations along a spherulite diameter. Figure 3 shows results for 1 and 4 wt % additive. This shows a uniform melt concentration, a boundary peak, a lower concentration within the spherulite, and a central dip. The resolution and sensitivity with this technique are poorer than with the optical microscopy. With every method, thin film crystallized samples and microtomed sections of bulk samples gave similar results.

Photomicrographic microdensitometry was used to generate quantitative distribution data for the additives. Calibration was conducted using standard quenched samples with a known additive concentration, photographed under the same conditions. Quenched melt regions, distant from growing spherulites, can also be used for internal calibration. We estimate that relative concentrations within any sample can be found to ±5%, but sample-to-sample errors are closer to ±20%.

Redistribution Model

A computer model of the partition and diffusion processes has been constructed for distribution analysis. This will be described in detail elsewhere (*10*). Assuming that we start with a polymer melt containing a uniform impurity concentration, growing crystallites at the spherulite edge will reject impurities into the surrounding melt and intercrystalline amorphous material. Redistribution at the individual lamellae will be too fine to observe, but it will lead to a local high impurity concentration at the growing edge of the spherulite. On this scale we can replace the spherulite's fine structure with a uniform solid phase. The rejection process is characterized by a partition coefficient, the ratio of the concentration of any impurity in the solid to that in the liquid at the growing

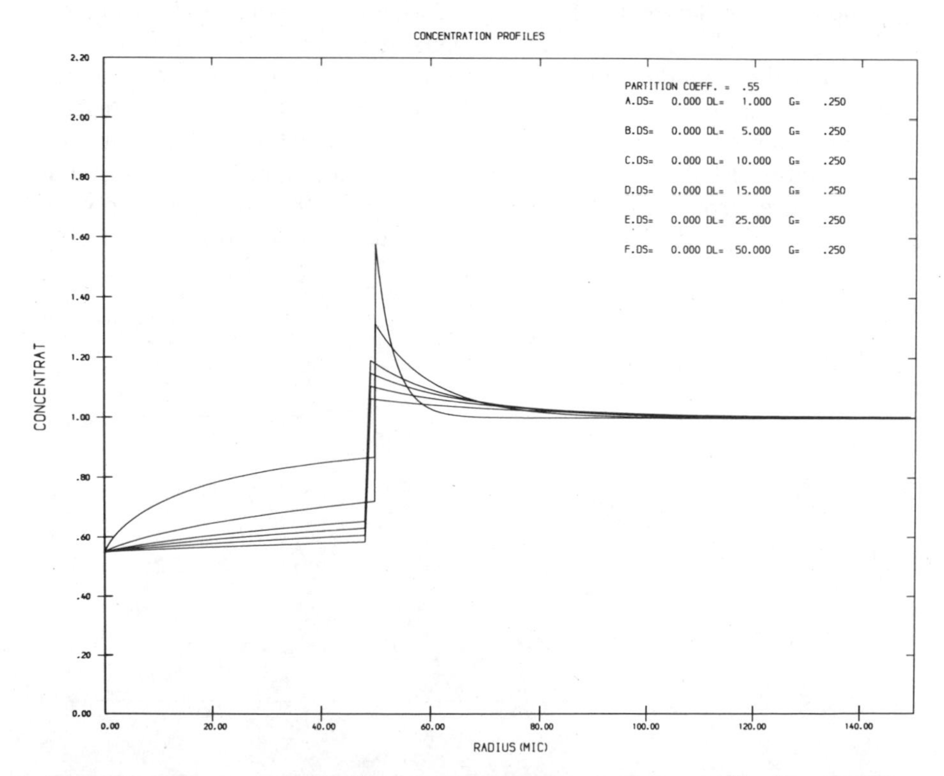

Figure 4. Computed distributions for samples partly crystallized at 125°C and quenched. DL = diffusion coefficient of additive in liquid μm^2 sec^{-1}; DS = back diffusion coefficient; G = spherulite growth rate μm sec^{-1}.

front. This will be equal to $(1 - x)$, where x is the spherulite crystallinity at the growth front. Under these circumstances, the model is similar to that used for normal freezing, a simple case of zone refining (*11*). The other necessary parameters are the impurity's diffusion coefficients in the melt and in the spherulite, the spherulite growth rate, and the final spherulite radius.

On this basis, impurity distribution curves can be calculated for partially or completely grown spherulites using an iterative procedure. Figure 4 shows a calculated set of curves for spherulites grown in polypropylene at 125°C for a number of melt diffusion rates, and a zero diffusion rate within the spherulite (back diffusion). The crystallinity of 45% was determined by scanning calorimetry to be that holding at the end of primary crystallization at 125°C in samples that were remelted without further cooling. This is equal to the crystallinity at the growth front. Figure 5 shows the fit obtained between the observed distribution in Figure 1 and the calculated one. The value of the diffusion coefficient in the melt is $7 \mu m^2 s^{-1}$. This is the right order of magnitude since a benzophenone of similar molecular weight has an extrapolated diffusion coefficient of $4 \mu m^2 s^{-1}$ in solid polypropylene at the same temperature (*12*). Although the boundary peak fit is good, there is no central dip on the calculated curve to correspond to that observed. Diffusion coefficients can be obtained by comparing observed and calculated peak heights and peak widths at half-height. Using this method for samples crystallized at temperatures from 120°–130°C, where the growth rate decreases by a factor of 7, the calculated diffusion coefficient changes from $6–10 \mu m^2 s^{-1}$.

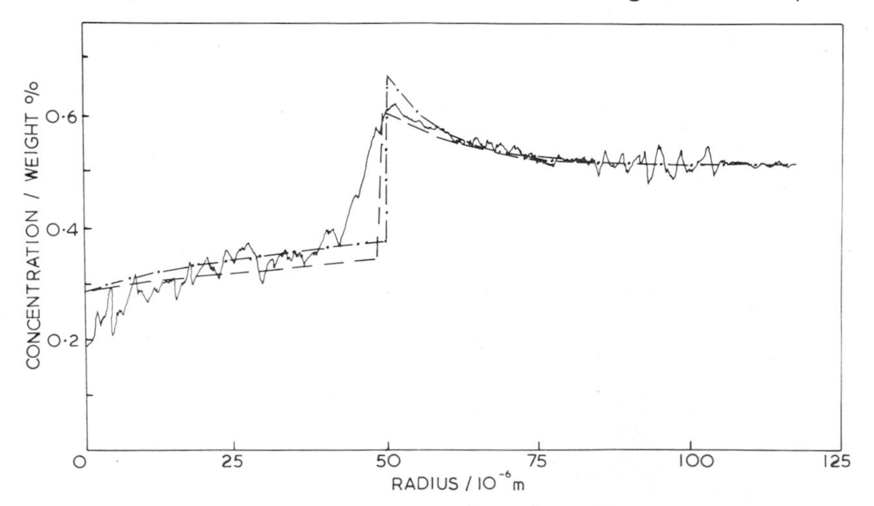

Figure 5. Observed and computed distributions for sample containing 0.5% Uvitex OB partly crystallized at 125°C. (– · –) DL = 5 μm^2 sec^{-1}, (– – –) DL = 10 μm^2 sec^{-1}.

This is support for the model, but it is not a way to determine diffusion coefficients since the errors are ±50%. This method is usable only if the ratio of diffusion coefficient to growth rate is in the range 10–100μm.

We have added a number of corrections to the simple model. There is no sound basis for estimating back diffusion coefficients because there are no liquid and solid diffusion measurements at the same temperature, but based on measurements through the melting range, a back diffusion coefficient of up to one-third of the melt value seems reasonable (*13*). Addition of this correction only slightly affects the interfacial peak, but it tends to flatten out the concentration gradients within the spherulite. For small spherulites a correction can be added for the apparent peak broadening when a curved interface is viewed through a thin film. A more important correction is necessary when diffusion fields of approaching spherulites overlap. This is difficult to calculate since the system lacks spherical symmetry, but it can be avoided by observing only spherulites that are well separated. Finally, impurities may redistribute by diffusion after quenching to take up a pattern characteristic of a crystallinity distribution within the quenched sample. If this crystallinity distribution reflected some other impurity—e.g., atactic polymer—it would be difficult to distinguish from the additive distribution. We believe that this is not the case since markedly different distributions and diffusion rates are obtained for different antioxidants.

The foregoing analysis of the peak in impurity concentration at the boundary of a growing spherulite establishes the validity of the normal freezing model. This model can easily be run to complete spherulite growth to produce a predicted final distribution which we will call the dynamic distribution. However, the discrepancy between the predicted and observed curves at the spherulite center has already been noted, and this leads one to consider a second coexisting equilibrium distribution.

Crystallinity Variation

The model so far fails to predict a concentration dip at the spherulite center. This can be resolved by observing samples crystallized without additive, where the additive is allowed to diffuse in later. In this case we would expect the additive to adopt a uniform distribution throughout the amorphous regions. Figure 6 compares the observed distributions in polypropylene samples crystallized completely at 130°C with Uvitex OB, crystallized completely with Uvitex OB followed by annealing for 7 days, and crystallized without additive, then immersed in a glycerol solution of the additive at 130°C. The distributions are essentially identical. Figure 7 shows the distribution in a fully crystallized sample. If an average crystallinity is determined for each sample, it is possible to

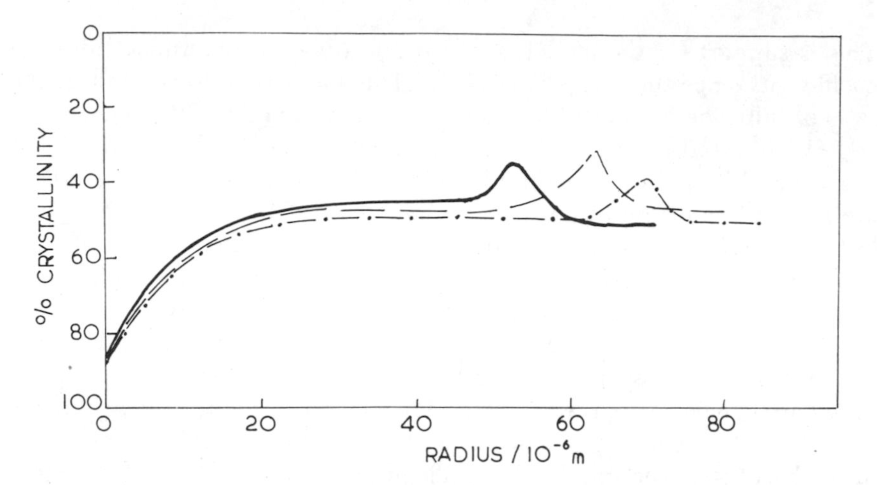

Figure 6. Observed distributions for samples fully crystallized at 130°C containing Uvitex OB, (——) crystallized for 2 hr, (- - -) crystallized and annealed for 7 days, (– · –) crystallized without additive and annealed in a solution of Uvitex OB in glycerol at 130°C

interpret the local additive concentrations as local amorphous contents so that these curves are plotted as crystallinity variations within the fully grown spherulites. This crystallinity variation has the general form of a redistribution curve, suggesting that it is caused by the rejection of species, presumably atactic or a stereoblock polymer, with a diffusion coefficient of about 0.1–1$\mu m^2 s^{-1}$ under these experimental conditions. Crystallinity varies from about 90% at the spherulite center, to 50% through most of the spherulite, to about 30% at the boundary. The central dip is more manifest than the boundary peak, but this is partly a product of the spherical geometry. This equilibrium distribution, with

Figure 7. Polypropylene section containing 0.1% Uvitex OB crystallized at 115°C, viewed by fluorescence. Shows α and β forms. Bar is 40 μm.

a uniform concentration of additive in the amorphous regions, would be expected in slowly crystallized or annealed samples. After rapid crystallization the dynamic distribution will be superimposed on this, giving a greater concentration of additive in the amorphous regions at the boundaries.

Thus far redistribution of wholly soluble impurities has been discussed. If the impurity or additive is insoluble (e.g., carbon black) or precipitates before the polymer crystallizes, it will be redistributed as solid particles. This effect has been studied for low molecular weight matrices (*14*) but not for polymers. If the additive precipitates after the polymer crystallizes, the distribution should remain similar to the completely soluble case. Partially oxidized or imperfect polymer chains can partly enter the crystal and have a higher partition coefficient. The diffusion coefficients for flexible molecules are generally greater than for rigid molecules of the same molecular weight. Thus, the diffusion coefficients of polymeric impurities should overlap those of the antioxidants and decrease with increasing molecular weight. The overall distribution pattern of polymeric additives is similar to those described here (*7*).

Effects of Impurity Distributions on Oxidation Behavior

In previous sections we have shown that the redistribution of additives at the spherulite boundaries during polymer crystallization leads to the additive's uneven distribution, whose form is determined by the kinetics of the growth rejection process. In time, this initial dynamic distribution should relax to an equilibrium form in which the noncrystalline polymer is uniformly permeated by the additive, whose distribution reflects that of the noncrystalline polymer. The relevance of these observations to oxidative degradation processes in semi-crystalline polyolefins is discussed in this section.

Oxidative degradation is confined to the amorphous regions of both polyethylene and polypropylene because the crystalline regions of both polymers are impermeable to oxygen (*1, 2, 3*). In addition there is reasonable evidence to suggest that oxidation does not occur uniformly throughout the amorphous regions. Oxidation measurements in both solution (*15, 16*) and solid phase (*17, 18*) systems have shown that chain scission during autoxidation of polypropylene occurs relatively infrequently compared with oxygen absorption, much of the absorbed oxygen appearing as low molecular weight volatile products. Adams (*19*) studied thermal and photochemical oxidation of polypropylene and found that embrittlement occurred when the number average molecular weight fell from 70,000–30,000. This drop is too small to account for embrittlement and implies that there is preferential degradation of vulnerable load-bearing chains during oxidation. Radical (*20*) and boundary (*21*) cracking of the spherulites have been observed during autoxidation of polypropylene thin films although boundary cracking appears to predominate (*22*). Similar conclusions can be drawn for polyethylene. Winslow et al. (*23*) found that linear polyethylene embrittles during thermal oxidation at an oxygen uptake of about 6 mlg^{-1}. This level of oxygen uptake would represent one scission for every 130 repeat units if each oxygen molecule absorbed led to one random chain scission. Other data, however, presented by Winslow and Matryek (*24*), suggest that the decrease in molecular weight at embrittlement is little more than a factor of two. Furthermore, the relatively low level of chain breaking is confirmed by the fact that the polyethylene sample recovered much of its toughness upon remolding. A similar recovery of toughness was observed by Kafavian (*25*). From previous work it seems reasonable to suggest that oxidative embrittlement arises from the preferential oxidation of vulnerable molecules within the polymer, these vulnerable molecules being either intercrystalline tie chains within the bulk of the spherulite or load-bearing chains within the boundary regions. Another factor that may be important is the effect of stress since crystallization can lead to boundary stress (*26*) and the oxidation rate of polypropylene is markedly increased by stress (*27*). Therefore, the spherulite boundaries are particularly vulnerable regions; their vulnerability can be increased by concentration of partially oxidized polymer, an effect which is discussed later. There is scope for a detailed reinvestigation of the oxidation locus in crystalline polymers.

We have demonstrated that slow polymer crystallization followed by annealing will lead to an equilibrium additive distribution in which an antioxidant will be dispersed uniformly in the amorphous phase of the polymer. If we accept that the spherulte boundaries are most vulnerable to oxidation, we should clearly concentrate the antioxidant in these regions as far as possible. The redistribution model suggests that the maximum boundary concentration effect will occur if the additive is rapidly diffusing and the polymer is quench-cooled after crystallization. However, it is unlikely that the boundary additive concentration could be increased by more than a factor of 2–4 over the equilibrium value. Although a rapidly migrating additive is likely to concentrate in the boundary regions, such additives also will diffuse rapidly back into the spherulite as the dynamic distribution relaxes to equilibrium; extrapolation of diffusion coefficients to room temperature is hazardous, but our present feeling is that the relaxation of the dynamic distribution can be expected to take place in a time scale for mobile additives, which is relatively short compared with the polymer's oxidative lifetime.

Where an additive is being used as a UV stabilizer, the situation is more complex since many of the additives commonly used operate by both chemical reaction and simple UV screening (28). The considerations outlined above will apply where chemical reaction is important. Given a typical absorber with a molar extinction coefficient of 10^4 l cm mol^{-1} at a uniform concentration of 0.1 wt %, the intensity of light absorbed at the absorption maximum is reduced to 10% of its surface value at a sample depth of 0.5 mm; this distance will be considerably larger at solar UV wavelengths where the extinction coefficients of most additives are well below their values at the absorption maximum. Distances of this order are at least 10 times greater than a typical spherulite radius. Therefore, redistribution will not significantly affect the action of a simple UV screening additive.

Within the limitations of simple, soluble, mobile additives, our results suggest that there is no way to improve significantly the spherulite boundary stabilization. However, the growth rejection process is expected to apply to any species which cannot enter the crystal phase, and it is relevant to consider the effects of redistribution of other impurities present in the polymer. Both thermal and photochemical oxidation of polyolefins are initiated by reactive functional groups (usually hydroperoxide or carbonyl groups) introduced into the polymer by partial oxidation during processing (29). Polymeric impurities, including partially oxidized polymer, are expected to be rejected by the growing polymer crystals and may be responsible for the observed intraspherulitic crystallinity variation. Rejection of polymeric impurities may have two effects. It may weaken the boundary regions or it may greatly increase their sensitivity to oxidation by concentrating initiation sites within the boundaries. In a polymer where there is no crystallinity variation with radius, partially oxidized polymer will be rejected to the boundaries, the effect decreasing with increasing crystallization rate. In polypropylene, it remains unclear whether partially oxidized polymer is boundary rejected or simply contributes to the observed crystallinity variation within the spherulites. However, any tendency to be boundary rejected must lead to a significant effect on stability, and although more detailed studies are required, we believe that redistribution of partially degraded polymer may exert a major influence upon oxidative embrittlement.

Acknowledgments

The authors thank I. C. I. Plastics Division for supporting this work via a CASE grant. They also thank A. Curson and D. G. M. Wood for helpful discussions and access to equipment and D. Back of Kings College, London, for use of the uv microscope. T. G. Ryan thanks the Science Research Council for the maintenance grant award.

Literature Cited

1. Hawkins, W. L., Matreyek, W., Winslow, F. H., *J. Polym. Sci.* (1959) **41**, 1.
2. Winslow, F. H., Aloisio, C. J., Hawkins, W. L., Matreyek, W., Matsuoka, S., *Chem. Ind. (London)* (1963) 533.
3. Hansen, R. H., *in* "Thermal Stability of Polymers," R. T. Conley, Ed., p. 153, Dekker, New York, 1970.
4. Billingham, N. C., Prentice, P., Walker, T. J., *J. Polym. Sci., Polym. Symp.*, (1976) **57**, 287.
5. Keith, H. D., Padden, F. J., Jr., *J. Appl. Phys.* (1964) **35**, 1270.
6. Barnes, W. J., Luetzel, W. G., Price, F. P., *J. Phys. Chem.* (1961) **65**, 1742.
7. Moyer, J. D., Ochs, R. J., *Science* (1963) **142**, 1316.
8. Frank, H. P., Lehner, H., *J. Polym. Sci., Polym. Symp.* (1970) **31**, 193.
9. Curson, A. D., *Proc. R. Micros. Soc.* (1972) **7**, 96.
10. Ryan, T. G., Calvert, P. D., *Polymer*, in press.
11. Pfann, W. G., "Zone Melting," 2nd ed., Wiley, New York, 1966.
12. Dubini, M., Cicchetti, O., Vicario, G. P., Bua, E., *Eur. Polymer J.* (1967) **3**, 473.
13. Klein, J., Briscoe, B. J., *Polymer* (1976) **17**, 481.
14. Uhlmann, D. R., Chalmers, B., Jackson, K. A., *J. Appl. Phys.* (1964) **35**, 2986.
15. Van Sickle, D. E., *J. Polym. Sci., Polym. Chem. Ed.* (1972) **10**, 355.
16. Bawn, C. E. H., Chaudri, S. A., *Polymer* (1968) **9**, 81.
17. Monaci, A., Lassari, P., Bernaducci, E., *Chem. Ind. (Milan)* (1963) **45**, 1337.
18. Mizutani, Y., Yamamoto, K., Matsuoka, S., Ihara, H., *Chem. High Polym. (Tokyo)* (1965) **22**, 97.
19. Adams, J. H., *J. Polym. Sci., Polym. Chem. Ed.* (1970) **8**, 1077.
20. Van Schooten, J., *J. Appl. Polym. Sci.* (1960) **4**, 122.
21. Inoue, M., *J. Polym. Sci.* (1961) **55**, 443.
22. Barish, L., *J. Appl. Polym. Sci.* (1962) **6**, 617.
23. Winslow, F. H., Hellman, M. Y., Matreyek, W., Stills, S. M., *Polym. Eng. Sci.* (1966) **6**, 273.
24. Winslow, F. H., Matreyek, W., *Polym. Prepr., Am. Chem. Soc.* (1964) **5**, 552.
25. Kafavian, G., *J. Polym. Sci.* (1957) **24**, 501.
26. Calvert, P. D., *Nature (London)* (1977) **268**, 321.
27. Czerny, J., *J. Appl. Polym. Sci.* (1972) **16**, 2623.
28. Guillory, J. P., Cook, C. F., *J. Polym. Sci., Polym. Chem. Ed.* (1971) **9**, 1529.
29. Rånby, B., Rabek, J. F., "Photodegradation, Photooxidation and Photostabilization of Polymers," Wiley, New York, 1975.

RECEIVED May 26, 1977.

Microscopic Mechanisms of Oxidative Degradation and Its Inhibition at a Copper–Polyethylene Interface

D. L. ALLARA and C. W. WHITE[1]
Bell Laboratories, Murray Hill, NJ 07974

Several techniques—including atomic absorption spectroscopy, Rutherford Back Scattering, and internal reflection infrared spectroscopy—have been applied to the study of solubilities and diffusion rates of the copper carboxylate salts $Cu(C_nH_{2n+1}CO_2)_2$ *with* n = *3, 7, 11, 17, and 29 in low density polyethylene at 90°C. Solubilities can increase by several orders of magnitude when the polyethylene is preoxidized in* O_2*. This observation is attributed to trapping of copper in the polymer matrix by ligand exchange between the diffusing salt and free carboxylic acid groups created by oxidative scission in the oxidized polymer. Calculations show that the observed solubilities and diffusion rates correlate well with postulated mechanisms of copper-catalyzed oxidation under typical conditions. Experiments with N,N′-diphenyloxamide, a typical copper deactivator, showed little effect of the deactivator on diffusion rates and indicate that surface and interface reactions are the critical inhibition processes rather than bulk phase scavenging of copper ions.*

Metal surfaces often have adverse catalytic effects on the rates of oxidative degradation (by molecular oxygen) of polymers, particularly polyolefins. Previous papers in this series have shown for the specific case of the O_2/polyethylene/copper system between 40°C and 90°C that copper carboxylate salts, initially formed at the interface, are re-

[1] Current address: Oak Ridge National Laboratories, Oak Ridge, TN 37830

0-8412-0381-4/78/33-169-273$05.00/1

sponsible for this catalysis (*1, 2, 3, 4*) and that extensive migration of these salts into the polyethylene matrix can occur during the oxidation (concentrations as high as 0.1*M* within 6000 Å of the interface). The corresponding activity of so-called "metal deactivators" in inhibiting the metal's catalytic effects is caused mostly by the reaction of these deactivator molecules with copper carboxylate salts whereby insoluble copper complexes of low catalytic activity are formed (*5*). These reactions can occur either at the copper/polyethylene interface or in the polyethylene matrix. The critical location for inhibition may be very dependent on the rate at which the copper carboxylate salts diffuse into and through the polymer. It is thus clear that a detailed knowledge of the solubilities and diffusion coefficients of appropriate copper salts in polyethylene is necessary for the determination of the fundamental chemical mechanisms of the copper-catalyzed oxidation and the related stabilization by deactivators.

Little published information is available on the fundamental details of diffusion and solution of metal salts in polymers, particularly polyolefins. Accordingly, the purpose of this study is to determine, as quantitatively as possible, appropriate solubility and diffusion parameters. We report results on the salts $Cu(O_2CR)_2$, where the alkyl R group varies from 2 to 29 carbons, in low density polyethylene at 90°C. The effects of the extent of reaction of the bulk polymer with oxygen were found to be important and were studied extensively. Some experiments were carried out with low concentrations of a typical copper deactivator, *N,N'*-diphenyloxamide, in an attempt to develop a better understanding of the corresponding inhibition mechanisms. A combination of analytical techniques was applied to the analysis of copper carboxylates in various depth regions of the polymer: Rutherford Back Scattering (RBS) (*3*) for the near surface regions (0–6000 Å), infrared internal reflection spectroscopy (IRS) (*1, 2*) for the deeper surface regions (up to several μm), and atomic absorption and x-ray fluorescence for bulk analyses. With the application of these various techniques we have attempted to characterize the diffusion of copper carboxylates in both the surface regions and bulk and to correlate these results as closely as possible.

Experimental

Sample Preparation. Polyethylene pellets were processed on a two-roll mill followed by pressing between clean aluminum sheets at 110°–125°C for one minute. The resultant films were preoxidized to varying extents by initially exposing them in air to ~ 5 Mrads of γ-radiation from a Co^{60} source followed by exposure to one atmosphere of oxygen gas in sealed vessels at 90°–100°C. The extent of oxidation was monitored by ir spectroscopy (*see below*). Two types of laminates were formed from ~ 1.5 cm × 1.0 cm polyethylene films about 0.025 cm in thickness: (a) a

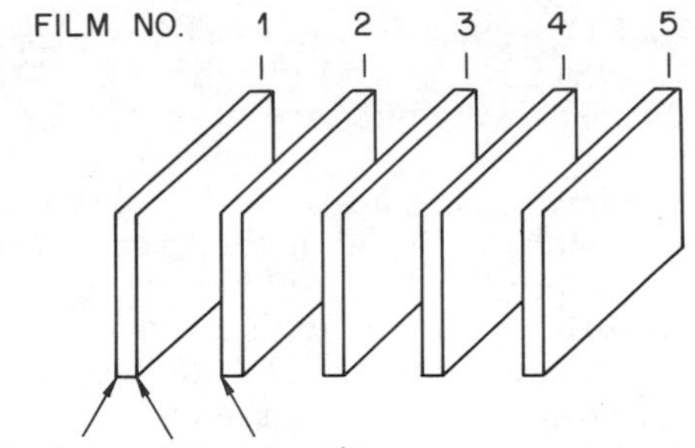

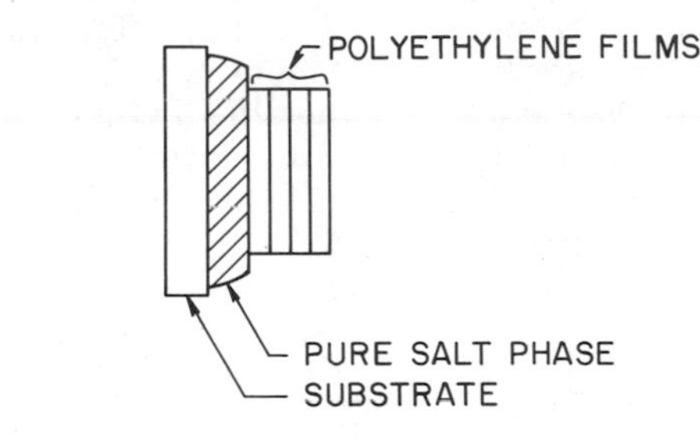

Figure 1. Configurations of typical diffusion samples

stack of films in which one or both outer films were loaded with various concentrations of a copper salt, and (b) a stack of films in which an outermost film surface was in contact with a smooth film of a copper salt supported on a polished silicon or a Teflon substrate. Details are shown in Figure 1. Both types of laminates were pressed between two stainless steel plates held by 4–6 screws and lined with Teflon sheets. The samples were pressed tightly enough to provide effectively complete interfacial contact without film deformation at the experimental temperatures. All diffusion–solubility experiments were run in an atmosphere of helium or argon in glass vessels sealed by viton O-rings. All runs were at 90.0° ± 0.05°C except for a few runs at 60.0° ± 0.1°C.

Starting Materials. Polyethylene, low density (0.92), was obtained from Union Carbide as additive-free pellets. Copper propionate, used only for liquid-phase solubilities, was obtained commercially and used without further purification. All other copper salts were synthesized as follows: equal equivalents of $Cu(OH)_2 \cdot CuCO_3$ (99+%, ROC/RIC Chemical Corp.) and the appropriate carboxylic acid were stirred in a minimum amount of xylene and heated to ~ 120°C under N_2 overnight. After reacting the xylene solution was diluted with additional xylene, reheated, and filtered hot to remove any copper oxides and/or carbonates. The xylene solution was cooled and the resultant precipitate collected by suction filtration and washed with additional xylene and then hexane. Additional recrystallization was done from hexane, isooctane, and/or xylene. In the case of the copper octanoate the solubility in xylene was quite high, so minimum amounts were used and recrystallization was done from hexane. Analyses for copper: $Cu(C_7H_{15}CO_2)_2$,

96.9%; $Cu(C_{11}H_{23}CO_2)_2$, 99.2%; $Cu(C_{17}H_{23}CO_2)_2$, > 98%; $Cu(C_{29}H_{59}CO_2)_2$, 98.8%; $Cu(Br(CH_2)_{10}CO_2)_2$, 97.8% (Br, 99.3% of theory). The starting carboxylic acids were obtained from Aldrich as the purest grades available.

Analysis of Films. After heating, the films in the laminate stack were separated for analysis by the methods described below. To ensure that excess surface impurities (e.g., salt) were removed, the film surfaces were usually washed repeatedly with concentrated HCl followed by actone and water rinses. Prior to RBS analysis the films were coated with several hundred angstroms of carbon.

Details of the Rutherford Back Scattering technique as applied to polyethylene have been described elsewhere (*3*). The IRS measurements were carried out using a Perkin Elmer 621 infrared spectrometer. A KRS-5 reflection element with dimensions 50 mm × 2 mm was used with a 45° angle of incidence. Further details can be found elsewhere (*1*). Most of the atomic absorption measurements were performed by the Fairfield Testing Laboratory, Fairfield, NJ. The x-ray fluorescence measurements were carried out on a GE-XRD6 spectrometer.

Liquid Phase Solubilities. The solubilities of copper carboxylates in octane (Aldrich, 99+%) and hexadecane (Chem Samples Co., 99.9+%) were determined by intermittently, over a period of several hours, adding small weighted portions of salt into continuously stirred, fixed weights of solvent, contained in large test tubes in a theromstated bath at 90.0°C (±0.05°), until visual inspection detected opacity in the solution caused by light scattering off insoluble particles. The solutions at this point were then heated above 90°C until they became clear again and then returned to the 90°C bath and allowed to come slowly to thermal equilibrium (over several hours). If the cloudiness returned, an upper limit to the solubility was thus determined. The method gave good results for all the salts except the octanoate which exhibited such a high solubility at 90°C that visual observation of small amounts of insoluble salt was very difficult.

Results

Solubilities. EXTRAPOLATION FROM LIQUID PHASE. Solubilities of various copper carboxylate salts in pure octane and hexadecane at 90°C are shown in Figure 2. The solubility of the octanoate salt is quite high but cannot be measured accurately because of experimental difficulties (discussed above). If one applies regular solution theory (*6*) with the assumption that the excess free energies of mixing (or alternatively the solvent–solute interaction parameters) are equal for a particular salt in an alkane medium, then the following equation can be derived (*7*):

$$\Phi_1 \propto e^{-(1-V_1/V_2)}$$

where Φ_1 = volume fraction of the solute in the solution = $N_1V_1/(N_1V_1 + N_2V_2)$; N_1,N_2 = number of moles of solute and solvent, respectively; and V_1,V_2 = molar volumes of solute and solvent. Thus a plot of

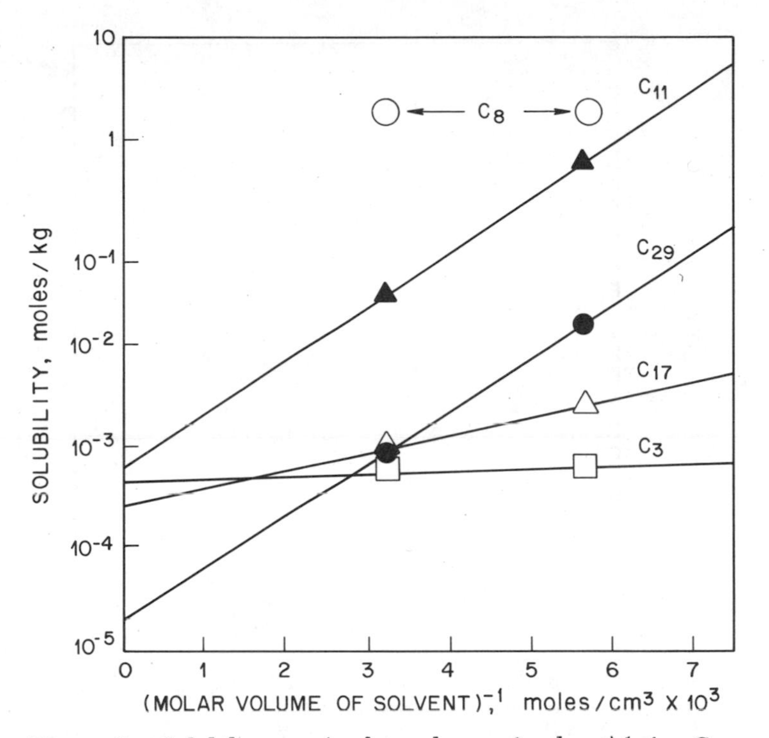

Figure 2. Solubility vs. (molar volume of solvent)$^{-1}$ for Cu-$(RCO_2)_2$ salts in hexadecane and octane at 90°C. R groups indicated on curve.

ln Φ_1 vs. $1/V_2$ should be linear for a fixed solute, and if one assumes that a polyethylene chain is infinite, viz., $V = \infty$, then the value of the intercept at $1/V_2 = 0$ represents the solubility in polyethylene. Since the concentration of solute is approximately proprotional to Φ, we have plotted log [solubility (mol/kg)] vs. reciprocal molar volume of octane and hexadecane in Figure 2. The molar volumes of the solvents at 90°C were determined in separate experiments by standard volumetric techniques.

DIRECT MEASUREMENT OF SOLUBILITY. The solubilities in polyethylene films were determined as follows. Samples consisting of stacks of unloaded films (generally two) in contact with pure-salt phase (*see* Figure 1) were heated at 90°C for various times. At sufficiently long times (several hundred hours or greater) the bulk copper analyses of successive films in a stack became nearly equal in concentration, and the concentration in the first film was then assumed equal to the solubility. The results are presented in Figure 3 along with the extrapolated values (divided by two, to correct for ~ 50% crystallinity in the solid polymer) from the liquid phase measurements. Solubilities are given for non-preoxidized films and for films preoxidized to $A = 0.08$ and $A = 0.41$–0.43,

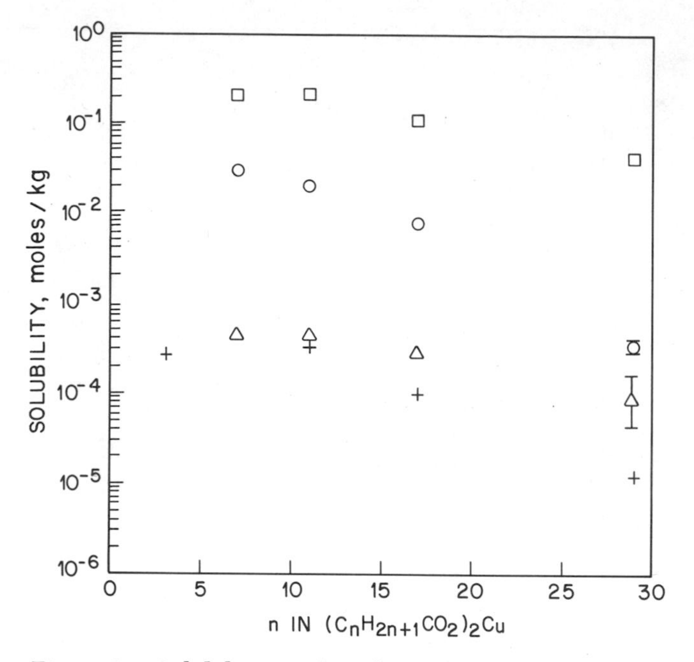

Figure 3. Solubility vs. chain length for $Cu(RCO_2)_2$ salts in preoxidized low density polyethylene at 90°C. □, A = *0.4;* ○, A = *0.08;* △, *unpreoxidized;* +, *extrapolated from liquid-phase experiments in* n-*alkanes.*

where A = absorbance (C=O, 1715 cm^{-1} band)/absorbance (CH_2, 1465 cm^{-1} band). Parallel measurements of the O_2 consumption during preoxidation indicates that A = 0.4 corresponds approximately to 30 cm^3 of O_2 per gram of polyethylene at STP. The extrapolated pure polyethylene and the non-preoxidized film values agree quite well except for the C_{29} salt for which a difference of up to an order of magnitude may exist. There was (unexplained) difficulty obtaining reproducible values for the C_{29} salt in non-preoxidized film samples, and the error bars for this point show the range of values. Although the solubility of the C_{29} salt in slightly preoxidized films appears too low, the value was quite reproducible, as shown by the error bars. The non-preoxidized films always exhibited trace amounts of oxidation as measured by small absorption in the 1715–1740 cm^{-1} region ($A < 0.01$). Most likely this arises because of processing and is generally unavoidable for unstabilized polyolefins. This trace oxidation does not appear to affect the solubilities since they agree quite well (except for C_{29}) with the liquid-phase extrapolated values. As an attempt to obtain a film with a minimum amount of preoxidation, films pre-loaded with 0.02 wt % ($\sim 2 \times 10^{-4}$ mol/kg) of a 2,4,6-trisubstituted phenol antioxidant (Irganox 1010) were used to measure solubilities. These films consistently gave solubilities larger than observed in unstabilized films. These results indicate that the copper salts react with the stabilizer.

Bulk Diffusion Rates. EXPERIMENTAL PROFILES. Diffusion profiles were generated from the atomic absorption analyses. Data for a variety of selected, typical experiments are given in Table I. Data are not given for non-preoxidized films because of the irreproducibility of the profile measurements (*see below*). Diffusion coefficients were calculated from the total copper absorbed in the laminate stacks using Equation 2 given below. With these diffusion coefficients values of the average concentration of copper for each film in a stack were calculated from Equation 3 below and are also included in Table I. Results with the C_7 salt were difficult to obtain because the melting point was sufficiently close to 90°C (91.5–93.0°C for a recrystallized sample) so that softening and flowing of the salt film around the pressed polyethylene laminate occurred during heating. The C_7 runs presented in Table I seemed the most free of this problem. Other carboxylate salts had higher melting points and presented no problem. The extreme limits of reproducibility of the results can be estimated by a direct comparison of runs 75-6 and 67-8. There were occasional exceptions to the expected monotonic decrease of copper concentration with increasing distance from the salt/polymer interface, e.g., run 75-6. Good interfacial contact seemed present in these samples (like all others) as judged by the clear appearance of the stacks after heating and the appearance on peeling the films apart. However, several other experimental problems which may have contributed to the scatter in the results were: (a) brittleness and occasional cracking of the highly preoxidized films which led to excessive salt located along the crack areas (samples were cut out avoiding these enriched areas as carefully as possible), (b) difficulty of removing all the surface phase with HCl/acetone wash, particularly in the C_{17} and C_{29} salts where fast penetration in and complete solubilization of the salt film by the wash solution were difficult to achieve, and (c) the inherent scatter in the atomic absorption measurements near the limit of detection, $\sim 10^{-5}$ mol/kg (in the polymer), for these particular film geometries. The latter was particularly a problem in non-preoxidized polyethylene, where the solubility limit is low ($\sim 9 \times 10^{-5}$ mol/kg), and represents an upper limit on the diffusion profile concentrations.

DIFFUSION EQUATIONS. Fick's second law can be solved for the special case of a layer of substance of constant (with time) concentration diffusing into a semi-infinite medium which initially contains no solute. Using the method of the Laplace transform (*8*) the solution yields an expression for the concentration in the medium:

$$C = C_s \operatorname{erfc}\left(\frac{x}{2\sqrt{Dt}}\right) \qquad (1)$$

Table I. Diffusion of Copper Carboxylates into Polyethylene[a]

Run	*Time (hr)*	n *in* $(C_nH_{2n+1}CO_2)$	*Preoxidation (A)*[c]	*Film Analyses*[b, d] *Film 1* *[Cu]*	*d*
69-5	20	7	0.43	8.0×10^{-2}	0.024
—	—	—	—	7.8×10^{-2}	—
75-2	73.5	7	0.43	0.12	0.030
—	—	—	—	0.11	—
75-1	73.5	7	0.08	1.1×10^{-2}	0.022
—	—	—	—	1.1×10^{-2}	—
75-3	74	11	0.43	8.7×10^{-2}	0.032
—	—	—	—	9.4×10^{-2}	—
75-11	97	11	0.08	4.8×10^{-3}	0.028
—	—	—	—	5.0×10^{-3}	—
62-9	20	17	0.41	4.4×10^{-2}	0.020
—	—	—	—	4.3×10^{-2}	—
63-4	65	17	0.41	5.8×10^{-2}	0.020
—	—	—	—	6.6×10^{-2}	—
67-5	100	17	0.08	6.7×10^{-3}	0.023
—	—	—	—	5.6×10^{-3}	—
75-8[e]	30	17	0.39	3.1×10^{-2}	0.021
—	—	—	—	3.1×10^{-2}	—
75-9[e]	68	17	0.39	7.7×10^{-2}	0.019
—	—	—	—	7.1×10^{-2}	—
62-7	20	29	0.41	2.9×10^{-3}	0.020
—	—	—	—	2.8×10^{-3}	—
67-8	214	29	0.43	1.5×10^{-2}	0.021
—	—	—	—	1.8×10^{-2}	—
62-6	1000	29	0.43	3.9×10^{-2}	0.020
—	—	—	—	3.3×10^{-2}	—
75-6	212	29	0.43	3.1×10^{-3}	0.027
—	—	—	—	5.9×10^{-3}	—

[a] Samples consisted of polyethylene films pressed against a salt film deposited on a silicon wafer.
[b] Calculated concentrations are based on the diffusion equation for a semi-infinite medium (*see text*) and are given immediately under the experimental values. Diffusion coefficients are calculated from the total copper absorbed by the laminate stack.

where C_s is the solubility in the medium, χ is the distance from the interface, D is the diffusion coefficient, and t is time. The semi-infinite criterion is satisfied in the present laminate samples provided that $C(\chi = l) << C(\chi = 0)$, where l = stack thickness. In most cases in Table I this is a valid approximation, and the inherent scatter in the data does not warrant more exact expressions. The amount of solute (per unit area) which is absorbed between χ and ∞ (approximated by l experimentally) is given by the integral

at 90°C: Experimental Analyses and Calculated Profiles[b]

Film Analyses[b, d]

Film 2		*Film 3*		*Film 4*		
[Cu]	*d*	*[Cu]*	*d*	*[Cu]*	*d*	*D* *(cm²/sec)*
9.3×10^{-4}	0.018	1.5×10^{-5}	0.022	—	—	—
2.3×10^{-5}	—	1.3×10^{-5}	—	—	—	1.0×10^{-9}
1.7×10^{-3}	0.024	$< 1 \times 10^{-5}$	0.028	$< 1 \times 10^{-5}$	0.027	—
1.6×10^{-2}	—	1.0×10^{-3}	—	1.6×10^{-6}	—	1.0×10^{-9}
2.4×10^{-4}	0.023	1.1×10^{-4}	0.023	2.7×10^{-5}	0.025	—
1.5×10^{-4}	—	3.4×10^{-8}	—	$< 10^{-10}$	—	2.0×10^{-10}
8.5×10^{-3}	0.032	1.9×10^{-3}	0.031	8.1×10^{-4}	0.032	—
3.6×10^{-3}	—	1.1×10^{-5}	—	2.3×10^{-9}	—	6.0×10^{-10}
7.9×10^{-5}	0.028	$< 10^{-5}$	0.025	$< 10^{-5}$	0.027	—
1.3×10^{-6}	—	$< 10^{-10}$	—	$<< 10^{-10}$	—	1.0×10^{-10}
7.0×10^{-4}	0.020	$< 4.6 \times 10^{-6}$	0.020	—	—	—
9.6×10^{-4}	—	8.5×10^{-7}	—	—	—	7.0×10^{-10}
1.5×10^{-2}	0.022	6.9×10^{-3}	0.017	—	—	—
1.2×10^{-2}	—	7.7×10^{-4}	—	—	—	7.0×10^{-10}
6.9×10^{-4}	0.020	—	—	—	—	—
1.6×10^{-3}	—	—	—	—	—	8.0×10^{-10}
5.2×10^{-4}	0.021	4.0×10^{-5}	0.020	$< 10^{-5}$	0.017	—
4.8×10^{-5}	—	$< 10^{-11}$	—	$<< 10^{-11}$	—	2.5×10^{-10}
8.0×10^{-3}	0.022	1.7×10^{-3}	0.017	7.2×10^{-4}	0.020	—
1.6×10^{-3}	—	1.7×10^{-3}	—	1.1×10^{-4}	—	8.0×10^{-10}
$< 5 \times 10^{-5}$	0.020	$< 5 \times 10^{-5}$	0.020	—	—	—
$< 10^{-10}$	—	$<< 10 \times 10^{-10}$	—	—	—	2.0×10^{-11}
3.4×10^{-3}	0.021	—	—	—	—	—
7.0×10^{-4}	—	—	—	—	—	9.0×10^{-11}
1.3×10^{-2}	0.018	1.0×10^{-2}	0.020	—	—	—
1.9×10^{-2}	—	8.9×10^{-3}	—	—	—	2.0×10^{-10}
5.4×10^{-4}	0.023	5.4×10^{-4}	0.026	1.5×10^{-3}	0.025	—
$< 10^{-10}$	—	$<< 10^{-10}$	—	—	—	1.5×10^{-11}

[c] A = absorbance (C=O, 1715–1740 cm^{-1})/absorbance (CH_2 1465 cm^{-1}); *see text.*
[d] [Cu] in units of mol/kg; d = film thickness in cm.
[e] Sample contains 0.1% (wt) *N,N′*-diphenyloxamide.

$$M_{\chi,\infty} = \int_{\chi}^{\infty} C(\chi)\,d\chi \qquad (2)$$

$$= 2C_s\sqrt{Dt}\,[\mathrm{ierfc}(\chi/2\sqrt{Dt})]$$

The amount absorbed by a film located between χ_1 and χ_2 is $M(\chi_1, \infty) - M(\chi_2, \infty) = M(\chi_1, \chi_2)$. The average concentration in the region χ_1,χ_2 is $M(\chi_1,\chi_2)/(\chi_2 - \chi_1) = \hat{C}(\chi_1,\chi_2)$. Equation 2 was solved by machine computation on a Honeywell 600 computer using a Rand approximation.

Table II. Surface Measurements of Diffusion of Copper Polyethylene Film[a]

Run	n *in* $(C_nH_{2n+1}CO_2)_2Cu$	*Temp (°C)*	*Time (hr)*
17-1	17	90	502
11-1	11	90	502
11-3	11	90	41
17-2	17	60	473

[a] A = 0.41, film thickness 0.025 cm; prior to analysis films washed with concentrated HCl followed by acetone and distilled water rinses.

The calculated values of $\hat{C}(\chi_1,\chi_2)$ are given in Table I under the experimental values.

Surface Region Measurements. LOCAL SURFACE REGIONS (0–6000Å). Rutherford Back Scattering (RBS) analysis was applied to a variety of individual films to determine local surface region copper profiles. The total concentration gradient at 0–6000 Å was obtained, but for convenience the results are discussed below in terms of < 1000-Å and ~ 6000-Å surface regions. It was originally hoped that the measured gradients could be used for calculation of diffusion coeffiicents, but in most cases this was not possible because of fast formation of pure salt phase at the polymer surfaces following cooling and separation of the laminate stacks.

Uniform distribution of copper is indicated for highly preoxidized polyethylene films at equilibrium with pure salt phase as shown in Table II. The variations of surface region concentrations from bulk solubilities are negligible relative to the combined errors of the experimental and analytical techniques. The results suggest that pure surface phase is mostly (but not completely) removed by the washing techniques (pure-salt phase concentration will be > 1.5 mol/kg). Removal of further salt by additional washings proved difficult. A large number of preoxidized and non-preoxidized films from multi-film laminates at shorter diffusion times were studied, and in all cases the RBS analyses (washed and unwashed samples) showed surface region copper concentrations higher than the average bulk concentrations but always less than the pure-salt phase concentrations. Thus even in films with expected steep concentration gradients small amounts of pure surface phase formed during or after the diffusion period (and persisted partially through washing), and the expected surface-to-surface gradients could not be measured. A phenomenon involving localized enrichment within the surface regions of the polymer during diffusion could also be considered in view

Carboxylates from Pure Salt Phase into a Preoxidized at 90°C and 60°C

[Cu], mol/kg

	Surface Concentrations by RBS			
	Pure Salt Interface		*Ambient Interface*	
Bulk Conc. by AA	*0–1000 Å*	*~ 6000 Å*	*0–1000 Å*	*~ 6000 Å*
0.11[b]	0.11	0.095	—[c]	—[c]
0.22[b]	0.29	0.19	0.50	0.41
0.068	0.19	0.095	—[c]	—[c]
0.0076	0.090	0.041	0.014	0.008

[b] Solubility limit.
[c] No result because of experimental problem.

of the incomplete efficiency of the washing operation but seems an unlikely possibility. Since the detection limit of the RBS is ~ 10^{-3} mol/L, well above the solubility limit (*see* Figure 2) for non-preoxidized films, no possibility exists of seeing anything but pure salt, surface phases in the latter films.

A number of laminate stacks in which non-preoxidized and preoxidized films were loaded with *N,N′*-diphenyloxamide, at concentrations of 3.8×10^{-3}, 3.8×10^{-4}, and 3.8×10^{-5} mol/kg (0.1–0.001 wt %), were run at 90°C for a variety of times and analyzed by RBS. Within the reproducibility of the experiments the general pattern of results is essentially the same as for non-loaded samples. An occasional run was observed to give significantly slower diffusion with the oxanilide loaded films than with unloaded films, but this effect was not reproducible. Furthermore, no corresponding differences of significance in the surface concentration patterns were seen with these samples. These results agree with the selected diffusion experiments reported in Table I.

Profiles of O and Br atoms were also measured in a few samples and will be discussed in later sections dealing with exchange reactions.

DEEP SURFACE REGIONS (< 2 μm). Internal reflection (IRS) infrared measurements of several selected films in diffusion stacks were performed to monitor the concentration of the carboxylate group. Interesting results were obtained from analysis of bands at ~ 1590–1630 and 1710–1720 cm^{-1}. The approximate penetration depth can be calculated (9) from the equation

$$d_p = \frac{\lambda_1}{2\pi(\sin^2\theta - n_{21}^2)^{1/2}} \qquad (3)$$

where d_p = distance at which the electric field amplitude is $1/e$ of the boundary surface value, λ_1 = wavelength/n_1, θ = angle of incidence,

Table III. Summary of Internal Reflection

Run No. Film and Interface	R *in* $(RCO_2)_2Cu$	*Time (hr)*	*Bulk [Cu] (mol/kg)*
43-1	C_7H_{15}	502	
1-0,1			0.049
-0,1[a]			
-1,2			
2-1,2			0.035
-1,2[b]			
-2,3[b]			
49-2	$C_{17}H_{35}$	984	
1-0,1			0.058
-1,2			
2-1,2			0.0098
-2,3			
49-OX-2	C_7H_{15}	1008	
1-0,1			0.058
-1,2			
2-1,2[c]			0.0092
-2,3[c]			
49-OX-1	C_7H_{15}	1008	
1-0,1[d]			0.054
-1,2[d]			
Standard film, 0.1*M* salt	C_7H_{15}	—	0.076
Standard film, 0.1*M* salt	C_7H_{15}	—	
51-OX-5	C_7H_{15}	65	
4-4,5[c]			0.0092
-3,4[c]			
62-OX-5	C_7H_{15}	20	
1-1,2			0.066
2-1,2			0.0018
-1,2[a]			
-2,3			

[a] Washed with HCl/acetone.
[b] Washed with acetone.

n_1 = refractive index of the reflection element, n_2 = refractive index of the polymer, and $n_{21} = n_2/n_1$. Using $n_2 \sim 1.5$ (our measured value) one calculates $d_p \sim 3$ μm (30,000 Å) for the above-mentioned bands. The types of samples which best showed the appearance of the 1710–1720 cm^{-1}

Infrared Measurements of 90°C Diffusion Films

$\left(\frac{ABS_{(1710\text{-}1720\ cm^{-1})}}{ABS_{(1456\ cm^{-1})}}\right)$ Pre-Diffusion	Post Diffusion	Post Diffusion; 1550–1650 cm^{-1}: $\left(\frac{ABS(\tilde{\nu}_{max})}{ABS(1465\ cm^{-1})}\right)$	$\tilde{\nu}_{max}$, cm^{-1}
~ 0	0.19	0.44	1590
~ 0	0.11	0.06	broad, ~ 1650
~ 0	0.04	weak	broad, ~ 1610
0.41	0.41	~ 0.04	broad, ~ 1610
0.41	0.42	~ 0.04	broad, ~ 1610
0.41	0.49	0.05	broad, ~ 1610
~ 0	0.19	0.54	1590
~ 0	~ 0	weak	broad, ~ 1610
0.54	0.49	~ 0.04	broad, ~ 1610
0.54	0.52	~ 0.03	broad, ~ 1610
~ 0	0.21	0.39	1590
~ 0	~ 0	~ 0.03	broad, ~ 1610
0.53	0.44	0.03	broad, 1620
0.53	0.46	weak	broad, 1600-1650
~ 0	0.24	0.41	1590
~ 0	0.05	~ 0.04	broad, 1610
~ 0	—	~ 0.24	1590
~ 0	—	0.05	broad, 1650
0.53	0.43	weak	broad, 1620
0.53	0.46	weak	broad, 1620
~ 0	0.04	~ 0.03	broad, ~ 1620
0.45	0.45	0.06	broad, ~ 1600
0.45	0.45	weak	broad, ~ 1600
0.45	0.49	~ 0	—

[c] Initially 0.01 wt % *N,N'*-diphenyloxamide in film.
[d] Initially 0.001 wt % *N,N'*-diphenyloxamide in film.

bands were stacks consisting of three preoxidized inner films sandwiched between two outer non-preoxidized reservoir films pre-loaded with copper carboxylate, ~ 0.1 mol/kg. [This concentration is obviously well above the solubility limit, and diffusion coefficients cannot be calculated since the boundary conditions (driving concentration) are not certain.] In these non-preoxidized outer films 1710–1720 cm^{-1} absorption is nearly absent initially, whereas in the preoxidized films the oxidation products exhibit these bands quite strongly. The results are presented in Table III.

The absorption between 1590 and ~ 1630 cm^{-1} can be attributed to the carboxylate ($R–CO_2^-$) group asymmetric stretching mode. A standard film of unoxidized polyethylene, nominally loaded with a $Cu(C_7H_{15}CO_2)_2$ concentration of ~ 0.01 mol/kg (actual analysis by AA, 0.008 mol/kg), showed a sharp band at 1590 cm^{-1} (Table III). This is attributed to a pure salt phase in the surface since only a broad weak absorption at 1600–1630 cm^{-1} is left after HCl/acetone rinses. The latter absorption can be taken as the "bulk" carboxylate band for ~ 0.01 mol/kg. No significant 1550–1630 cm^{-1} absorption was observed in the unloaded (inner) films prior to diffusion. After diffusion the outer surface (0,1) of the outer (1 or 5) films showed the expected sharp 1590 cm^{-1} band of pure salt phase, whereas the corresponding inner surface (1,2) gave only a very broad absorption with a maximum at ~ 1650 cm^{-1}. Since the outer films were not preoxidized, the solubility of salt is quite low, and diffusion to inner films appears to deplete the excess concentrations rapidly and permanently at the 1,2 interface of the 1 and 5 films. RBS analysis of these films gave a corresponding decrease in Cu at the 1,2 interface, several-fold below the bulk level. After diffusion the inner films (both 1,2 and 2,3 interfaces) showed the appearance of broad, bulk carboxylate absorption varying between 1610 and 1650 cm^{-1}. Quantitative characterization was difficult because of the broadness and background absorption by the KRS5 crystal surfaces which became rapidly contaminated, but intensity estimates are given in Table III. The *N,N'*-diphenyloxamide loaded films showed no major differences in the carboxylate region spectra except for the 20-hr run, 62-OX-5, which showed significant carboxylate (assigned on basis of 1620-cm^{-1} band) accumulation at the 1,2 interface of inner film 2. In comparison, the corresponding opposite 2,3 surface of film 3 exhibits no detectable carboxylate. On the basis of this one result no significant conclusion can be reached. *N,N'*-diphenyloxamide exhibits characteristic bands at 1555 and 1650 cm^{-1}, but the lower limit for detection of these in a polyethylene matrix under our conditions was found to be < 0.01 mol/kg, larger than the highest concentration of the loaded films (0.1 wt %, ~ 4×10^{-3} mol/kg).

The absorption at 1710–1720 cm^{-1} which appears after diffusion in the outer films is typical of long-chain, aliphatic carboxylic acids (*10*). The absorption is considerably more intense at the ambient (0,1) interface than at the 1,2 interface. These bands are caused, at least in part, by surface phases since washing diminishes their intensity (*see* run 43-1). The inner films are preoxidized and show strong absorption in the 1710–

1720 cm^{-1} region caused by oxidation of the polymer. This absorption is generally unaffected by diffusion of salt and is not removed by washing.

Exchange Experiments. To test whether or not the initial copper carboxylate salt was transported through the polymer matrix as an integral molecular unit, an experiment was run using the bromine-tagged salt $Cu(Br(CH_2)_{10}CO_2)_2$. Diffusion was carried out at 90°C using a five-film stack of two non-preoxidized outer films (1 and 5) preloaded with ~ 0.005 mol/kg salt and three preoxidized inner (2, 3, and 4) unloaded films (*see* Figure 1). The results are given in Table IV. It is obvious that the distribution of copper is far from equilibrium, whereas the bromine is completely in an equilibrium distribution. No significant material has been lost judging from the mass balance (85% based on the summed film contents). Since there is no reason to assume that the C–Br bond has been broken under these conditions, it must be concluded that the carboxylate group has been dissociated in some manner from the cupric ion and transported rapidly throughout the polymer.

Discussion

Solubilization Mechanism. All our experimental results support the proposal that the correlation between degree of preoxidation in the polyethylene and solubility is caused by exchange reactions:

$$Cu(RCO_2)_2 \rightleftarrows Cu(RCO_2)_2(m) \quad (4)$$

$$Cu(RCO_2)_2(m) + P\text{–}CO_2H \rightleftarrows (P\text{–}CO_2)Cu(RCO_2) + RCO_2H(m) \quad (5)$$

$$(P\text{–}CO_2)Cu(RCO_2) + P\text{–}CO_2H \rightleftarrows (P\text{–}CO_2)_2Cu + RCO_2H(m) \quad (6)$$

Reaction 4 is the dissolution of pure salt into the polymer matrix phase, designated by m. (Equation 4 is probably an oversimplification of the actual molecular species involved. Copper carboxylates tend to dimerize in nonpolar media, and one must also consider monomer–dimer equilibrium and the accompanying diffusion coefficients of each species.) $P\text{–}CO_2H$ is a polyethylene chain containing a carboxylic acid group produced during oxidation of the polymer. The released acid, $RCO_2H(m)$, is free to diffuse and would be expected to migrate rapidly to the surface to form a pure salt phase when the matrix solubility is exceeded. The latter phenomenon is indicated by our exchange experiments and surface analyses (although most of the phase change may occur on cooling to ambient). For an oxygen consumption (at 100°C) of 1.34×10^{-3} mol per gram of polymer (30 cm^3/g at STP) if each mole of O_2 produced one

Table IV. Diffusion of $(Br(CH_2)_{10}CO_2)_2Cu$

	Analyses	
	Film 1	*Film 2*
[Cu][b], mol/kg	5.4×10^{-2}	3.5×10^{-3}
[Br][c], mol/kg	3.5×10^{-2}	3.2×10^{-2}
[Br]/[Cu]	0.65	9.1

[a] Outer films (1 and 5) unpreoxidized, loaded with ~0.006 mol/kg salt; inner films (2, 3, and 4) preoxidized (A = 0.41) and initially salt-free.
[b] Analysis by atomic absorption.

oxygenated functional group, there would be approximately one functional group per 50 $-CH_2-$ units in the polyethylene. A significant fraction of the groups should be free carboxylic acid (*11*). IRS ir analysis of standard films of $C_{17}H_{35}CO_2H$ ($\lambda_{max} = 1710\ cm^{-1}$ for C=O stretch) with absorbance ($1710\ cm^{-1}$)/absorbance ($1465\ cm^{-1}$) = 0.4 corresponds roughly to $[-CO_2H] = 2 \times 10^{-4}$ mol per gram of polymer or equivalently one $-CO_2H$ group per 360 $-CH_2-$ units. From the above two calculations one can estimate that there is probably one CO_2H unit per several hundred $-CH_2-$ units or a concentration on the order of ~ 0.1 mol/kg. The latter is roughly the observed solubility of Cu for the highly preoxidized samples in Figure 2. For the lightly preoxidized samples where $A \sim 0.08$, the $-CO_2H$ concentration is expected to be $\lesssim 2 \times 10^{-2}$ mol/kg since at earlier oxidation times CO_2H is a more minor oxidation product (*11*). The solubility results in Figure 3 agree roughly with this calculation except for the apparent low solubility of the C_{29} salt. The non-preoxidized polyethylene and hexadecane–octane solution measurements give the solubilities of salt in a pure $-CH_2-$ medium (Figure 2).

In all samples there is a trend of decreasing solubility with increasing size of R. This effect is not explained in a straightforward manner on the basis of the above discussion.

Diffusion Mechanisms. FICK'S LAW DIFFUSION. The Fick's law diffusion coefficients given in Table I are useful as indicators of the general diffusion rates. These data point out the general trend of decreasing diffusion rate with increasing size of R. However, the fit of the data for Equation 2 is not good enough to substantiate a diffusion mechanism based on boundary conditions and assumptions implicit in Equations 1 and 2. In fact, for oxidized polyethylene films in which exchange processes 5 and 6 can accompany the diffusion of $Cu(RCO_2)_2(m)$ it is clear that the actual transport of copper ions through the polymer will be a complex function involving the forward and reverse rate constants corresponding to Equations 5 and 6 at the rates of transport of the various accompanying species. Analytical solutions of such a complex system

into Polyethylene[a] for 95.5 hr at 90°C

Analyses			*Calculated[d] Initial Values for Film 1 (or 5)*
Film 3	*Film 4*	*Film 5*	
1.1×10^{-4}	3.7×10^{-3}	5.4×10^{-2}	5.8×10^{-2}
3.7×10^{-2}	—	—	9.8×10^{-2}
336	—	—	1.7 (theory 2.0)

[a] Analysis by x-ray fluorescence.
[d] Calculated from mass balance.

are not available (*see* Ref. 8, Ch. 14). However, if the rate constants for processes in Equations 4, 5, and 6 can be assigned, reasonable values should be obtainable from machine-computed numerical solutions.

Implications for the Copper-Catalyzed Oxidation of Polyolefins. CORRELATION OF CONTRIBUTING MECHANISMS. The susceptibility of a polyolefin/copper composite to enhanced oxidation rates will depend critically on the rate of dissolution of copper ions from the copper(oxide) surface and the subsequent rates at which catalytically active concentrations of copper ions can be established within significantly large regions of the polymer matrix. Dissolution of copper with formation of interfacial salt layers appears to occur at very early times (*4*) (during the induction period) and can be described by overall reactions such as Equation 7:

$$Cu_2O + 4RCO_2H + \tfrac{1}{2}O_2 \rightarrow 2Cu(RCO_2)_2 + 2H_2O \qquad (7)$$

where RCO_2H is an acid formed from oxidative scission of the polymer chain. Initially this acid is formed via non-copper-catalyzed oxidation. Previous studies (*3*) indicate that several-fold acceleration of oxidation rate occurs when copper concentrations $> 10^{-3}$ mol/kg can build up in regions of the polymer of 1–4 μm in depth from the metal interface for polymer films of $\sim$ 0.006-in. thickness.

With the above in mind it is of interest to calculate the rate and extent of copper diffusion and to correlate these results with the time scale of the induction period and autocatalytic stages of the oxidation in order to determine how reasonable the proposed mechanisms are. This calculation is carried out below for conditions representative of a typical polyethylene/copper sample oxidized at 90°C.

To calculate the rate and extent of diffusion one must know: (a) the size of the R group in the interfacial salt layer (Equation 7), and (b) the extent of bulk oxidation of the polymer, i.e., the concentration of free carboxylic acid groups. An earlier study (*1*) indicates that R has $<$ 100

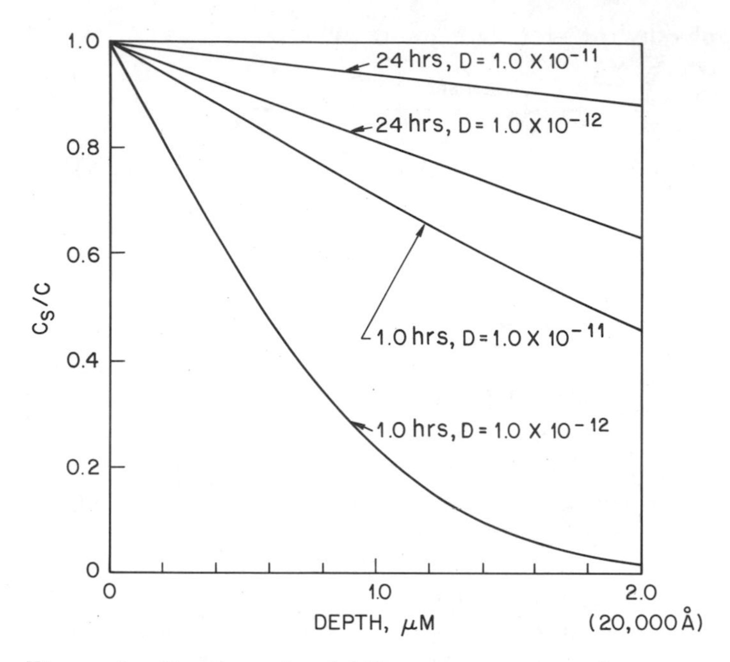

Figure 4. Fraction of solubility concentration of copper attained as a function of depth for a diffusing salt as calculated from Equation 1 (see text)

carbons in the later stages of the degradation. It is thus reasonable to use the C_{29} salt as an approximate model with the realization that the calculated diffusion rates and solubilities will be on the high side (*see* Table I and Figure 3). With respect to the second point, the calculation below is done for an early stage of oxidation, during the induction period, and later initial autocatalytic stage where bulk $-CO_2H$ contents have been estimated for both stages.

Using a Fick's law diffusion coefficient lower limit of $\sim 1 \times 10^{-11}$ cm^2/sec (Table I), values of copper concentration/copper solubility (C/C_s) as a function of distance from the copper interface (χ) were calculated from Equation 1 and are presented in Figure 4. The induction period of a typical low-density polyethylene/copper metal (e.g., gauze or powder) sample at 90°C is $\sim$ 50–100 hr (*12*). For a state of bulk oxidation of approx. several cm^3 of O_2 (at STP) per gram of polymer, appropriate for the middle of the induction period, the solubility for copper will correspond to our data for $A \sim 0.08$ (Figure 2). Thus for the C_{29} salt Figure 4 indicates that over a 24-hr period 88% of the solubility limit, $\sim 3 \times 10^{-4}$ mol/kg, will be attained at a depth of 2.0 μm, a value great enough to create observable catalytic effects. The appearance of these copper ions will accelerate the oxidation and production of acid groups allowing increased copper concentrations to accumulate quickly. This calculation, of course, assumes that a sufficiently extensive interfacial carboxylate phase exists to act as a continuous reservoir (and to satisfy Equation 1), but a simple calculation shows that only a layer of pure salt < 10 Å in thickness is sufficient; since much thicker films are observed, the assumption is valid. As the oxidation progresses and more acid groups are formed, the solubility of salt increases until at the typical endpoint of the induction period (~ 10 cm^3 of O_2 (STP) per gram of polymer) the solubility is probably $\sim 10^{-2}$ mol/kg (estimate from Figure 2, with this extent of oxidation approximately one-third of the $A = 0.4$ curve). At this point the [Cu] at 2 μm is near 10^{-2} mol/kg and a several-hunded-Å layer of interfacial salt would be necessary to supply this amount of copper ions. The latter thickness has in fact been observed (*1*). A more conservative set of calculations could be based on a diffusion coefficient of 10^{-12} cm^2/sec (e.g., for $(C_nH_{2n+1}CO_2)_2Cu$ where $n > 29$; *see* Table I). The plot in Figure 4 shows that the [Cu] at 2 μm is only 28% less than the $D = 1 \times 10^{-11}$ cm^2/sec case; thus the estimates and conclusions above are still quite applicable.

MECHANISMS OF INHIBITION. A number of diffusion experiments were run in which various concentrations of *N,N'*-diphenyloxamide (< 0.1 wt % or $\sim 3 \times 10^{-3}$ mol/kg) were loaded in the polyethylene films. The great majority of runs showed essentially no effect of the additive on the diffusion rate, nor were any unusual surface phases reproducibly noted (*see* Tables I and III for typical results). Previous work (*2, 5*) has indicated that the inhibition effect of a deactivator may be caused by both surface and homogeneous scavenging effects. On the basis of the present results we conclude that the major effect of the deactivator involves surface–interface reactions rather than bulk scavenging mechanisms. The former may consist of poisoning of active surface sites on the Cu_2O/Cu film (*13*) and/or conversion of an interfacial copper carboxylate layer to a relatively inert phase of insoluble copper complex (*5*). Work is in progress to separate these mechanisms further.

Acknowledgment

The authors wish to acknowledge the helpful discussions of R. J. Roe and the cooperation of S. Vincent in obtaining the x-ray fluorescence measurements.

Literature Cited

1. Allara, D. L., Chan, M. G., *J. Colloid Interface Sci.* (1974) **47**, 697.
2. Chan, M. G., Allara, D. L., *Polym. Eng. Sci.* (1974) **4**, 12.

3. Allara, D. L., White, C. W., Meek, R. L., Briggs, T., *J. Polym. Sci., Polym. Chem. Ed.* (1976) **14**, 93.
4. Allara, D. L., *in* "Characterization of Metal and Polymer Surfaces," L. H. Lee, Ed., Vol. 2, p. 193, Academic, New York, 1977.
5. Allara, D. L., Chang, M. G., *J. Polym. Sci., Polym. Chem. Ed.* (1976) **14**, 1857.
6. Hildebrand, J. H., Prausnitz, J. M., Scott, R. L., "Regular and Related Solutions," Van Nostrand Reinhold, New York, 1970.
7. Roe, R. J., Bair, H. E., Gieniewski, C., *J. Appl. Polym. Sci.* (1974) **18**, 843.
8. Crank, J., "The Mathematics of Diffusion," Ch. 2, Clarendon, Oxford, 1975.
9. Harrick, N. J., "Internal Reflection Spectroscopy," Ch. 2, Interscience, New York, 1967.
10. Bellamy, L. J., "The Infrared Spectra of Complex Molecules," Ch. 10, Wiley, New York, 1975.
11. Cheng, H. N., Schilling, F. C., Bovey, F. A., *Macromolecules* (1976) **9**, 363; and references therein.
12. Allara, D. L., unpublished data.
13. Allara, D. L., Roberts, R. F., *J. Catal.* (1976) **45**, 54.

RECEIVED May 12, 1977.

Stabilization of Poly(fluoroalkoxyphosphazene) (PFAP) Elastomer against Thermal Degradation

G. S. KYKER and J. K. VALAITIS

Ceneral Research Laboratories, The Firestone Tire and Rubber Company, Akron, OH 44317

Poly[(trifluoroethoxy)(octafluoropentoxy)(X)phosphazene] (X = cross-link site) terpolymer exhibits a loss in molecular weight at temperatures above 149° that is detrimental to mechanical properties (1, 2). This degradation reaction was studied using solution viscosity, gel permeation chromatography, and weight loss measurements on samples aged in contact with air in the temperature range 135°–200°C. The mechanism of degradation is consistent with a random chain scission at weak sites along the polymer chain. Weight loss measurements on PFAP at high temperatures indicate that depolymerization is very slow at 135°–175°C but fairly rapid at 200°C. The addition of small amounts of bis(8-oxyquinolate)zinc(II) or magnesium(II) to PFAP results in a large decrease in the rate of chain scission. These zinc(II) complexes are postulated to deactivate the weak sites on the PN chain by forming a thermally stable complex.

Poly(fluoroalkoxyphosphazenes)(PFAP)(I) are semi-inorganic polymers that contain a phosphorus nitrogen chain. Pendant fluoroalkoxy groups are attached to the phosphorus to impart hydrolytic stability and hydrocarbon resistance.

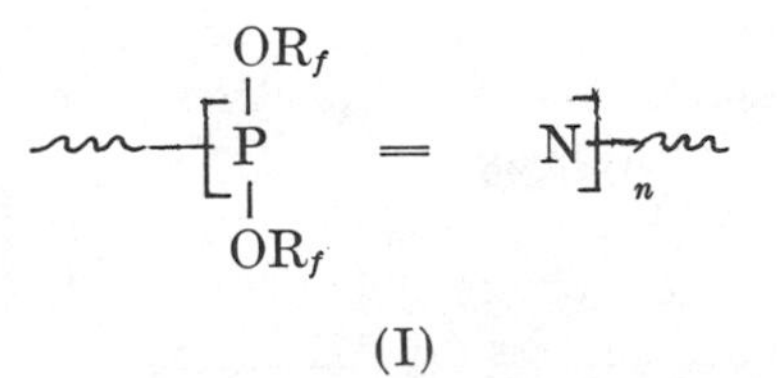

(I)

Highly crystalline plastics are obtained when both pendant fluoroalkoxy groups are of the same or similar structure, e.g., trifluoroethoxy

0-8412-0381-4/78/33-169-293$05.00/1

(3). However, when the two pendant groups are of dissimilar structure, amorphous fluoroelastomers are realized (4).

The PFAP(I) selected for this study is an amorphous fluoroelastomer that consists of pendant trifluoroethoxy and octafluoropentoxy groups (II). A small quantity of cross-link site was incorporated also to facilitate vulcanization via conventional methods, i.e., organic peroxides, sulfur-accelerator, and radiation (high-energy electrons).

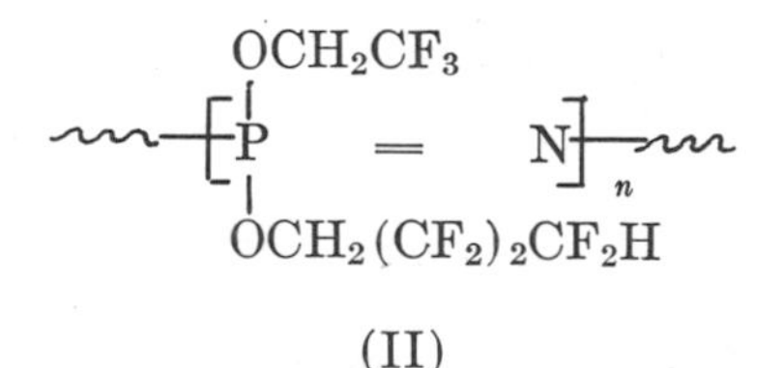

(II)

Preliminary studies indicated that PFAP(II) undergoes a loss in molecular weight at temperatures above 149°C (*1, 2*). The degradation process was followed by monitoring the change in solution viscosity of samples aged at 150°C in air. The viscosity exhibited a rapid loss during the first 200 hr, then leveled off to a slower rate.

Metal dialkyldithiocarbamates and Group IA and IIA oxides, hydroxides, and carbonates retarded the rate of thermal degradation (*2*). In general, acid scavengers and basic compounds retarded the rate of thermal degradation. Subsequent studies revealed that bis(8-oxyquinolate)zinc(II) and magnesium(II) were effective stabilizers for preventing thermal degradation of PFAP(II) (*5*).

Several investigators have reported studies on the thermal degradation of poly(alkoxy-, aryloxy-, and fluoroalkoxyphosphazene) homopolymers that are generally crystalline plastics (*6–13*). The early investigations involved only qualitative estimates of mechanical strength retention on aging at high temperatures; no degradation mechanism was discussed (*6*). Allcock presented theoretical calculations which predicted that poly(phosphazenes) should depolymerize to cyclic oligomers at high temperatures in order to relieve steric interactions between pendant groups (*7, 8*).

MacCallum and Tanner studied the thermal degradation of both poly(aryloxy- and fluoroalkoxyphosphazenes) by thermogravimetric (TGA) and solution viscosity measurements (*9*). The thermal stabilities indicated by TGA are misleading since they represent weight loss owing to volatilization of cyclic oligomers. Major degradation of the PN chain probably occurs at much lower temperatures than indicated by the temperature at which weight loss becomes rapid. The changes in solution viscosity of aged poly[bis(phenoxy)phosphazene] were interpreted to result from depolymerization to cyclic oligomers and the mechanism of degradation was stated to be of the depolymerization type. This data appears to be more consistent with a random chain scission type of degradation.

G. Allen et al. (*10*) studied the thermal degradation of several poly(aryloxy- and fluoroalkoxyphosphazenes) by use of isothermal TGA and retention of solution viscosity. Thermal stability was determined by the temperature at which the samples exhibited 10% weight loss. No mechanism was presented for the thermal degradation.

Allcock studied the thermal degradation of poly[bis(trifluoroethoxy)-phosphazene] via gel permeation chromatography and solution viscosity measurements on samples aged at 150°–300°C (*11*) in sealed, evacuated tubes. The mechanism of degradation was presented as a two-step process: (1) chain scission at weak links and (2) depolymerization to cyclic oligomers. Allcock also reported similar studies on the thermal breakdown of poly[bis(phenoxy)phosphazene] and arrived at the same degradation mechanism (*12*).

Hagnauer and LaLiberte (*13*) recently studied the degradation of poly[bis(*m*-chlorophenoxy)phosphazene] via gel permeation chromatography and solution viscosity measurements on samples aged at 165°C. The degradation mechanism was postulated to be random chain scission at weak links in the polymer backbone. No evidence was obtained for a depolymerization-type mechanism.

No studies have been published on the thermal degradation of poly(fluoroalkoxyphosphazene) (PFAP) elastomers. In this paper we report the first study on the mechanism of the thermal degradation of a PFAP elastomer (II). The bis(8-oxyquinolate)zinc(II) stabilization of PFAP(II) against thermal degradation also is reported and a mechanism for stabilization is proposed.

Experimental Section

Materials. The synthesis of poly[(trifluoroethoxy)(octafluoropentoxy)(X)phosphazene] (X = crosslink site) terpolymer has been described in previous publications (*1, 14*). Bis(8-oxyquinolate)zinc(II) was synthesized according to the procedure described in an earlier publication (*5*).

Viscosity Measurements. Solution viscosity data were obtained with the use of a Cannon–Fenske viscometer (size 50) at 25°C. Solutions were prepared in acetone (0.30 g/100 mL).

Gel Permeation Chromatography (GPC). Measurements were made with the use of a Waters Associates Model 100 GPC equipped with four stainless steel columns (4 ft $\times$ ⅜ in.) containing Styragel of nominal pore sizes 10_4, 10_5, 10_6, and 10_7 Å. Dimethylformamide was used as a solvent at 85°C and a flow rate of 0.7 mL/min. PFAP(II) samples were injected at a concentration of 0.12 wt %. Approximate calibration of the columns was accomplished by means of fractionated samples of PFAP(II) of known weight average molecular weight (M_w) (via light scattering measurements). A log (M_w) vs. elution count calibration curve was used to determine the M_w of PFAP(II) samples. Accurate values of number average molecular weights (M_n) for PFAP(II) samples were difficult to obtain via osmometry because of diffusion of low molecular weight materials through the membrane. Therefore, the GPC columns were calibrated in M_w units. M_n of aged PFAP(II) samples should actually be used for kinetic plots, but M_w also should be valid.

Addition of Bis(8-oxyquinolate)zinc(II) to PFAP(II). PFAP(II) (25 g) was dissolved in reagent acetone to give a viscous solution. Then 0.25, 0.50, and 0.75 g of bis(8-oxyquinolate)zinc(II) was stirred in reagent tetrahydrofuran (50 mL) to give a bright yellow solution (not completely soluble). This solution (containing undissolved solid) was mixed with the PFAP(II) in acetone at 23°C and stirred for 100 hr. The PFAP(II) samples containing 1, 2, and 3 wt % of bis(8-oxyquinolate)-zinc(II) were purified by coagulation of the PFAP(II) in distilled water. These samples were dryed to constant weight at 60°C in a vacuum oven.

A homogeneous reaction of PFAP(II) with bis(8-oxyquinolate)-zinc(II) was achieved by dissolving both materials in warm dimethylformamide (60°C). The PFAP(II) sample was purified in the same manner as described above.

Thermal Degradation Techniques. PFAP(II) samples (5.0 g) were contained in small glass Petri dishes and placed in forced-air ovens maintained at 135°, 149°, 177°, or 200°C ± 2°C for the specified time intervals. The weight loss of these samples was monitored as a function of aging time. Small portions of these samples were dissolved in acetone for viscosity measurements and in dimethylformamide for GPC studies.

Mechanical Mixing of PFAP(II) Compounds. PFAP(II) O-ring seal formulations were mixed in a Brabender internal mixer equipped with a mixer head of 85-mL volume and cam-type blades. The mix order was as follows: (1) PFAP(II), (2) silica (Quso WR82, silane-treated surface) (3) magnesium oxide and (4) bis(8-oxyquinolate)zinc(II). This compound then was placed on a small two-roll rubber mill (2 in. × 6 in.) and dicumyl peroxide was mixed in at 57°C.

Mechanical Testing of PFAP(II) O-ring Seal Compounds. Cut-ring stress–strain specimens were cut from slabs (6 in. × 6 in. × 0.050 in.) vulcanized in a press for 30 min at 177°C. These cut-ring specimens were aged in a forced-air oven at 200°C for the specified time intervals (ASTM D-573). Stress–strain measurements on unaged and aged cut-ring specimens were carried out by use of an Instron 1130 (ASTM D-412).

Results and Discussion

Investigation of the Thermal Degradation of PFAP(II). PFAP(II) was aged in air at varying time intervals (0–700 hr) at 135°, 149°, 177°, and 200°C. The change in solution viscosity with aging time at these temperatures is shown in Figure 1. The viscosity exhibits an initial rapid decrease, then levels off and approaches a constant value which appears

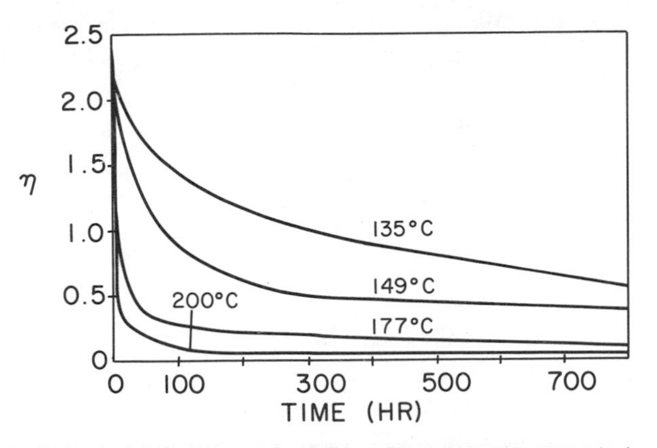

Figure 1. Change in solution viscosity of PFAP(II) with time at different temperatures

to be a function of the temperature. The initial rate of viscosity loss increases as the temperature is raised from 135° to 200°C and shorter time intervals are required to approach the plateau viscosity. This type of viscosity–time profile is consistent with a random chain scission process at weak links in the PFAP(II) backbone. The PFAP(II) initially contains a large number of weak links that undergo thermal cleavage so the initial rate is quite rapid. As the concentration of weak links decreases, the rate of scission is diminished concurrently so that eventually all weak sites have been decomposed and only a small change in viscosity occurs with additional aging (*11*). This type of viscosity–time profile has been reported for poly(phosphazene) plastics aged at comparable temperatures (*10, 11, 12, 13*).

Depolymerization to cyclic oligomers could be occurring concurrent with chain scission or may be initiated by the scission process itself at the ends of the lower molecular weight chains. The formation of depolymerization products (PFAP(II) cyclic oligomers) would not be detected as a change in the solution viscosity which is more dependent on high molecular weight PFAP(II).

However, depolymerization to cyclic oligomers could be detected by monitoring the weight loss of PFAP(II) at high temperatures (135°–200°C) providing the oligomers are volatile. The wt % PFAP(II) volatilized is plotted as a function of aging time at 135°, 149°, 177°, and 200°C in Figure 2. In the temperature range 135°–177°C, PFAP(II) exhibits an initial rapid weight loss during the first 50 hr followed by a leveling off to a very slow rate. This initial weight loss is attributed to the volatilization of occluded cyclic oligomers carried over from the synthesis of PFAP. The differences in initial weight losses at different temperatures arises from the different vapor pressures of the mixture of cyclic oligomers. Only cyclic trimer and tetramer may vaporize at 135°C, while at 149° and 177°C higher molecular weight cyclics also would be expected to vaporize. The very slight amount of material volatilized between 50 and 764 hr at 135°–177°C indicates that essentially no depolymerization to cyclic oligomers is occurring. If only nonvolatile, high molecular weight cyclics were produced, this would not be a valid conclusion; however, this is very unlikely.

At 200°C the initial volatilization rate does decrease after 50 hr, but not to the extent observed at lower temperatures. The total weight loss after 764 hr is 40%. This weight loss is attributed to the formation of cyclic oligomers produced by depolymerization of PFAP(II). After 50 hr the chain scission degradation process has lowered the molecular weight to ca. 450 repeat units. Thereafter, the depolymerization mechanism is viewed as an unzipping reaction initiated at the ends of the low molecular weight chains.

Gel permeation chromatograms of PFAP(II) samples aged for varying times at 135°, 149°, 177,° and 200°C are depicted in Figure 3. The bottom scale is elution volume, while the top scale is $\overline{M}_w$ obtained by calibration of the columns with fractionated samples of PFAP(II). The molecular weight distribution of unaged PFAP(II) is very broad with $\overline{M}_w$'s ranging from $0.2–30 \times 10^6$. As PFAP(II) is aged at 135°C, the GPC curves shift slowly to lower molecular weights with increasing aging times. Similar results are obtained when PFAP(II) is aged at 149°, 177°, and 200°C. The extent of chain scission increases as the temperature is raised from 135° to 200°C at the same aging times. The GPC curves also become more narrow with increasing aging times at 177° and 200°C.

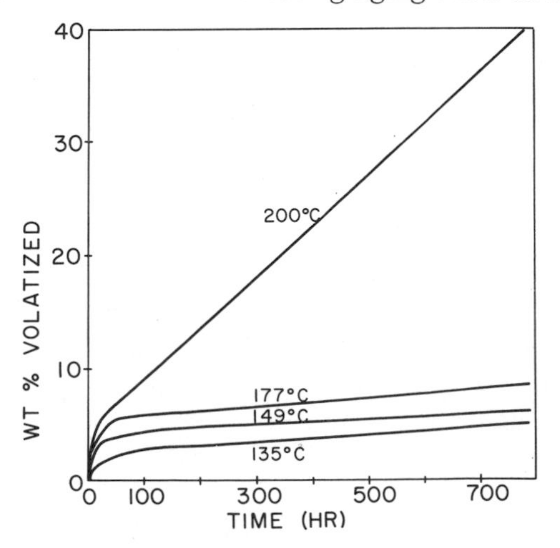

Figure 2. Wt % of PFAP(II) volatilized vs. time at different temperatures

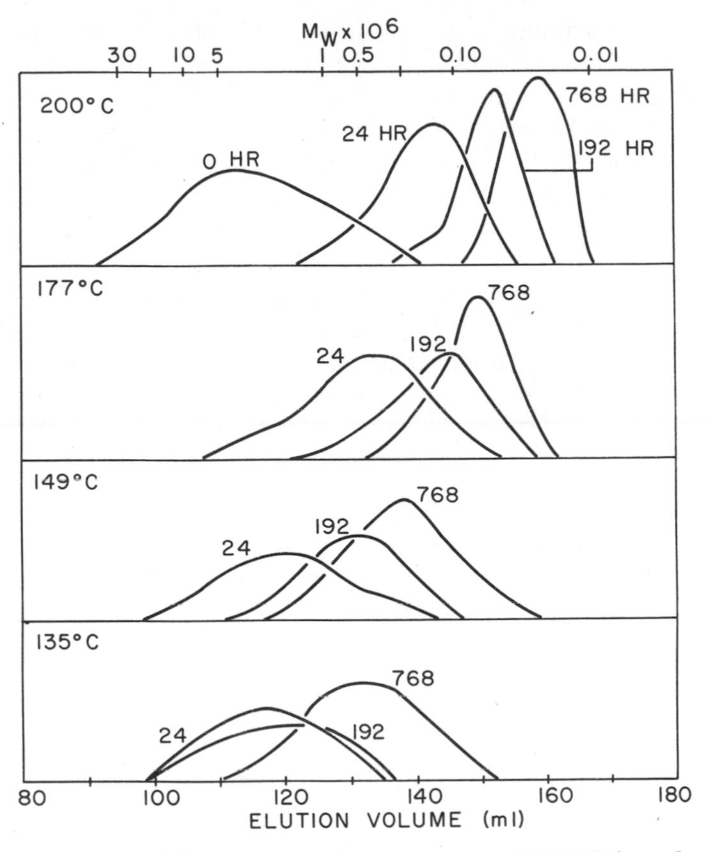

Figure 3. Gel permeation chromatograms of PFAP(II) aged at different temperatures

This lowering of molecular weight and narrowing of molecular weight distribution with increasing exposure to high temperatures are characteristic of a random chain scission degradation mechanism. If the main degradation process consisted of depolymerization to cyclic oligomers, the GPC curve of the high molecular weight polymer would decrease in area while undergoing only a very slight shift to lower molecular weight. Concurrently, a GPC trace for cyclic oligomers should appear in the low molecular weight region and the area of this peak should increase as a function of aging time. This would be valid only if the cyclic oligomers were nonvolatile under the aging conditions, i.e., 135°–200°C in a forced-air oven. These oligomers have sufficient volatility under these aging conditions to exhibit considerable weight losses (Figure 2). Therefore, integration of this low molecular weight GPC peak would probably be meaningless. The GPC curve for cyclic oligomers occurs at an elution volume of ca. 180 mL and overlaps with an impurity peak eluting in the same region. Therefore, it was not possible to obtain accurate integration of the area of the cyclic oligomer GPC curve to determine the residual concentration. However, the GPC of unaged PFAP did indicate the presence of small amounts of cyclic oligomers (from synthesis) and is thus consonant with the weight loss data obtained on PFAP(II) samples at high temperature (Figure 2). Unfortunately, it is not possible to determine if depolymerization is occuring concurrent or subsequent to chain scission by the GPC analysis of PFAP(II) samples aged under the conditions stated in this paper. If PFAP(II) samples were aged in a closed system to prevent volatilization of cyclic oligomers and the low molecular weight GPC peak could be resolved from the solvent impurities peak, it should be possible to ascertain if depolymerization is occuring concurrent or subsequent to chain scission. The PFAP(II) samples would also have to be free of residual cyclic oligomers from the synthesis.

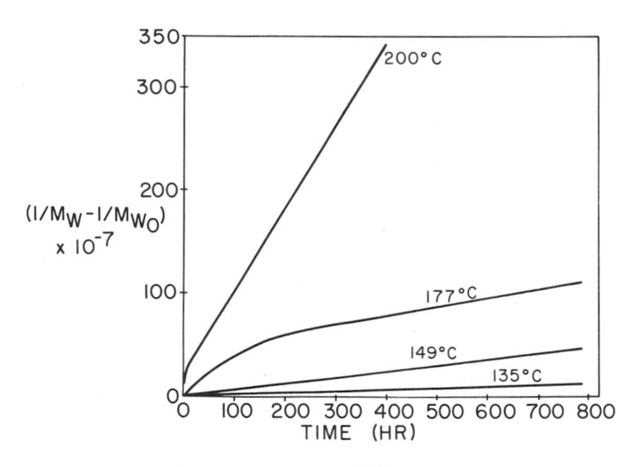

Figure 4. Kinetic plots for $\overline{M}_w$ loss in PFAP(II) at selected temperatures

The reciprocal weight average molecular weight of aged PFAP(II) ($1/M_w$) minus that of the unaged sample ($1/\overline{M}_{w_0}$) is plotted vs. aging time at 135°, 149°, 177°, and 200°C in Figure 4. These plots exhibit a constant rate of chain scission with aging time at 135° and 149°C. However, at 177°C the rate of chain scission is quite rapid in the initial 200 hr; it then levels off to a slower rate. Similar results are obtained at 200°C except that the initial rapid rate lasts only 10 hr, then decreases slightly to a slower rate. These types of plots have been observed for other polymers which degrade by a random chain scission at weak links (*15*). Small amounts of weak links may be present in PFAP(II) which undergoes scission at 135°–149°C but the predominate degradation path appears to be random scission (without weak links) similar to that observed for polystyrene (*13*). Higher temperatures, 177°–200°C, are required to produce scission at weak links in the PN chain. For a purely random scission, the following equation should apply:

$$1/\overline{M}_w - 1/\overline{M}_{w_0} = Kt$$

However, if weak links are present the equation becomes:

$$1/\overline{M}_w - 1/\overline{M}_{w_0} = Kt + b$$

where b is the ordinate intercept obtained by extrapolation of the linear portion of the curves back to 0 time and is interpreted as the concentration of weak links (Figure 4) (*15*). The rate constant is calculated from the linear portions of these plots. The values of this rate constant, K, and the concentration of weak sites, b, are summarized in Table I. The rate of chain scission is fairly slow at 135°C with the rate increasing ca. fivefold when the temperature is increased to 177°C. However, ca. a tenfold increase in rate of chain scission is realized when the temperature is increased from 177° to 200°C.

Table I. Rate Constants for Chain Scission and Weak Link Concentration in PFAP(II)

T(°C)	*K (Moles Chain Scission g^{-1} HR^{-1})* $\times 10^{-9}$	b *(Weak Link Concentration)* $\times 10^{-6}$
135	1.7	0
149	5.9	0
177	8.6	4.5
200	80.0	2.0

These data suggest that several types of weak links may be situated in a random manner along the PN backbone. Different types of weak links would be expected to undergo scission at different temperatures as is indicated by the plots in Figure 4. The value of b, the weak link concentration, indicates that these sites are located at $\overline{M}_w$ intervals of ca. $2.2 \times 10_5$ which corresponds to ca. 587 repeat units. These weak links are thought to be of the following structures (*11, 12, 13*):

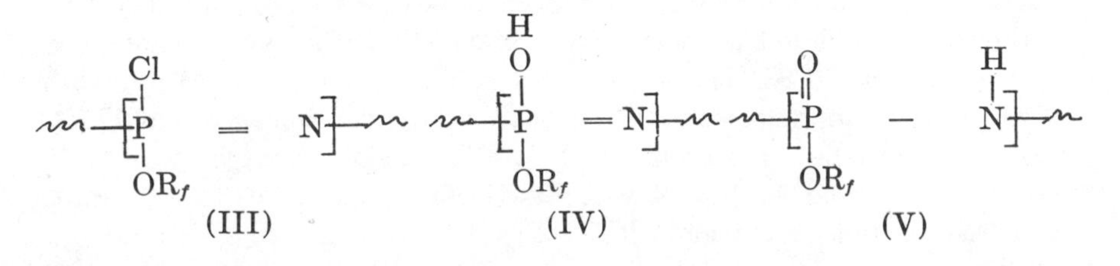

Groups of type (III) result from incomplete nucleophilic substitution of the PCl moieties of the precursor poly(dichlorophosphazene). These groups are latent sites for hydrolysis to produce groups of types (IV) and (V). The most likely weak link is thought to be the phosphazane (V) which is produced by the rearrangement of the phosphazene moiety (IV). This phosphazane moiety (V) would not be expected to undergo thermal PN bond breaking in the temperature range used in the current studies (135°–200°C). Chain scission most likely occurs by attack of water, acids, or hydroxy-containing compounds (BOH) at the phosphazane moiety (V) to produce lower molecular weight polymers with active end groups (VI) and (VII).

$$\sim\!\!\sim\!\!-\underset{OR'_f}{\overset{OR_f}{P}}\!=\!N\!-\!\left[\underset{OR_f}{\overset{O}{\overset{\|}{P}}} - \overset{H}{N}\right]\!-\!\underset{OR'_f}{\overset{OR_f}{P}}\!=\!N\!-\!\!\sim\!\!\sim + BOH \rightarrow \sim\!\!\sim\!\underset{OR'_f}{\overset{OR_f}{P}}\!=\!N - \underset{OR_f\ \ OB}{\overset{O}{\overset{\|}{P}}} + H_2N - \underset{OR'_f}{\overset{OR_f}{P}} = N\!\sim\!\!\sim$$

(V) (VI) (VII)

$$B = H, R_f, \sim\!\!\sim\!\!-\underset{OR_f}{\overset{|}{P}} = N\!-\!\!\sim\!\!\sim$$

In addition to the scission of the phosphazane group by traces of water and acids, residual POH moieties in the polymer could produce chain scission via either inter- or intramolecular reactions.

One approach to improving the thermal stability of PFAP(II) would be to deactivate the labile moieties (III and IV) which are precursors to the phosphazane weak links (V). Improvements in the thermal stability of PFAP(II) have been realized by treatment with acid scavengers and basic compounds (*2*). These results are considered consonant with the proposed mechanism of chain scission at the phosphazane (V) site.

Stabilization of PFAP(II) with Bis(8-oxyquinolate)zinc(II). Preliminary studies indicated that bis(8-oxyquinolate)zinc(II) and magnesium(II) slowed down the rate of molecular weight loss in PFAP(II) (*5*). This stabilization effect also was observed in reinforced PFAP(II) vulcanizates and is of commercial significance. The retention of mechanical properties of PFAP(II) O-ring seal formulations upon prolonged aging (>600 hr) at high temperatures (177°–200°C) is improved greatly by the addition of 1–2 wt % bis(8-oxyquinolate)zinc(II) (*5*). In order to elucidate the mechanism by which bis(8-oxyquinolate)zinc(II) stabilizes PFAP(II) against thermal degradation, samples of PFAP(II) containing selected levels of this stabilizer were aged at 177°C in air for short time intervals (32 hr). These aged samples were subsequently analyzed by use of solution viscosity and GPC to determine the extent of molecular weight loss.

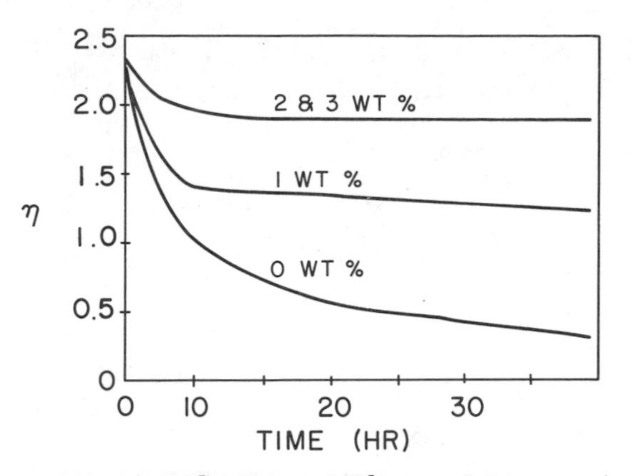

Figure 5. Change in solution viscosity of PFAP(II) (with 0, 1, 2, and 3 wt % bis(8-oxyquinolate)zinc(II) with time at different temperatures

The plots of solution viscosity vs. aging time at 177°C for pure PFAP(II) and samples containing 1, 2, and 3 wt % of bis(8-oxyquinolate)zinc(II) are depicted in Figure 5.

The pure PFAP(II) exhibits a rapid decrease in viscosity from 2.35–0.38 dl/g over a span of 32 hr at 177°C. In contrast, the PFAP(II) samples containing 1, 2, and 3 wt % of stabilizer exhibit viscosity decreases to 1.20, 1.88, and 1.93, respectively, under the same conditions. These plots indicate that the zinc complex reacts with the weak links (III and IV) to deactivate them and thus prevent chain scission. The addition of the stabilizer in excess of 2 wt % doesn't appear to improve the thermal stability. Evidently the zinc complex has deactivated all of the weak links that are capable of reaction under these conditions. The fact that the PFAP(II) containing 2 and 3 wt % stabilizer still exhibits a small initial decrease in viscosity to 1.93 indicates that a small amount of weak links have not been deactivated.

The GPC curves of PFAP(II) containing 0 and 1 wt % bis(oxyquinolate)zinc(II) are depicted in Figure 6. Only short aging times at 177°C were investigated since most of the molecular weight loss occurs during this time interval. After 32 hr at 177°C the molecular weight of PFAP(II) decreases from 5.0×10^6 to 0.3×10^6, while that of the stabilized PFAP(II) is lowered from 10×10^6 to 2.7×10^6.

The reaction between PFAP(II) and the zinc complex evidently results in a type of chain extension or complexation since the molecular weight shows an increase from 5 to 10×10^6. The GPC curves of unaged PFAP(II) containing 2 and 3 wt % bis(8-oxyquinolate)zinc(II) are shown in Figure 7 along with plots of samples aged 16 and 32 hr at 177°C. With 2 wt % of this stabilizer, the molecular weight decreases

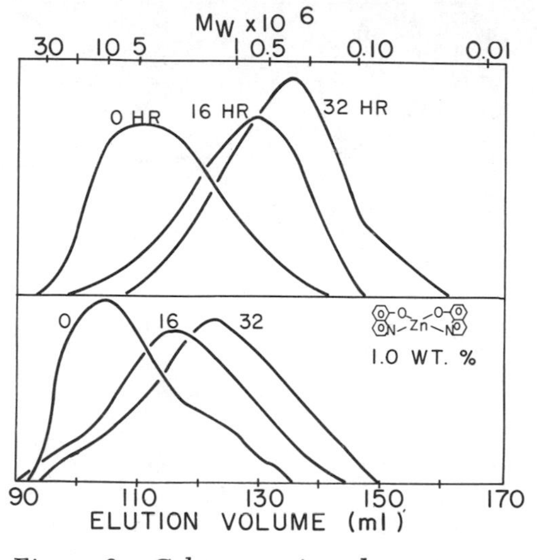

Figure 6. Gel permeation chromatograms for PFAP(II) (with 0 and 1 wt % bis(8-oxyquinolate)zinc(II)) aged at 177°C

from 10 to 3 and 2×10^6 upon aging 16 and 32 hr, respectively, at 177°C. With 3 wt % of bis(8-oxyquinolate)zinc(II), the molecular weight of PFAP(II) decreases from 6 to 4×10^6. Bis(8-oxyquinolate)zinc(II) decreases the amount of chain scission in PFAP(II) through reaction

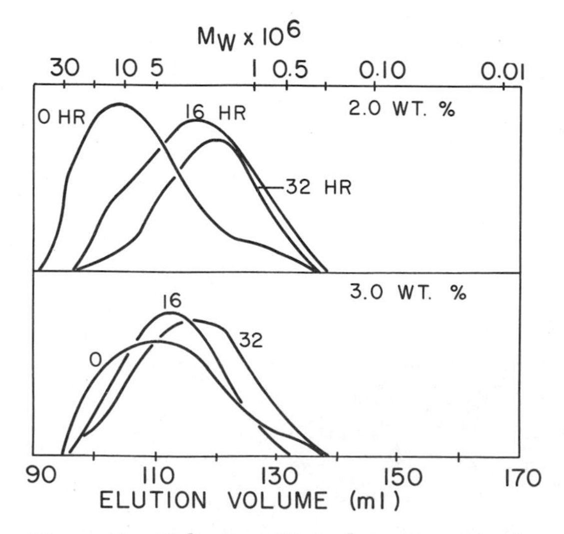

Figure 7. Gel permeation chromatograms for PFAP(II) (with 2 and 3 wt % bis(8-oxyquinolate)zinc(II)) aged at 177°C

with the weak links (III) and (IV) to form a thermally and hydrolytically stable complex which blocks the phosphazene–phosphazane rearrangement.

The plots of $(1/\overline{M}_w - 1/\overline{M}_{w_0})$ vs. aging time at 177°C for pure and stabilized PFAP(II) are depicted in Figure 8. The rate of chain scission is quite rapid in the pure PFAP(II) and shows no evidence of leveling off at the short time interval studied. The addition of increasing levels of the zinc complex results in a concurrent decrease in the rate of chain scission. The rate constants are summarized in Table II. The addition of 3 wt % of bis(8-oxyquinolate)zinc(II) results in approximately a tenfold reduction in the rate of chain scission.

A possible mechanism for the stabilization of PFAP(II) with bis(8-oxyquinolate)zinc(II) involves complexation of this zinc compound with POH moieties (IV) to form thermally and hydrolytically stable sites (VIII).

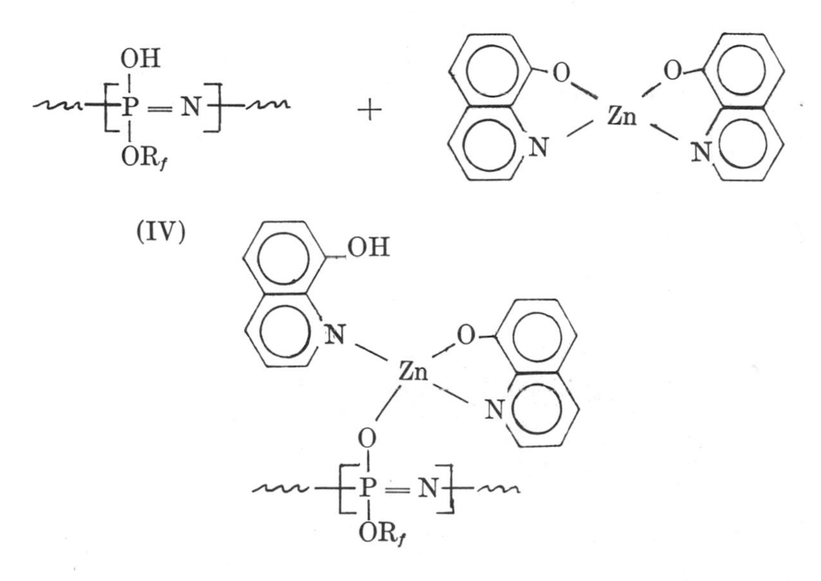

The reaction is considered to be a ligand displacement in which the oxygen atom of the 8-oxyquinolate ligand is displaced by the oxygen atom of the P–O group. The zinc complex now is bonded covalently to the PFAP(II) via a P–O–Zn bond that is reported to be hydrolytically and thermally stable (*16*). This complexation of bis(8-oxyquinolate)-zinc(II) to the P–O moiety blocks the rearrangement to a phosphazane-type structure which is postulated to be the weak link in the PFAP(II) chain.

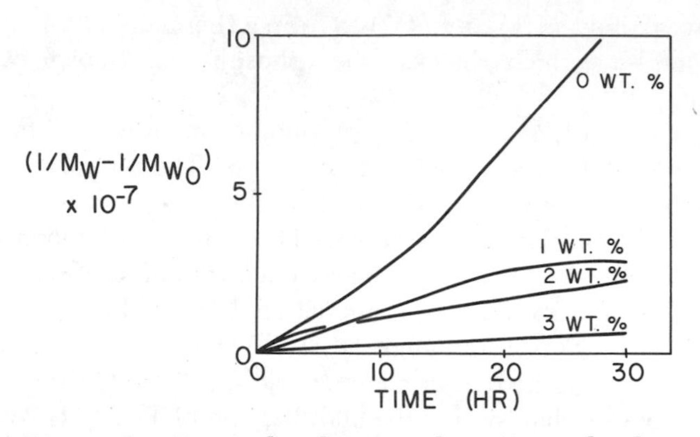

Figure 8. First-order kinetic plots for molecular weight loss in PFAP(II) (with 0, 1, 2, and 3 wt % bis(8-oxyquinolate)zinc(II)) aged at 177°C

Table II. Rate Constants for Chain Scission (177°C) in PFAP(II) Stabilized with Bis(8-oxyquinolate)zinc(II)

Bis(8-oxyquinolate)zinc(II) (wt %)	*K (Moles Chain Scission g^{-1} Hr^{-1}) $\times 10^{-8}$*
0	4.1
1	0.8
2	0.8
3	0.4

Table III. PFAP Formulations

Formulation	*A*	*B*	*C*
PFAP(II)	100	100	100
Quso WR82[a]	30	30	30
Stan Mag ELC[b]	6	6	6
$(C_9H_6NO)_2Zn$	0	3	3[c]
Dicumyl peroxide	0.8	0.8	0.8
Stress–Strain			
Cure 15 min, 170°C			
σ_{50} (psi)	340	275	298
σ_{100} (psi)	—	939	967
σ_f (psi)	1125	1270	1303
ϵ_f (%)	93	126	126

[a] Silane-treated precipitated silica.
[b] Magnesium oxide.
[c] PFAP(II) treated with 3 wt % $(C_9H_6NO)_2Zn$ in dimethylformamide (homogeneous).

In order to evaluate the commercial importance of bis(8-oxyquinolate)zinc(II) stabilization of PFAP, this stabilizer was evaluated in a standard PFAP O-ring formulation (Table III). Compound A should be considered as a control since no stabilizer was added. Compound B is identical in composition to A except that 3 wt % of stabilizer was added during the mixing of B on a rubber mill. Compound C was the same as A except the PFAP was treated with 3 wt % stabilizer in DMF (homogeneous) prior to mixing the compound. The stress–strain properties of cured specimens of B and C were essentially identical but appear to be of a lower cross-link density than A.

Reinforced PFAP vulcanizates, prepared from compounds A, B, and C, were aged for varying times at 200°C. Stress–strain measurements were made on both the unaged and aged samples to determine the extent of degradation. The stress at 50% elongation or 50% modulus (σ_{50}) was selected as the most critical property for an O-ring compound to retain upon heat aging.

The retention of 50% modulus as a function of aging time at 200°C for reinforced PFAP vulcanizates A, B, and C is shown in Figure 9. It is clearly illustrated that the incorporation of bis(8-oxyquinolate)zinc(II) results in a significant improvement in retention of σ_{50}. After 300 hr the control retained 27% of the initial σ_{50}, while B retained 57% and C retained 65%.

These plots of the retention of σ_{50} as a function of aging time are quite similar to those observed for the change in viscosity vs. aging time (Figure 5). Therefore, the retardation of chain scission to decrease molecular weight loss results in a slowing down of the rate of σ_{50} loss as would be expected.

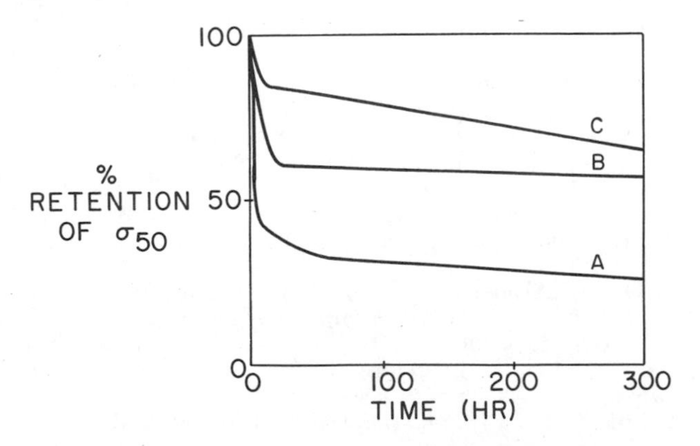

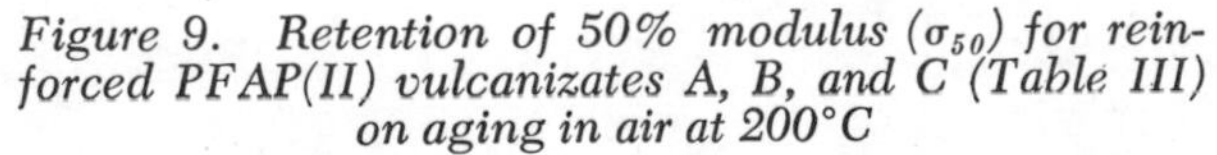
Figure 9. Retention of 50% modulus (σ_{50}) for reinforced PFAP(II) vulcanizates A, B, and C (Table III) on aging in air at 200°C

Conclusions

Poly[(trifluoroethoxy)(octafluoropentoxy)(X)phosphazene] (X = cross-link site) PFAP(II), undergoes a loss in molecular weight at elevated temperatures (135°–200°C) that is detrimental to mechanical properties. The mechanism of molecular weight loss was found to be random chain scission at weak links in the PFAP(II) chain. These weak links are postulated to be phosphazane moieties which would be attacked by water, acids, and POH to produce PN chain scission.

It is proposed that bis(8-oxyquinolate)zinc(II) blocks the formation of these weak links by complexing to residual POH moieties that are the precursor to the phosphazene–phosphazane rearrangement. Incorporation of this zinc complex in both PFAP(II) gum and reinforced PFAP(II) vulcanizates results in large improvements in thermal stability.

Acknowledgments

The authors are grateful to M. L. Stayer, T. A. Antkowiak, and T. C. Cheng for the synthesis of the fluoroelastomer used in this study. We also wish to thank The Firestone Tire and Rubber Company which made this research possible and encouraged the publication of this study.

Literature Cited

1. Kyker, G. S., Antkowiak, T. A., *Rubber Chem. Technol.* (1974) **32**, 4761.
2. Kyker, G. S., Beckman, J. A., Halasa, A. F., Hall, J. E., "Stabilized Phosphonitrile Elastomers," U.S. Patent No. **3,843,596** (1974).
3. Allcock, H. R., "Phosphorus–Nitrogen Compounds," Academic Press, New York, 1972.
4. Rose, S. H., *J. Polym. Sci., Part B* (1968) **6**, 837.
5. Kyker, G. S., "Metal 2-Hydroxyquinolate Complexes as Stabilizer for oly-(phosphazenes) to Inhibit Thermal Degradation at Elevated Temperatures," U.S. Patent No. **3,867,341** (1975).
6. Mirhez, M. E., Henderson, J. F., *J. Macromol. Chem.* (1966) **1**, 187.
7. Allcock, H. R., *Inorg. Chem.* (1966) **5**, 1320.
8. Allcock, H. R., *J. Polym. Sci., Part A-1* (1967) **5**, 355.
9. MacCallum, J. R., Tanner, J., *J. Macromol. Sci., Chem.* (1970) **A4**(2), 481.
10. Allen, G., Lewis, C. J., Todd, S. M., *Polymer* (1970) **77**, 44.
11. Allcock, H. R., Cook, W. J., *Macromolecules* (1974) **7**, 284.
12. Allcock, H. R., Moore, G. Y., Cook, W. J., *Macromolecules* (1974) **7**(5), 571.
13. Hagnauer, G. L., LaLiberte, B. R., *J. Appl. Polym. Sci.* (1976) **20**, 3073.
14. Tate, D. P., *J. Polym. Sci., Polym. Symp.* (1974) **48(C)**, 33.
15. Cameron, G. G., MacCallum, J. R., "Reviews in Macromolecular Chemistry," G. B. Butler, K. F. O'Driscol, Eds., Ch. 8, Marcel Dekker, New York, 1967.
16. Rose, S. H., Block, B. P., *J. Polym. Sci.*, Av (1966) **4**, 583.

RECEIVED May 12, 1977.

25

Recent Fundamental Developments in the Chemistry of Poly(vinyl chloride) Degradation and Stabilization

W. H. STARNES, JR.

Bell Laboratories, Murray Hill, NJ 07974

Poly(vinyl chloride) degradation and stabilization are briefly reviewed emphasizing significant fundamental work that has been published during the past four years. Nonoxidative thermal degradation and its prevention are the principal areas of coverage, although a few comments are made also about degradation in the presence of oxygen. Topics surveyed include the mechanism of nonoxidative thermal dehydrochlorination, initiation of degradation from polymer defect sites and normal monomer units, effects of tacticity on polymer stability, stabilization by organotins, metal soaps, and miscellaneous compounds, and stabilization by chemical modification.

The great commercial importance of poly(vinyl chloride) (PVC), together with its high tendency to decompose under processing and end-use conditions, have stimulated much research on PVC degradation and stabilization mechanisms. Earlier work in this field has been summarized in many comprehensive reviews; thus this survey is concerned only with developments that have occurred since the most recent reviews of this type were written (*1, 2, 3, 4, 5*). Nonoxidative thermal degradation and its prevention are the principal areas discussed because most of the published fundamental studies pertain to these areas. Oxidative degradation also is considered here (although very briefly), but no attempt has been made to update recent surveys of work on UV-initiated decomposition (*3, 6, 7*). Apart from these limitations, literature coverage in this chapter is reasonably complete but not exhaustive.

0-8412-0381-4/78/33-169-309$05.00/1

Nonoxidative Thermal Degradation

General Considerations. Nonoxidative thermal degradation of PVC at temperatures ranging from about 100° to 200°C is primarily a dehydrochlorination process that produces conjugated polyene sequences in the polymer chains (*1, 2, 3, 4, 5*). These sequences impart undesirable color to the polymer, even at very low levels of dehydrochlorination (*1, 2, 3, 4, 5*), and they also cause mechanical and electrical properties to deteriorate when they have been formed to a sufficient extent (*1, 2, 3, 5*). Thus, fundamental studies relating to PVC degradation chemistry have been concerned with such things as the nature, number, and relative stabilities of the sites where dehydrochlorination begins, the effects of various factors on the subsequent steps responsible for polyene growth and its termination and, of course, the detailed mechanism of the overall degradation process (*1, 2, 3, 4, 5*). Recent work in these areas will now be assessed briefly.

Initiation Sites. Approaches to this problem have included theoretical calculations on model structures and measurements of the stabilities of model compounds in the vapor and liquid phases (*1, 2, 3, 4, 5*). Recent computations by Valko et al. have been concerned with the cis elimination of HCl from possible irregular polymer structures (*8, 9*), from the normal saturated structure (*8, 9*), and from ethyl chloride (*10*). The HCl-catalyzed dehydrohalogenation of ethyl chloride has been examined also by computation (*11*), and the point has been made that the uncatalyzed dehydrochlorination is not forbidden by orbital symmetry if it involves a nonbonding electron pair of the halogen atom (*9, 10, 12, 13*). The most sophisticated theoretical treatment to appear in this area thus far has been by Hiberty (*13*), who performed ab initio SCF calculations on the unimolecular dehydrochlorination of ethyl chloride using STO-3G and 4-31G basis sets. The process was predicted to occur via a syn transition state which is planar but not rectangular (*13*); its geometry resembles that of the triangular array that was suggested recently by Kwart and Stanulonis (*14*) in order to accommodate experimental evidence for hydrogen tunneling during a related ether thermolysis.

The normal head-to-tail monomer units of PVC frequently are considered to be thermally stable at temperatures higher than those at which nonoxidative degradation of the polymer begins (*1, 2, 3, 4, 5*). Thus many workers have argued that this type of degradation starts from polymer defect sites exclusively (*1, 2, 3, 4, 5*). However, the possibility of simultaneous initiation from the normal units now is receiving increased attention (*3, 15–19*) and is supported (though not proved) by the increasing number of initiation points as the degradation proceeds (*20, 21*). Attempts have been made to dissect the overall degradation rate into contributions from normal units and defect structures (*3, 15–19*), but various authors have disagreed on the details of the separation. Certain conclusions reached in this regard would seem to be highly improbable; for example, dehydrochlorination of the normal units is said to involve an Arrhenius A factor of only $10^{3.3}$ and an activation energy that is lower by some 12–16 kcal/mol than the activation energy required for dehydrochlorination of internal chloroallylic groups (*15*). Furthermore, a theoretical scheme involving initiation from normal, allylic, and tertiary halide structures is claimed to agree well with experimental data (*16*) despite a current lack of conclusive evidence for the presence of any groups of the latter type (*see* below).

Recent research at Bell Laboratories has shown that defect sites play a dominant role under some conditions. This conclusion is based on the discovery that stability enhancements ranging up to a factor of 9 can be produced by the prior reaction of PVC with a number of organotins (*22, 23, 24, 25*). These enhancements are associated with the incorporation of only a few organotin ligands into the polymer (*22, 23, 24, 25*), and they cannot be ascribed to morphological changes since they are observed in solution (*25*) as well as in bulk (*22, 23, 24, 25*). Several other possible explanations are excluded by the available facts, and it now seems that the stability enhancements must result from the chemical deactivation of defect structures (*22, 23, 24, 25*). However, the reason for the residual instability of the modified polymers is not known at the present time (*22, 23*).

Enhanced stability caused by chemical modification has been observed also by other researchers (*see* below), although much of their work is difficult to interpret for reasons that were discussed previously (*22*).

Head-to-head (H–H) PVC, prepared by adding chlorine to *cis*-polybutadiene, appears to be less stable than the ordinary head-to-tail polymer (*26*). The ordinary polymer itself is said to contain 3–3.5 H–H linkages per 1000 carbons (*26*), but these results were obtained with an iodometric titration technique (*26*) that needs to be corroborated.

Instability arising from oxygen-containing structural defects in PVC is suggested by the decreased stabilities observed for polymers prepared in the presence of oxygen (*27*). This preparation method introduces polyperoxide moieties that can decompose into HCHO, HCl, and CO (*28*). However, the importance of oxygenated structures remains unclear for polymers prepared under normal conditions. Thermal dehydrochlorination of PVC can, of course, be accelerated greatly by introducing molecular oxygen into the surrounding environment (*1, 2, 4, 29*), but the rate soon falls to its original (nonoxidative) value when the oxygen is removed (*1, 4, 29*).

Tertiary chloride groups are expected to have relatively low stabilities (*1–5, 9*), and recent work with vinyl chloride/2-chloropropene copolymers suggests that only 0.1–0.2 mol % of such groups would account for the instability of PVC (*30*). Possible routes to tertiary Cl's include chain transfer to polymer and copolymerization with unsaturated moieties at the ends of chains (*30*). However, the latter path requires terminal $CH_2{=}C(Cl)-$ groups and thus is not consistent with the work of Enomoto (*31*), which suggests that the β-hydrogens of vinyl chloride are involved in the process of chain transfer to monomer. Thus, the presence of tertiary Cl at branch points formed by chain transfer to polymer currently seems more likely, and indeed, recent ^{13}C NMR studies on $LiAlH_4$- (*32*) and n-Bu_3SnH-reduced (*33, 34*) polymers have shown that PVC contains a small number of long branches. On the other hand, the ^{13}C NMR spectrum of a polymer reduced with $LiAlD_4$ indicates that the major branch structure of PVC is $-CH_2CH(Cl)CH(CH_2Cl)CH_2CH(Cl)-$ (*35*), a grouping containing no tertiary halogen.

The presence of tertiary Cl also could not be confirmed in another recent study (*36*), which involved attempts to determine the total labile halogens content of PVC by UV analysis after phenolysis. Since the number of labile halogens found was inversely proportional to polymer molecular weight, the authors (*36*) suggested that most of these halogens occur in allylic end-group structures (*see* below).

The occurrence of pendant methyl groups in $LiAlH_4$-reduced PVC (*32, 33, 34, 35*) has been supported now by pyrolysis gas chromatography (*37*), and results obtained with this technique have suggested also that a few ethyl branches are present (*37*). However, ^{13}C NMR studies on reduced PVC (*32, 33, 34, 35*) have not confirmed the latter conclusion yet. In view of these results and other observations summarized above, the presence of tertiary halogen in PVC still remains to be verified or disproved convincingly.

Olefinic unsaturation may be an important source of PVC's thermal instability (*1, 2, 3, 4, 5*), and several recent attempts have been made to determine the polymer's internal double-bond content by measuring the changes in molecular weight resulting from oxidative double-bond scission (*15, 38, 39, 40*). Correlations of stability with the number of internal double bonds found were observed in some cases (*15, 38, 39*) but not in others (*40*), and such correlations may be precluded by the presence of other labile groups (*38*).

The total number of double bonds in PVC was measured recently by bromination (*41*), but again no relationship was found between analytical results and thermal stabilities (*41*). However, the stability data used in this work (*41*) are difficult to interpret, since they were obtained in the presence of air on samples containing basic metal salts.

NMR techniques using signal accumulation appear to hold the greatest promise for the conclusive identification and enumeration of the unsaturated structures in PVC, and some studies of this type using pulsed Fourier-transform ^{1}H NMR have now appeared (*42, 43*). In this work (*42, 43*) the resonances ascribed to vinyl protons could not be assigned unambiguously, but the results did suggest the presence of several types of isolated olefinic groups that were associated with allylic halogen. Total double-bond contents found by ^{1}H NMR (*43*) were about (1.2–2.5)/1000 C's; values estimated by other methods are (0.7–1.5)/1000 C's (bromination) (*41*) and (0.2–1.0)/1000 C's (phenolysis) (*36*).

Thermal degradation of PVC at 180°C under nitrogen causes the vinyl proton signals to disappear when only 1–2% of the available HCl has been lost (*43*). This result can be attributed to signal broadening for protons in conjugated polyene structures (*43*), and thus it seems to constitute direct evidence for the involvement of olefinic defect sites in an early stage of the degradation process.

Consideration of the above results and others described in the literature (*1, 2, 3, 4, 5*) suggests that much more work is needed before the instabilities caused by the various structural defects in PVC can be sorted out and measured quantitatively.

Effects of Tacticity on Polyene Growth. The rate of the nonoxidative thermal dehydrochlorination of PVC is increased by the presence of syndiotactic sequences, although there seems to be no significant effect of tacticity on the overall activation energy (*44, 45*). These observations are consistent with a slow initiation step whose activation energy is tacticity-independent, followed by rapid polyene growth that is facilitated by the syndiotactic arrangement (*44, 45*). In keeping with this hypothesis, highly syndiotactic polymers have yielded abnormally long polyenes (*44, 45*). Similar results have been reported for copolymers containing syndiotactic vinyl chloride sequences (*46, 47*). Increases in polyene length with increasing syndiotacticity have also been reported for PVC dehydrochlorinations induced by UV irradiation (*48*).

Enhanced dehydrochlorination rates of syndiotactic sequences may result from faster propagation of polyene growth along favored (*tt*) conformations (*44, 45, 48*). However, decreased rates of polyene termination could play an important role also, since the all-trans polyenes thought to result from syndiotactic sequence degradation (*48, 49*) would be less likely to end their growth by undergoing intramolecular Diels–Alder cyclizations (*49*). In any event, these interesting tacticity effects clearly merit further attention.

Thermal scission of cyclohexadiene structures formed via the intramolecular Diels–Alder route (*49, 50*) constitutes an attractive mechanism for the formation of benzene (*49, 50*), a substance that appears among the pyrolysis products of PVC at temperatures of 180–200°C (*50*) and above (*49*). It is therefore of considerable interest to know whether benzene results from an intramolecular reaction path. Strong evidence for such a route has been reported now by O'Mara (*51*), who found that the pyrolysis of a mixture of perdeuterated and nondeuterated PVC gave C_6D_6 and C_6H_6 in considerable quantity, whereas only small amounts of benzenes of intermediate isotopic composition were formed.

Mechanism of Nonoxidative Thermal Dehydrochlorination. This subject is still very controversial, with various workers being in favor of radical, ionic, or "molecular" (concerted) paths. Recent evidence for a radical mechanism has been provided by studies of decomposition energetics (*52*), the degradation behavior of PVC–polystyrene (*53*) or PVC–polypropylene (*54*) mixtures, and the effects of radical traps (*54*). Evidence for an ionic mechanism comes from solvent effects (*55*) and studies of the solution decomposition behavior of a model allylic chloride (*56*). Theoretical considerations (*57, 58*) also suggest that an ionic (E1) path is not unreasonable. Other model compound decompositions have been interpreted in terms of a concerted process (*59*), but differences in solvent effects led the authors to conclude that PVC degrades via a different route (*59*).

Nolan and Shapiro (*19*) found that triphenylmethane and hexamethylbenzene reduced the rate of PVC dehydrochlorination and raised its activation energy by 26–27 kcal/mol. These results were interpreted in terms of a dual mechanism involving "molecular" elimination from ordinary monomer units, followed by a radical chain process that was thought to have been inhibited by the two hydrocarbon additives used (*19*). In a related study, Figge and Findeiss (*60*) found an activation energy increase of about 27 kcal/mol for PVC stabilized with an organotin mercaptide. These workers suggested that the organotin destroyed reactive defect structures and that the higher activation energy was required for "statistical" decomposition (*60*) (a term presumably intended to describe HCl loss from ordinary monomer units).

Observations concerning the kinetic effects of HCl have now come full circle with a report of the last remaining permutation: HCl is said to inhibit dehydrochlorination under certain experimental conditions (*55*)! However, the inhibition effect is small and may have resulted from HCl addition to double bonds in the closed degradation vessel used (*55*). Under most conditions, HCl exhibits strong dehydrochlorination catalysis (*1, 2, 3, 4*), which modern workers usually prefer to interpret in terms of an ionic path (*1, 2, 3, 55*). Attempts to quantify the catalytic effect of HCl have been made recently (*3, 61*), and dehydrochlorination proceeding from chloroallylic groups has been suggested to be the only catalyzed step (*61*).

Oxidative Degradation

Although PVC degradation in the presence of air has obvious practical significance, PVC oxidation has not been an especially popular field from the standpoint of basic research. A possible reason is that the chemistry involved is undoubtedly quite complex, particularly at higher temperatures where nonoxidative dehydrochlorination occurs simultaneously. However, the situation can be simplified by oxidizing at lower temperatures with known rates of radical generation, and an interesting study along these lines was described recently by Decker (*62*), who carried out PVC oxidation at 25°C using γ-irradiation for initiation.

In Decker's report (*62*), the identified oxidation products and their yields are: hydroperoxide (24–39%), aldehyde plus ketone (14–19%), alcohol (15–24%), carboxylic acid (7–10%), and carbon monoxide (2–4%) (the yields are based on O_2 absorbed at different radiation dose rates). Hydrogen chloride was also formed in significant quantity: ~ 0.5–0.66 mol of HCl per mol of O_2 consumed. The suggested mechanism (*62*) involves chain propagation via an essentially exclusive attack of peroxy radicals on the hydrogens of chloromethylene groups, a postulate which is not inconsistent with the photooxidation results of Kwei (*63, 64*). However, absolute rate data for some related model compounds (*65, 66*) suggest that the CH_2 and CHCl groups of PVC may react with peroxy radicals near 25°C at rates that are mutually competitive. Furthermore, exclusive $RO_2\cdot$ attack on CHCl's seems unable to account for the presence of appreciable numbers of alcohol moieties, since under such circumstances the OH's would belong to geminal chlorohydrin functions [$-CH_2CCl(OH)CH_2-$] (*62*), whose decomposition into ketone and HCl should occur rapidly and irreversibly. In any event, under the conditions used (*62*), the oxidizability of PVC (per initiating radical) was greater than that of polyethylene or polypropylene (*62*).

PVC degradation during processing is, of course, an important technological problem, and Scott et al. (*67*) have shown that rapid oxidation of the polymer occurs at very short mastication times. This early oxidation produces peroxides and is accompanied by the rapid appearance of unsaturation; both processes are faster at 170°C than at 210°C (*67*). The temperature effect suggests that the initial degradation reactions start from radicals formed by mechanochemical backbone scission (*67*). This interpretation is supported by data on mixing torque and by the decreased initial degradation rates found in the presence of an ester lubricant (*67*).

Oxidative bleaching of colored polyenes in PVC is a well-established phenomenon, and a quantitative kinetic study of this reaction has appeared now in an important paper by Nagy et al. (*68*). In this work (*68*), disappearance rates of conjugated polyenes containing 5–11 double bonds were measured at 40°–70°C using constant rates of radical initiation. The individual polyenes were destroyed at rates that were directly proportional to their lengths, a result that was shown only to be caused by differences in the number of contiguous double bonds, rather than by variations in their intrinsic reactivity (*68*). However, the reactivity (oxidizability) of a given double-bond unit was much greater for these polyenes than for conjugated dienes or isolated olefins (*68*). Since no evidence was obtained for the conversion of long polyenes into shorter ones during an autoxidation, the authors suggested that the polyenes underwent an initial rate-determining radical attack that was followed by rapid intramolecular propagation steps involving polyenylperoxy radical cyclization (*68*). This mechanism bears a formal resemblance to the mechanism of polypropylene autoxidation (*69, 70*).

Stabilization

With Organotins. In recent years the ability of organotins and metal soaps (*see* below) to prevent the appearance of color in degrading PVC has been ascribed most frequently to chemical reactions leading to the deactivation of defect sites. Very strong evidence for the operation of such a mechanism, at least in the case of the organotins, has been provided now by the chemical stabilization effects whose discovery (*22, 23, 24, 25*) was noted above. However, these effects do not rule out the concurrent operation of other mechanisms in "technological" situations (*22, 23*).

The defect site deactivation theory is supported also by recent studies that have confirmed the chemical binding of organotin ligands to the polymer upon heating (*22–25, 60, 71, 72*). Measurements of the rate of incorporation of sulfur-containing groups into PVC during heating with an organotin mercaptide have been used to estimate the number of allylic halogens in the polymer (*71*), and these measurements (and ancillary experiments) have been considered also to provide evidence for addition reactions of polyenes with free mercaptan formed from the stabilizer and HCl (*71*). During thermal treatment, sulfur-containing groups derived from organotin mercaptides are incorporated preferentially into PVC fractions having low molecular weights (*60, 72*), suggesting that the reactive sites may be concentrated at the ends of polymer chains (*see* above).

Carbon radicals generated by the polymerization of methyl methacrylate do not undergo displacement reactions with the Sn–S, Sn–O, Sn–C, or Sn–Cl bonds of typical organotin stabilizers (*73*). However, PVC stabilization caused by scavenging of chlorine atoms by organotins has not been ruled out experimentally (*73*). Furthermore, the ability of some organotins to undergo S_H2 displacements with alkoxy radicals (*74*) suggests that similar displacements may occur during the oxidative degradation of PVC containing organotins of appropriate structure.

Attempts have been made to correlate the stabilization properties of organotins with their reactivity toward HCl (*75, 76*), and the stabilizing effectiveness of bis(4-ketopentyl)tin bis("isooctyl" mercaptoacetate) has been found to be significantly greater than that of the corresponding di-*n*-butyltin compound (*77*). This difference was attributed to intramolecular coordination of tin with the C=O groups of bis(4-ketopentyl)-tin dichloride, an effect that should make this dichloride a weaker dehydrochlorination catalyst than $n\text{-Bu}_2\text{SnCl}_2$ (*77*).

It has been thought for some time that ligand exchange between organotins and the chlorides formed from them in situ may well play an important role in the overall stabilization process (*22, 78*). Some exchange reactions of this type have been investigated quantitatively now by ^{1}H and ^{13}C NMR techniques (*79, 80*), and the results have been discussed in terms of actual stabilization behavior (*79, 80*).

With Metal Soaps. The well-known stabilization synergism observed with mixtures of barium and cadmium soaps has been reaffirmed recently (*81*) and interpreted in terms of the usual mechanism (*1, 5*) involving replacement of labile halogen by the carboxylate groups associated with Cd(II), followed by regeneration of the cadmium soap via ligand exchange with the barium carboxylate (*81*). In a somewhat related study using 4-chloro-2-hexene as a model compound, Hoang et al. (*82*) found that the model could be esterified by reaction with zinc stearate or calcium stearate and that the esterification rates were much faster when zinc chloride was present. Mixtures of the two stearates exhibited synergistic esterification effects which were ascribed, not to ligand exchange, but to the formation of bimetallic complexes whose ability to catalyze dehydrochlorination and diene oligomerization was said to be less than that of $ZnCl_2$ (*82*).

Reactivity toward HCl has been used to rationalize the stabilization behavior of metal soaps and other metallic compounds (*83*), and the color-stabilizing properties of metal soaps in heated PVC have been shown to be influenced strongly by complementary color development (*84, 85, 86*).

With Miscellaneous Compounds. Several mechanistic studies have been performed recently on reactions of "auxiliary" stabilizers with 4-chloro-2-hexene, both in the presence and in the absence of metal salts (*82, 87, 88, 89, 90*). The stabilizers used were phosphites (*82, 87*), phosphines (*82, 87*), epoxides (*88*), 2-phenylindole (*89*), and the bis(β-aminocrotonate) ester of 1,4-butanediol (*90*). Important mechanistic features delineated in this work included substitutive removal of allylic halogen and modification of the activities of acidic catalysts by complexation phenomena (*82, 87, 88, 89, 90*).

Support has been claimed recently for an alternative theory of epoxide stabilization which postulates that epoxides merely assist the transport of HCl to primary stabilizers (metallic bases) via intermediate chlorohydrins (*91*).

By Chemical Modification. Stabilization by chemical pretreatment with organotins (*22, 23, 24, 25*) has been discussed already. Pretreatment studies with other reagents have been reviewed briefly (*22*), and improved stability is now also said to result from the prior reaction of PVC with maleic anhydride (*92*). Kennedy's interesting work on stabilization reactions of organoaluminums (*93*) has been continued by a report of enhanced stability (compared with that of ordinary PVC) for poly(vinyl chloride-g-styrene) obtained from the Et_2AlCl-induced reaction of PVC and styrene monomer (*94*). Selective destruction of defect sites seems a plausible explanation for this result (*94*), but improved stability has been observed also for the PVC parts of PVC–polystyrene blends (*53, 95, 96*) and for the PVC moieties of poly(vinyl chloride-g-styrene)s prepared by alternative methods (*95, 96, 97*).

Concluding Remarks

In recent years progress has been made toward a mechanistic understanding of PVC degradation and stabilization processes. Nevertheless, many of the major problems in this field remain unsolved at the present time. It is to be hoped that modern techniques, in particular those used to determine molecular structure, eventually will allow these problems to be solved in unambiguous ways.

Addendum

This section summarizes the results of very recent work which is described in papers that were not available when this review was prepared originally.

^1H NMR studies, using signal accumulation, on low molecular weight fractions of commercial PVC are considered to have provided evidence for the presence of $-CH_2CH{=}CHCH_2Cl$ and $-CH_2CH_2CH(Cl)CH_2Cl$ moieties at the chain ends (*98*). The latter group was believed to arise via HCl addition to the former structure (*98*), but the data implying the existence of the latter group are suggestive rather than conclusive. Double-bond contents estimated from this work were about (1–1.5)/1000 C's (*98*), a range of values comparable with those obtained by earlier workers (*36, 41, 43*).

The allylic chlorine contents of some commercial PVC's have now been determined by a radiochemical technique involving exchange with $SO^{36}Cl_2$ (*99*). Values found were in the range of (0.60–0.80)/1000 C's, but the method was thought to be incapable of detecting allylic halogen associated with conjugated double bonds (*99*).

The entire defect-structure problem has been discussed at length by Suzuki (*100*), who reaffirms the common belief that the ordinary PVC monomer units are stable at processing temperatures (*100*). A recent review by Carrega (*101*) on PVC characterization contains information pertaining to branches, double bonds, and polymer tacticity.

The relative instability of syndiotactic (crystalline) PVC has been supported by new observations (*102, 103*). However, the instability was said to be associated with decreases in the activation energy (*103*), a finding that does not agree with the conclusions of previous researchers (*44, 45*).

The work communicated in Ref. *19* has been published now in full (*104*). This study (*104*) shows that the nonoxidative dehydrochlorination of PVC can be inhibited by anthracene to some extent.

A comprehensive review by Wirth and Andreas (*105*) discusses the mechanisms of action of most of the common thermal and UV stabilizers for PVC. Special attention is given (*105*) to the organotin mercaptides, which are suggested to function as antioxidants, HCl scavengers, and deactivators of labile groups by ligand exchange or via the specific complexation of allylic chloride moieties. The latter explanation was invoked in order to account for the authors' finding that $R_{4-n}SnCl_n$ compounds (R = octyl, n = 0–3) can reduce the rate of thermal dehydrochlorination (*105*). However, it seems that some consideration should be given also to the possibility that these substances can scavenge HCl by oxidative addition

$$R_{4-n}SnCl_n + 2HCl \rightarrow [R_{4-n}SnCl_{n+2}]^{=} + 2H^{+} \ (106)$$

or dealkylation

$$R_{4-n}SnCl_n + HCl \rightarrow RH + R_{4-(n+1)}SnCl_{n+1} \ (76, 107, 108).$$

In any event, the author's report (*105*) of stabilization by (octyl)$SnCl_3$ is very surprising indeed, since alkyltin trichlorides appear ordinarily to induce a catastrophic degradation of the polymer (*76*).

Treatment with organotin stabilizers improves the color of thermally degraded PVC (*100, 105*). This effect was attributed to the destruction of complexes formed from polyenes and HCl (*100, 105*) and, in the case of the organotin mercaptides, to the addition of free mercaptan to polyene double bonds (*105*). Other recent work with organotins includes a systematic study of the stabilization effects observed with various R groups in a series of $R_2Sn(SCH_2CO_2R')_2$ (R′ = "isooctyl") compounds (*109*).

The work of Iida et al. (*84, 85, 86*) on stabilization by complementary color development has been extended now to include the use of various color-masking agents that apparently can prevent the formation of excessive amounts of highly colored polyene–metal complexes (*110*).

By means of computer calculations, an attempt has been made to analyze a set of experimental stabilization data in order to extract the individual rate constants for various reactions that may occur during stabilization with metal salts (*111*). Ethyl β-aminocrotonate, a representative of an "auxiliary" stabilizer type, has been shown to react with a model allylic halide (1-bromo-2-butene) via alkylation on C-α rather than nitrogen (*112*).

Stabilization by chemical modification has been reviewed recently (*100*), and details have been published on a new modification technique involving pretreatment with aqueous ethanol (*100*). This method improves the static and dynamic stabilities of PVC by factors ranging up to 3–3.5; its effectiveness is believed to derive from the deactivation of labile halogen sites by solvolytic displacements (*100*).

Literature Cited

1. Ayrey, G., Head, B. C., Poller, R. C., *J. Polym. Sci., D* (1974) **8**, 1.
2. Mayer, Z., *J. Macromol. Sci., C* (1974) **10**, 263.
3. Braun, D., *Degradation Stab. Polym., Proc. Plenary Main Lect. Int. Symp., 1974* (1975) 23.
4. David, C., *Compr. Chem. Kinet.* (1975) **14**, 78.
5. Nass, L. I., *Encycl. PVC* (1976) **1**, 271.
6. Owen, E. D., *ACS Symp. Ser.* (1976) **25**, 208.
7. Rånby, B., Rabek, J. F., "Photodegradation, Photooxidation, and Photostabilization of Polymers," pp. 192–195, Wiley–Interscience, New York, 1975.
8. Valko, L., Kovařík, P., *J. Phys. Chem.* (1976) **80**, 19.
9. Valko, L., Tvaroška, I., Kovařík, P., *Eur. Polym. J.* (1975) **11**, 411.
10. Tvaroška, I., Klimo, V., Valko, L., *Tetrahedron* (1974) **30**, 3275.
11. Tvaroška, I., Bleha, T., Valko, L., *Polym. J.* (1975) **7**, 34.
12. Goddard III, W. A., *J. Am. Chem. Soc.* (1972) **94**, 793.
13. Hiberty, P. C., *J. Am. Chem. Soc.* (1975) **97**, 5975.
14. Kwart, H., Stanulonis, J. J., *J. Am. Chem. Soc.* (1976) **98**, 5249.
15. Minsker, K. S., Lisitsky, V. V., Berlin, A. A., Kolinski, M., Vymazal, Z., *Second International Symposium on Poly(vinyl chloride), Preprints,* 269–272 (Lyon–Villeurbanne, France, July 5–9, 1976).
16. Tvaroška, I., *Second International Symposium on Poly(vinyl chloride), Preprints,* 301–304 (Lyon–Villeurbanne, France, July 5–9, 1976).
17. Razuvaev, G. A., Troitskii, B. B., Troitskaya, L. S., *Second International Symposium on Poly(vinyl chloride), Preprints,* 261–264 (Lyon–Villeurbanne, France, July 5–9, 1976).

18. Troitskii, B. B., Dozorov, V. A., Minchuk, F. F., Troitskaya, L. S., *Eur. Polym. J.* (1975) **11**, 277.
19. Nolan, K. P., Shapiro, J. S., *J. Chem. Soc., Chem. Commun.* (1975) 490.
20. Abbås, K. B., Sörvik, E. M., *J. Appl. Polym. Sci.* (1975) **19**, 2991.
21. Abbås, K. B., *Second International Symposium on Poly(vinyl chloride), Preprints,* 277–280 (Lyon–Villeurbanne, France, July 5–9, 1976).
22. Starnes, Jr., W. H., Plitz, I. M., *Macromolecules* (1976) **9**, 633, 878.
23. Starnes, Jr., W. H., Plitz, I. M., *Polym. Prepr., Am. Chem. Soc., Div. Polym. Chem.* (1975) **16** (2), 500.
24. Plitz, I. M., Starnes, Jr., W. H., Hartless, R. L., *Polym. Prepr., Am. Chem. Soc., Div. Polym. Chem.* (1976) **17** (2), 495.
25. Plitz, I. M., Willingham, R. A., Starnes, Jr., W. H., *Macromolecules* (1977) **10**, 499.
26. Mitani, K., Ogata, T., Awaya, H., Tomari, Y., *J. Polym. Sci., A1* (1975) **13**, 2813.
27. Garton, A., George, M. H., *J. Polym. Sci., A1* (1974) **12**, 2779.
28. Bauer, J., Sabel, A., *Angew. Makromol. Chem.* (1975) **47**, 15.
29. Henson, J. H. L., Hybart, F. J., *J. Appl. Polym. Sci.* (1973) **17**, 129.
30. Berens, A. R., *Polym. Eng. Sci.* (1974) **14**, 318.
31. Enomoto, S., *J. Polym. Sci. A1* (1969) **7**, 1255.
32. Abbås, K. B., Bovey, F. A., Schilling, F. C., *Makromol. Chem., Suppl.* (1975) **1**, 227.
33. Starnes, Jr., W. H., Hartless, R. L., Schilling, F. C., Bovey, F. A., ADV. CHEM. SER. (1978) **169**, 324.
34. Starnes, Jr., W. H., Hartless, R. L., Schilling, F. C., Bovey, F. A., *Polym. Prepr., Am. Chem. Soc., Div. Polym. Chem.* (1977) **18** (1), 499.
35. Bovey, F. A., Abbås, K. B., Schilling, F. C., Starnes, Jr., W. H., *Macromolecules* (1975) **8**, 437.
36. Robilă, G., Buruiană, E. C., Caraculacu, A. A., *Eur. Polym. J.* (1977) **13**, 21.
37. Ahlstrom, D. H., Liebman, S. A., Abbås, K. B., *J. Polym. Sci., A1* (1976) **14**, 2479.
38. Abbås, K. B., Sörvik, E. M., *J. Appl. Polym. Sci.* (1976) **20**, 2395.
39. Braun, D., Quarg, W., *Angew. Makromol. Chem.* (1973) **29/30**, 163.
40. Lindenschmidt, G., *Angew. Makromol. Chem.* (1975) **47**, 79.
41. Boissel, J., *J. Appl. Polym. Sci.* (1977) **21**, 855.
42. Buruiană, E. C., Robilă, G., Bezdadea, E. C., Bărbîntă, V. T., Caraculacu, A. A., *Eur. Polym. J.* (1977) **13**, 159.
43. Caraculacu, A., Bezdadea, E., *J. Polym. Sci., A1* (1977) **15**, 611.
44. Martínez, G., Millán, J., Bert, M., Michel, A., Guyot, A., *Second International Symposum on Poly(vinyl chloride), Preprints,* 293–296 (Lyon–Villeurbanne, France, July 5–9, 1976).
45. Millán, J., Madruga, E. L., Martínez, G., *Angew. Makromol. Chem.* (1975) **45**, 177, and references cited therein.
46. Millán, J., Guzmán Perote, J., *Eur. Polym. J.* (1976) **12**, 299.
47. Malhotra, S. L., Hesse, J., Blanchard, L.-P., *Polymer* (1975) **16**, 269.
48. Mitani, K., Ogata, T., *J. Appl. Polym. Sci.* (1974) **18**, 3205.
49. Chang, E. P., Salovey, R., *J. Polym. Sci., A1* (1974) **12**, 2927.
50. Tüdös, F., Kelen, T., Nagy, T. T., Turcsányi, B., *Pure Appl. Chem.* (1974) **38**, 201.
51. O'Mara, M. M., *Pure Appl. Chem.* (1977) **49**, 649.
52. Gupta, V. P., St. Pierre, L. E., *J. Polym. Sci., A1* (1973) **11**, 1841.
53. Dodson, B., McNeill, I. C., *J. Polym. Sci., A1* (1976) **14**, 353.
54. Papko, R. A., Pudov, V. S., *Polym. Sci. USSR (Engl. Transl.)* (1974) **16**, 1636.
55. Zafar, M. M., Mahmood, R., *Eur. Polym. J.* (1976) **12**, 333.

56. Hoang, T. V., Michel, A., Pichot, C., Guyot, A., *Eur. Polym. J.* (1975) **11**, 469.
57. Haddon, R. C., Starnes, Jr., W. H., ADV. CHEM. SER. (1978) **169**, 333.
58. Haddon, R. C., Starnes, Jr., W. H., *Polym. Prepr., Am. Chem. Soc., Div. Polym. Chem.* (1977) **18** (1), 505.
59. Varma, I. K., Grover, S. S., *Makromol. Chem.* (1974) **175**, 2515.
60. Figge, K., Findeiss, W., *Angew. Makromol. Chem.* (1975) **47**, 141.
61. Abdullin, M. I., Malinskaya, V. P., Kolesov, S. V., Minsker, K. S., *Second International Symposium on Poly(vinyl chloride), Preprints,* 273–276 (Lyon–Villeurbanne, France, July 5–9, 1976).
62. Decker, C., *J. Appl. Polym. Sci.* (1976) **20**, 3321.
63. Kwei, K.-P. S., *J. Polym. Sci., A1* (1969) **7**, 237.
64. Kwei, K.-P. S., *J. Polym. Sci., A1* (1969) **7**, 1075.
65. Chenier, J. H. B., Tremblay, J. P.-A., Howard, J. A., *J. Am. Chem. Soc.* (1975) **97**, 1618.
66. Hendry, D. G., Mill, T., Piszkiewicz, L., Howard, J. A., Eigenmann, H. K., *J. Phys. Chem. Ref. Data* (1974) **3**, 937.
67. Scott, G., Tahan, M., Vyvoda, J., *Chem. Ind. (London)* (1976) 903.
68. Nagy, T. T., Kelen, T., Turcsányi, B., Tüdös, F., *J. Polym. Sci., A1* (1977) **15**, 853.
69. Chien, J. C. W., Vandenberg, E. J., Jabloner, H., *J. Polym. Sci., A1* (1968) **6**, 381.
70. Niki, E., Decker, C., Mayo, F. R., *J. Polym. Sci., A1* (1973) **11**, 2813.
71. Alavi–Moghadam, F., Ayrey, G., Poller, R. C., *Eur. Polym. J.* (1975) **11**, 649.
72. Alavi–Moghadam, F., Ayrey, G., Poller, R. C., *Polymer* (1975) **16**, 833.
73. Ayrey, G., Head, B. C., Poller, R. C., *J. Polym. Sci., A1* (1975) **13**, 69.
74. Davies, A. G., Scaiano, J. C., *J. Chem. Soc., Perkin Trans. 2* (1973) 1777.
75. Rockett, B. W., Hadlington, M., Poyner, W. R., *J. Appl. Polym. Sci.* (1973) **17**, 3457.
76. Rockett, B. W., Hadlington, M., Poyner, W. R., *J. Appl. Polym. Sci.* (1974) **18**, 745.
77. Abbas, S. Z., Poller, R. C., *Polymer* (1974) **15**, 543.
78. Klemchuk, P. P., ADV. CHEM. SER. (1968) **85**, 1.
79. Parker, R. G., Carman, C. J., ADV. CHEM. SER. (1978) **169**, 363.
80. Parker, R. G., Carman, C. J., *Polym. Prepr., Am. Chem. Soc., Div. Polym. Chem.* (1977) **18** (1), 510.
81. Braun, D., Hepp, D., *Angew. Makromol. Chem.* (1975) **44**, 131.
82. Hoang, T. V., Michel, A., Guyot, A., *Eur. Polym. J.* (1976) **12**, 337.
83. Wypych, J., *J. Appl. Polym. Sci.* (1976) **20**, 557.
84. Iida, T., Nakanishi, M., Gotō, K., *J. Appl. Polym. Sci.* (1975) **19**, 235.
85. Iida, T., Nakanishi, M., Gotō, K., *J. Appl. Polym. Sci.* (1975) **19**, 243.
86. Iida, T., Gotō, K., *Second International Symposium on Poly(vinyl chloride), Preprints,* 281–284 (Lyon–Villeurbanne, France, July 5–9, 1976).
87. Hoang, T. V., Michel, A., Pham, Q. T., Guyot, A., *Eur. Polym. J.* (1975) **11**, 475.
88. Hoang, T. V., Michel, A., Guyot, A., *Eur. Polym. J.* (1976) **12**, 347.
89. Hoang, T. V., Michel, A., Guyot, A., *Eur. Polym. J.* (1976) **12**, 357.
90. Michel, A., Hoang, T. V., Guyot, A., *Second International Symposium on Poly(vinyl chloride), Preprints,* 324–327 (Lyon–Villeurbanne, France, July 5–9, 1976).
91. Wypych, J., *J. Appl. Polym. Sci.* (1975) **19**, 3387.
92. Varma, I. K., Patnaik, S., Sahoo, J. C., *Second International Symposium on Poly(vinylchloride), Preprints,* 309–312 (Lyon–Villeurbanne, France, July 5–9, 1976).
93. See, inter alia, Kennedy, J. P., Ichikawa, M., *Polym. Eng. Sci.* (1974) **14**, 322.

94. Kennedy, J. P., Nakao, M., *Second International Symposium on Poly(vinyl-chloride), Preprints,* 61–64 (Lyon–Villeurbanne, France, July 5–9, 1976).
95. Yamakawa, S., Stannett, V., *J. Appl. Polym. Sci.* (1974) **18,** 2177.
96. McNeill, I. C., Neil, D., Guyot, A., Bert, M., Michel, A., *Eur. Polym. J.* (1971) **7,** 453.
97. Guyot, A., Bert, M., Michel, A., Spitz, R., *J. Polym. Sci., A1* (1970) **8,** 1596.
98. Pétiaud, R., Pham, Q.-T., *Makromol. Chem.* (1977) **178,** 741.
99. Buruiană, E. C., Bărbîn,tă, V. T., Caraculacu, A. A., *Eur. Polym. J.* (1977) **13,** 311.
100. Suzuki, T., *Pure Appl. Chem.* (1977) **49,** 539.
101. Carrega, M. E., *Pure Appl. Chem.* (1977) **49,** 569.
102. Ellinghorst, G., Hummel, D. O., *Angew. Makromol. Chem.* (1977) **63,** 167.
103. Ellinghorst, G., Hummel, D. O., *Angew. Makromol. Chem.* (1977) **63,** 183.
104. Nolan, K. P., Shapiro, J. S., *J. Polym. Sci., C* (1976) **55,** 201.
105. Wirth, H. O., Andreas, H., *Pure Appl. Chem.* (1977) **49,** 627.
106. Marks, G. C., Benton, J. L., Thomas, C. M., *SCI Monogr.* (1967) **26,** 204.
107. Druesedow, D., Gibbs, C. F., *Nat. Bur. Stand. (U.S.), Circ.* (1953) **525,** 69.
108. Minsker, K. S., Fedoseyeva, G. T., Zavarova, T. B., Krats, E. O., *Polym. Sci. USSR (Engl. Transl.)* (1971) **13,** 2544.
109. Poller, R. C., ADV. CHEM. SER. (1976) **157,** 177.
110. Iida, T., Kataoka, N., Ueki, N., Gotō, K., *J. Appl. Polym. Sci.* (1977) **21,** 2041.
111. Prochaska, K., Wypych, J., *J. Appl. Polym. Sci.* (1977) **21,** 2113.
112. Cointet, P. d., Pigerol, C., Mendes, E., *Eur. Polym. J.* (1977) **13,** 531.

RECEIVED June 9, 1977.

26

Reductive Dehalogenation with Tri-*n*-butyltin Hydride: A Powerful New Technique for Use in Poly(vinyl chloride) Microstructure Investigations

W. H. STARNES, JR., R. L. HARTLESS, F. C. SCHILLING, and F. A. BOVEY

Bell Laboratories, Murray Hill, NJ 07974

*Poly(vinyl chloride) (PVC) reacts with tri-*n*-butyltin hydride in inert solvents containing a thermal free-radical source (azobisisobutyronitrile) to give quantitative yields of a polymeric product containing hydrogen in place of halogen. Using ^{13}C NMR and IR spectroscopy, the structures of reduced polymers prepared in this way are compared with those of PVC that have been reduced with lithium aluminum hydride. The results show that, in contrast to the $LiAlH_4$ reductions, reactions with *n-Bu_3SnH are relatively clean and probably do not involve dehydrochlorination. The data also indicate that the carbon skeletal dislocations responsible for the appearance of methyl branches in $LiAH_4$- or *n-Bu_3SnH-reduced PVC must occur exclusively during polymerization rather than during the reduction steps.*

The thermal stability of poly(vinyl chloride) (PVC), as well as its physical properties in bulk and in solution, are influenced by the presence of defect structures such as long- and short-chain branches and internal or terminal double bonds (*1–8*). The identification and enumeration of these structures, with the aid of modern instrumental analytical techniques, can be facilitated by chemically converting the polymer into a hydrocarbon of related constitution. A method for effecting such a transformation was described in the 1950s by Cotman (*9, 10, 11*), who used lithium aluminum hydride as a chemical reducing agent (Reaction 1). This technique soon gained general acceptance, and in the interven-

0-8412-0381-4/78/33-169-324$05.00/1

$$(\mathrm{CH_2\overset{\overset{\displaystyle Cl}{|}}{C}H})_n \xrightarrow{\mathrm{LiAlH_4}} (\mathrm{CH_2CH_2})_n \tag{1}$$

ing years it has been used by many workers whose principal interest has been to determine the number of branches in the reduced polymer (and, therefore, in the starting PVC) by IR spectroscopy (*12, 13*). Recently, the value of this approach to the defect structure problem was confirmed and enhanced by the research of Abbås (*12*), Bovey (*14*), and their collaborators, who used Fourier transform ^{13}C NMR spectroscopy to examine $LiAlH_4$-reduced materials.

Despite its extensive utilization, the $LiAlH_4$ reduction technique has serious disadvantages. These include long reaction times [e.g., ranging from one (*12, 15*) to three weeks (*16*)], incomplete halogen removal (*9–13, 15–17*), and a lack of reproducibility (*12, 15*). Moreover, explosions frequently occur when attempts are made to circumvent these difficulties by carrying out the reaction at high temperatures under pressure (*15*), and side reactions supervene (*see* below) which make it difficult to relate the microstructure of the reduced polymer to that of the starting PVC. Nevertheless, it appeared to us that reduction followed by spectroscopic examination could still be a viable approach to PVC microstructure elucidation if a better method were developed for carrying out the reduction step.

Consideration of the chemistry of many reducing agents suggested that an organotin hydride might be useful. The reductive dehalogenation of simple organic halides by triorganotin monohydrides (Reaction 2) was

$$R'X + R_3SnH \rightarrow R'H + R_3SnX \tag{2}$$

$$X = Cl, Br, I;\ R' = \text{alkyl, aryl};\ R = \text{alkyl, aryl}$$

an extensively studied reaction (*18, 19, 20*) that was applicable to *sec*-alkyl monochlorides that could be regarded as rough models for PVC (*18, 20*). The mechanism of this reduction was well understood and generally recognized as a free-radical chain process involving $R_3Sn\cdot$ and $R'\cdot$ intermediates (*18, 19, 20, 21*) (Reactions 3 and 4). Furthermore,

$$R_3Sn\cdot + R'X \rightarrow R_3SnX + R'\cdot \tag{3}$$

$$R'\cdot + R_3SnH \rightarrow R'H + R_3Sn\cdot \tag{4}$$

absolute rate constants reported for the individual propagation and termination steps (*21*) indicated that long kinetic chain lengths were to be expected for *sec*-alkyl chloride reductions (of at least several hundred (*21*), depending on conditions), and that the small amounts of termination occurring during these reductions would involve $R_3Sn\cdot$ radicals almost exclusively (*21*). The absence of terminations involving two $R'\cdot$ radicals or one $R'\cdot$ and one $R_3Sn\cdot$ seemed advantageous for PVC since the introduction into reduced polymer of tin-containing groups, cross-links, or double bonds—formed via these alternative termination routes—would thereby be precluded. Other data (*19, 20, 21, 22, 23*) suggested that PVC reduction by an R_3SnH of appropriate structure would not be subject to undue complications from side reactions such as irreversible intermolecular addition of radicals to internal double bonds and hydrogen abstraction from the polymer or possible solvents. Finally, it seemed that the reduction should occur at temperatures where PVC would not undergo appreciable thermal dehydrochlorination (*1–4, 20, 21*). Encouraged by these considerations, we undertook to develop a new method for PVC reduction using an organotin hydride reagent.

Results and Discussion

For reasons of economy and commercial availability, the specific reagent selected was tri-*n*-butyltin hydride; the Experimental section gives details of the procedure developed for its use. The first stage of this procedure requires overnight heating of the polymer at ca. 80°C under nitrogen in a 2-methyltetrahydrofuran (2-MeTHF) solution containing a small amount of azobisisobutyronitrile (for thermal initiation) and a 20 mol % excess of *n*-Bu_3SnH. During this treatment, the reduced polymer precipitates from the reaction mixture. After a simple work-up involving filtration, solvent washing, Soxhlet extraction with methanol, and drying, one obtains an essentially quantitative yield of snow-white polymer which usually contains ca. 2–4 wt % of halogen (lower values are sometimes achieved) and is ordinarily suitable for direct examination by ^{13}C NMR or other spectral techniques. For further reduction, this material can be subjected to a second reduction step involving overnight heating at ~ 90°C under nitrogen in a xylene solvent mixture containing additional initiator and organotin. Precipitation of the polymer with methanol, followed by a work-up similar to that described above, gives a snow-white product that ordinarily contains no chlorine at all and is again recovered almost quantitatively.

This reduction technique has not necessarily been carried to its ultimate stage of development; it is quite likely that the procedure could be improved. Nevertheless, this technique already seems far superior to the $LiAlH_4$ method with respect to reaction velocity, safety aspects, and the reproducible production of a polymer that has been extensively reduced. Further evidence for the superiority of the new technique has been provided by ^{13}C NMR studies, which will now be briefly described.

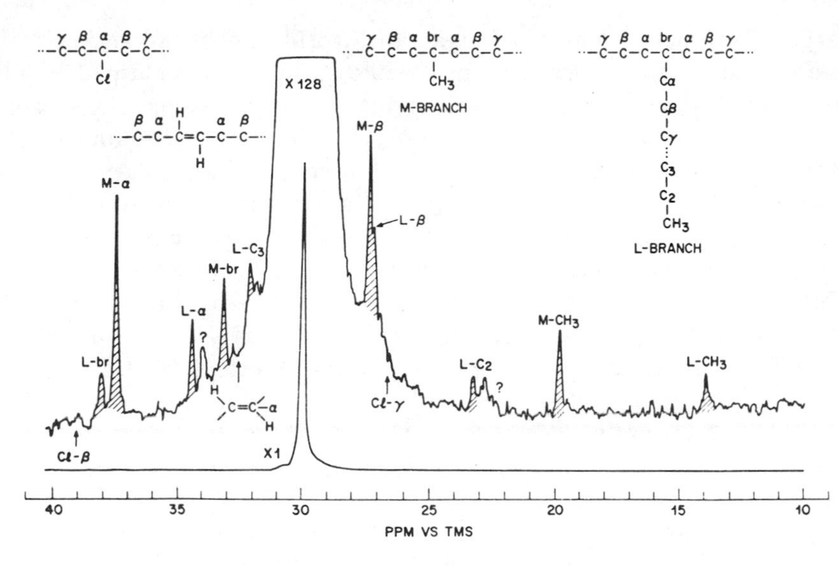

*Figure 1. 25-MHz ^{13}C NMR spectrum of PVC reduced with tri-*n*-butyltin hydride* (see *text for details*)

Previous workers (*17*) have used IR spectroscopy to demonstrate the presence of *trans*-alkene linkages in $LiAlH_4$-reduced PVC, and it has been argued that these groups must have been present in the starting polymer (*17*) and are the main source of its thermal instability (*4, 17*). Moreover the ^{13}C NMR spectra of PVCs reduced with $LiAlH_4$ (*12*) or $LiAlD_4$ (*14*) display a resonance at 32.7 ppm (chemical shifts in this paper are referred to tetramethylsilane (TMS)) that can be ascribed to methylene carbons α to internal trans unsaturation (*14*). However, the intensity of this resonance implies a much greater internal double bond content for the starting PVC than can be detected by ozonolysis followed by molecular weight measurements using gel permeation chromatography (*14*). Thus the possibility of double bond formation during reduction obviously requires consideration (*14*).

Figure 1 shows the ^{13}C NMR spectrum of the polymer obtained by subjecting a sample of PVC to a "two-pass" n-Bu_3SnH reduction. The PVC used was identical to that used in the work of reference (*12*), and the absence of resonances at 26.7 and 39.1 ppm (corresponding to carbons β and γ to isolated secondary halogen (*12*)) demonstrates that the reduction was essentially complete (the carbon α to chlorine, if present, would have appeared at 63.7 ppm.) (An erroneous chemical shift value was previously reported for this carbon (*12*)). However, the α-methylene resonance at 32.7 ppm now has vanished entirely. This resonance also failed to appear in the spectrum of the polymer obtained from the same starting PVC after a "one-pass" organotin hydride treatment, and in both cases the absence of internal trans olefinic groups was confirmed by IR inspections. Complete destruction of internal *trans*-alkene moieties by intermolecular reactions during n-Bu_3SnH reductions seems unlikely in view of literature information pertaining to certain related systems (*18–23*) and the apparent absence of n-Bu_3Sn- groups from polymers reduced with the organotin (*see* below). Double bond destruction via intramolecular carbon radical addition is not ruled out entirely, but the ^{13}C NMR data suggest that the structures expected to result from such addition (1,2-disubstituted alicyclic rings) (*18, 19, 21, 23–26*) are present in the n-Bu_3SnH-reduced polymers to only a minor extent, if at all. Thus, the double bonds detected in PVC that has been reduced with $LiAlH_4$ (or $LiAlD_4$) are probably mostly an artifact of the reduction method—at least for cases in which the specific method of Refs. *12* and *14* has been used (further details concerning this method were recently published (*27*)).

Another point of interest concerning $LiAlH_4$-reduced polymers relates to the probable origin of the one-carbon (methyl) branches. ^{13}C NMR data have shown that a considerable amount of hydrogen scrambling occurs near these branches when deuterium-labeled PVC is reduced with $LiAlH_4$ according to the procedure of Ref. *12* and *14* (*28*). This scrambling constitutes permissive evidence for the presence of carbenium ion intermediates and suggests that the methyl branches themselves may be formed, at least in part, by carbenium ion rearrangements during reduction. However, the methyl branch frequency calculated from the relative intensity of the "M-α" resonance in Figure 1 (*see* Experimental section) is equivalent to the frequency computed in an identical manner from the spectrum of the polymer prepared by reducing the same type of PVC with lithium aluminum hydride (*12*) (the actual frequency value is 2.8 ± 0.2 CH_3 groups per 1000 carbons). A possible rationale for these results is that carbenium ion rearrangements occurred to the same extent during both reduction reactions, but such a coincidence seems improbable since the tin compounds present in the n-Bu_3SnH system are probably weaker Lewis acids (therefore, much less effective rearrangement catalysts) than the aluminum chloride formed from $LiAlH_4$. Nevertheless, in order to investigate this possibility, a sample of PVC was pretreated by refluxing it overnight in the 2-MeTHF mother liquor obtained from a prior organotin hydride reduction. The polymer was then recovered and reduced with the organotin in the usual way. Since the ^{13}C NMR spectrum of the reduced polymer showed that the pretreatment had caused no significant change in the number of methyl branches, we conclude that these branches were present in the starting PVC itself (apparently as -CH_2Cl's (*12, 13, 14*)). This conclusion was confirmed by carrying out n-Bu_3SnH and $LiAlH_4$ reductions on another PVC sample of slightly different constitution, whose branch content after reduction by either reagent was 2.0 ± 0.2 CH_3's per 1000 carbons.

Additional ^{13}C NMR studies have shown that $LiAlH_4$ reductions of PVC can lead to other problems which are more serious than those already described. These reductions are frequently incomplete, as already noted; and in a typical case of this type where the residual content of isolated secondary halogen was approximately 0.7 wt %, the ^{13}C NMR spectrum of the reduced polymer (polymer A) provided no clear-cut evidence for the presence of "L-br" or "L-α" carbons (for nomenclature, *see* Figure 1), even though strong NMR evidence for the presence of long branches in $LiAlH_4$-reduced PVC had been obtained previously (*12*). Moreover, the spectrum of polymer A contained strong resonances at 19.2, 26.6, 34.6, and 39.3 ppm whose origins were quite obscure (assignments for some of these resonances will be discussed in a subsequent paper (*29*)). For comparison, the virgin PVC used to obtain these results was subjected to a "one-pass" organotin hydride reduction to obtain reduced polymer (polymer B) whose isolated secondary halogen content ($\sim$ 0.4 wt %) was roughly comparable with that of polymer A. The ^{13}C NMR spectrum of polymer B was recorded, and it contained well-defined resonances at the "L-br" and "L-α" positions. However, the unexplained resonances at 19.2, 26.6, 34.6, and 39.3 ppm were absent entirely, and in other respects this spectrum was similar to that of Figure 1.

Concluding Remarks

In view of the foregoing observations, it would seem that $LiAlH_4$ reduction must now be regarded as an unsatisfactory tool for microstructure investigations on PVC. Although this procedure did yield the "correct" number of methyl branches in the examples cited above, recent work suggests that the method can fail even in this respect (*29*). All aspects of the ^{13}C NMR spectra of polymers reduced with n-Bu_3SnH are not yet completely explained (e.g., note the resonance at 33.9 ppm, which also appears in the spectra of $LiAlH_4$-reduced materials), but from the data in hand it is clear that the organotin hydride reduction technique holds great promise for use in PVC microstructure determinations. Further studies in this area are in progress (including some labeling experiments with n-Bu_3SnD), and extensions of the new reduction technique to other polymers are planned.

Experimental

Materials. Tri-n-butyltin hydride (Ventron), mixed xylenes (bp 137°–144°C, MC&B reagent grade), methanol (Baker electronic grade), and azobisisobutyronitrile (AIBN, Aldrich) were used as received.

2-Methyltetrahydrofuran (2-MeTHF, Aldrich) was distilled under nitrogen and subjected to column chromatography on neutral alumina immediately before use. The polymer whose spectrum is displayed in Figure 1 was prepared from a PVC sample (Nordforsk S-54) that has been described previously (*12*).

Tri-*n*-butyltin Hydride Reductions. These reactions were carried out in a round-bottomed flask equipped wtih a Teflon-coated magnetic stirring bar, a reflux condenser, and an inlet tube for inert gas introduction below the liquid level. In a typical experiment, a solution of PVC (4.0 g), n-Bu_3SnH (22.0 g, 1.2 mol/mol of monomer units), and AIBN (0.10 g, 0.010 mol/mol of monomer units) in 350 mL of 2-MeTHF was stirred and heated under gentle reflux (ca. 80°C) for approximately 24 hr with continuous nitrogen bubbling. The solid that precipitated was recovered by suction filtration and washed in succession with fresh 2-MeTHF and methanol. Soxhlet extraction of the product with methanol for 24 hr, followed by drying under vacuum at ca. 50°C, afforded 1.9 g (yield, approximately 100%) of reduced PVC as a snow-white powder whose chlorine content ranged from about 2–4 wt % in typical cases. Further reduction could be effected by heating this "one-pass" reduced material for an additional 24 hr at 90°C in a nitrogen-bubbled, mixed-xylenes solution (250 mL) containing 4.0 g (6.5 mol/mol of monomer units) of n-Bu_3SnH and 0.050 g (0.14 mol/mol of monomer units) of AIBN (the mol ratios refer to starting polymer having a chlorine content of 4 wt %). The solution was then added slowly with rapid stirring to ~ 600 mL of methanol to precipitate the polymer, which was recovered by filtering, washed with fresh methanol, extracted, and dried in the manner described above. By these operations reduced polymer was again obtained in essentially quantitative yield (~ 1.8 g) as a snow-white powder whose halogen content was 0.0–0.4 wt %.

Polymers prepared by either the "one-pass" or "two-pass" reduction procedure were generally suitable for NMR analysis without further purification. However, these polymers occasionally contained small amounts of impurities that caused undesirable broadening of spectral lines. In some cases these impurities could be removed by filtering the NMR solutions through 5-μ Teflon filters. In other cases, satisfactory purification was achieved by redissolving the polymers in mixed xylenes, adding a few drops of concentrated hydrochloric acid, stirring until there was no visible turbidity (10–15 min), and then precipitating the polymers into methanol that also contained a small amount of concentrated HCl. Following their recovery by suction filtration, these polymers were processed in the usual way (*see* above).

^{13}C NMR Measurements. The NMR spectra were observed with a Varian XL-100 spectrometer modified for pulse Fourier transform spectroscopy and interfaced with a Nicolet model 1080 computer. The protons were decoupled from the carbon nuclei using a random noise decoupling field. Free induction decays were stored in 8K computer locations using a dwell time of 200 μsec, i.e., a spectral window of 2500 Hz. The pulse width was 23 μsec (for a 90° pulse), and the pulse interval was 3.0 sec. Hexamethyldisiloxane was used as an internal reference (2.0 ppm vs. TMS), and the internal deuterium lock signal was provided by *p*-dioxane-d_8. All spectra represent accumulations of approximately 20,000 scans.

Reduced PVCs were observed as 33% (w/v) solutions in 1,2,4-trichlorobenzene at 110°C. At this temperature, T_1 for the normal backbone carbons is approximately 1.3 sec. Methyl branch frequencies were measured quantitatively by comparing the intensities of the "M-α" carbon resonances (*see* Figure 1) with those of the carbon resonances arising from the normal backbone methylenes. The absence of tri-*n*-butyltin groups from reduced polymers prepared with *n*-Bu_3SnH was inferred from the absence of a resonance at ca. 9.1 ppm arising from *n*-butyl methylene carbons α to tin (*30*).

Acknowledgments

The authors acknowledge the very valuable contributions of K. B. Abbås, who provided samples of $LiAlH_4$-reduced PVC and the original polymers used to prepare them, and they thank S. M. Vincent, D. J. Freed, and J. P. Luongo for analytical assistance.

Literature Cited

1. Ayrey, G., Head, B. C., Poller, R. C., *J. Polym. Sci., Macromol. Rev.* (1974) **8**, 1.
2. Mayer, Z., *J. Macromol. Sci., Rev. Macromol. Chem.* (1974) **10**, 263.
3. David, C., *Compr. Chem. Kinet.* (1975) **14**, 78.
4. Braun, D., "Degradation and Stabilization of Polymers," G. Geuskens, Ed., pp. 23–41, Wiley, New York, 1975.
5. Abbås, K. B., Sörvik, E. M., *J. Appl. Polym. Sci.* (1976) **20**, 2395.
6. Starnes, Jr., W. H., ADV. CHEM. SER. (1978) **169**, 309.
7. Starnes, Jr., W. H., *Am. Chem. Soc., Div. Polym. Chem., Preprints* (1977) **18** (1), 493.
8. Frissell, W. J., "Encyclopedia of PVC," Vol. 1, L. I. Nass, Ed., pp. 257–269, Dekker, New York, 1976.
9. Cotman, Jr., J. D., *Ann. N. Y. Acad. Sci.* (1953) **57**, 417.
10. Cotman, Jr., J. D., *J. Am. Chem. Soc.* (1955) **77**, 2790.
11. Cotman, Jr., J. D., U. S. Patent **2,716,642** (1955).
12. Abbås, K. B., Bovey, F. A., Schilling, F. C., *Makromol. Chem., Suppl.* (1975) **1**, 227.
13. Rigo, A., Palma, G., Talamini, G., *Makromol. Chem.* (1972) **153**, 219.
14. Bovey, F. A., Abbås, K. B., Schilling, F. C., Starnes, Jr., W. H., *Macromolecules* (1975) **8**, 437.
15. Carrega, M., Bonnebat, C., Zednik, G., *Anal. Chem.* (1970) **42**, 1807.
16. Baijal, M. D., Wang, T. S., Diller, R. M., *J. Macromol. Sci., Chem.* (1970) **4**, 965.
17. Braun, D., Schurek, W., *Angew. Makromol. Chem.* (1969) **7**, 121.
18. Kuivila, H. G., *Synthesis* (1970) **2**, 499.
19. Kuivila, H. G., *Acc. Chem. Res.* (1968) **1**, 299.
20. Kuivila, H. G., *Adv. Organomet. Chem.* (1964) **1**, 47.
21. Carlsson, D. J., Ingold, K. U., *J. Am. Chem. Soc.* (1968) **90**, 7047.
22. Kuivila, H. G., Sommer, R., *J. Am. Chem. Soc.* (1967) **89**, 5616.
23. Ingold, K. U., *Free Radicals* (1973) **1**, 37.
24. Walling, C., Cioffari, A., *J. Am. Chem. Soc.* (1972) **94**, 6059.
25. Beckwith, A. L. J., Moad, G., *J. Chem. Soc., Chem. Commun.* (1974) 472.

26. Beckwith, A. L. J., Blair, I., Phillipou, G., *J. Am. Chem. Soc.* (1974) **96,** 1613.
27. Abbås, K. B., Sörvik, E. M., *J. Appl. Polym. Sci.* (1975) **19,** 2991.
28. Schilling, F. C., Bovey, F. A., Abbås, K. B., unpublished observations.
29. Starnes, Jr., W. H., Plitz, I. M., Schilling, F. C., Bovey, F. A., unpublished data.
30. Mitchell, T. N., *Org. Magn. Reson.* (1976) **8,** 34.

RECEIVED July 1, 1977.

Molecular Orbital Theory of Polyenes Implicated in the Dehydrochlorination of Poly(vinyl chloride)

II. Electronic Structures, Equilibrium Geometries, and Energetics of the Ground States of Polyenyl Cations and Neutral Polyenes

ROBERT C. HADDON and WILLIAM H. STARNES, JR.
Bell Laboratories, Murray Hill, NJ 07974

The geometries, energetics, and charge densities of linear polyenes and polyenyl cations are studied with ab initio SCF MO theory using the STO-3G and 4-31G basis sets; particular attention is given to the possible intermediacy of these conjugated species in the propagative dehydrochlorination of PVC. Polyenyl cations are found to retain C_{2v} *symmetry, with the positive charge being principally concentrated at the center of unsaturation. The conjugation energy of these cations is large at short chain lengths but asymptotically approaches a constant value at high degrees of unsaturation. Our results suggest, for high degrees of unsaturation, (i) the feasibility of the E1 dehydrochlorination mechanism in PVC, (ii) an explanation for the arrest of propagative dehydrochlorination, and (iii) a mechanism for the prevention of color with stabilizers.*

It is known that the principal pathway for nonoxidative thermal degradation of poly(vinyl chloride) (PVC) involves the evolution of HCl and the accumulation of unsaturation in the polymer chains (*1, 2, 3, 4*), as shown in Equation 1:

$$-(CH_2\text{–}CHCl)_n- \xrightarrow{\Delta} -(CH{=}CH)_n- + n\,HCl \qquad (1)$$

The mechanism of this process is not well understood, however, and a number of alternative possibilities have been suggested. One of the

0-8412-0381-4/78/33-169-333$07.50/1

more popular of these is the E1 or ionic mechanism (*2*) (Equation 2), which involves the intermediacy of a polyenyl cation:

$$-(CH{=}CH)_n\text{–}CHCl\text{–}CH_2\text{–} \rightleftharpoons -(CH)^+_{2n+1}\text{–}CH_2\text{–},\ Cl^- \tag{2a}$$

$$-(CH)^+_{2n+1}\text{–}CH_2\text{–},\ Cl^- \rightarrow -(CH{=}CH)_{n+1}\text{–} + HCl \tag{2b}$$

It is this mechanism which forms the primary focus of the present study, and we shall be particularly concerned with the effects of unsaturation on the electronic structures, equilibrium geometries, and energetics of the polyenyl cations and the polyenes derived therefrom. Some of the more practical aspects of the degradation and stabilization of PVC will then be discussed in terms of our results. The question of initiation by defect structures will be considered elsewhere, and we shall be mainly concerned in this work with the propagative or "unzipping" phase of the mechanism.

We wish to emphasize at the outset that we have not attempted to prove the correctness of the E1 mechanism for PVC dehydrochlorination. What we have tried to do, instead, is to use molecular orbital theory in order to (a) determine whether or not the E1 mechanism is a reasonable possibility, and to (b) elucidate theoretical principles which can serve as a guide for future experimental work in the PVC degradation/stabilization field. Some preliminary results of this study were published in Part 1 of this series (*5*).

Theoretical Method

Single determinant ab initio LCAO–SCF molecular orbital theory is used throughout this study (*6, 7*). Molecular geometries were optimized with the minimal STO–3G basis set (*8, 9*), and where possible the energy of the final structure was recalculated with the extended 4–31G basis set (*9, 10*). Such a procedure has been shown to provide a reasonable description of the structures and energies of neutral organic molecules (*11*) and carbocations (*12*). The potential surface for the concerted elimination of hydrogen chloride from ethyl chloride also has been studied successfully (*13*) with this technique.

Owing to the size of the molecules considered in this paper, it would have been impractical to optimize fully the geometry of each molecule. Instead we adopted a set of standard geometries (Table I) which were chosen to approximate the STO–3G equilibrium values. Thus, all bond angles and C–H bond lengths were fixed at standard values, but the C–C bond lengths were optimized fully in many cases by parabolic interpolation (*14*). Total energies obtained in the calculations are reported in Appendix 1, whereas the experimental heats of formation

Table I. Standard STO–3G Bond Lengths and Bond Angles[a]

Type	*Bond(s)*	*Bond Length (Å) or Bond Angle (degrees)*
Single Bond	C4–H	1.09
	C3–H	1.08
	C4–C4	1.54
	C4–C3	1.52
	C3–C3	1.485
	C4–Cl	1.77[b]
Double Bond	C3–C3	1.325
Bond Angle	–C4–	109.47
	–C3–	120

[a] Chosen to approximate STO–3G equilibrium values. In the notation X*n*, *n* refers to the connectivity (number of bonded neigbors) of the element X (*96*).
[b] Ref. *13*.

which were utilized for comparison purposes are collected in Appendix 2. The theoretical energies refer to fixed nuclei, so that comparisons actually should be made with heats of formation at 0°K with corrections for zero-point vibrations (*15, 16, 17*). However, the vibrational frequencies are not available for many of the molecules considered in this study, so we have followed standard practice (*12, 15, 16*) and utilized the 298°K ΔH_f values for comparison purposes [the ΔH_f's (0°K) are included where available to provide some indication of the errors incurred in this approximation]. Differences involving total energies or heats of formation are denoted by ΔE and are referred to as energy changes—such quantities are related most closely to changes in enthalpy.

Table II. Observed and Calculated Bond Lengths for Linear Polyenes

Linear Polyene	*Bond Length (Å)*				*Method*	*Ref.*
	1–2	*2–3*	*3–4*	*4–5*		
L_2	1.330				GPED	*97*
	1.306				STO–3G	*18*[a]
	1.309				STO–3G	This Work[b]
	1.316				4–31G	*18*[a]
L_4	1.337	1.483			GPED	*98*
	1.344	1.467			GPED	*99*
	1.341	1.463			GPED	*100*
	1.313	1.488			STO–3G	*19*[a, b]
L_6	1.337	1.458	1.368		GPED	*101*
	1.319	1.488	1.327		STO–3G	*102*[b]
L_8	1.319	1.487	1.329	1.483	STO–3G	This Work[b]

[a] Full geometry optimization.
[b] Partial geometry optimization (carbon–carbon bond lengths only), this work.

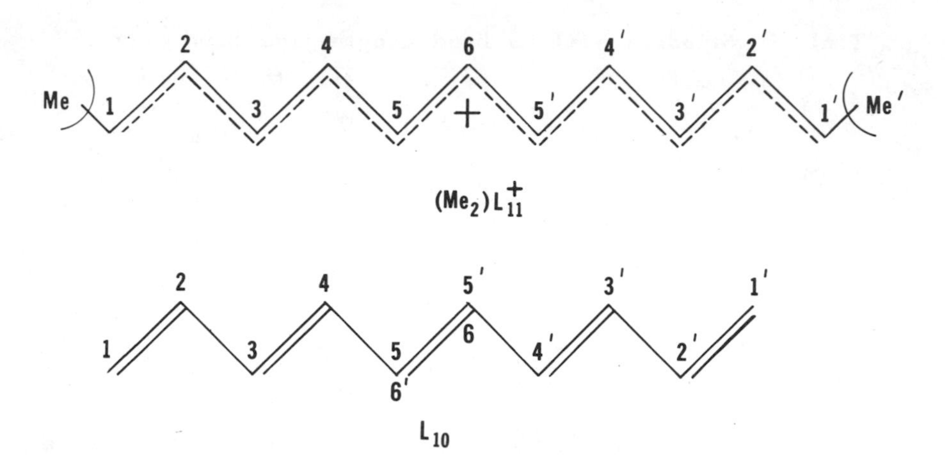

Figure 1. Numbering system and nomenclature used for the linear polyenyl cations and polyenes

Results

Polyenes. GEOMETRIES. The calculated bond lengths are compared with the available experimental values [which are from gas-phase electron diffraction (GPED) studies] in Table II (with the numbering scheme exemplified in Figure 1). It is reassuring to note that our partial geometry optimizations lead to C–C bond lengths in good agreement with those found in the fully optimized structures reported for ethylene (*18*) and butadiene (*19*). We find that the C–C double bond lengths are underestimated by 2–3%, whereas the conjugated single bonds are overestimated by about the same amount. Such behavior is well established now as a systematic error in STO–3G structure calculations (*20*). However, trends in the bond lengths of conjugated molecules usually are well described by this basis set (*20*), and while the slight tendency toward bond equalization at the center of the polyene chains is reproduced correctly by the theory, it is surprising to find that the magnitude of this effect is underestimated (*see* Appendix 3).

ENERGIES. The conjugation energies of the linear polyenes are assessed by means of the isodesmic reaction shown in Table III. As may be seen from Figure 2, the conjugation energy is positive (conjugation is favored) and shows a linear dependence on the number of double bond units in the polyene sequence. This behavior is in agreement with suggestions that energies of linear polyenes may be derived from a sum of characteristic (conjugated) single and double carbon–carbon bond energies (*21, 22*). The most recent experimental value (*23*) for the torsional barrier separating the trans and cis isomers of butadiene is 7.2

Table III. Conjugation Energies for Linear Polyenes (kcal/mol)

$$L_n + 2\left(\frac{n}{2} - 1\right) CH_4 \rightarrow \frac{n}{2} CH_2{=}CH_2 + \left(\frac{n}{2} - 1\right) CH_3CH_3 \qquad \Delta E_n$$

	ΔE_n			$(\Delta E_n - \Delta E_{n-2})$		
Polyene	*STO-3G*	*4-31G*	*Expt*[a]	*STO-3G*	*4-31G*	*Expt*[a]
L_2	0	0	0	—	—	—
L_4	12.3	11.2	14.2	12.3	11.2	14.2
L_6	25.2	21.1		12.9	9.9	
L_8	38.2			13.0		
L_{10}	51.2			13.0		

[a] Calculated from data in Appendix 2.

$$L_n + 2(n/2-1)\ CH_4 \longrightarrow n/2\ CH_2{=}CH_2 + (n/2-1)\ CH_3CH_3$$

e.g. [hexatriene structure] $+ 4\ CH_4 \longrightarrow 3\ CH_2{=}CH_2 + 2\ CH_3CH_3 \quad (n=6)$

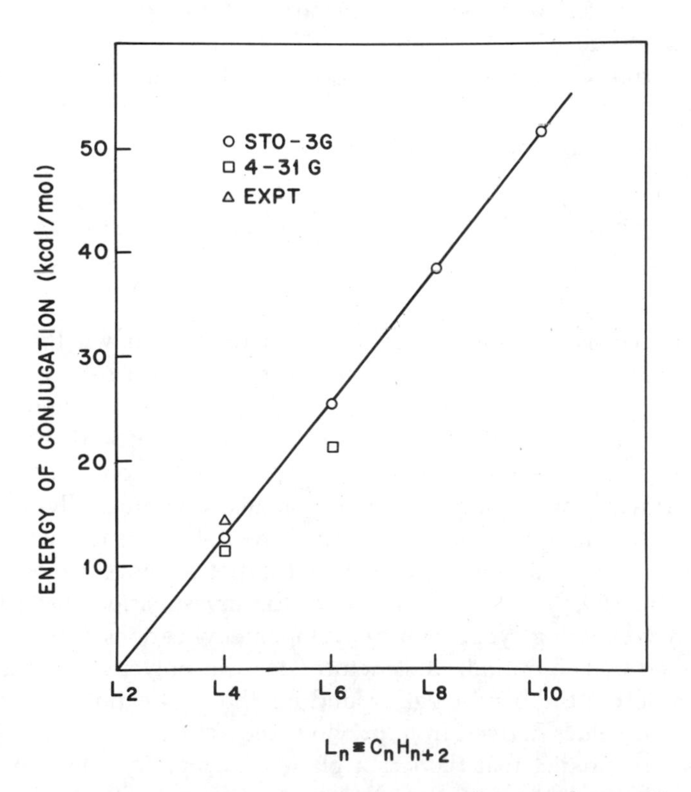

Figure 2. Energy of conjugation for linear polyenes ($L_n \equiv C_nH_{n+2}$)

kcal/mol [as against theoretical estimates (*24*) of 5.6 (STO–2G) and 6.7 (STO–3G) kcal/mol]. Therefore, it is reasonable to attribute the remainder of the butadiene conjugation energy (ca. 5 kcal/mol) to a change in carbon–carbon σ-bond energy with connectivity (*21*) [that is, from the conversion of $C(sp^3)–C(sp^3)$ to $C(sp^2)–C(sp^2)$]. This type of argument is presumably also valid for the polyenes with higher sequence lengths.

Polyenyl Cations. The first theoretical treatment of polyenyl systems was carried out by Coulson using the HMO method (*25*). This study (*25*) and others (*26–34*) which neglect both electron repulsion and configuration interaction necessarily lead to the prediction of bond equalization (apart from end effects) in polyenyl cations, radicals, and anions as well as in the neutral polyenes (*see* Appendix 3). However, results obtained upon introduction of configuration interaction (*35–47*) suggest that these molecules will be subject to a second-order Jahn–Teller distortion to a state of lower symmetry [this is entirely analogous to the Peierls instability in one-dimensional metals (*48*)]. Very long sequence lengths were suggested as being necessary for the onset of such a distortion (*39, 40, 43–45*). Most authors (*37–45*) have favored a B_2-type distortion (within C_{2v} symmetry) of the polyenyl framework, according to Equation 3:

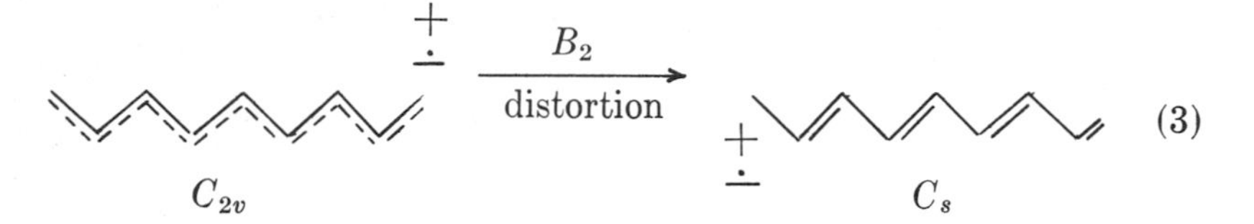

This point is reconsidered in Appendix 3, where it is shown that even within the simple HMO treatment, there are at least two serious candidates for the ground-state structures of polyenyl systems. There have been few calculations of sufficient sophistication to describe these molecules correctly (*12, 49–59*), and most of these calculations have been concerned primarily with the first member of this series: the allyl cation.

Most of the limited amount of experimental evidence which is available on the ground states of polyenyl cations comes from NMR solution studies (*60, 61, 62*), and even here the investigations have been hampered by competing cyclization to cyclopentenyl cations (*60, 61, 62*).

GEOMETRIES. The optimized structures for the polyenyl cations are shown in Table IV; the bond length found for the allyl cation is in good agreement with values derived from previous theoretical studies (*51–59*). Our calculations predict that there is a plane of symmetry which bisects the polyenyl chains at the central carbon atom, perpendicular to the path of conjugation. We find no evidence for the distortion shown in Equation 3,

Table IV. STO–3G Calculated Bond Lengths for Linear Polyenyl Cations

	Calculated Bond Lengths (Å)					
Cation	*1–2*	*2–3*	*3–4*	*4–5*	*5–6*	*Ref.*
L_3^+	1.385					57[a,b]
L_5^+	1.353	1.428				This Work[b]
L_7^+	1.339	1.448	1.391			This Work[b]
L_9^+	1.332	1.461	1.371	1.415		This Work[b]
L_{11}^+	1.33	1.47	1.34	1.45	1.391	This Work[b]

[a] Full geometry optimization.
[b] Partial geometry optimization (carbon–carbon bond lengths only), this work.

although our calculations, together with the analysis presented in Appendix 3, suggest that at very long chain lengths the distortion brought about by the positive charge will be only weakly pinned to the polyenyl skeleton and may have quite a high mobility that is limited only by the end effects which are dominant at the short sequence lengths treated in this study (Table IV). This point may be illustrated nicely by a consideration of the terminal 1–2 bond lengths. The theoretical bond distance for the allyl cation is 1.385 Å, precisely the bond length given for benzene by the STO–3G basis set (*63*), and it is apparent that this result must correspond to a fully delocalized partial double bond. On proceeding to longer chain lengths, the terminal bond is seen to contract until its length is close to the characteristic value found for conjugated double bonds in the neutral polyenes (Table II). The same trend (in a reverse sense) occurs with the penultimate 2–3 bond lengths, which expand with increasing sequence length toward the characteristic conjugated single-

Table V. STO–3G π-Orbital Atomic Charge Densities

	Charge Densities (−e)						
Cation	*1*	*2*	*3*	*4*	*5*	*6*	*Me*
L_1^+	+1.0						
L_3^+	+.533	−.066					
L_5^+	+.347	−.073	+.452				
L_7^+	+.247	−.066	+.364	−.092			
L_9^+	+.185	−.056	+.294	−.092	+.339		
L_{11}^+	+.138	−.047	+.227	−.087	+.319	−.098	
$Me_2L_1^+$	+.832						+.165
$Me_2L_3^+$	+.501	−.104					+.131
$Me_2L_5^+$	+.356	−.101	+.415				+.114
$Me_2L_7^+$	+.271	−.095	+.345	−.101			+.103
$Me_2L_9^+$	+.223	−.091	+.294	−.103	+.323		+.087
Me_2L_{2n}	———	———	−.02 – +.02	———	———	———	+.067

Figure 3. Idealization of resonance forms found for linear polyenyl cations

bond value. Although he did not report bond lengths, the rotational barriers calculated by Baird (*54*) for the lower polyenyl cations with the NDDO approximation lend themselves to the pattern of bond lengths shown in Table IV and the interpretation presented above.

CHARGE DENSITIES. The STO–3G π-orbital charge densities are shown in Table V, and it should be stressed that these are taken from a Mulliken population analysis of the π-orbitals alone—the total atomic charge densities are distributed more evenly owing to polarization of the σ-framework. It is interesting to note that the alternation in charge densities which was found in the early semiempirical SCF MO π-electron treatments (*49, 50, 64*) still persists at this higher theoretical level. In the zero differential overlap approximation, it can be shown that the sign of the charge densities at the even-numbered carbon atoms depends on a delicate balance between the electron repulsion integrals for this atom and its neighbors (*64*). Apparently the even-numbered carbon atoms prefer to achieve a charge of the opposite polarity to that of their nearest neighbors, thereby minimizing this electron repulsion, but at the expense of increased electron interaction at the odd-numbered carbon atoms, which must bear a greater charge density as a result of this polarization. Given the propensity of the Hartree–Fock SCF MO theory to overestimate the importance of ionic contributions to the wave function and the finding that configuration interaction reduces the importance of such terms (*65*), the final resolution of the charge alternation question will have to await the calculation of correlated wave functions for some of these systems.

Nevertheless, a definite pattern does emerge, and it is apparent that two types of polyenyl cation may be distinguished (Figure 3). The theoretical bond lengths and charge densities are in agreement with this interpretation, but it must be remembered that the diagrams of Figure 3 are merely a convenient representation of a rather complex bonding situation which involves a gradual transition from delocalized to localized bonding. A particularly clear distinction between the $4N+1$ and $4N+3$ types (Figure 3) is apparent from the charge densities at the central carbon atom, but both categories are seen to terminate with a double bond and a positively charged carbon atom. It is apparent that the lower charge densities at the terminal atoms at longer sequence lengths stem from a combination of two factors: (i) there are simply more atoms over which to spread the charge; (ii) the dispersal of charge over more distant atoms (from the center) is damped by the gradual increase in bond alternation. This latter point has been considered in a number of different contexts for the $4N+1$ polyenyl type (*5, 41, 42*). Adopting the nomenclature of Part 1 (*5*), it may be shown that while the *i*th coefficient of the nonbonding molecular orbital (which determines charge and spin densities) for the $4N+1$ polyenyl systems is given by

$$C_i = \left(-\frac{\beta_1}{\beta_2}\right)^i \left(\frac{\beta_2^2 - \beta_1^2}{\beta_2^2 + \beta_1^2}\right)^{1/2},$$

the corresponding quantity for the $4N+3$ types takes the form

$$C_i = \left(-\frac{\beta_1}{\beta_2}\right)^i \left(\frac{\beta_2^2 - \beta_1^2}{2\beta_2^2}\right)^{1/2},$$

at long chain lengths.

The gross trends shown by the charge densities of the α,ω-dimethylpolyenyl cations parallel those of the parent ions. The charge densities of the out-of-plane hydrogen atoms of the methyl groups drop rapidly with increasing unsaturation and asymptotically approach values that are characteristic of the neutral dimethylpolyenes.

There is an interesting reversal shown by the charge densities at the terminal carbon atoms: For L_1^+ and L_3^+ the charge density at this position is decreased upon α,ω-dimethyl substitution, whereas the reverse behavior is apparent in the case of L_5^+, L_7^+, and L_9^+. These results can be rationalized in the following manner. For L_1^+ and L_3^+ the localized representations (Figure 3) do not include double bonds at the termini, and the electron demand is such that these positions do experience a reduction in charge densities on substitution. On the other hand, for L_5^+, L_7^+, and L_9^+, where terminal double bonds are present in the localized representation, the main function of the methyl substituents is to raise the energy of these double bonds (thus improving their donor properties) by repulsion of filled shells (*66*). As a result, the terminal carbon atom bears an excess positive charge because of the increased availability of electron density from the terminal double bond. A further consequence

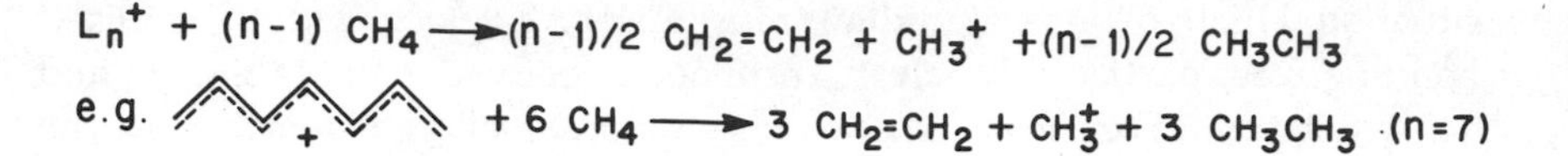

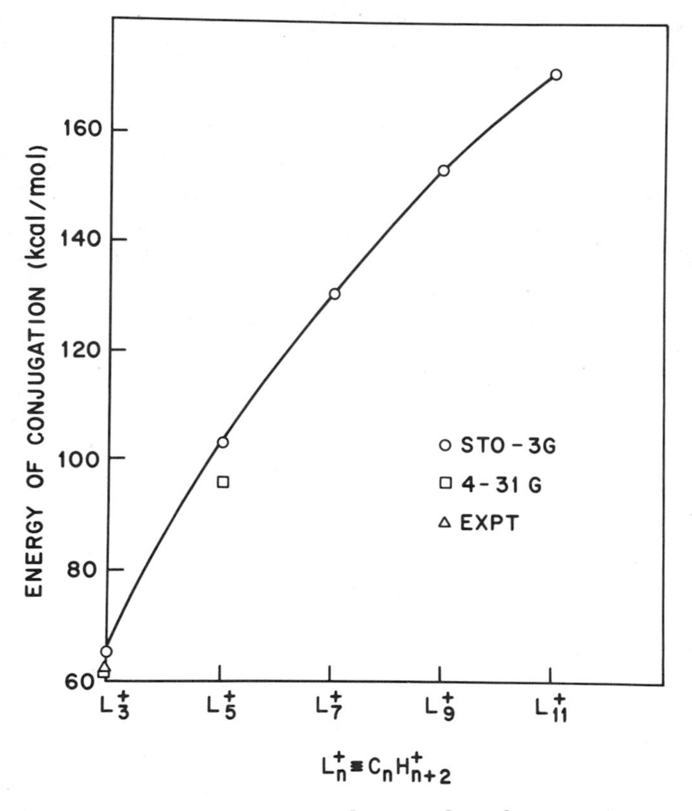

Figure 4. Energy of conjugation for linear polyenyl cations ($L_n^+ \equiv C_nH_{n+2}^+$)

of these effects is a more even distribution of positive charge throughout the polyenyl skeleton (charge densities at the center are reduced) owing to the presence of more polarizable terminal groups.

NMR chemical shifts of the α,ω-tetramethylpolyenyl cations (*60, 61, 62*) provide qualitative support for the trends in calculated charge densities noted above.

Energies. In the case of the polyenyl cations, it is logical to assess the conjugation energies by means of the isodesmic reaction shown as Equation 4 (in analogy to the neutral polyenes).

$$L_n^+ + (n-1)CH_4 \rightarrow$$
$$(n-1)/2\ CH_2{=\!=}CH_2 + CH_3^+ + (n-1)/2\ CH_3CH_3 \qquad (4)$$

Table VI. Conjugation Energies for Linear Polyenyl Cations (kcal/mol)

$$L_n^+ + (n-1)CH_4 \rightarrow$$
$$(n-1)/2CH_2{=\!=}CH_2 + CH_3^+ + (n-1)/2CH_3SH_3 \qquad \Delta E_n$$

Polyenyl Cation	ΔE_n			$(\Delta E_n - \Delta E_{n-2})$		
	STO–3G	*4–31G*	*Expt*[a]	*STO–3G*	*4–31G*	*Expt*[a]
L_3^+	64.7	61.5	62	64.7	61.5	62
L_5^+	102.8	95.4		38.1	34.0	
L_7^+	130.2			27.4		
L_9^+	152.3			22.1		
L_{11}^+	170.6			18.3		

[a] Calculated from data in Appendix 2.

It is not surprising to find that the energy change for this reaction is large and positive (Figure 4 and Table VI), as the reaction proceeds with loss of both conjugation and charge dispersal.

An alternative comparison is provided by Reaction 5, where the conjugation energy of the neutral polyene is included on the right-hand side of the equation.

$$L_n^+ + CH_4 \rightarrow MeL_{n-1} + CH_3^+ \qquad (5)$$

This hydride abstraction reaction provides a good index of the effect of unsaturation on the stability of carbocations (Figure 5 and Table VII). Previous work with the basis sets used here has shown that the present theory can provide a reasonable description of the energy changes in this type of reaction (*12*). The allyl cation has been treated already by Pople and coworkers (*57*) with these basis sets, and their results are very

Table VII. Conjugation Energies for Linear Polyenyl Cations (kcal/mol)

$$L_n^+ + CH_4 \rightarrow MeL_{n-1} + CH_3^+ \qquad \Delta E_n$$

Polyenyl Cation	ΔE_n			$(E_n - \Delta E_{n-2})$		
	STO–3G	*4–31G*	*Expt*[a]	*STO–3G*	*4–31G*	*Expt*[a]
L_3^+	62.4	59.8	57	62.4	59.8	57
L_5^+	88.1	81.7		25.7	21.9	
L_7^+	102.6			14.5		
L_9^+	111.6			9.0		
L_{11}^+	116.9			5.3		

[a] Calculated from data in Appendix 2.

similar to those obtained here; with *d*-functions on carbon (6–31G* basis set), the energy change for the allyl cation (Equation 5) is computed to be 53.0 kcal/mol (57). The experimental value for allyl and the 4–31G results for allyl and pentadienyl suggest that the energy change for Reaction 5 is overestimated by the STO–3G basis set to the extent of ca. 10–15%.

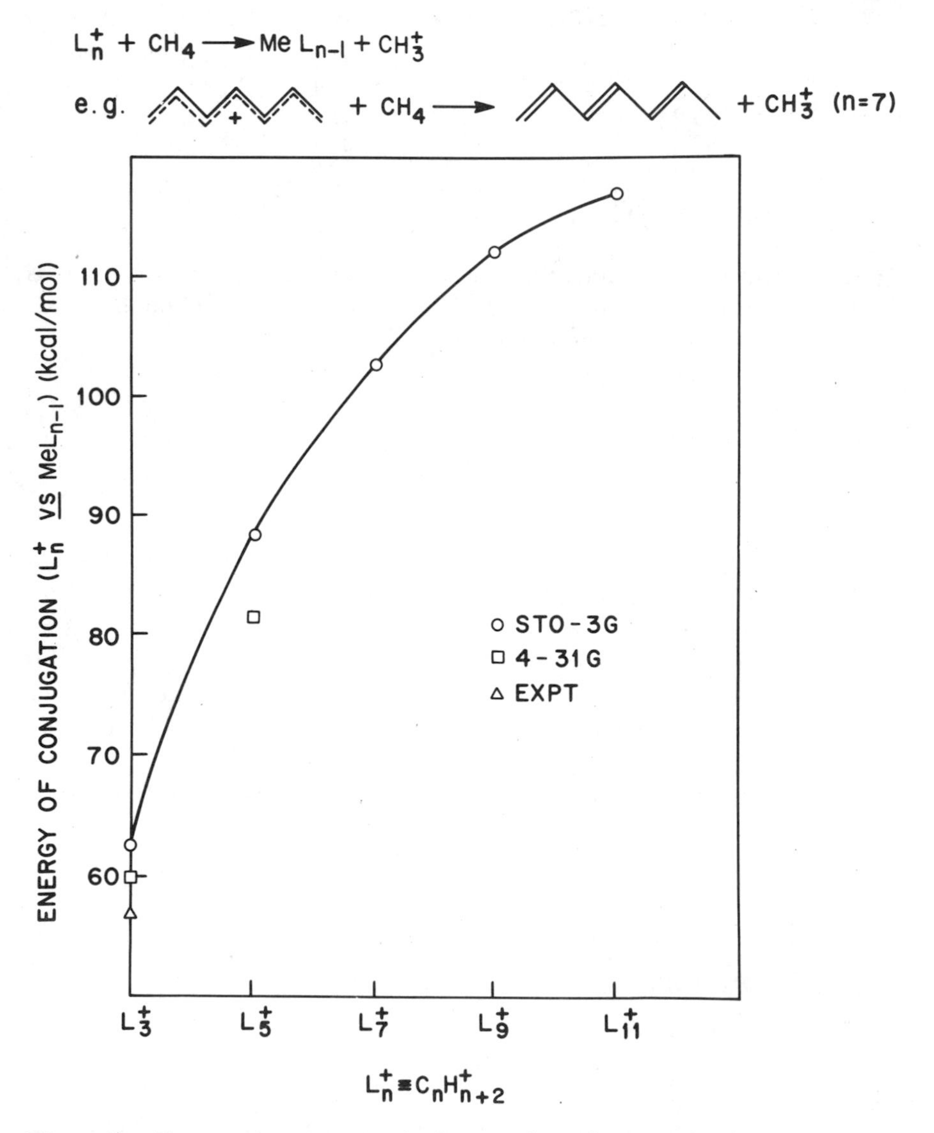

Figure 5. Energy of conjugation for linear polyenyl cations (L_n^+) with reference to 1-methylpolyenes (MeL_{n-1})

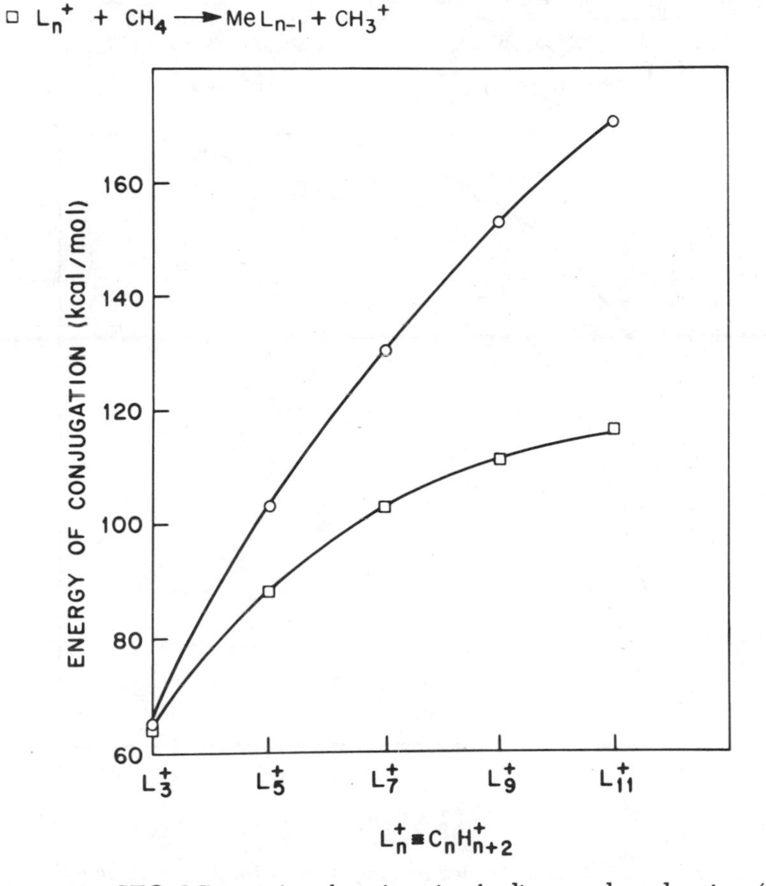

Figure 6. STO–3G energies of conjugation for linear polyenyl cations ($L_n^+ \equiv C_nH_{n+2}$)

It is clear from Figure 6, where the effects of unsaturation on conjugation energies are contrasted for the above reactions, that the energy change for the second process (Equation 5) is rapidly approaching a constant value, beyond which increases in the degree of unsaturation will have no effect. As may be seen from Figure 7 and Table VIII, the α,ω-dimethyl substitution pattern is very effective in stabilizing positive charge at shorter sequence lengths, but even here the conjugation energies (as defined in Equation 6) asymptotically approach about

$$\alpha,\omega\text{-}Me_2L_n^+ + CH_4 \rightarrow \alpha,\omega\text{-}MeEtL_{n-1} + CH_3^+ \quad (6)$$

the same value found previously in the absence of substitution. It may be seen from Figure 7 that we find α,ω-dimethyl substitution in polyenyl

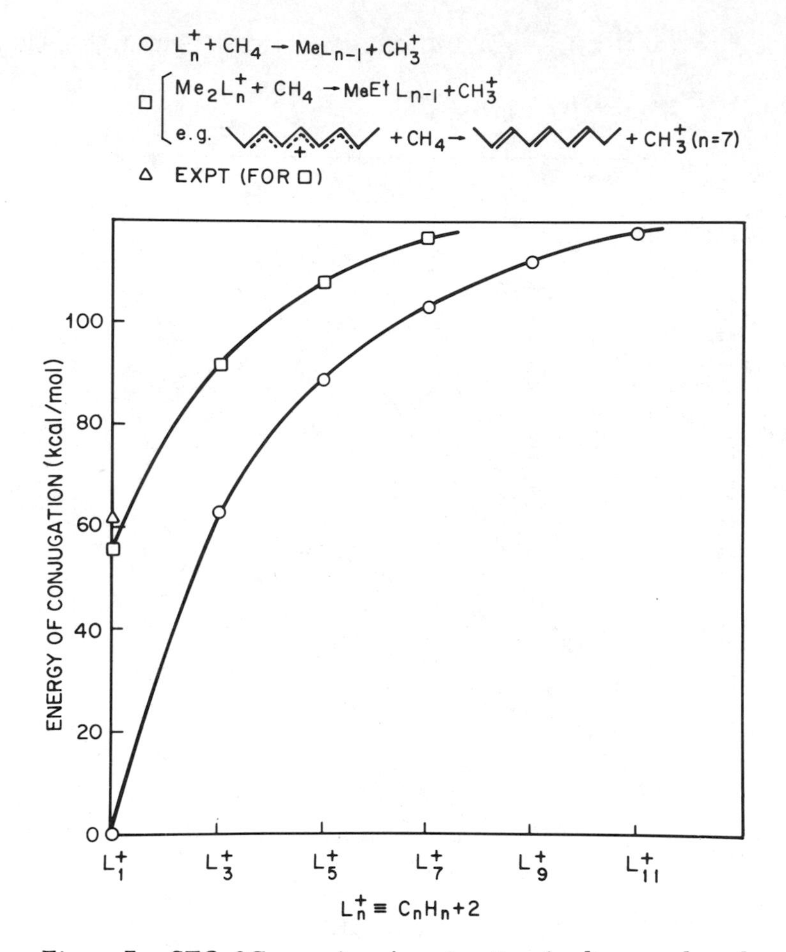

Figure 7. STO–3G energies of conjugation for linear polyenyl and α,ω-dimethylpolyenyl cations with reference to neutral polyenes

Table VIII. Conjugation Energies for Linear α,ω-Dimethyl Polyenyl Cations (kcal/mol)

$$\alpha,\omega\text{-}Me_2L_n^+ + CH_4 \rightarrow \alpha,\omega\text{-}MeEtL_{n-1} + CH_3^+ \qquad \Delta E_n$$

Polyenyl Cation	ΔE_n			$(\Delta E_n - \Delta E_{n-2})$	
	STO–3G	*4–31G*	*Expt*[a]	*STO–3G*	*4–31G*
$Me_2L_1^+$	53.7	50.9	61	—	—
$Me_2L_3^+$	91.7	86.8		38.0	35.9
$Me_2L_5^+$	107.4			15.7	
$Me_2L_7^+$	116.5			9.1	

[a] Calculated from data in Appendix 2.

cations to be about as effective as the addition of one double bond unit in the stabilization of positive charge, in agreement with previous experimental (*60*) and theoretical studies (*12, 57, 67, 68*).

The above pattern is in agreement with our analysis of the molecular structures and charge densities of the unsubstituted polyenyl cations (previous sections).

The positive charge acts much like a defect site in the polyene, and all of the properties of the system respond to its presence. The modifications are particularly noticeable at the locus of maximum positive charge, away from which the effects taper off; although each property appears to respond at a slightly different rate. We find that the effect of the positive charge on properties (at long chain length) decreases in the order: charge densities > conjugation energies > molecular structure.

Discussion

Heterolytic Gas-Phase Dissociation Energies. Given the nature of the first step of the E1 mechanism proposed for the dehydrochlorination reaction of PVC (Equation 2a), it would have been particularly useful to examine directly the effect of unsaturation on the energy for heterolytic cleavage of the C–Cl bond (Equation 7).

$$R\text{–}Cl \rightarrow R^{+} + Cl^{-} \qquad (7)$$

Experimental and theoretical results for the energy of this reaction are shown in Table IX for some representative cases. It may be seen that the STO–3G basis set provides a rather poor description of this process. This is to be expected, since Equation 7 is very far from being an isodesmic reaction, and the minimal basis set is fully saturated by the chloride ion. The values given by the 4–31G basis set are considerably better, but it is apparent that the theoretical description of Equation 7 is vastly inferior to the treatment of isodesmic reactions presented previously. In a subsequent section we show that these latter results may be used indirectly to provide information on Reaction 7. It is clear from the results presented in Table IX, however, that the energy for heterolytic cleavage of the C–Cl bond in the gas phase is enormous in comparison with the energies associated normally with chemical reactions, although as shown below this picture is modified drastically in the solution phase.

Much of the energy associated with Reaction 7 comes from the work required to dissociate opposite charges to infinity: that is, the energy required to take the components from ion pairs to free ions. In fact, this latter process is not necessarily a prerequisite for the occurrence of Reaction 2b and appears unlikely under the prevailing conditions of

Table IX. Energy of Heterolytic Gas-Phase Dissociation of Organic Chlorides (kcal/mol)

$$RCl \rightarrow R^+ + Cl^-$$

Group (R)	*Expt*[a]	*STO–3G*	*4–31G*
CH_3	197.3	290.2	213.6
CH_3CH_2	162.4	260.9	186.8
$CH_2{=}CHCH_2$	142.9	230.0	155.5

[a] Calculated from data in Appendix 2.

poly(vinyl chloride) degradation (*see* below). However, the gas-phase energies of organic ion pairs are difficult to estimate, and from the theoretical standpoint, optimization of a C–Cl bond leads automatically to the structure of the covalent chlorocarbon. [Where possible the following nomenclature has been utilized: tropenyl chloride (covalent species) and tropenylium chloride (ionic species) (*69*). However, the former term is used in generic context.] Tropenyl chloride, however, offers a unique opportunity for this type of calculation, since in this case we may idealize an ion pair by constraining the chlorine atom to lie on the sevenfold axis of the ring (*70*). The results of such calculations are shown in Figure 8. The geometry of the covalent chloride was estimated from experimental structure determinations of cyclohepta-

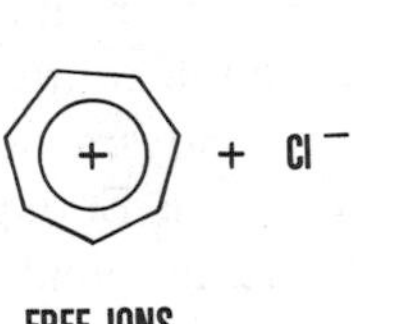

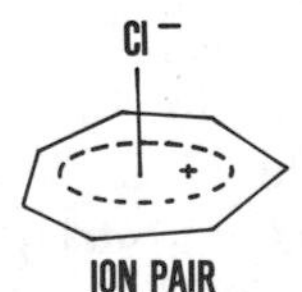

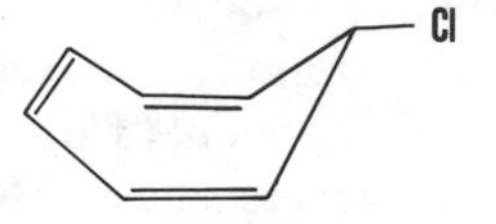

Figure 8. STO–3G energies for tropenyl chloride

triene derivatives (*71, 72*). The planar tropenylium cation ring was optimized independently in D_{7h} symmetry to give a calculated C–C bond length of 1.397 Å for the free ion. Optimization of the apical ring-center-to-Cl distance in the ion pair (with the ring geometry fixed) led to a value of 2.780 Å (C–Cl distances of 3.214 Å). The electronic structure of this species conforms well with the notion of an ion pair—the Mulliken population analysis finds 18.000 electrons on chlorine, with most of the positive charge located in the π-system of the tropenylium ring. Although Table IX shows that the energies cannot be trusted in detail, the results presented in Figure 8 do indicate that most of the energy involved in dissociating tropenyl chloride in the gas phase is lost in the electrostatic work required to separate an ion pair into its components.

Solution-Phase Ionization of Tropenyl Chloride and Triphenylmethyl Chloride. Insofar as thermal degradation is concerned, the critical phase in PVC processing occurs in the melt, which may reach temperatures in the vicinity of 200°C (*73*) over a period of some minutes. The dielectric constant of PVC (3–4 at 25°C, 11–12 at 140°C) (*74*) is comparable with that of small chlorocarbons (*75*). Of the organic chlorides whose ionization has been studied in such solvents, tropenyl chloride and triphenylmethyl chloride seem to have received the most attention.

TRIPHENYLMETHYL CHLORIDE. The ionization of triphenylmethyl chloride (Ph_3CCl) in 1,2-dichloroethane (dielectric constant of 10.4 at 25°C) (*75*) has not been measured directly, but an extrapolation from experimental data (*76, 77*) allows the following estimates:

$$Ph_3CCl \overset{K_1}{\rightleftharpoons} Ph_3C^+, Cl^- \text{ (ion pair)} \qquad K_1 = 6 \times 10^{-5} \tag{8}$$

$$Ph_3CCl \overset{K_2}{\rightleftharpoons} Ph_3C^+ + Cl^- \text{ (free ions)} \qquad K_2 = 1 \times 10^{-10} \tag{9}$$

In other words, the dissociation of Ph_3CCl in a solvent of similar dielectric constant to PVC occurs quite readily at room temperature and, providing the equilibria are rapidly established, might seem to provide some analogy for Reaction 2a. Unfortunately, little thermodynamic data are available to provide the necessary calibration, although the following energy changes can be estimated from data in Appendix 2:

$$Ph_3CCl \text{ (solid)} \rightarrow Ph_3C^+(g) + Cl^-(g) \qquad \Delta E = 138 \text{ kcal/mol} \tag{10}$$

$$Ph_3C^+(g) + CH_4(g) \rightarrow Ph_3CH \text{ (solid)} + CH_3^+(g) \qquad \Delta E = 77 \text{ kcal/mol} \tag{11}$$

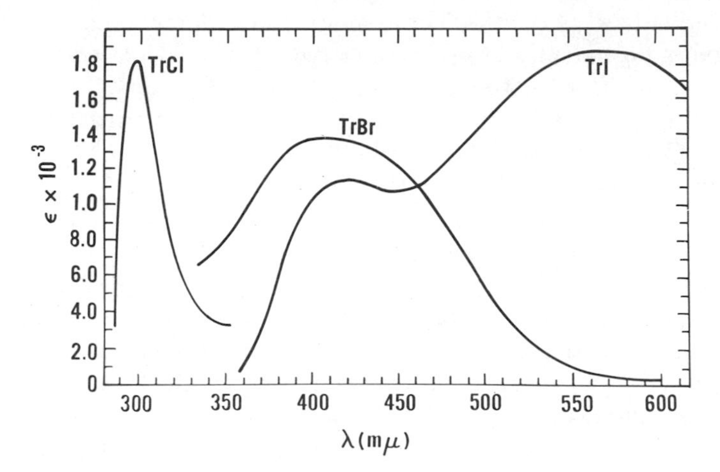

Figure 9. Charge–transfer absorption bands of tropenylium halides in methylene chloride (115)

Owing to the phase changes, the energies of Reactions 10 and 11 represent upper and lower limits, respectively, for the occurrence of these reactions (wholly) in the gas phase. Indeed, the true gas-phase energy changes might differ from these values by as much as 30 kcal/mol.

TROPENYL CHLORIDE. Tropenyl chloride is thought to exist completely in the form of an ion pair in methylene chloride (*69*) (dielectric constant of 8.9 at 25°C) (*75*). Charge-transfer spectra of some tropenyl halides (*69*) are shown in Figure 9. In a solvent such as acetonitrile (dielectric constant of 37.5 at 20°C) (*75*), these absorptions are absent, and the tropenyl halides are assumed to be dissociated into solvent-separated ion pairs which are in equilibrium with free ions (as determined by conductance measurements) (*69*). Thermodynamic data (Appendix 2) allow the following gas-phase energy change (between tropenylium cation and cycloheptatriene) to be estimated:

$$C_7H_7^+ + CH_4 \rightarrow C_7H_8 + CH_3^+ \qquad \Delta E = 113 \text{ kcal/mol} \qquad (12)$$

Now this process is analogous to Reactions 5 (Table VII and Figure 5) and 6 (Table VIII and Figure 7). If we assume that the C–Cl heterolytic bond dissociation energies in tropenyl chloride and the polyenyl chlorides parallel the C–H bond strengths toward hydride abstraction, then it is reasonable to use the above results to extrapolate the behavior of the polyenyl chlorides into the solution phase. On this basis it seems clear that the establishment of equilibrium in Reaction 2a in solution is eminently reasonable—bearing in mind that Reaction 1 occurs in the melt and that we require only a kinetic amount of the ionized chloride for the occurrence of Reaction 2b. It is more difficult to establish the minimum value of n (Equation 2) for which the E1 mechanism will become operative, although on the basis of a comparison between Table VIII and ΔE for Equation 12, $n = 1$ or 2 would not seem to be unreasonable. Perhaps in concluding this section we should point out that there are certain objections to using tropenyl chloride as a model for the polyenyl chlorides. The most important of these is the unique geometry available to the ion pair, with the chloride ion lying on the sevenfold axis of the tropenylium ring: a situation which might be expected to provide an ideal coulombic interaction. Nevertheless, as pointed out above, tropenyl chloride does indeed form solvent-separated ion pairs and free ions in organic solvents, so it must be assumed that the electrostatic attraction is not enormously important in a comparative sense.

Ion-Pair Mechanism for the Thermal Dehydrochlorination of Poly(vinyl chloride). Clearly there is nothing in our theoretical analysis of the electronic structures and energetics of the polyenyl cations which would preclude their participation in Reaction 2 as either transition states or intermediates. However, it remains to establish the values of n at which the pathway of Reaction 2 will be competitive with other reaction processes. On the basis of our comparison (*see* above) with the ionization of tropenyl chloride, it is apparent that kinetic amounts of contact and perhaps even solvent-separated ion pairs will be available at n values (Equation 2) of about 1 or 2. However, in view of our previous arguments, it seems unlikely that free ions will ever achieve a high concentration in PVC. Given the fact that proton abstraction by chloride ion (Equation 2b) is necessary to complete the elimination

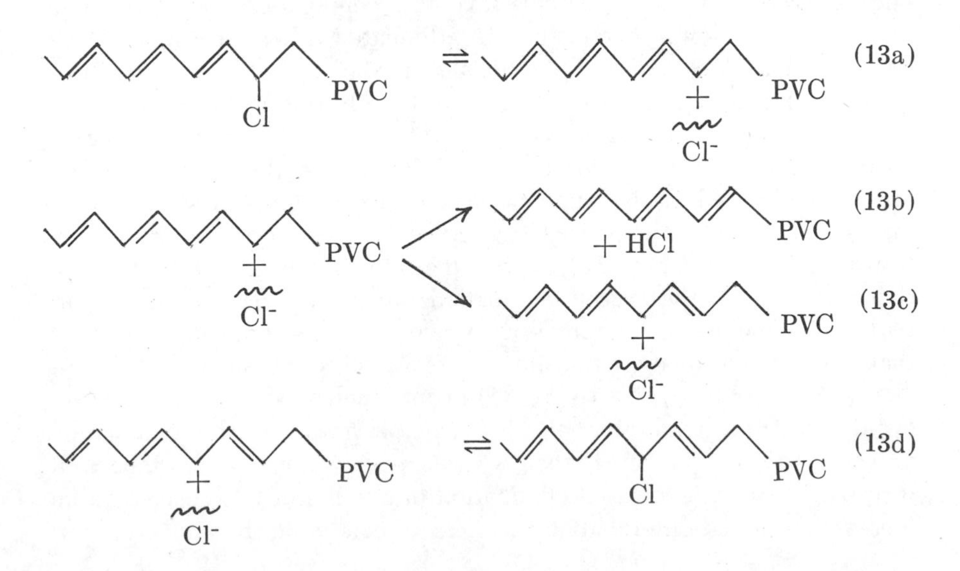

reaction, the high reactivity (toward other substrates) associated normally with free ions may be less important in this particular instance. Accordingly, in our reconsideration of the ionic dehydrochlorination mechanism (Equation 13), we have formulated the transition states/intermediates as ion pairs of unspecified nature. We also note here that the literature contains experimental evidence which has been interpreted in terms of the existence of carbenium chloride ion pairs in the thermally degraded polymer (*78, 79, 80*), together with prior speculation on ion pair mechanisms for the dehydrochlorination reaction per se (*2, 78, 80*).

Our results clearly suggest that the forward step of Reaction 13a is favored by unsaturation. In Reaction 13b we show the normal completion step in an E1 elimination reaction, whereas in Reaction 13c we take cognizance of our finding that charge is better established at the center of unsaturation rather than at the termini. With increasing stabilization of the ion pair by unsaturation, migration of the ion pair toward the center of the chain may become a competitive process; and, of course, once the chloride ion is no longer in the neighborhood of a terminal methylene group, proton abstraction (Reaction 13b) will become increasingly unlikely. As we noted previously, the charge density at the methylene protons falls quite quickly with increasing unsaturation, and the concomitant kinetic trend is obviously unfavorable for elimination. This effect seems capable of providing at least a partial explanation of the finding that polyene sequences apparently do not grow beyond a certain length in thermally degrading PVC (*1, 2, 3*). Previous explanations of this observation which have relied on resonance destabilization of the product polyenes (*2, 3, 81–84*) clearly are discounted by our finding (Table III, Figure 2) of a favorable conjugation energy in these materials that shows no tendency to attenuate with increasing polyene size. Furthermore, unlike the rationale of Marks and coworkers (*80*), which was couched in terms of resonance stabilization of the polyenyl cations (a concept supported by our results), our analysis points toward a kinetic effect. In other words, our results suggest that if the polyenes are indeed formed via E1 elimination, then cessation of polyene growth may be related to an enhanced free energy of activation for the proton transfer step (Equation 2b), rather than to any unfavorable thermodynamic trend in the overall elimination process (Equation 2). Admittedly, this conclusion may represent an oversimplification, since it is clear that polyene termination may involve side reactions such as Friedel–Crafts alkylation (*80*), inter- (*84, 85*) or intramolecular (*86*) Diels–Alder cyclization, and readdition of HCl (i.e., reversal of the elimination process) (*87, 88, 89*). Nevertheless, even under these circumstances, it is apparent that a decreased elimination rate will still be reflected in an increased termination probability (unless, of course, in the unlikely event

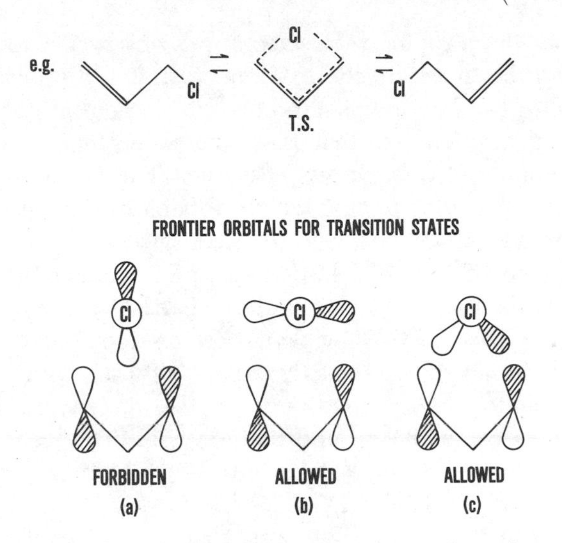

Figure 10. Concerted migration of polyenylic chloride

that the rates of the terminating side reactions are reduced concurrently to a comparable extent).

It is interesting to note that it may also be rather difficult to fix the position of the chlorine substituent in the covalent PVC polyene structures because there are concerted pathways available for [1, 3] sigmatropic halogen shifts. The available transition states for this type of process are shown in Figure 10, and these possibilities are quite general insofar as the degree of unsaturation is concerned. Preliminary calculations suggest that representation (b) of Figure 10 constitutes the best description of the important orbitals. Gas-phase studies of this type of rearrangement (*90*) led to an activation energy of about 40 kcal/mol for the conversion of γ,γ-dimethylallyl chloride to the α,α-isomer (*90*).

Implications of the Ion-Pair Dehydrochlorination Mechanism for Poly(vinyl chloride) Stabilization. The best thermal stabilizers for PVC are believed to function (at least in part) by replacing labile halogens with stabilizer ligands that are less susceptible to removal during thermolysis (*1, 4*). Therefore, if our conclusion concerning ion pair migration to the centers of long polyenyl cations is correct, it is clear that the most effective color-preventing stabilizers will be substances which can trap these cations rapidly in reactions whose regioselectivities are controlled kinetically. Under such circumstances, ligand incorporation will tend to bisect the polyene chromophores with saturated –CHX– units (X = a stabilizer ligand) and will thus tend to prevent the appearance of color in a very efficient manner [in analogy with a speculation of Marks, et al. (*80*) concerning the structures of polyenyl cation, SnX_6^{-2} ion pairs]. On the other hand, the trapping regioselectivities observed with less reactive stabilizers will be more likely to depend upon the overall reaction thermodynamics (i.e., upon the stabilities of the neutral polyenes formed in the trapping reaction), so that these stabilizers may tend to react at the cation termini and thus be less effective. The overall situation is, in fact, somewhat analogous to that which obtains in the well-known addition of HBr to 1,3-butadiene (*91*). In that system (*91*), kinetic control affords mostly the 1,2-type of addition since Br^- reacts most rapidly with the cationic carbon bearing the highest partial positive charge (*92*) (Equation 14a) [our STO–3G calculation on the 1-methylallyl cation finds π-orbital atomic charge densities of $+ .546$ (C^1) and $+ .480$ (C_3)]. Conversely, reaction under conditions of thermodynamic control (i.e., under conditions conducive to addition reversibility) leads to the more stable 1,4-addition product (Equation 14b). Similar considerations should apply to other systems in which conjugated (allylic-type) carbocations react with nucleophiles at rates that are not controlled by rates of diffusion.

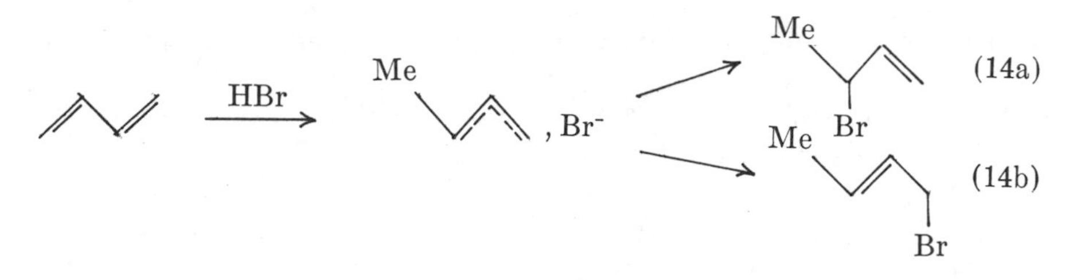

At the present time, it is not known whether any of the common PVC stabilizers do, indeed, react preferentially at the centers of polyenyl cations. However, it would seem that this possibility should be amenable to experimental tests.

Appendix 1

Total Energies

Designation	*Formula*	*Geometry*	*Basis*	*Energy (a.u.)*	*Ref.*
	CH_4	STO-3G	STO-3G	−39.72686	*103*[a]
		STO-3G	4-31G	−40.13976	*103*[a]
	CH_3CH_3	STO-3G	STO-3G	−78.30618	*18*[a]
		STO-3G	4-31G	−79.11582	*18*[a]
L_2	C_2H_4	Std	STO-3G	−77.07228	
		STO-3G	STO-3G	−77.07265	[b]
		STO-3G	STO-3G	−77.07396	*18*[a]
		Std	4-31G	−77.92100	
		STO-3G	4-31G	−77.92088	[b]
		STO-3G	4-31G	−77.92188	*18*[a]

Designation	*Formula*	*Geometry*	*Basis*	*Energy (a.u.)*	*Ref.*
L_4	C_4H_6	Std	STO-3G	−153.01658	
		STO-3G	STO-3G	−153.02036	[b]
		STO-3G	STO-3G	−153.02036	*19*
		Std	4-31G	−154.69620	
		STO-3G	4-31G	−154.69568	[b]
		STO-3G	4-31G	−154.69906	*19*
L_6	C_6H_8	Std	STO-3G	−228.96189	
		STO-3G	STO-3G	−228.96198	[b]
		Std	4-31G	−231.47189	
		STO-3G	4-31G	−231.47160	[b]
L_8	C_8H_{10}	Std	STO-3G	−304.90737	
		STO-3G	STO-3G	−304.90753	[b]
L_{10}	$C_{10}H_{12}$	Std	STO-3G	−380.85290	
α-MeL_2	C_3H_6	Std	STO-3G	−115.65536	
		STO-3G	STO-3G	−115.66030	*20*
		Std	4-31G	−116.89975	
		STO-3G	4-31G	−116.90459	*20*
α-MeL_4	C_5H_8	Std	STO-3G	−191.59976	
		Std	4-31G	−193.67463	
α-MeL_6	C_7H_{10}	Std	STO-3G	−267.54508	
α-MeL_8	C_9H_{12}	Std	STO-3G	−343.49057	
α-MeL_{10}	$C_{11}H_{14}$	Std	STO-3G	−419.43610	
$CH_3CH_2CH_3$	C_3H_8	Std	STO-3G	−116.88513	
		STO-3G	STO-3G	−116.88642	*20*
		Std	4-31G	−118.09212	
		STO-3G	4-31G	−118.09360	*20*
α,β-MeEtL_2	C_5H_{10}	Std	STO-3G	−192.81671	
		Std	4-31G	−194.85365	
α,δ-MeEtL_4	C_7H_{12}	Std	STO-3G	−268.76160	
α,ζ-MeEtL_6	C_9H_{14}	Std	STO-3G	−344.70704	
L_1^+	CH_3^+	STO-3G	STO-3G	−38.77948	*103*[a]
		STO-3G	4-31G	−39.17129	*103*[a]
L_3^+	$C_3H_5^+$	STO-3G	STO-3G	−114.80735	[b]
		STO-3G	STO-3G	−114.80953	*57*
		STO-3G	4-31G	−116.02652	[b]
		STO-3G	4-31G	−116.02511	*57*
L_5^+	$C_5H_7^+$	STO-3G	STO-3G	−190.79279	[b]
		STO-3G	4-31G	−192.83796	[b]
L_7^+	$C_7H_9^+$	STO-3G	STO-3G	−266.76116	[b]
L_9^+	$C_9H_{11}^+$	STO-3G	STO-3G	−342.72110	[b]
L_{11}^+	$C_{11}H_{13}^+$	STO-3G	STO-3G	−418.67493	[b]
$Me_2L_1^+$	$C_3H_7^+$	Std	STO-3G	−116.02328	[c]
		STO-3G	STO-3G	−116.02765	*67*
		Std	4-31G	−117.20481	[c]
		STO-3G	4-31G	−117.20864	*67*

Designation	*Formula*	*Geometry*	*Basis*	*Energy (a.u.)*	*Ref.*
$Me_2L_3^+$	$C_5H_9^+$	Std	STO-3G	−192.01547	[c]
		STO-3G	STO-3G	−192.01548	[b]
		Std	4-31G	−194.02349	[c]
		STO-3G	4-31G	−194.02353	[b]
$Me_2L_5^+$	$C_7H_{11}^+$	Std	STO-3G	−267.98518	[c]
$Me_2L_7^+$	$C_9H_{13}^+$	Std	STO-3G	−343.94538	[c]
	$C_7H_7^+$	STO-3G	STO-3G	−265.66771	*104*[a]
		STO-3G	STO-3G	−265.66789	[b]

[a] Full geometry optimization.
[b] Partial geometry optimization (carbon–carbon bond lengths only).
[c] Carbon–carbon partial double bonds taken from corresponding unsubstituted polyenyl cation optimization.

Appendix 2

Experimental Heats of Formation (kcal/mol)

Molecule	ΔH_f*(298°K)*	*Ref.*	ΔH_f*(0°K)*	*Ref.*
CH_4	−17.88	*105*	−15.970	*106*
CH_3CH_3	−20.24	*105*	−16.523	*106*
$CH_3CH_2CH_3$	−24.8	*107*	−19.482	*106*
L_2	12.49	*105*	14.515	*106*
L_4	26.33	*108*		
MeL_2	4.88	*107*	8.468	*106*
L_1^+	260	*105*		
MeL_1^+	219	*105*		
$Me_2L_1^+$	192	*109*		
L_3^+	226	*110*		
$C_7H_7^+$	209	*111*[a]		
C_7H_8	44.5	*108*[b]		
CH_3Cl	−20.63	*112*		
CH_3CH_2Cl	−26.7	*112*		
$CH_2{=}CHCH_2Cl$	−0.15	*112*		
Ph_3C^+	240	*113*		
Ph_3CH (s)	39.2	*112*		
Ph_3CCl (s)	43.4	*112*		
Cl^-	−83.2	*114*		

[a] Tropenylium cation.
[b] 1,3,5-Cycloheptatriene.

Appendix 3

It is well known that the ground-state molecular structures of π-electron systems are usually well described by molecular orbital theory, even at the one-electron level (*93*). In the Hückel Molecular Orbital (HMO) Theory, the relationship between structure and theory arises from the calculated bond orders, which have been found to show a good correlation with experimentally determined conjugated C–C bond lengths (*93*). It was this dependence which led Lennard–Jones (*26, 27*), Penney (*28*), and Coulson (*25*) to suggest that the C–C bond lengths at the center of a long polyene chain would tend to a common value (bond equalization). Subsequently it was realized that configuration interaction becomes very important in such a situation, as the energy gap between the occupied and vacant orbitals vanishes in this approximation (*35–47*). Inclusion of configuration interaction reestablishes the energy gap, with the introduction of bond alternation throughout the polyene, and it is now known that bond equalization does not occur at the center of polyenes (*94*).

The situation with polyenyl systems is less well defined. In the case of the linear polyenes, it was the experimental observation that the wavelength of maximum absorption at long sequence length tends toward a finite limit, instead of increasing steadily, which first suggested that the extant ideas about bond equalization might be incorrect (*35*). At the present time, electronic spectra have been determined experimentally (*61, 95*) for various α,ω-substituted polyenyl cations with up to 13 carbon atoms in the path of conjugation. The absorption maxima among homologous series show a linear dependence on the number of double bonds with no sign of convergence to a finite limit (*61, 95*). This result generally has been interpreted in favor of a bond-equalized structure.

Theoretical treatments of polyenyl systems have often focused on radicals, and it generally has been assumed that the distortion coordinate will correspond to that depicted in Equation 3 (*25–30*). In addition, it has been suggested that very long sequence lengths would be necessary for the onset of a distortion (*37–45*).

In this section we briefly reconsider polyenyl systems with the HMO method (the present treatment is valid for cation, radical, and anion) in order to delineate the candidate distortions. As previously noted, the bond lengths of conjugated π-electron systems usually are dealt with via the bond order (P_{ij}). Where it seems that configuration interaction may be important (pseudo Jahn–Teller effect, second-order bond fixation), this level of theory is patched up by a consideration of the bond–bond polarizability matrix ($\pi_{ij,kl}$) (*47*). The eigenvalues of this matrix provide a measure of the energy gain accompanying distortions, whereas the eigenvectors indicate the symmetry (*47*). The quantities discussed above are collected in Table X for L_6 and $L_7^{+,\cdot,-}$; and it may be seen there that while the symmetry of L_6 in the first-order approximation also is preserved by the eigenvector of the highest eigenvalue, this is not true in the case of $L_7^{+,\cdot,-}$. In fact, A_g in C_{2h} and B_2 in C_{2v} both correspond in simple bond length alternation (Equation 3), whereas A_1 in C_{2v} corre-

Table X. First- and Second-Order Bond Fixation in Hexatriene (L_6) and Heptatrienyl ($L_7^{+,\cdot,-}$)

	L_6			$(L_7^{+,\cdot,-})$		
	Bond Order	*Eigenvectors of* $\pi_{ij,kl}$		*Bond Order*	*Eigenvectors of* $\pi_{ij,kl}$	
Bond	P_{ij}	*1*	*2*	P_{ij}	*1*	*2*
1–2	.8711	.3143	.3431	.8155	.3039	.4138
2–3	.4834	−.5439	−.6183	.5449	−.4115	−.5556
3–4	.7849	.4592	0.0	.6533	.4882	.1418
4–5	.4834	−.5439	.6183	.6533	−.4882	.1418
5–6	.8711	.3143	−.3431	.5449	.4115	−.5556
6–7	—	—	—	.8155	−.3039	.4138
Eigenvalues of $\pi_{ij,kl}$	—	.7244	.4066	—	1.0617	.5466
Symmetry (Point Group)	A_g (C_{2h})	A_g (C_{2h})	B_u (C_{2h})	A_1 (C_{2v})	B_2 (C_{2v})	A_1 (C_{2v})

sponds to the structure shown in Figure 3. Therefore, it is apparent that the structural situation found for the polyenyl systems is far more complicated than that of the neutral polyenes. If the C_{2v} symmetry of the polyenyl systems is to be preserved, then the second highest eigenvalue of $\pi_{ij,kl}$ would have to be the controlling distortion, a requirement which fies in the face of previous experience where the highest eigenvalue is dominant (*47*). Nevertheless, this problem must be addressed with more powerful techniques than the HMO theory if these questions are to be resolved satisfactorily. As we already have remarked (*see* Results section), our ab initio STO–3G calculations find no evidence for the B_2 distortion in polyenyl cations at short chain lengths. However, it is appropriate to point out that a B_2 distortion with $\Delta r = 0.01$ Å in L_9^+ raises the energy by only ~ 0.14 kcal/mol. Clearly this vibrational mode has a very low force constant, presumably as a result of the situation detailed above. Under such circumstances it is obviously desirable for the present results to be further tested with correlated wave functions. Given the fact that the lowest unoccupied molecular orbital (LUMO) in polyenes and the second LUMO in polyenyl cations (the lowest is nonbonding) lead to distortions via the partial bond orders which are opposed diametrically to those found in the ground state, it seems reasonable to suppose that the most favorable doubly excited configurations will act so as to depress the extent of single-determinant bond alternation (*see* Results).

Literature Cited

1. Ayrey, G., Head, B. C., Poller, R. C., *J. Polym. Sci., Macromol. Rev.* (1974) **8**, 1.
2. Mayer, Z., *J. Macromol. Sci., Rev. Macromol. Chem.* (1974) **10**, 263.
3. Braun, D., *Degradation Stab. Polym., Proc. Plenary Main Lect. Int. Symp., 1974* (1975) 23.
4. Starnes, Jr., W. H., ADV. CHEM. SER. (1978) **169**, 309.
5. Haddon, R. C., Starnes, W. H., Jr., *Polym. Prepr., Am. Chem. Soc., Div. Polym. Chem.* (1977) **18**(1), 505.
6. Hall, G. G., *Proc. R. Soc. London* (1951) **A205**, 541.
7. Roothaan, C. C. J., *Rev. Mod. Phys.* (1951) **23**, 69.
8. Hehre, W. J., Stewart, R. F., Pople, J. A., *J. Chem. Phys.* (1969) **51**, 2657.
9. Hehre, W. J., Lathan, W. A., Ditchfield, R., Newton, M. D., Pople, J. A., Gaussian 70, QCPE Program No. 236.
10. Ditchfield, R., Hehre, W. J., Pople, J. A., *J. Chem. Phys.* (1971) **54**, 724.
11. Lathan, W. A., Curtiss, L. A., Hehre, W. J., Lisle, J. B., Pople, J. A., *Prog. Phys. Org. Chem.* (1974) **11**, 175.
12. Radom, L., Poppinger, D., Haddon, R. C., "Carbonium Ions," Vol. 5, G. A. Olah, P. v. R. Schleyer, Eds., Chapter 38, Wiley, New York, 1976.
13. Hiberty, P. C., *J. Am. Chem. Soc.* (1975) **97**, 5975.
14. Newton, M. D., Lathan, W. A., Hehre, W. J., Pople, J. A., *J. Chem. Phys.* (1970) **52**, 4064.
15. Snyder, L. C., *J. Chem. Phys.* (1967) **46**, 3602.
16. Snyder, L. C., Basch, H., *J. Am. Chem. Soc.* (1969) **91**, 2189.
17. Hehre, W. J., Ditchfield, R., Radom, L., Pople, J. A., *J. Am. Chem. Soc.* (1970) **92**, 4796.
18. Lathan, W. A., Hehre, W. J., Pople, J. A., *J. Am. Chem. Soc.* (1971) **93**, 808.
19. Hehre, W. J., Pople, J. A., *J. Am. Chem. Soc.* (1975) **97**, 6941.
20. Radom, L., Lathan, W. A., Hehre, W. J., Pople, J. A., *J. Am. Chem. Soc.* (1971) **93**, 5339.
21. Dewar, M. J. S., *Chem. Soc., Spec. Publ.* (1967) **21**, 177.
22. Hess, B. A., Schaad, L. J., *J. Am. Chem. Soc.* (1971) **93**, 305.
23. Carreira, L. A., *J. Chem. Phys.* (1975) **62**, 3851.
24. Poppinger, D., *Chem. Phys.* (1976) **12**, 131.
25. Coulson, C. A., *Proc. R. Soc. London* (1938) **A164**, 383.
26. Lennard-Jones, J. E., *Proc. R. Soc. London* (1937) **A158**, 280.
27. Lennard-Jones, J. E., Turkevich, J., *Proc. R. Soc. London* (1937) **A158**, 297.
28. Penney, W. G., *Proc. R. Soc. London* (1937) **A158**, 306.
29. Longuet-Higgins, H. C., *J. Chem. Phys.* (1950) **18**, 265.
30. Oseen, D., Flewwelling, R. B., Laidlaw, W. G., *J. Am. Chem. Soc.* (1968) **90**, 4209.
31. Fabian, J., Hartmann, H., *J. Signalaufzeichnungsmaterialen* (1974) **2**, 457.
32. Fabian, J., Hartmann, H., *Theor. Chim. Acta* (1975) **36**, 351.
33. Fabian, J., Hartmann, H., *J. Molec. Struct.* (1975) **27**, 67.
34. Fabian, J., Hartmann, H., *J. Signalaufzeichnungsmaterialen* (1976) **4**, 101.
35. Kuhn, H., *J. Chem. Phys.* (1948) **16**, 840.
36. Dewar, M. J. S., *J. Chem. Soc.* (1952) 3544.
37. Ooshika, Y., *J. Phys. Soc. Jpn.* (1957) **12**, 1388.
38. Ooshika, Y., *J. Phys. Soc. Jpn.* (1957) **12**, 1246.
39. Labhart, H., *J. Chem. Phys.* (1957) **27**, 957.
40. Longuet–Higgins, H. C., Salem, L., *Proc. R. Soc. London* (1959) **A251**, 172.
41. Ooshika, Y., *J. Phys. Soc. Jpn.* (1959) **14**, 747.
42. Pople, J. A., Walmsley, S. H., *Mol. Phys.* (1962) **5**, 15.

43. Tsuji, M., Huzinga, S., Hasino, T., *Rev. Mod. Phys.* (1960) **32,** 425.
44. Hanna, M. W., McLachlan, A. D., Dearman, H. H., McConnell, H. M., *J. Chem. Phys.* (1962) **37,** 361.
45. Hanna, M. W., McLachlan, A. D., Dearman, H. H., McConnell, H. M., *J. Chem. Phys.* (1962) **37,** 3008.
46. Salem, L., "Molecular Orbital Theory of Conjugated Systems," Chapter 8, Benjamin, New York, 1967.
47. Binsch, G., Heilbronner, E., Murrell, J. N., *Mol. Phys.* (1966) **11,** 305.
48. Peierls, R. E., "Quantum Theory of Solids," Oxford University Press, London, 1955.
49. Brickstock, A., Pople, J. A., *Trans. Faraday Soc.* (1954) **50,** 901.
50. Hush, N. S., Pople, J. A., *Trans. Faraday Soc.* (1955) **51,** 600.
51. Clark, D. T., Armstrong, D. R., *Theor. Chim. Acta* (1969) **13,** 365.
52. Peyerimhoff, S. D., Buenker, R. J., *J. Chem. Phys.* (1969) **51,** 2528.
53. Billingsley, F. P., Trindle, C., *J. Phys. Chem.* (1972) **76,** 2995.
54. Baird, N. C., *Tetrahedron* (1972) **28,** 2355.
55. Bodor, N., Dewar, M. J. S., Lo, D. H., *J. Am. Chem. Soc.* (1972) **94,** 5303.
56. Shanshal, M., *J. Chem. Soc., Perkin Trans. 2* (1972) 335.
57. Radom, L., Hariharan, P. C., Pople, J. A., Schleyer, P. v. R., *J. Am. Chem. Soc.* (1973) **95,** 6531.
58. Levin, G., Goddard, W. A., Huestis, D. L., *Chem. Phys.* (1974) **4,** 409.
59. Bingham, R. C., Dewar, M. J. S., Lo, D. H., *J. Am. Chem. Soc.* (1975) **97,** 1294.
60. Deno, N. C., "Carbonium Ions," Vol. 2, G. A. Olah, P. v. R. Schleyer, Eds., Chapter 18, Wiley, New York, 1970.
61. Sorensen, T. S., "Carbonium Ions," Vol. 2, G. A. Olah, P. v. R. Schleyer, Eds., Chapter 19, Wiley, New York, 1970.
62. Deno, N. C., Haddon, R. C., Nowak, E. N., *J. Am. Chem. Soc.* (1970) **92,** 6691.
63. Radom, L., personal communication, 1975.
64. Salem, L., "Molecular Orbital Theory of Conjugated Systems," Chapter 2, Benjamin, New York, 1967.
65. Craig, D. P., "Nonbenzenoid Aromatic Compounds," D. Ginsburg, Ed., p. 1, Interscience, New York, 1959, and references cited therein.
66. Haddon, R. C., *Aust. J. Chem.* (1977) **30,** 1.
67. Radom, L., Pople, J. A., Buss, V., Schleyer, P. v. R., *J. Am. Chem. Soc.* (1972) **94,** 311.
68. Hariharan, P. C., Radom, L., Pople, J. A., Schleyer, P. v. R., *J. Am. Chem. Soc.* (1974) **96,** 599.
69. Harmon, K. M., "Carbonium Ions," Vol. 4, G. A., Olah, P. v. R. Schleyer, Eds., Chapter 29, Wiley, New York, 1973.
70. Sundaralingham, M., Chwang, A. K., "Carbonium Ions," Vol. 5, G. A. Olah, P. v. R. Schleyer, Eds., Chapter 39, Wiley, New York, 1976.
71. Traetteberg, M., *J. Am. Chem. Soc.* (1964) **86,** 4265.
72. Davis, R. E., Tulinsky, A., *J. Am. Chem. Soc.* (1966) **88,** 4583.
73. Brighton, C. A., Marks, G. C., Benton, J. L., *Encycl. Polym. Sci. Technol.* (1971) **14,** 363.
74. *Polym. Handb., 2nd Ed., 1975,* p. V-43.
75. Riddick, J. A., Bunger, W. B., "Organic Solvents," Wiley, New York, 1970.
76. Franklin, J. L., "Carbonium Ions," Vol. 1, G. A. Olah, P. v. R. Schleyer, Eds., Chapter 2, Wiley, New York, 1968.
77. Freedman, H. H., "Carbonium Ions," Vol. 4, G. A. Olah, P. v. R. Schleyer, Eds., Chapter 28, Wiley, New York, 1973.
78. Onokuza, M., Asahina, M., *J. Macromol. Sci., Rev. Macromol. Chem.* (1969) **3,** 235.
79. Figge, K., Findeiss, W., *Angew. Makromol. Chem.* (1975) **47,** 141.

80. Marks, G. C., Benton, J. L., Thomas, C. M., *SCI Monogr.* (1967) **26**, 204.
81. Braun, D., *Pure Appl. Chem.* (1971) **26**, 173.
82. Berlin, A. A., Yanovskii, D. M., Popova, Z. V., "Aging and Stabilization of Polymers," M. B. Neiman, Ed., p. 175, Consultants Bureau, New York, 1968.
83. Winkler, D. E., *J. Polym. Sci.* (1959) **35**, 3.
84. Abbås, K. B., Laurence, R. L., *J. Polym. Sci., Polym. Chem. Ed.* (1975) **13**, 1889.
85. Oleinik, É. P., Vasileiskaya, N. S., Razuvaev, G. A., *Bull. Acad. Sci. USSR, Div. Chem. Sci.* (1968) 472.
86. Tüdös, F., Kelen, T., Nagy, T. T., Turcsányi, B., *Pure Appl. Chem.* (1974) **38**, 201.
87. Guyot, A., Bert, M., *J. Appl. Polym. Sci.* (1973) **17**, 753.
88. Kelen, T., Bálint, G., Galambos, G., Tüdös, F., *J. Polym. Sci., Part C* (1971) **33**, 211.
89. Mayo, F. R., *Polym. Prepr., Am. Chem. Soc., Div. Polym. Chem.* (1971) **12**(2), 61.
90. See, inter alia, Maccoll, A., *Chem. Rev.* (1969) **69**, 33.
91. Morrison, R. T., Boyd, R. N., "Organic Chemistry," 3rd Ed., pp. 271–273, Allyn and Bacon, Boston, 1973.
92. See, inter alia, Streitwieser, Jr., A., "Molecular Orbital Theory for Organic Chemists," p. 379, Wiley, New York, 1962, and references cited therein.
93. Salem, L., "Molecular Orbital Theory of Conjugated Systems," Chapter 3, Benjamin, New York, 1967.
94. Sly, W. G., *Acta Crystallogr.* (1964) **17**, 511.
95. Olah, G. A., Pittman, C. U., Symons, M. C. R., "Carbonium Ions," Vol. 1, G. A. Olah, P. v. R. Schleyer, Eds., Chapter 5, Wiley, New York, 1968.
96. Pople, J. A., Beveridge, D. L., "Approximate Molecular Orbital Theory," p. 110, McGraw–Hill, New York, 1970.
97. Kuchitsu, K., *J. Chem. Phys.* (1966) **44**, 906.
98. Almenningen, A., Bastiansen, O., Traetteberg, M., *Acta Chem. Scand.* (1958) **12**, 1221.
99. Haugen, W., Traetteberg, M., *Acta Chem. Scand.* (1966) **20**, 1726.
100. Kuchitsu, K., Fukuyama, T., Morino, Y., *J. Mol. Struct.* A*1967–1968*Q **1**, 463.
101. Traetteberg, M., *Acta Chem. Scand.* (1968) **22**, 628.
102. Haddon, R. C., unpublished data (1975).
103. Lathan, W. A., Hehre, W. J., Curtiss, L. A., Pople, J. A., *J. Am. Chem. Soc.* (1971) **93**, 6377.
104. Hehre, W. J., *J. Am. Chem. Soc.* (1974) **96**, 5207.
105. Franklin, J. L., Dillard, J. G., Rosenstock, H. M., Herron, J. T., Draxl, K., "Ionization Potentials, Appearance Potentials, and Heats of Formation of Gaseous Ions," NSRDS-NBS-26, National Bureau of Standards, Washington, D.C.
106. Rossini, F. D., Pitzer, K. S., Arrett, R. L., Braun, R. M., Pimentel, G. C., "Selected Values of Physical and Thermodynamic Properties of Hydrocarbons and Related Compounds," Carnegie, Pittsburgh, 1953.
107. Cox, J. D., Pilcher, G., "Thermochemistry of Organic and Organometallic Compounds," Academic, New York, 1970.
108. Benson, S. W., Criuckshank, F. R., Golden, D. M., Haugen, G. R., O'Neal, H. E., Rodgers, A. S., Shaw, R., Walsh, R., *Chem. Rev.* (1969) **69**, 279.
109. Lossing, F. P., Semchuk, G. P., *Can. J. Chem.* (1970) **48**, 955.
110. Lossing, F. P., *Can. J. Chem.* (1971) **49**, 357.
111. Vincow, G., Dauben, H. J., Jr., Hunter, F. R., Volland, W. V., *J. Am. Chem. Soc.* (1969) **91**, 2823.
112. Stull, D. R., Westrum, E. F., Simke, G. C., "The Chemical Thermodynamics of Organic Compounds," Wiley, New York, 1969.

113. Franklin, J. L., "Carbonium Ions," G. A. Olah, P. v. R. Schleyer, Eds., Vol. 1, Chap. 2, Wiley, New York, 1968.
114. "CRC Handbook of Chemistry and Physics," R. C. Weast, Ed., p. E-55, Chemical Rubber Co., Cleveland, OH, 1973.
115. Harmon, K. M., Cummings, F. E., Davis, D. A., Diestler, D. J., *J. Am. Chem. Soc.* (1962) **84**, 3349.

RECEIVED June 13, 1977.

28

The Ligand Exchange Reaction of Some Dialkyltin Dimercaptides and Dicarboxylates with Dialkyltin Dichlorides

Observation by ^{1}H and ^{13}C NMR. Implications for Poly(vinyl chloride) Stabilization

RICHARD G. PARKER and CHARLES J. CARMAN

The B F Goodrich Company, Research and Development Center,
9921 Brecksville Road, Brecksville, OH 44141

The existence of a ligand exchange reaction between dialkyltin dimercaptides and dicarboxylates (typical poly(vinyl chloride) stabilizers) and dialkyltin dichlorides is demonstrated by ^{1}H and ^{13}C NMR. Dibenzyl- and dibutyltin dimercaptides undergo the reaction instantaneously and quantitatively at 25°C to yield dialkylchlorotin mercaptides. Dialkyltin dicarboxylates and dimethyltin dimercaptides appear to participate in an equilibrium exchange reaction with dialkyltin dichlorides in which the individual components are not identifiable by NMR at 25°C. These results are discussed as they pertain to poly(vinyl chloride) stabilization.

The mechanism of poly(vinyl chloride) (PVC) stabilization has been studied for many years and the area continues to be an active one (*1, 2, 3, 4*). The organotin stabilizers have been of special interest. Klemchuk (*5*), working in part to confirm the stabilization theory of Frye and Horst (*6*), studied the reaction of a PVC defect-site model, 4-chloropent-2-ene, with dibutyltin dilaurate and dibutyltin di(dodecyl mercaptide), and concluded that dibutylchlorotin derivatives were among the stabilizing species. The dibutylchlorotin compounds (3) were postu-

0-8412-0381-4/78/33-169-363$05.00/1

lated to arise through a ligand exchange reaction (I) between the stabilizer (**1**) and dibutyltin dichloride (**2**). In these model compound

$$\underset{\mathbf{1}}{Bu_2SnL_2} + \underset{\mathbf{2}}{Bu_2SnCl_2} \rightarrow \underset{\mathbf{3}}{2\ Bu_2SnClL} \quad \text{(I)}$$

studies dibutyltin dichloride was added to the reaction mixtures. In the PVC it would be formed most likely by reaction of the stabilizer with two equivalents of hydrogen chloride during the early stages of PVC decomposition. Some dibutylchlorotin compound (**3**) could be formed also by reaction with one equivalent of hydrogen chloride at this time. The dibutylchlorotin compounds were not isolated, nor was the existence of the exchange reaction (1) established firmly. Using ^1H and ^{13}C NMR, we wish to summarize our observations concerning the ligand exchange reaction and to point out the possible implications of this reaction in PVC stabilization. NMR has been used to study ligand exchange reactions involving organotin compounds in other systems (*7, 8*).

Experimental Section

Dialkyltin Compounds. All of the dialkyltin dimercaptides and dicarboxylates were prepared by reaction of the appropriate dialkyltin oxide (Alfa Inorganics) with the desired mercaptan or carboxylic acid. The reactions were carried out in refluxing benzene or toluene using a Dean–Stark water separator. The yields are quantitative and products require no purification (purity confirmed by ^1H and ^{13}C NMR). The dialkyltin dichlorides were obtained commercially (Alfa Inorganics) and used as received.

NMR Spectra. The 60-MHz ^1H NMR spectra were obtained on a Varian A-60 or T-60 spectrometer using the solvents and concentrations listed in the tables.

The ^{13}C NMR spectra were obtained on a Varian XL-100–15 (25.16 MHz) or a Brüker HX–90E (22.62 MHz) using the solvents and concentrations listed in the tables. Broad-band ^1H decoupling was used during the collection of 8K data points over sweep widths of 5 or 6 KHz. The pulse widths represented flip angles of 25°–30°. Excellent S/N ratios were obtained after 100–400 transients. The T_1 values of the carbonyl carbons in Compounds **7** and **9** were obtained at 30°C on the Varian XL-100–15 using a 90° pulse, a 300-Hz sweep width, a 60-sec repetition rate, and the inversion–recovery method (*9*).

Results and Discussion

We first studied the dibenzyltin di(benzyl mercaptide)/dibenzyltin dichloride system owing to the expected simplicity of its 60-MHz ^1H spectra. The methylene resonances in the starting materials are distinct singlets with differing chemical shifts. The experimental results are presented in Table I.

Table I. 60-MHz ^1H NMR Data for the Dibenzyltin Di(benzyl mercaptide)/Dibenzyltin Dichloride System

$$4 = \underset{A}{(C_6H_5\text{–}CH_2)_2}Sn\underset{B}{(S\text{–}CH_2\text{–}C_6H_5)_2}$$

$$5 = \underset{C}{(C_6H_5\text{–}CH_2)_2}SnCl_2$$

$$6 = \underset{D}{(C_6H_5\text{–}CH_2)_2}Sn\underset{E}{(S\text{–}CH_2\text{–}C_6H_5)}Cl$$

Experiment No.		*Concentrations (M) in DMSO-d_6*			*^1H Chemical Shifts of Methylene Protons in ppm (TMS)*				
		4	*5*	*6*	*A*	*B*	*C*	*D*	*E*
1		0.0932	—	—	2.78	3.76			
2		—	0.1442	—			3.11		
3	initial	0.0932	0.1442	—					
	final[a]	0	0.0510	0.1864			3.11	2.87	3.88
4	initial	0.0932	0.0721	—					
	final[a]	0.0211	0	0.1442	2.78	3.75		2.87	3.89

[a] Calculated from Reaction I stoichiometry.

When **4** was mixed with excess dibenzyltin dichloride (Experiment 3), the methylene proton resonances (A and B) were not observed, but two new lines (D and E) were observed at lower field. When excess **4** was mixed with dibenzyltin dichloride (Experiment 4), all four methylene resonances (A, B, D, and E) appeared in the NMR spectrum. This eliminates the possibility that D and E arise as a result of A and B being solvent-shifted by addition of the polar component dibenzyltin dichloride. The spectral features present in Experiments 3 and 4 were unchanged after one hour at 100°C. That the new spectrum observed in Experiments 3 and 4 was that of **6** was rationalized in the following way. Protons that are in close proximity to electron attracting groups or atoms, such as chlorine, are deshielded electronically and their resonances appear at lower field (higher frequency) than they would in the absence of the electron attracting moiety. Therefore, the methylene resonances (A and B) in **4** should be shifted to lower field in **6** (D and E). The exchange reaction was quantitative at 25°C based on the integrated areas of the resonances in Experiments 3 and 4.

Other investigators (*10*) have reported ligand exchange effects on the tin–proton coupling constants ($J_{Sn\text{-}C\text{-}H}$) for dimethyltin derivatives. We found that in the dibenzyltin system (Table I) the coupling constant increased as the number of chlorines on tin increased. We found also

Table II. Tin–Proton Coupling Constants for the Dibenzyltin Di(benzyl mercaptide)/Dibenzyltin Dichloride System

Compound		$J^a_{Sn\text{-}C\text{-}H}$	$J^a_{Sn\text{-}S\text{-}C\text{-}H}$
$(C_6H_5CH_2)_2Sn(SCH_2C_6H_5)_2$	(4)	75.3	30.8
$(C_6H_5CH_2)_2SnCl_2$	(5)	141.5	—
$(C_6H_5CH_2)_2Sn(SCH_2C_6H_5)Cl$	(6)	89.3	43.4

[a] Coupling constants are the average of the ^{117}Sn- and ^{119}Sn-proton couplings expressed in Hz.

that the two-bond coupling ($J_{Sn\text{-}C\text{-}H}$) was larger than the three-bond coupling ($J_{Sn\text{-}S\text{-}C\text{-}H}$). These data are summarized in Table II. Although the ligand exchange effects alter the coupling constants we did not try to use them in interpreting the course of the exchange reaction.

Our attention turned next to a widely used type of commercial stabilizer, dibutyltin di-isooctylthioglycolate. The ligand exchange reaction in the dibutyltin di-isooctylthiogylcolate/dibutyltin dichloride system was studied by both ^1H and ^{13}C NMR. In the ^1H experiments the resonance of the methylene group in the thioglycolate moiety ($-S-CH_2-COO-$) was used to follow the reaction. When Compounds 7 and 8 (Table III, Experiments 6, 7, and 8) were mixed at 25°C in deutero-

Table III. 60-MHz ^1H NMR Data for the Dibutyltin Di-isooctylthioglycolate/Dibutyltin Dichloride System

$$7 = Bu_2Sn(S-\underset{F}{CH_2}-COO-\underset{G}{CH_2}-C_7H_{15})_2$$

$$8 = Bu_2SnCl_2$$

$$9 = Bu_2Sn(S-\underset{H}{CH_2}-COO-\underset{I}{CH_2}-C_7H_{15})Cl$$

Experiment No.	*Concentrations M in $CDCl_3$* 7	8	9	*^1H Chemical Shifts of Methylene Protons in ppm (TMS)* F	G	H	I
5	0.0732			3.42	4.15		
6 initial	0.0732	0.0916	0				
final[a]	0	0.0184	0.1464			3.59	4.25
7 initial	0.1084	0.0907	0				
final[a]	0.0177	0	0.1814	3.42	4.15	3.60	4.23
8 initial	large	0.0907	0				
final[a]	excess	0	0.1814	3.41	4.15	3.59	4.24

[a] Calculated from Reaction I stoichiometry.

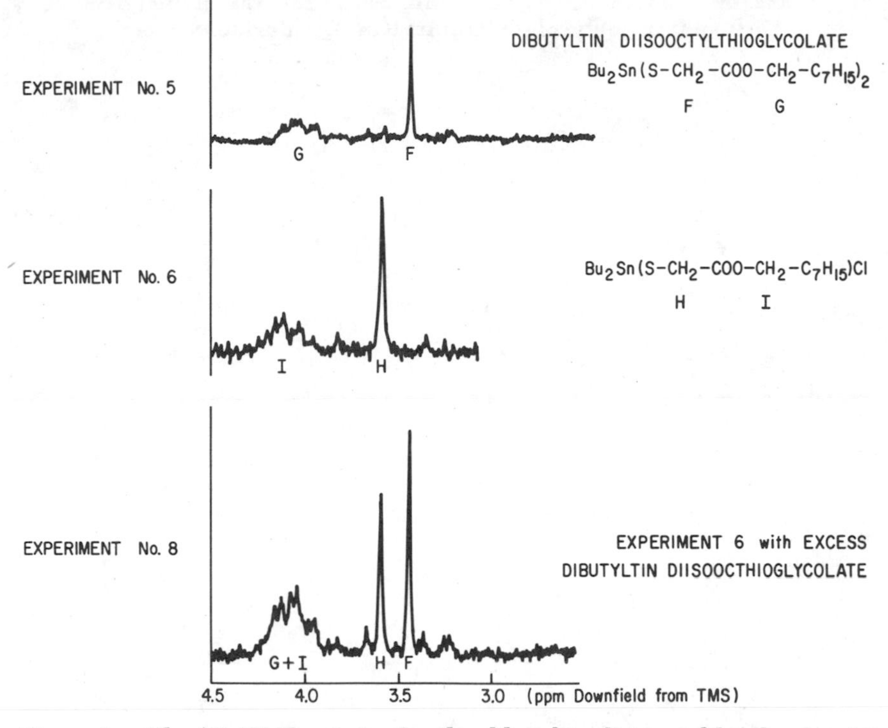

Figure 1. The ^1H NMR spectra for the dibutyltin di-isooctylthioglycolate/dibutyltin dichloride system (Table III)

chloroform, the exchange reaction occurred instantaneously and quantitatively. The data are presented in Table III and Figure 1.

Analyses of the ^1H-decoupled ^{13}C NMR spectrum of the commercial sample of dibutyltin di-isooctylthioglycolate used in this study indicates that its thioglycolate precursor is produced from a mixture of C_8 aliphatic alcohols and thioglycolic acid ($HSCH_2COOH$). Consequently, the aliphatic portion of its ^{13}C spectrum (0–60 ppm) is extremely complex. However, only one carbonyl carbon resonance is present (line width at half height = 4.7 Hz), and its chemical shift (173.9 ppm) is separated widely from those of the aliphatic carbons. The carbonyl group in these compounds apparently is not subject to the larger substituent effects well-documented for alkanes (*11*) and other classes of compounds. Thus, the carbonyl carbon resonance was used to monitor the reaction.

The ^{13}C spectra of the samples containing less than a stoichiometric amount of dibutyltin dichloride (Table IV, Experiments 10 and 11) contain two resonances in the carbonyl region—one at 173.9 ppm caused the stabilizer and the other at 179.2 ppm owing to the product (*9*) of the exchange reaction. When excess dibutyltin dichloride is present the

Table IV. 25.16 MHz ^{13}C NMR Data for the Dibutyltin Di-isooctylthioglycolate/Dibutyltin Dichloride System

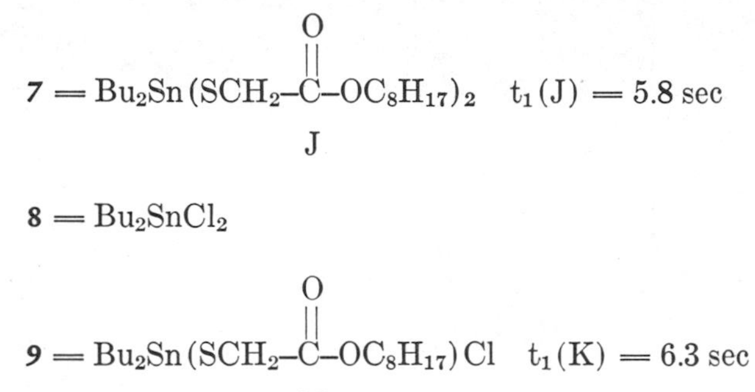

Experiment No.	Concentrations (mM) in $CDCl_3$			^{13}C Chemical Shifts of Carbonyl Groups in ppm (TMS)	
	7	8	9	J	K
9	7.7	—	—	173.9	
10 initial	7.7	2.5	—		
final[a]	5.2	0	5.0	174.0	179.1
11 initial	8.4	5.0	—		
final[a]	3.4	0	10.0	173.9	179.2
12 initial	10.3	10.3	—		
final[a]	0	0	20.6	—	179.2
13 initial	5.6	10.6	—		
final[a]	0	5.0	11.2	—	179.2

[a] Calculated from Reaction I stoichiometry.

carbonyl resonance from *7* is not observed. The results of these experiments are presented in Table IV and Figure 2.

The concentrations of Components *7* and *9* in Experiment 10 were verified by integration of the carbonyl resonances. In order to do this it was necessary to measure the spin lattice relaxation times (T_1) (*12*) of the carbonyl carbons. These were 5.8 sec in *7* and 6.3 sec in *9*. The spectrum in Experiment 10 was then reaccumulated, using a 60-sec repetition rate, and integrated. The calculated and observed integral ratios were within the limits of the experimental error.

From these results we conclude that the ligand exchange reaction (I) exists and occurs instantaneously and quantitatively at 25°C in the organotin mercaptide stabilizer systems examined.

A possible alternative interpretation of these results is that dibutyltin dichloride catalyzes the interconversion of the cis and trans forms of *7* reported by Stapfer and Herber (*13*). These forms arise because of one of the ester groups being bidentate (tin bonded to both sulfur and carbonyl oxygen) resulting in a trigonal bipyramidal configuration about tin. We do not believe that we are observing this isomer interconversion because of the stoichiometry observed (integrated ^{1}H and ^{13}C NMR spectra) and the observation of this phenomenon in the case of the dibenzyltin di(benzyl mercaptide) system (no carbonyl present).

We also have obtained ir spectra on the diisooctylthioglycolate system in the absence of solvent. The carbonyl bands of each component can be seen and identified clearly (*7* (C=O) at 1740 cm^{-1} and *9* (C=O) at 1680 cm^{-1}). The relative intensities of the two bands vary with the initial concentration of dibutyltin dichloride.

At least one exception in the organotin mercaptides has been noted. In the dimethyltin di(benzyl mercaptide)/dimethyltin dichloride system, regardless of the relative concentrations used, the ^{1}H NMR spectra consisted of broad lines and did not contain any resonances characteristic of the excess component. At 25°C, the reaction appears to be one involving a rapid equilibrium exchange of groups (*14*). However, at about −10°C, the resonances of the individual components can be observed owing to a slow rate of exchange (*15*).

Examination of the dibutyltin dilaurate/dibutyltin dichloride system by ^{13}C NMR gave an unexpected result. No evidence was found for the existence of a complete exchange reaction in this case. The only effect noted in the spectra as the concentration of dibutyltin dichloride increased was a linear, downfield shift of the laurate carbonyl resonance

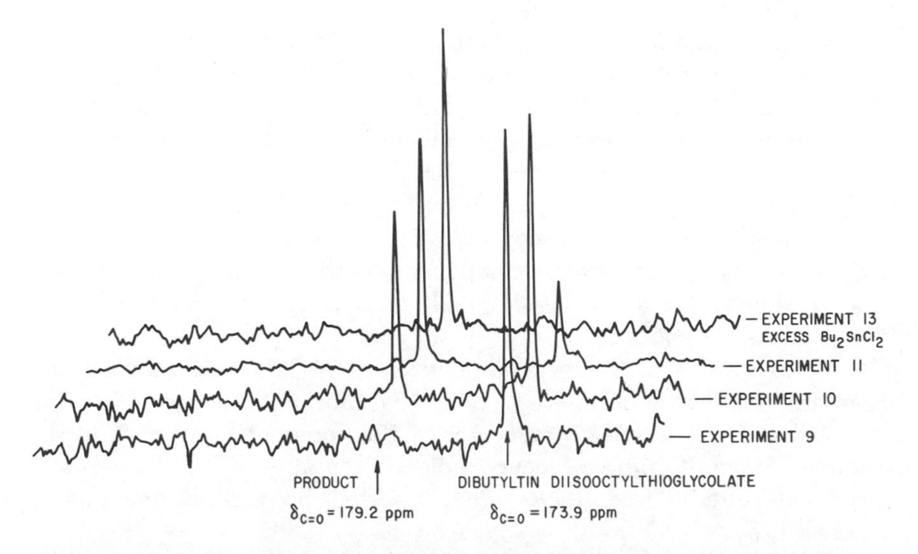

Figure 2. The ^{13}C NMR spectra of the carbonyl region for the dibutyltin diisooctylthioglycolate/dibutyltin dichloride system (Table IV)

and some scrambling of the resonances in the rather complex alkyl region. This system was studied also between 25° and 250°C by ir spectroscopy. No evidence for a complete exchange reaction was noted.

There are two plausible explanations for the dibutyltin dilaurate case: (1) the only association between the two components is some type of simple, nonequilibrating complexation which is the cause of the downfield shift of the laurate carbonyl resonance, or (2) an equilibrium exchange reaction (II) is occurring and increasing the concentration

$$\underset{\mathbf{10}}{Bu_2Sn(O_2CR)_2} + Bu_2SnCl_2 \leftrightarrows 2\ \underset{\mathbf{11}}{Bu_2SnCl(O_2CR)} \qquad (II)$$

of dibutyltin dichloride drives the equilibrium toward the dibutylchlorotin species (**11**). The carbonyl resonance in **11** is expected to be downfield from that of **10** owing to electron withdrawal by chlorine. The carbonyl chemical shift observed is then an average dependent on the relative concentrations of dibutyltin dilaurate and dibutylchlorotin laurate.

To resolve this question we prepared dibutyltin dibutyrate [$Bu_2Sn(O_2CCH_2CH_2CH_3)_3$] and examined its ^{13}C NMR spectrum in the presence of various amounts of dibutyltin dichloride. The same response of the carbonyl resonance was noted as in the case of dibutyltin dilaurate, but the simplicity of the alkyl resonances enabled us to determine that there were no dibutyltin dichloride resonances present when it was in excess.

An even simpler system composed of dibutyltin diacetate and dibutyltin dichloride was examined by both ^{13}C NMR and low-temperature ^1H NMR because of the report that dibutylchlorotin acetate formed immediately when the reagents were mixed in carbon tetrachloride (*16*). The ^{13}C spectra of the pure diacetate and the mixed system were obtained using conditions suitable for quantitative analysis. In deuterochloroform, at 30°C, over a wide range of concentrations, no evidence could be found for the presence of the excess component or the exclusive formation of dibutylchlorotin acetate. The ^1H NMR spectra gave similar results and the spectra were unchanged between 50° and −50°C.

We conclude that the reaction between dibutyltin dicarboxylates and dibutyltin dichloride near 30°C is an equilibrium reaction (II); the chemical shifts observed are averages of the components present in the reaction. Since the process occurs rapidly even at −50°C, it must be a very facile one. These results are supported by the observations of Mitchell (*17*).

The implications for PVC stabilization in the behavior of mercaptides vs. carboxylates in this reaction are not clear at the present time but they undoubtedly reflect certain important aspects of the tin–ligand bonding. The overall efficiency of an organotin stabilizer may be highly dependent on the type of bonding present in each class of stabilizer.

Gol'dshtein et al. (*18*) proposed that the ability of simple tin mercaptides to undergo ligand exchange reactions was caused by $d_\pi - p_\pi$ conjugation in the Sn–S bonds. They proposed also that tin alkoxides form 1:1 complexes with tin chlorides because of the lack of $d_\pi - p_\pi$ conjugation in the Sn–O bond. No definite conclusions were stated regarding tin carboxylates. Starnes (*19*) has reported that dibutyltin dimethoxide is not a very efficient PVC stabilizer.

If the dialkylchlorotins, R_2SnClL, are the actual or most active stabilizing species in PVC being processed in the presence of organotin stabilizers, then it might be expected that deliberate, initial introduction of these materials into PVC might lead to a lengthening of the induction period during decomposition owing to more efficient initial stabilization. Starnes (*20*) has noted some interesting results concerning the increased stability of PVC when it is chemically stabilized prior to being subjected to decomposition, by tin mercaptides in the presence of dibutyltin dichloride.

When stabilizers prepared by mixing **7** and **8** were added to PVC (Table V), no general improvement was noted in the results of an oven

Table V. Oven Aging Data and HCl-Evolution Data for PVC Stabilized with Dibutyltin Di-isooctylthioglycolate/ Dibutyltin Dichloride

Recipe: Geon 103 PVC, 100 g; dibutyltin di-isooctylthioglycolate (Bu_2SnL_2), 1.30 g, 2 mmol; polyethylene processing aid, 3.0 g.

		Resultant Stabilizer System			*HCl Evolution Data*	
Experiment No.	*Bu_2SnCl_2 Added (mmol)*	*Bu_2SnL_2 (mmol)*	*Bu_2SnLCl[a] (mmol)*	*Time to Black[b] (min)*	*Induction Period[c] (min)*	*Stabilizer Exhaustion[d] (min)*
14	none	2.0	—	145	53	89
15	0.5	1.5	1.0	145	42	88
16	1.0	1.0	2.0	155	27	87
17	2.0	0.0	4.0	150	13	89

[a] Calculated from exchange Reaction I stoichiometry.
[b] At 170°C in an air-circulated oven.
[c] HCl evolution measured conductometrically at 180°C in air; time for recorded curve to rise a standard height.
[d] Time at which the slope of the HCl-evolution curve becomes more positive and there is a well-defined break point in the curve.

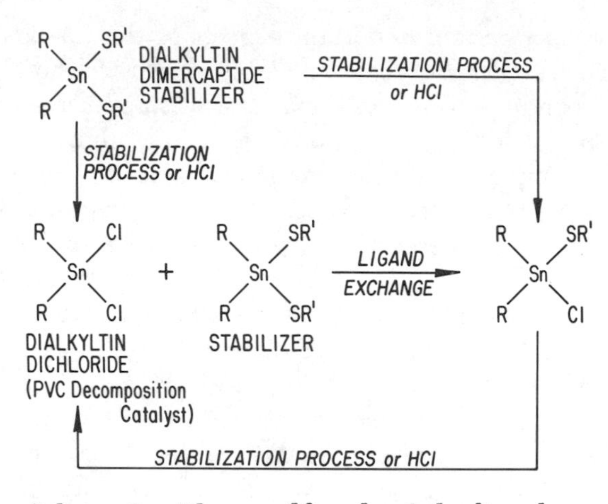

Scheme 1. The possible role of the ligand exchange reaction (I) in PVC stabilization by dialkyltin dimercaptides

aging test or in an HCl evolution test compared with pure 7. In fact, the induction period is shortened considerably even though the same molar amounts of mercaptide are available to participate in stabilization or react with hydrogen chloride. However, the concentration of dibutyltin dichloride builds up because the exchange reaction can no longer occur to the same extent because of stabilizer depletion. The dibutyltin dichloride, a Lewis acid, acts as a catalyst for the decomposition of PVC in Experiments 15, 16, and 17 (Table V) (*1*). Similar effects were noted when the samples listed in Table V were subjected to milling at 180°–185°C (the samples failed in the order 17, 16, 15, and 14, decreasing initial concentration of dibutyltin dichloride) as evidenced by the rate of color development and eventual debanding. Thus, the exchange reaction (I) may play an important role in PVC stabilization by organotin mercaptides through preventing the early build up of dibutyltin dichloride. An overview of this process is presented in Scheme 1.

These exchange reaction phenomena may be general for all types of organotin PVC stabilizers. In the cases where an equilibrium exchange reaction is observed (dimethyltin dimercaptides and dialkyltin dicarboxylates), the equilibrium at PVC processing temperatures may be far toward the dialkylchlorotin species and thus the mechanism is still operative.

Literature Cited

1. Ayrey, G., Head, B. C., Poller, R. C., *J. Polym. Sci. Macromol. Rev.* (1974) **8**, 1.
2. David, C., *Compr. Chem. Kinet.* (1975) **14**, 78.

3. Nass, L. I., "Encyclopedia of PVC," L. I. Nass, Ed., Vol. 1, pp. 271–293, Dekker, New York, 1976.
4. Starnes, W. H., Jr., *Am. Chem. Soc., Div. Polym. Chem., Polym. Prepr.* (1977) **18**(1), 493.
5. Klemchuk, P. P., in "Stabilization of Polymers and Stabilizer Processes," ADV. CHEM. SER. (1968) **85**, 1–17, especially p. 10.
6. Frye, A. H., Horst, R. W., Paliobagis, M. A., *J. Polym. Sci., A* (1964) **2**, 1801, and references therein.
7. Rockett, B. W., Hadlington, M., Poyner, W. R., *J. Appl. Polym. Sci.* (1973) **17**, 315.
8. Kawasaki, Y., *Org. Mag. Reson.* (1970) **2**, 165.
9. Freeman, R., Hill, H. D. W., *J. Chem. Phys.* (1971) **54**, 3367.
10. Kitching, W., *Tetrahedron Lett.* (1966) **31**, 3689.
11. Carman, C. J., Tarpley, A. R., Jr., Goldstein, J. H., *Macromolecules* (1973) **6**, 719.
12. Levy, G. C., Nelson, G. L., "Carbon-13 Nuclear Magnetic Resonance for Organic Chemists," Chap. 9 and p. 5, Wiley–Interscience, New York, 1972.
13. Stapfer, C. H., Herber, R. H., *J. Organometal. Chem.* (1974) **66**, 425.
14. Van Den Berghe, E. V., Van Der Kelen, G. P., *J. Organometal. Chem.* (1974) **72**, 65.
15. Dorsch, J. L., B F Goodrich Co., Chemical Division, Avon Lake, OH, private communication.
16. Davies, A. G., Harrison, P. G., *J. Chem. Soc., C* (1967) 298.
17. Mitchell, T. N., *J. Organometal. Chem.* (1973) **59**, 189.
18. Gol'dshtein, I. P., et al., *Bull. Acad. Sci. USSR* (1967) (10), 2115.
19. Starnes, W. H., Jr., Plitz, I. M., Hartless, R. L., *Am. Chem. Soc., Div. Polym. Chem., Polym. Prepr.* (1976) **17**(2), 495.
20. Starnes, W. H., Jr., Plitz, I. M., *Macromolecules* (1976) **9**(4), 633.

RECEIVED May 12, 1977.

29

Stabilizer Consumption during Processing of Poly(vinyl chloride)

TRAN VAN HOANG, ALAIN MICHEL, ANDRÉ REVILLON, MICHEL BERT, ALAIN DOUILLARD, and ALAIN GUYOT

CNRS, Laboratoire des Matériaux Organiques, 69626 Villeurbanne, France

Although an estimate can be obtained from the induction period in thermal dehydrochlorination experiments, true measurements of the stabilizers' consumption during processing of PVC have only been carried out recently in the case of tin stabilizers. Techniques have been developed suited to the case of zinc–calcium and lead stabilizers; they have been applied to samples processed under industrial conditions either by extrusion (bottles and wires) or by calendering (sheets). Sophisticated dehydrochlorination methods show that the distribution of stabilizers in bottles is not regular. Coulometric titration allows the measurement of the chloride ions of either $ZnCl_2$ or $CaCl_2$. The distinction between the cations is more difficult and polarographic titration as well as atomic absorption experiments failed. Liquid chromatography in THF is able to separate epoxidized soyabean oil; in addition, α-phenylindole can be observed. The formation of carboxylic acid from metal carboxylates is studied by acidimetric titration. In calendering, most of the stabilizer consumption seems to occur during the premixing steps, including the alimentation of the machine by extrusion. Finally, in a lead formulation for cable insulation, the consumption of basic lead diphosphate may be measured using X-ray diffraction.

The new interest for recycling plastics material raises a question as to the amount of stabilizer remaining available in the material after processing. This knowledge also would be helpful in predicting the durability of a fabric upon weathering. An indirect estimate of this residual amount may be obtained from dehydrochlorination studies

0-8412-0381-4/78/33-169-374$05.00/1

(DHC); a test based on DHC has been proposed in Europe (*1*). However, very little has been published about the direct measurement of stabilizer consumption. In the case of tin stabilizer, an analytical method for measuring the tin derivative has been developed by Neubert and Wirth (*2*). It is based on alkylation after organomagnesium compounds' reactions with halides, oxides, carboxylates, and thioesters of tin derivatives followed by gas liquid chromatographic analysis. Some results have been presented at the Second International Symposium on PVC by Wirth (*3*). At the same symposium, Stepek, Vymazal, and Czako (*4*) presented a few results on coulometric titration of chlorides formed after processing and thermal degradation of PVC stabilized using Ba–Cd or Zn–Ca simplified formulations.

In this paper, we present a few methods of studying the consumption of organometallic as well as organic compounds of PVC stabilizer formulations used in making bottles (Ca–Zn formulation) or insulated cables (lead formulation) through extrusion and calendered sheets (Ca–Zn formulation). These methods include coulometric titration of chlorides (improvement of the Stepek–Vymazal–Czako method), potentiometric titration of carboxylic acids, liquid chromatographic determination of organic compounds, and x-ray titration of lead phthalate. Some DHC experiments using differential conductimetric determination of HCl (*5*) are carried out for comparison. Atomic absorption titration of calcium and zinc chlorides is also possible, but the preparation of solutions is very time-consuming, so the method has not been included here. The results may be compared with those from coulometric titration, which is a much simpler method. Applications of these methods to the study of the stabilization mechanism through the consumption of stabilizers in simplified formulations during mastication in a Brabender plastograph is presented in another chapter in this symposium (*6*).

Experimental

Materials. A set of 18 experimental bottles were prepared using the following recipe: (in phr): PVC (Kwert 57), 100; impact polymeric additive (Afcochoc R4011 Rhone–Poulenc), 10; processing aid polymeric additive (Acryloid R K 12ON Rohm and Haas), 2; calcium stearate, 0.24; zinc ethylhexanoate, 0.22; epoxidized soyabean oil (ESO), 3; α-phenylindole, 0.8; glycol monostearate (GMS), 1; and finally polyethylene wax, 0.2. The ingredients were mixed in a dry blending machine (Henschel) under the conditions indicated in Table I. All the powdered ingredients were introduced at room temperature and the liquids (zinc ethylhexanoate, ESO, GMS) occasionally were introduced at higher temperatures (bottles 13–18).

The extrusion was carried out in a machine with a rate of 30 Kg/hr. Bottles 1, 4, 7, 10, 13, and 16 were obtained using the fresh dry blend. A 30% recycling rate was used to make bottles 2, 5, 8, and 11. For bottles

Table I. Stabilizer Consumption in Bottles

Bottle No.	*Incorporation Temp. (°C)*		*Motor Speed (rpm)*	*Limiting Temp. (°C)*
	Solids	*Liquids*		
1				
2	r.t.	r.t.	3000	125
3				
4				
5	r.t.	r.t.	1500	125
6				
7				
8	r.t.	r.t.	3000	80
9				
10				
11	r.t.	r.t.	1500	80
12				
13				
14	r.t.	r.t.	3000	125
15				
16[a]				
17[a]	r.t.	r.t.	3000	125
18[a]				

[a] For bottles 16, 17, and 18, the processing aid has been introduced at 60°C during

3, 6, 9, 12, 15, and 18, a 100% recycling rate was adopted. A 30% recycling rate = 30% used PVC and 70% virgin PVC whereas a 100% recycling rate = 100% used PVC and no remaining virgin PVC.

A calendered sheet was obtained with the following recipe (in phr): PVC (Solvic 258 RD), 100; impact polymeric additive (Kane ACE B22, Kaneka Fushi), 8; processing aid (Modarez APVC 3, Protex), 1; calcium stearate (Rousselot), 0.17; zinc octoate (10% concentration, Polytitan), 0.05; epoxy costabilizer (Paraplex G62 Rohm and Haas), 6; phosphite costabilizer (Polygard, Chevassus), 0.5; α-phenylindole (Sapchim), 0.3; internal lubricants: hydroxystearic acid (Loxiol G15, Henkel), 1.5 and C16–C18 alcohols (Stenol PC, Henkel), 0.3; and optical agents β-phthalocyanine (Sandoz), 0.0006 and (Rubin Graphtol T2BP, Sandoz), 0.0006.

A 150-kg dry blend was prepared by mixing PVC with polymeric additives, internal lubricants, optical agents, zinc and calcium stearates, and α-phenylindole, starting at 1500 rpm and then increasing 300 rpm to reach a temperature of 80°C. At that point, the costabilizer and zinc octoate were added and the stirring speed increased to 3000 rpm, so that the temperature reached 120°C. The dry blend was placed in the cooling vessel that had been cooled under stirring to 30°C and then sieved (400 μ). The first sample (C1) was taken. An extrusion step was carried out using an Eickhoff PWE 200K machine working at 24 rpm with an extrusion temperature of 178°C. A second sample (C2) was taken at this point. The extruded material dropped directly into a calendering

According to Their Processing Conditions

Recycling Rate	*Consumption from Cl^- Ions % (our results)*		*Liberated Acid %*	*Epoxidized Oil Consumption (%)*
	Wall	*Bottom*		
0	51	44	39	53
30	54	46	39	46
100	65	64	42	53
0	59	60	49	46
30	65	65	42	38
100	76	76	42	38
0	62	61	32	38
30	56	49	34	38
100	70	62	49	46
0	42	37	49	69
30	67	67	51	46
100	76	75	56	30
0	52	44	46	46
30	49	54	46	38
100	77	71	54	30
0	49	49	49	38
30	50	52	54	30
100	70	65	55	38

the cooling cycle.

machine (Berstorff). The temperatures of the four cylinders were 192, 189, 193, and 187°C, respectively, and the following stretching step was in the range of 160–170°C with a final linear speed of 18 m/min. A final sample (C3) was taken from the calendered sheet.

Two other sets of samples have been taken from a recipe for cable insulation and for an ignifugated protective envelope of an electric cable. The recipe for cable insulation includes (in phr): PVC (Kwert 70), 100; ditridecyl phthalate, 34; octyldecyl trimellitate, 17; calcium carbonate covered with stearic acid ,14; calcinated kaolin, 6; dibasic lead phthalate, 8; bisphenol A, 0.05.

The recipe for the protective envelope is (in phr): PVC (Kwert 70), 100; propylene glycol polyadipate, 30; chlorinated paraffin (52% Cl), 20; ditridecyl phthalate, 15; dibasic lead phthalate, 10; epoxidized soyabean oil, 2; bisphenol A, 0.05; Sb_2O_3, 14; calcium carbonate covered with stearic acid, 15; carbon black, 2.

In both cases, a dry blend has been prepared in a Henschel machine (limiting temperature, 80°C; blending time, 10 min). A Werner machine was used for gelification (8 min at 120 rpm; final temperature 200°C) followed by a roll mill treatment (4 min; final temperature 170°C) in order to obtain a sheet.

Finally, extrusion has been carried out using an Andouart machine (∅ 69 mm; L/D = 20) at 25, 45, or 60 rpm. The temperatures of the machines' various zones were: filling zone, 160°C; melting zone, 165°C;

metering zone, 170°C; head, 180°C; die, 190°C. The temperatures of the extruded polymer were 165, 170, and 175°C, respectively; the extrusion times were 230, 100, and 70 sec., respectively for the three extrusion rates.

Coulometric Titration of Chloride Ions. The apparatus (Tacussel, Lyon) is composed of a coulometric generator (GCU) and a detection unit (DPA). Ag^+ ions are formed continuously from the generator. Chloride ions waiting to be titrated cause the precipitation of these ions. After their total consumption, the current increases to a preselected point of the detector unit which automatically stops the titration reaction. The amount of Ag^+ ions consumed is given by the generator unit.

The solvent medium we chose is a mixture of tetrahydrofuran (THF) (120 mL) and methanol (30 mL), allowing a homogeneous dissolution of PVC as well as $CaCl_2$ and $ZnCl_2$. NH_4NO_3 (500 mg) and HNO_3 (10*N* (1 mL)) are added to this mixture as electrolytic carriers. A 1% wt solution of polymer either in the same solvent mixtures or in DMF is prepared independently and, according to the amount of chloride ions, between 100 μL and 3 mL of this polymer solution are added to the electrolytic solution.

Preliminary studies of either $CaCl_2$ or $ZnCl_2$ solution (or their mixture) have shown that the maximum error is about 2%, and the sensitivity is such that about 10 nmol of Cl^- ions can be detected.

Potentiometric Titration of Carboxylic Acids. The apparatus (Tacussel Titrimat) is composed of a measuring unit (TAT4), a potentiometric recorder (EPL 1B), an electro addition unit for a solution $10^{-1}N$ of tetra-*n*-butyl ammonium hydroxide, $CH_3OH–CH_3CH_2CH_2OH$, and a Derivol unit that gives the differential curve of potentiometric titration as a function of the volume of added base. The reference electrode is glass, and the measuring electrode is a calomel electrode, both of which are full of dimethylformamide (DMF).

The polymer sample is dissolved in the same solution (4 THF, 1 CH_3OH) as that for coulometric titration, and between 10–20 mL of this solution is added to the titration cell, where 40 mL of DMF have been introduced already. Before each titration, a blank experiment is carried out with a mixture of the THF and CH_3OH solvents without any polymer. The sensitivity is such that about 4 μmol of stearic acid can be titrated with an error of about 6%. When the polymer is dissolved in dimethylformamide, the titration solution in the cell is a mixture of THF and methanol.

Liquid Chromatography Analysis of the Organic Compounds. A high-pressure apparatus built in the laboratory has been used. It is fitted with both a Waters R401 refractive index detector and a Chromatonix 220 UV detector. Three columns of microstyragel (100, 500, and 10000 Å) are used. The carrier solvent is THF. The technique has been applied to bottle samples dissolved in THF (4% in wt % concn). Before injection, solutions are filtered on Millipor (0.5 μ), because some components (calcium stearate, polyethylene wax, and impact modifier polymer) are not soluble in THF. 2 mL of the solution are injected in each analysis. Typical chromatograms are illustrated in Figure 1. PVC, together with the processing aid additive, are eluted first. The polymer peak is used as an internal standard. The epoxidized soyabean oil (ESO) is well separated, followed by a small peak of lubricant (glycol mono-

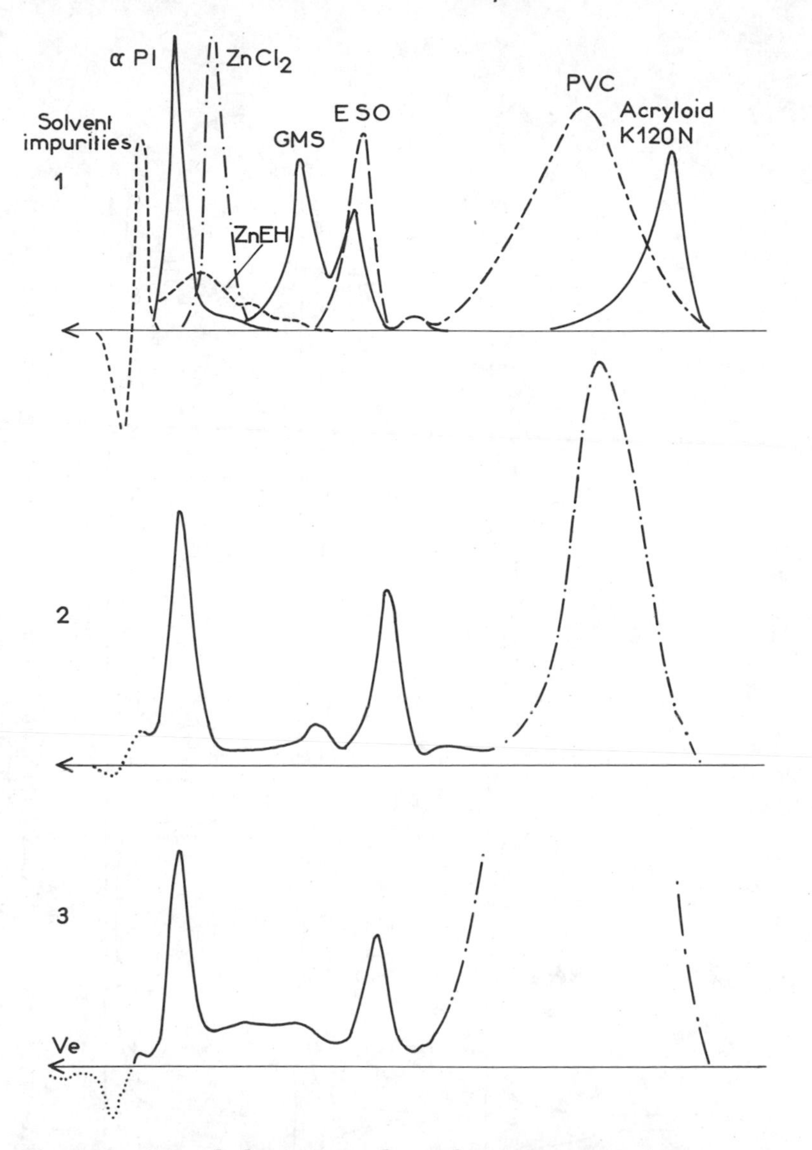

Figure 1. Liquid chromatography with refractometric detection of: (1) the individual components, (2) a fresh bottle formulation (mixture of components without processing), (3) bottle #7.

stearate). Then a complex absorption involves a part of the lubricant, the zinc octoate, the zinc chloride that might be formed, and finally the α-phenylindole. This last compound is the only one giving a strong UV absorption. In some cases, the polymer gives a small UV absorption.

X-Ray Diffraction. This technique has been used for lead formulation where lead phthalate and calcium carbonate show major x-ray peaks. Because of the morphology of the lead phthalate (platelets), oriented

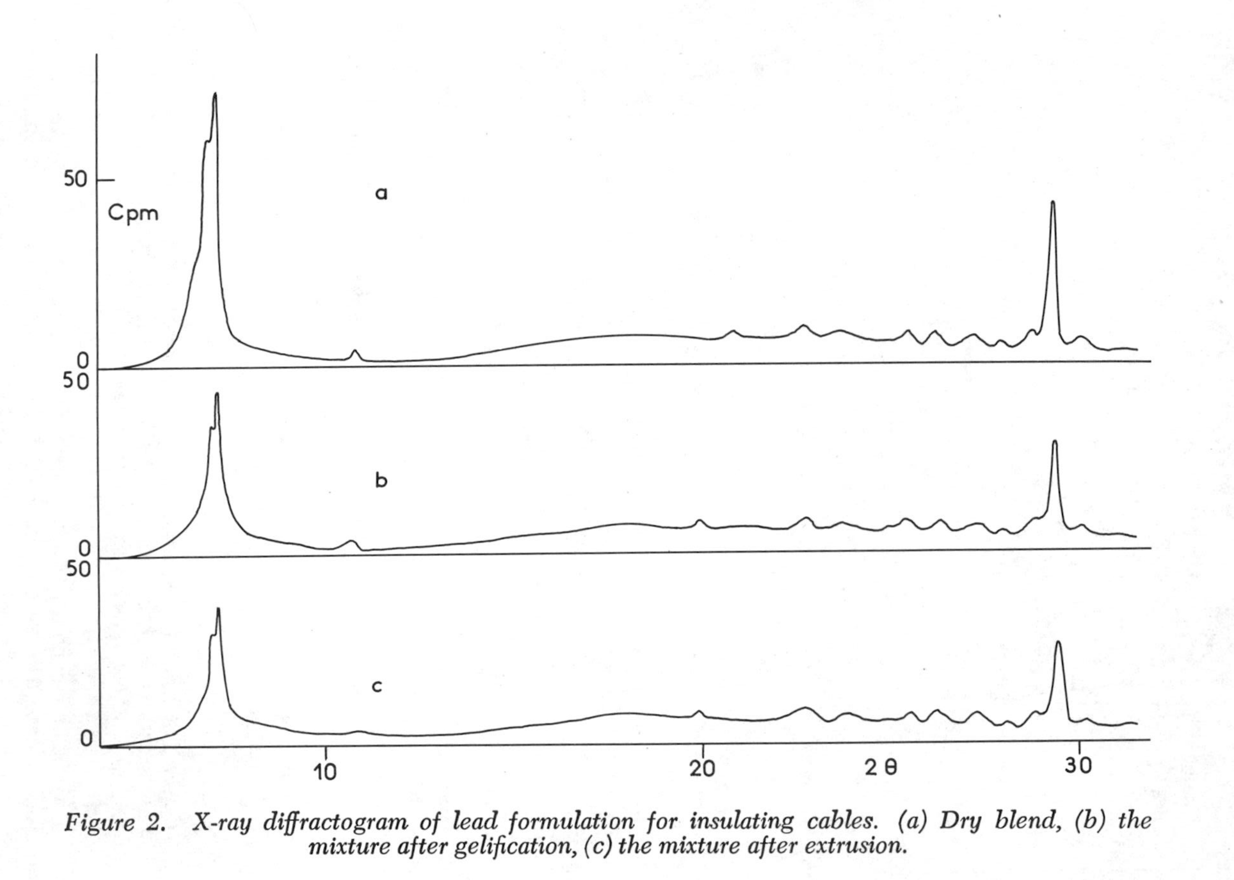

Figure 2. X-ray diffractogram of lead formulation for insulating cables. (a) Dry blend, (b) the mixture after gelification, (c) the mixture after extrusion.

structures are produced by the extrusion and cause a finer diffraction peak. In order to make comparable measurements for all the samples, they have been ground in a coffee mill in the presence of dry ice, in order to obtain fine particles (0.1–0.5 mm). A plain sample is obtained by sintering in an aluminum plate support. The diffractograms are obtained from a Siemens apparatus using the Ni-filtered CuKα band, the aperture of the beam being 0.25°. The step-by-step method (2 mn) has been chosen. In order to avoid too much degradation upon x-ray irradiation, the whole spectra has been taken rapidly from 4° to 50° by steps of 0.25° except for two intervals between 6.9° and 7.3° (lead phthalate), steps of 0.02° and between 29° and 29.2° (calcium carbonate), steps of 0.1°. Typical diffractograms are illustrated in Figure 2. The amounts of residual stabilizer are obtained after planimetric measurement of the area under the peaks, checking for the reproducibility of the continuous underlying halo, and for identical position of the peak with the authentic pure compounds.

Dehydrochlorination. The differential conductimetric apparatus described previously (5) has been used.

Results and Discussion

The results of chloride ions and acidic groups titration, together with that of liquid chromatographic analysis for the 18 bottles are reported in Table I. The consumption percent from Cl^- ions is the percentage of calcium + zinc compounds transformed into chlorides from carboxylates. The percent of acidic groups is based on the hypothesis that the carboxylate and ester groups, including those from the lubricant (GMS) but not those from the processing aid polymer, can be transformed into carboxylic acids from reaction with HCl. Of course, such a reaction is completed only if HCl displaces the ester groups fixed on the polymer after substitution of carboxylates for allylic chlorine atoms.

At first it appears that there are large differences in the same bottle, depending on the position of the sample. Generally the consumption increases when the thickness of the sample decreases. A parallel study (coulometric titration and DHC) has shown that in the thick parts (bottom and top) the results are rather reproducible, while in the thin parts (wall) there is a large dispersion. The results from coulometric titration are reported in Table II. Other results from DHC are shown in Figure 3. The dispersion is large for a bottle made using fresh material (B7)

Table II. Metal Carboxylates Consumption (%) in Bottle. Dispersion According to the Position of the Sample.

Bottle No.	*Wall*					*Bottom*		*Top*	
10	47	42	49			37	38	44	44
14	57	41	52.4	52.1	51.5	—		—	
15	78	74	83			71	71	77	77
18	73.4	61	54.5	77.5	56	—		—	

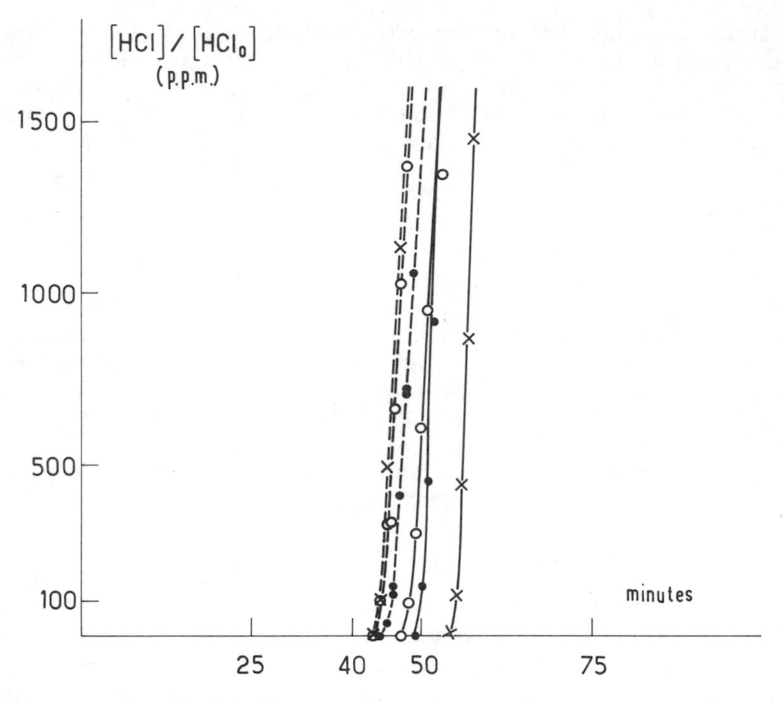

Figure 3. Dehydrochlorination at 178°C under nitrogen for three samples of the top part. Bottle #7 (——); bottle #3 (· · · ·).

although the samples were taken from a thick part (top). The reproducibility is better for a bottle made using a 100% recycling rate. When the samples are taken from the wall of a bottle made with fresh material, the dispersion of the DHC results is very large (Figure 4). Of course, when the recycling rate increases, the consumption also increases. All of the methods agree on that point.

Owing to the great complexity of the chromatogram, it is difficult to interpret the results of liquid chromatography. Only two products are clearly visible—epoxidized soyabean oil and α-phenylindole. A large consumption of the first one is noted, while α-phenylindole seems to be only slightly consumed. However the results concerning this product, which may be poorly separated from the zinc compounds, are not clear and have not been included in Table I. A slight UV signal is observed in some cases for the polymer. It probably results from the substitution of α-phenylindole for the allylic chloride generated in the polymer (7), in competition with the same reaction for epoxy compounds (8) and metal carboxylates (9).

The three samples (C_1, C_2, and C_3) taken from the calendering machine have been studied by coulometric titration and dehydrochlorination. Taking into account the calcium stearate and the zinc stearate and octoate, the consumptions are 25% in the dry blend, 38% after the extrusion step, and 54% for the calendered sheet, respectively. The results from DHC are shown in Figure 5. For the dry blend (Cl), it is clear that the stabilizer is not yet well-distributed in the sample because the slope of the curve is the lowest: that means that a part of the HCl produced may escape from the sample without reacting with the remaining stabilizer. For the extruded sample (C2) the slope is higher so that the stabilizer has reacted efficiently with HCl in the polymer mass. The induction time (74 min) may be used as a measure of the remaining stabilizer: the time is shorter for the final sample (59 min) indicating an extra consumption during calendering. For samples with lead formulation, the coulometric and acidimetric titration show that there are only very limited amounts of Cl^- ions or acid groups soluble in the THF–methanol mixture. Because of the presence of plasticizers that are partly volatilized and condensed and that may react with HCl, the response of DHC measurements is highly dispersed and probably erratic.

The results from x-ray diffraction are reported in Table III as percent of consumption of lead phthalate, calcium carbonate in both formulations, and antimony oxide in the formulation of the protective envelope. Most of the consumption of the lead phthalate (and of antimony oxide) are observed when producing the sheet although the major consumption of

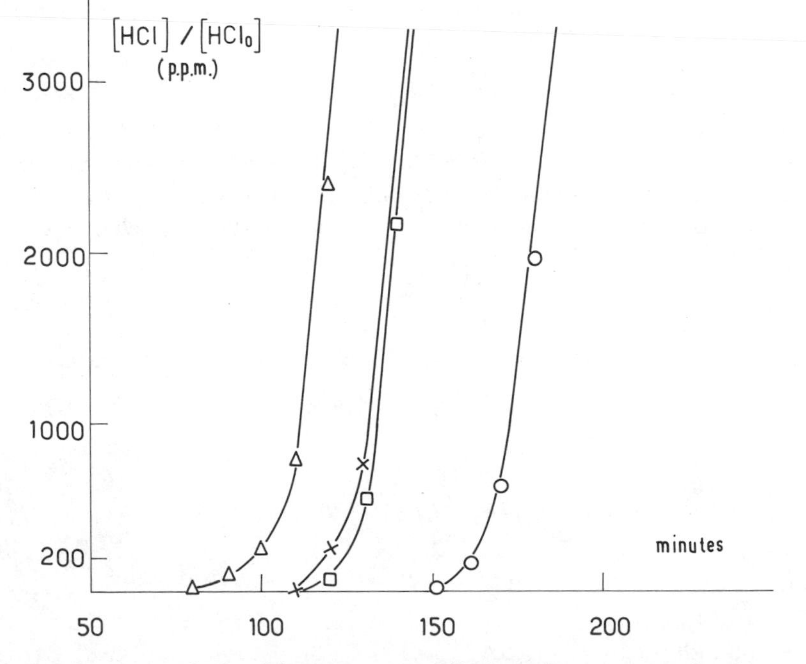

Figure 4. Dehydrochlorination at 170°C under nitrogen for four samples in the wall of bottle #7

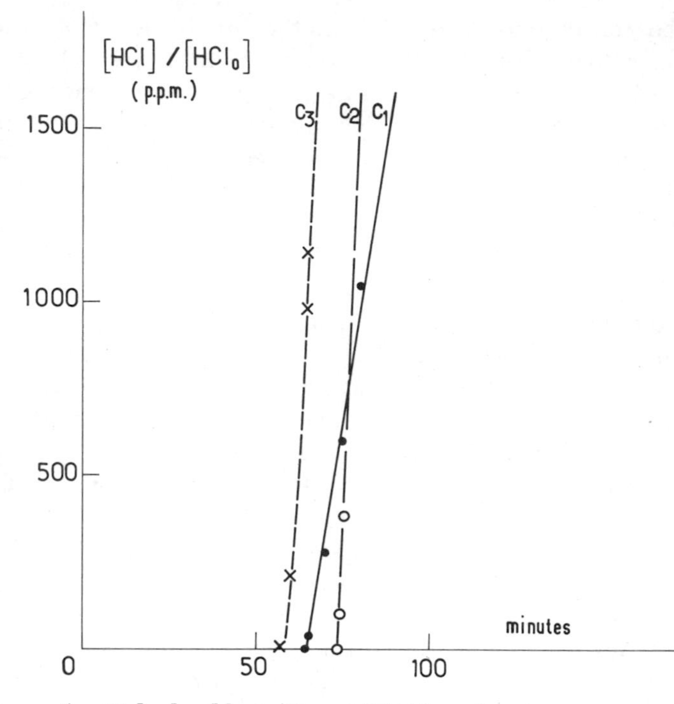

Figure 5. Dehydrochlorination at 184°C under nitrogen atmosphere of samples from the calendering operation

calcium carbonate is observed during the extrusion step. It seems that for the insulating formulation, extrusion at 45 rpm gives the smaller stabilizer consumption while it is clear for both formulations that extrusion at 60 rpm leads to a larger consumption. However, it must be noted that the x-ray results do not correspond obviously to a chemical consumption of the products: some of them can be dissolved or melted progressively.

Table III. X-Ray Diffraction. Results for Lead Formulation

		% Consumption After the Steps				
				Extrusion		
Product *Insulating formulation*	*phr*	*Dry Blend*	*Sheet*	*25 rpm*	*45 rpm*	*60 rpm*
lead phthalate	8	9	46	57	51	62
calcium carbonate	14	3	39	61	60	64
Protective Envelope						
lead phthalate	10	9	36	43	42	51
calcium carbonate	15	2	21	51	50	52
antimony oxide	14	2	17	21	21	21

Acknowledgments

The authors are indebted to M. Marois (Seprosy Lagnieu), M. Brive and Lemaure (Silec, Montereau), and M. Nivière and M. Boussand (BAT Taraflex, Tarare) who processed the bottles, cables, and calendered sheets, respectively. Grant 75–7–0128 obtained from the DGRST (Delégation Générale à la Recherche Scientifique et Technique) is acknowledged.

Literature Cited

1. Fougea, D., Ghaleb, M., Revirand, G., Pacelli, E., *Cah. Cent. Sci. Tech. Batim.* (1971) **142**, 1070.
2. Neubert, G., Wirth, O., *Z. Anal. Chem.* (1975) **273**, 19.
3. Wirth, H. O., Andreas, H., *Pure and Appl. Chem.* (1977) **49**, 627.
4. Stepek, J., Vymazal, Z., Czako, E., *J. Macromol. Sci., A*, in press.
5. Guyot, A., Bert, M., *J. Appl. Polym. Sci.* (1973) **17**, 753.
6. Michel, A., Van Hoang, T., Guyot, A., ADV. CHEM. SER. (1978) **169**, 386.
7. Van Hoang, T., Michel, A., Guyot, A., *Eur. Polym. J.* (1976) **12**, 357.
8. Van Hoang, T., Michel, A., Guyot, A., *Eur. Polym. J.* (1976) **12**, 347.
9. Van Hoang, T., Michel, A., Guyot, A., *Eur. Polym. J.* (1976) **12**, 337.

RECEIVED May 12, 1977.

30

Synergistic Mechanisms between Zinc, Calcium Soaps, and Organo Compounds in Poly(vinyl chloride) Stabilization

A. MICHEL, TRAN VAN HOANG, and A. GUYOT

CNRS, Laboratoire des Matériaux Organiques, 79, bd du 11 novembre 1918, 69626 Villeurbanne Cedex, France

Owing to metal chlorides titration by the coulometric method, and carboxylic acid titration by the potentiometric method, it is possible to follow the metal soaps consumption during thermomechanical heat treatments. This new technique provides a better understanding of the stabilization mechanisms of PVC with the calcium–zinc system, and offers a better explanation of synergistic effects between metal soaps and secondary stabilizers such as epoxidized soya-bean oil, α-phenylindole, and butanediol-β-aminocrotonate. The influence of these last stabilizers on zinc chloride formation enables us to classify them into short- and long-term stabilizers.

At a temperature above 80°C, poly(vinyl chloride) eliminates hydrogen chloride and allylic chlorinated structures appear, with 4-chloro-2-hexene being considered as a model. At the processing temperature (180–200°C), the main problem of poly(vinyl chloride) stabilization is preventing the zip dehydrochlorination that induces discoloration and cross-linking of the polymer.

In the zinc–calcium formulations, metal soaps are associated with secondary stabilizers such as alkylphosphites, epoxy compounds, α-phenylindole, or β-aminocrotonate esters that bring a synergistic effect. The first attempt to study the mechanisms behind this effect has been the reaction of all of these compounds with 4-chloro-2-hexene.

Zinc stearate (*1*) reacts with 4-chloro-2-hexene to yield hexenyl stearate and zinc chloride, which is a strong catalyst for this esterification reaction as well as for the degradation reaction that yields hexadiene and

0-8412-0381-4/78/33-169-386$05.00/1

hydrogen chloride. This last reaction is favored only when zinc carboxylate is consumed. On the contrary, calcium carboxylates do not react with 4-chloro-2-hexene without zinc chloride. From all of the studies with model compound (*1, 2*), the synergism between zinc and calcium soaps is explained by an exchange reaction between the ligands of calcium carboxylate and zinc chloride in order to regenerate zinc carboxylate which then esterifies allylic chlorine atoms. This exchange reaction is equilibrated and complexes can exist between calcium stearate or calcium chloride and zinc chloride, thereby reducing its prodegradent activity. Finally, in the presence of an HCl excess, the hexenyl stearate (*1*) is not stable at a temperature as low as 60°C and regenerates 4-chloro-2-hexene and stearic acid. This reaction can occur in the polymer when the stabilizer is consumed and allylic chlorine atoms are formed, which explains the catalytic effect of degradation.

For the synergistic mixture composed of metal soaps and organic compounds as secondary stabilizers, the studies with model compound gave the first explanation of synergism mechanisms. Alkyl phosphites (*3*) and epoxy compounds (*4*) react with 4-chloro-2-hexene through an O-alkylation reaction with formation of phosphonate ester and α-chloroether, respectively. Zinc chloride strongly catalyzes the esterification and etherification reactions, but phosphonate esters or ethers are not stable in the presence of HCl at a temperature as low as 60°C. The alkyl phosphites and epoxy compounds also have the capacity of binding hydrogen chloride evolved from the model compound or from the polymer. However, their main property in the synergism probably is caused by their reaction with the zinc chloride (*4, 5*) which aids in the inhibition of its activity in the dehydrochlorination of the polymer.

α-Phenylindole (*6*) and butanediol-β-aminocrotonate (*7*) react with allylic chlorine atoms through carbon-alkylation, or nitrogen-alkylation, respectively. These reactions are catalyzed with zinc chloride, but the substitution products are not sensitive to HCl at low temperatures and the stabilization reaction is not reversible. β-amino-crotonate esters can bind HCl to form NH_4Cl, and it can react with zinc chloride. Zinc chloride also forms a complex with NH_4Cl that strongly decreases its catalytic activity in the dehydrochlorination. Considering the above studies, it is possible to derive some conclusions about PVC from the results concerning the model compound by using a Brabender plasticorder that records the polymer viscosity during its heating. It is possible to define the stabilizer's action time (T_A) for PVC. This parameter represents the time in minutes or seconds between gelation and cross-linking of the polymer (between the maxima of gelation and cross-linking peaks on the plastogram). This parameter includes the influence of sta bilizers and their by-products, and varies with the stabilizer amount. It is then possible to compare its variation with the esterification, etherification, and carbon-alkylation yield of 4-chloro-2-hexene vs. the stabilizer molar fraction in synergistic mixtures.

Using new methods (7) for following the transformation of metal soaps in the polymer, this chapter studys the synergism obtained from chloride ions between zinc and calcium stearate in the absence or presence of secondary stabilizers during PVC thermomechanical treatments in the Brabender plasticorder or in a rolling mill.

Experimental

To follow the consumption of zinc and calcium carboxylates during heating time in the Brabender plasticorder at 190°C, the following recipe was used: PVC, 30 g; dioctylphthalate, 9 g; zinc or calcium stearate, 6.6 phr; for the synergistic mixture zinc stearate, 3.3 phr; and calcium stearate, 3.3 phr. These amounts of stabilizers are about three or four times larger than the quantity used in industrial conditions to facilitate the methods of titration. The plasticorder is stopped occasionally, and from all the samples only one part is placed in solution in tetrahydrofuran for titration of chlorides and stearic acid.

On the rolling mill, heat treatments were performed without plasticizer at 170°C for 4 min with the following recipe: zinc or calcium stearate, 6.6 phr; for their synergistic mixture: zinc stearate, 3.3 phr; calcium stearate, 3.3 phr; and stearic acid, 0.33 phr, as an external lubricant. Then the sheets were heated in a stove at 180°C in the presence of nitrogen; the carboxylate consumption was monitored during this treatment.

Heat treatments also were performed on a rolling mill at 170°C for the binary synergistic mixture in the presence of zinc carboxylates and organocompounds with the following recipe: zinc stearate, 6.6 phr; epoxidized soya-bean oil from Ciba Geigy (iodine index 5.8; acid index 2 and oxiran index 6.1), 5.53 or 16 phr; α-phenylindole (stabilizer 3621*N* of Sapchim Fournier Cimag), 2.13 phr; dibutyl-β-aminocrotonate ester (stabilizer 5420 of Sapchim Fournier Cimag), 2.7 phr. The amount of secondary stabilizer is such that its molar amount is equal to that of zinc stearate. Then the sheets were heated at 180°C in a stove in the presence of nitrogen; zinc chloride titration was carried out occasionally.

During heat treatments, zinc and calcium stearates react with hydrogen chloride that has evolved from the polymer or with labile chlorine atoms to liberate zinc and calcium chlorides. These compounds are titrated by a coulometric method in tetrahydrofuran (7) using a generator of Ag^+ ions that are precipitated in the presence of chloride ions. When zinc and calcium stearates are used together, both chlorides are titrated at the same time, and they are not differentiated.

Stearic acid titration is performed by a potentiometric method in dimethylformamide and in the presence of tetra-*n*-butyl ammonium hydroxide (7). The consumption of carboxylates vs. time is deduced from chloride and stearic acid titration according to the stoichiometry of Reaction 1 where Me is Zn or Ca and R is the stearic group.

$$Me(OCOR)_2 + 2\ HCl \rightarrow MeCl_2 + 2\ RCOOH \qquad (1)$$

Results and Discussion

Zinc and Calcium Stearate Synergism. With the Brabender plasticorder, it is possible to follow the viscosity variations of a polymer because of physical or chemical modifications. The shape of the typical plastograms for PVC plasticized and stabilized with calcium stearate ($Ca(OCOR)_2$) (6.6 phr), zinc stearate ($Zn(OCOR)_2$) (6.6 phr), and their mixture (3.3 phr both stearates), are shown in Figures 1, 2, and 3, respectively (– – –). Because of degradation, the first peak corresponds to the gelation phase and the second peak to the cross-linking phase. Each numbered point of the plastogram corresponds to the end of an experiment.

In the presence of calcium stearate, discoloration increases with the heating time after gelation and cross-linking appears in area 8–9 of the plastogram (Figure 1) after a 35–40 min heating time. For this stabilizing system, the action time of stabilizer T_A is about 52 min.

In the presence of zinc stearate, discoloration does not appear during gelation and fusion (area 1–5 on the plastogram, Figure 2), but darkening develops quickly after a 12–15 min heating time; and the polymer cross-linking strongly increases the viscosity simultaneously. Under these

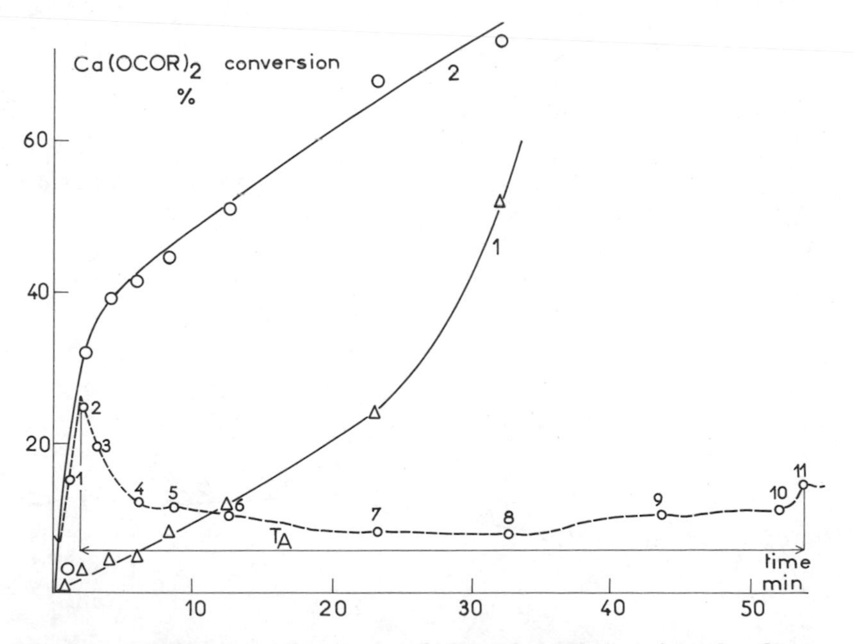

Figure 1. Viscosity evolution of plasticized PVC (– – –) and calcium stearate transformation during heating in the Brabender plasticorder at 190°C from chloride ions titration (Curve 1) and stearic acid titration (Curve 2). PVC, 30 g; DOP, 30 phr, calcium stearate, 6.6 phr.

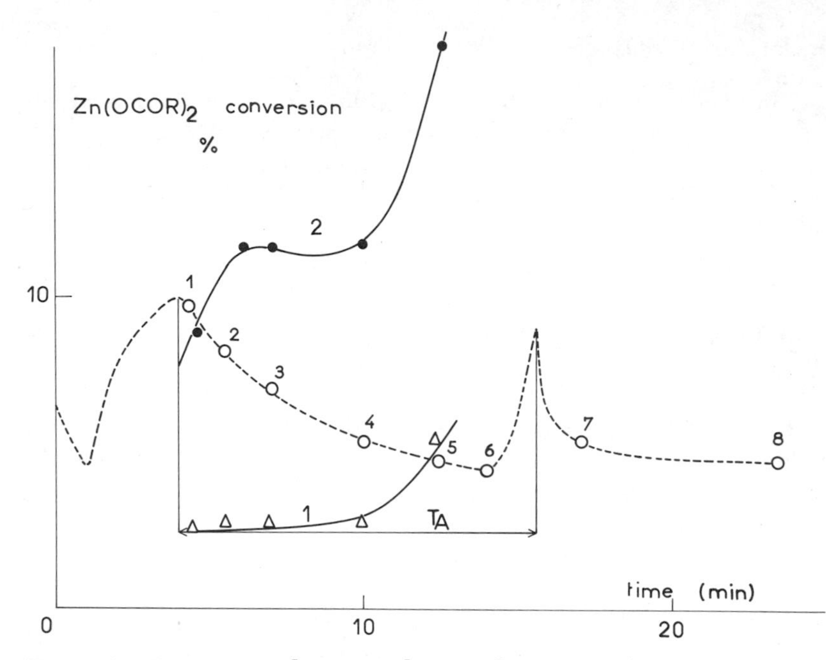

Figure 2. Viscosity evolution of plasticized PVC (---) and zinc stearate transformation during heating in the Brabender plasticorder at 190°C from chloride ions titration (Curve 1) and stearic acid titration (Curve 2). PVC, 30 g; DOP, 30 phr; zinc stearate, 6.6 phr.

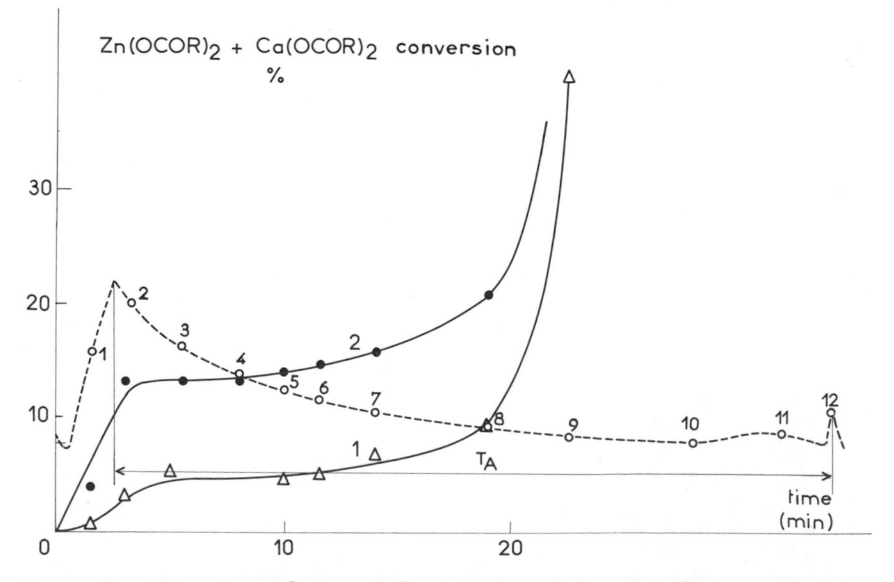

Figure 3. Viscosity evolution of plasticized PVC (---) and zinc and calcium stearate transformation during heating in the Brabender plasticorder at 190°C from chloride ions titration (Curve 1) and from stearic acid titration (Curve 2). PVC, 30 g; DOP, 30 phr; zinc stearate, 3.3 phr; calcium stearate, 3.3 phr.

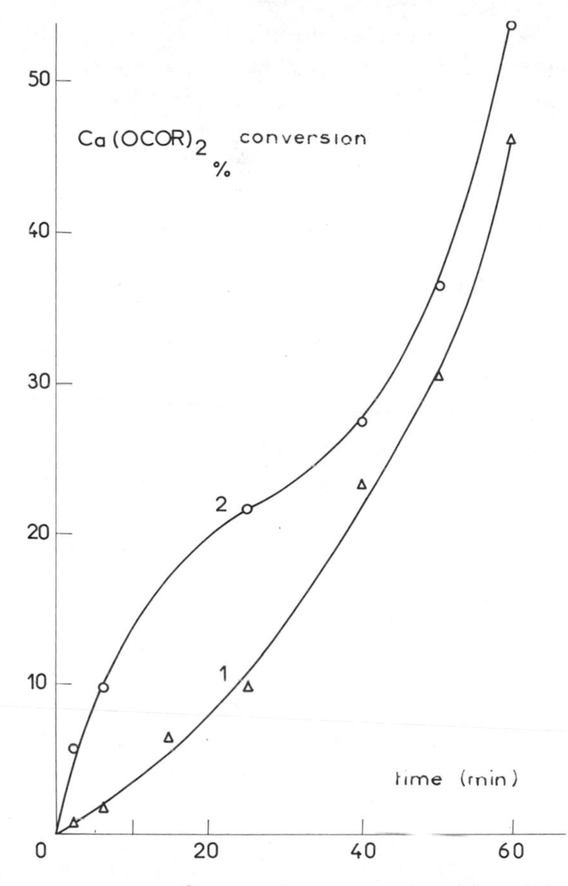

Figure 4. Calcium stearate transformation from chloride ions titration (Curve 1) or from stearic acid titration (Curve 2) in the sheets of rigid PVC heated in a stove at 180°C under nitrogen atmosphere

conditions, T_A = 14 min, 30 sec. For the zinc and calcium stearate mixture, discoloration develops for a heating time longer than that for zinc stearate only, but T_A (about 32 min) (plastogram on Figure 3) is smaller than that for the stabilizing system with only calcium stearate.

For each stabilizing system we have drawn the total amount of transformed carboxylates versus heating time from metal chloride titration (Curve 1) or stearic acid titration (Curve 2), according to the stoichiometry of Reaction 1 (Figures 1, 2, and 3). Curves 1 and 2 would be superposed on each figure if the stabilizer consumption was caused only by Reaction 1; however, experimentally they are not.

The consumption of calcium stearate is proportional to the mixing time at 190°C, but the amount of liberated stearic acid is more important than the amount of calcium chloride during polymer gelation (Figure 1). This result suggests that the amount of HCl evolved from the polymer during the gelation phase is more important than during the melting phase. HCl evolved from the polymer can react with calcium stearate in two steps according to Reactions 2 and 3.

$$Ca(OCOR)_2 + HCl \rightarrow ClCaOCOR + RCOOH \tag{2}$$

$$ClCa(OCOR) + HCl \rightarrow CaCl_2 + RCOOH \tag{3}$$

The same results are observed during heat treatments on sheets in a stove at 180°C in nitrogen atmosphere.

Figure 4 shows the transformation of calcium stearate vs. heating time without plasticizer. The difference between the amount of stearic acid and calcium chloride is less important at the beginning of heating, and Curves 1 and 2 tend to superpose from 50% conversion. The deviation between calcium chloride and stearic acid during both heat treatments in the Brabender plasticorder or in a stove can be caused partially by dioctylphthalate degradation at 190°C. This plasticizer degradation was not studied in detail, but with gas chromatography analysis we at least showed the presence of three products after heating at 190°C.

For zinc stearate, the situation is very different (Figure 2). The stabilizer transformation during heat treatments in the Brabender plasticorder appreciated from $ZnCl_2$ titration is very weak during the gelation and melting phases and is constant until the beginning of cross-linking. The transformation appreciated from stearic acid titration is a little more important, confirming the liberation of HCl during the gelation phase. The amounts of stearic acid and zinc chloride increase quickly, and at the same time polymer cross-linking and discoloration appear. When the polymer is melting and when the temperature is higher than the melting point of zinc stearate, this compound substitutes allylic chlorine atoms according to Reactions 4 and 5.

$$\sim CHCl{-}CH_2{-}CH{=}CH{-}\underset{\displaystyle Cl}{\underset{|}{CH}}{-}CH{-}CH_2\sim \quad + Zn(OCOR)_2 \longrightarrow$$

$$\sim CHCl{-}CH_2{-}CH{=}CH{-}\underset{\displaystyle O{=}C{-}R}{\underset{|}{\underset{O}{\underset{|}{CH}}}}{-}CH_2\sim \quad + ClZnOCOR \tag{4}$$

$$\sim\text{CHCl–CH}_2\text{–CH}{=}\text{CH–CH(Cl)–CH}_2\sim \quad + \text{ClZnOCOR} \longrightarrow$$

$$\sim\text{CHCL–CH}_2\text{–CH}{=}\text{CH–CH(O–C(=O)–R)–CH}_2\sim \quad + \text{ZnCl}_2 \qquad (5)$$

As soon as $ZnCl_2$ appears in Reaction 5, its prodegradent activity prevails and HCl evolved from the polymer can react with the unreacted zinc stearate or the chloro organozinc compound to liberate stearic acid. HCl also can react with the ester group grafted on the polymer according to Reaction 6.

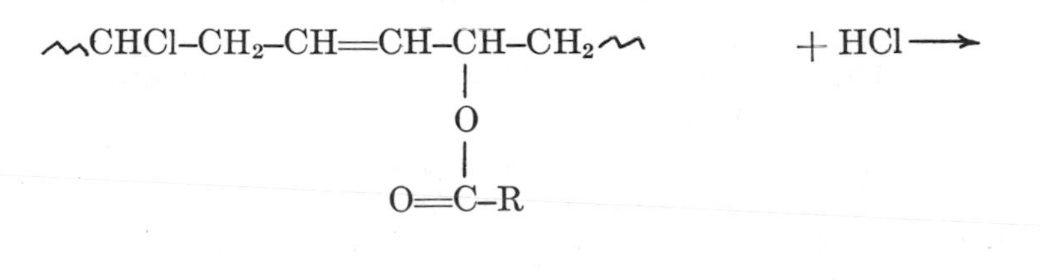

$$\sim\text{CHCl–CH}_2\text{–CH}{=}\text{CH–CH(Cl)–CH}_2\sim \quad + \text{RCOOH} \qquad (6)$$

This reaction was clarified with hexenyl stearate (*1*) at 60°C during the reaction between zinc stearate and 4-chloro-2-hexene in tetrahydrofuran. During this reaction, zinc stearate reacts immediately with 2 mol of model compound, and kinetic laws do not suggest the presence of ClZnOCOR as an intermediate product in the substitution reaction. Nevertheless, such an organometallic compound was suggested by Frye and Horst (*9*) and also by Anderson and McKenzie (*10*) who assumed that such compounds accounted for the ligand exchange between cadmium chloride and barium carboxylate.

For zinc and calcium-stearate mixture, there is always an excess of transformed stearate from stearic acid titration with respect to metal chlorides titration during gelation of plasticized PVC in the Brabender plasticorder (area 1–3 on the plastogram, and Curves 1 and 2 on Figure

3). However, without zinc stearate, the amount of stearic acid or chlorides remains practically constant until ~ 20-min heating time. In this phase of heating, the substitution reaction is the most important reaction and calcium stearate exchanges its ligands with chloroorganozinc compounds or $ZnCl_2$ to regenerate zinc stearate according to the following reactions:

$$ClZnOCOR + Ca(OCOR)_2 \rightarrow Zn(OCOR)_2 + ClCaOCOR \quad (7)$$

$$ClCa(OCOR) + ClZnOCOR \rightarrow CaCl_2 + Zn(OCOR)_2 \quad (8)$$

$$ZnCl_2 + Ca(OCOR)_2 \rightarrow Zn(OCOR)_2 + CaCl_2 \quad (9)$$

The last two reactions do not account for stearic acid formation and Reaction 9 assumes that the exchange rate is very high, which is difficult to admit in a polymeric phase even if it is melted.

After a 20-min heating time, free chlorides appear quickly and the polymer viscosity decreases slightly (area 6–10 on the plastogram (Figure 3), which proves the presence of free calcium chlorides which do not induce dehydrochlorination and cross-linking of PVC. In fact, both will appear in the presence of free zinc chloride. When exchange reactions with calcium stearate and substitution reactions are completed, HCl can evolve from the polymer and react with ClCaOCOR according to Reaction 3, and also with an ester group grafted on the polymer according to Reaction 6; this may be important in area 11–12 of the plastogram.

The previous results are confirmed during heat treatments on the sheets without plasticizer in a stove at 180°C, and then stabilized with the zinc–calcium synergistic mixture. Table I summarizes the chloride and stearic acid percentages that are liberated during heating with respect to the initial amount of stearates.

Table I. Chlorides and Stearic Acid Titration during Heating of the PVC Sheets in a Stove at 180°C

Time (Min.)	*Metal Chlorides (%)*	*Stearic Acid (%)*
0	0.5	30
2	1.25	31
6	1.50	43.3
15	9.5	45.7
20	28.7	71.6
25	insoluble	insoluble

These experiments show mainly that there is an insufficient amount of chlorides with respect to stearic acid during manufacturing of the sheets on the rolling mill, confirming HCl evolution during the gelation phase. In a stove, the chlorides appear after a 15-min heating time and the polymer cross-linking appears only after 25 min. This proves that the chlorides that appear first are calcium chlorides.

Calcium stearate acts only as an acceptor of hydrogen chloride when it is used alone and this reaction develops into two steps with formation of a chloro organocalcium compound (ClCaOCOR). This action prevents the HCl catalytic activity in the degradation of the polymer but not the formation of polyene sequences (discoloration) as soon as it is heated. On the contrary, zinc stearate quickly participates in the substitution reaction of allylic chlorine atoms with the formation of a chloro organozinc compound (ClZnOCOR), whose main role in the presence of calcium stearate is to exchange its chlorine atom for a stearic ligand and to regenerate zinc stearate and a chloro organocalcium compound inactive in the dehydrochlorination. Owing to coulometric titration of chloride ions during PVC heating, it was possible to confirm the theory of the ester exchange between zinc chloride and calcium soaps in order to explain the synergism between zinc and calcium carboxylates. A strong correlation exists between the studies of synergism mechanisms in solution with the model compound and in the polymer. It is believed that in solution, the intermediate chloro organocalcium and chloro organozinc compounds were not characterized in studies on the model compound because the equilibrium was reached easily and quickly.

Synergistic Mixture: Zinc Stearate and Organo Compounds as Secondary Stabilizers. From the previous results, it is clear that in order to prevent PVC cross-linking and to increase the action time of stabilizer (previously defined with the Brabender plasticorder (*1, 3, 4, 6, 7*)), it is necessary to impede zinc chloride accumulation in the polymer.

To study the influence of secondary stabilizers on zinc chloride formation, we have prepared PVC sheets on a rolling mill with the different binary mixtures zinc stearate and α-phenylindole, epoxidized soya-bean oil, or β-aminocrotonate esters, and we have followed the zinc chloride formation by coulometric titration of chloride ions. In Figure 5 we have combined the percentage of zinc chloride that may be liberated with respect to the initial amount of zinc stearate in the presence of α-phenylindole (Curve 1), epoxidized soya-bean oil (Curves 2 and 3), and butanediol-β-aminocrotonate (Curve 4) during heating in a stove at 180°C.

These results show that α-phenylindole does not influence zinc chloride formation and that it may catalyze the substitution reaction of allylic chlorine atoms on the polymer as it catalyzes the esterification reaction of 4-chloro-2-hexene (*6*). But its accumulation quickly implies discoloration. From these experiments, it is clear that α-phenylindole is a short-term stabilizer that only impedes initial coloring. Nevertheless, the darkening develops as quickly as the cross-linking. These results account also for T_A variations, implying that the time prior to cross-linking is not very

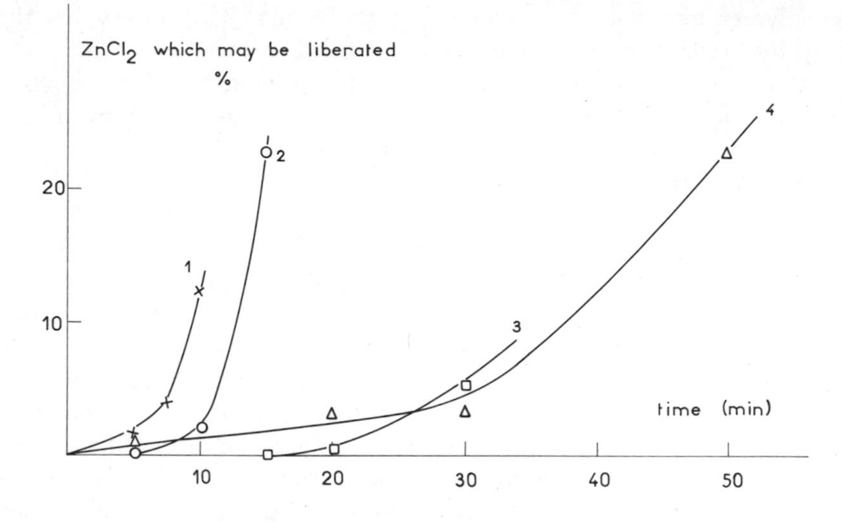

Figure 5. Zinc chloride formation in the sheets of rigid PVC heated in a stove at 180°C under nitrogen atmosphere and in the presence of a secondary stabilizer. (1) α-phenylindole, 2.13 phr; (2) epoxidized soya bean oil, 5.53 phr; (3) epoxidized soya bean oil, 16 phr; (4) butanediol-β-aminocrotonate, 2.7 phr.

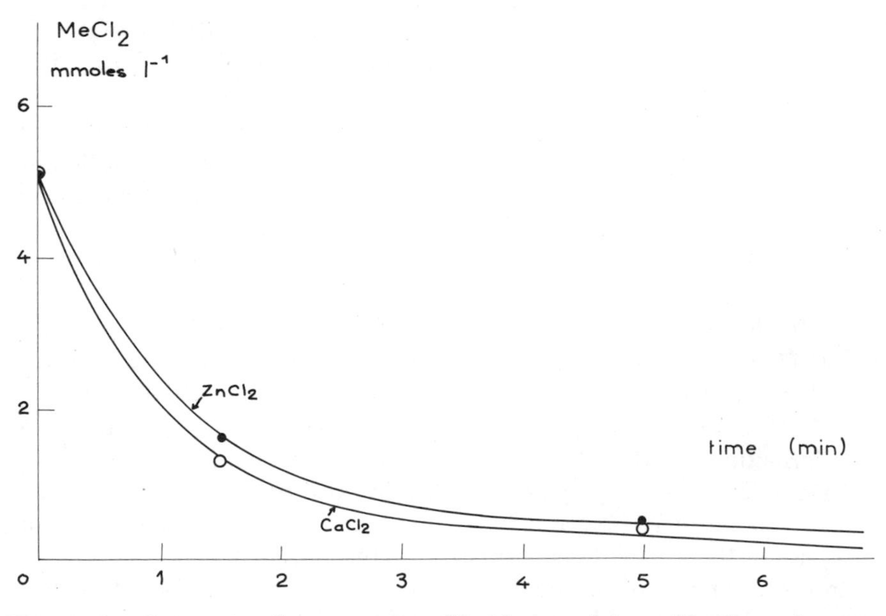

Figure 6. Interaction between zinc chloride or calcium chloride and epoxidized soya-bean oil at 180°C

sensitive to the molar fraction of α-phenylindole either in a binary stabilizing system with zinc stearate or in a tertiary system with zinc stearate and epoxidized soya bean oil as previously considered (*6*).

On the contrary, for the zinc stearate–epoxidized soya bean oil binary system, there is an inhibition time in zinc chloride formation (Curves 2 and 4) that is proportional to the amount of epoxy compound; the zinc chloride formation rate is inversely proportional to this amount. With this stabilizing system, a yellowing appears from the beginning of heating and progresses with the time of heating. Zinc chloride consumption is correlated with the power of this metal chloride to initiate the polymerization of epoxidized compounds (*4*). The reaction between these compounds and metal chlorides seems to be general, as evidenced by Figure 6, which shows the disappearance of zinc chloride or calcium chloride followed by coulometric titration in pure epoxidized soya bean oil at 180°C. From liquid chromatography experiments, it is possible to conclude that epoxidized soya bean oil is consumed and grafted on the polymer during heat treatments. Therefore, epoxy compound synergism binds hydrogen chloride evolved from the polymer, substitutes allylic chlorine atoms, and inhibits zinc chloride in its prodegradent activity. For these reasons, the compounds are long-term stabilizers; however, they do not prevent initial discoloration.

Finally, butanediol-β-aminocrotonate also delays zinc chloride formation (Curve 4), and with the same molar ratio it is more efficient than epoxidized soya-bean oil, although a small quantity of chlorides is formed at the beginning of the heat treatment. From studies with 4-chloro-2-hexene (*7*), these chlorides may be chlorohydrate, ammonium chloride, and zinc chloride. These compounds not only catalyze the substitution reaction of allylic chlorine atoms but also form a strong complex with ammonium chloride. On the chromatogram of gel permeation chromatography, there is a shoulder on the high molecular weight part of the polymer after a 50-min heating time confirming the reaction between poly(vinyl chloride) and β-aminocrotonate esters. Butanediol-β-aminocrotonate is a long-term stabilizer and delays the polymer cross-linking, preventing free chlorides formation. This property accounts for the increasing action time of stabilizer (T_A) with its molar fraction in the synergistic mixture with zinc stearate during thermomechanical treatments in the Brabender plasticorder.

Conclusions

Knowing the quantity of metal chlorides and carboxylate acid liberated during thermomechanical heat treatments permits a better understanding of the stabilization mechanisms of metal soaps. This study confirms that the action time of stabilizer T_A includes not only the primary influence of stabilizers, but also the influence of their by-products. Therefore, this parameter enables us to establish a correlation between the model compounds and PVC. The ability of a secondary stabilizer to permit or prevent zinc chloride evolution defines a long- or short-term stabilizer. In this first case, the yellowing appears quickly from the onset of heating and it develops slowly into an orange and brown color. In the second case, the discoloration does not develop immediately, but the darkening does.

Acknowledgments

We wish to thank A. Revillon for the liquid chromatography experiments.

Literature Cited

1. Van Hoang, T., Michel, A., Guyot, A., *Eur. Polym. J.* (1976) **12**, 337.
2. Onozuka, M., Asahina ,M., *J. Macromol. Sci., A* (1969) **2**, 235.
3. Van Hoang, T., Michel, A., Pham, Q. T., Guyot, A., *Eur. Polym. J.* (1975) **11**, 475.
4. Van Hoang, T., Michel, A., Guyot, A., *Eur. Polym. J.* (1976) **12**, 347.
5. Briggs, G., Wood, N. F., *J. Appl. Polym. Sci.* (1971) **15**, 25.
6. Van Hoang, T., Michel, A., Guyot, A., *Eur. Polym. J.* (1976) **12**, 357.
7. Van Hoang, T., Michel, A., Guyot, A., *J. Macromol. Sci., Prepr.*, International Symposium on PVC, 2nd (Lyon, France) in press.
8. Van Hoang, T., Michel, A., Bert, M., Revillon, A., Douillard, A., "Stabilization and Degradation of Polymers," *Am. Chem. Soc., Prepr.* (New Orleans, March, 1977).
9. Frye, A. H., Horst, R. W., *J. Polym. Sci.* (1960) **45**, 1.
10. Anderson, D. F., McKenzie, D. A., *J. Polym. Sci.* (1970) **8**, 2905.

RECEIVED May 12, 1977.

Antioxidative Properties of Phenyl-Substituted Phenols

Their Role in Synergistic Combinations with β-Activated Thioethers

C. R. H. I. DE JONGE, E. A. GIEZEN, F. P. B. VAN DER MAEDEN, W. G. B. HUYSMANS, W. J. DE KLEIN, and W. J. MIJS

Akzo Research Laboratories, Corporate Research Department, Arnhem, The Netherlands

Phenyl-substituted phenols and sulfides show remarkable antioxidant synergism in polypropylene at 180°C, i.e., close to processing conditions. In our investigations into phenyl-substituted phenols we synthesized potential end-use antioxidants of this type. Oven-aging at 80°, 120°, and 140°C was carried out with polyproylene films. The decay of the phenyl-substituted phenolic antioxidants was measured during the induction period using UV spectrophotometry to predict the time to embrittlement. In contrast to other chain-breaking antioxidants, the concentration of the phenyl-substituted phenolic antioxidant remained constant even after severe processing. These observations agree with the results of model compounds in polypropylene, which show that the phenyl-substituted phenol is regenerated continuously at higher temperatures (150°–200°C) by the sulfide or oxidation products of the sulfide.

Most organic polymers are highly susceptible to oxidative degradation and therefore require the addition of stabilizers to provide protection during processing and end-use. Stabilizers of extreme importance for polyolefins include phenolic antioxidants as chain-breaking antioxidants and sulfur compounds as hydroperoxide decomposers. Combinations of these two types of antioxidants often show marked synergistic effects (*1, 2, 3*). In spite of their importance little is known about the structural requirements for optimum interaction between phenolic antioxidants and peroxide decomposers.

0-8412-0381-4/78/33-169-399$06.75/1

Scheme 1

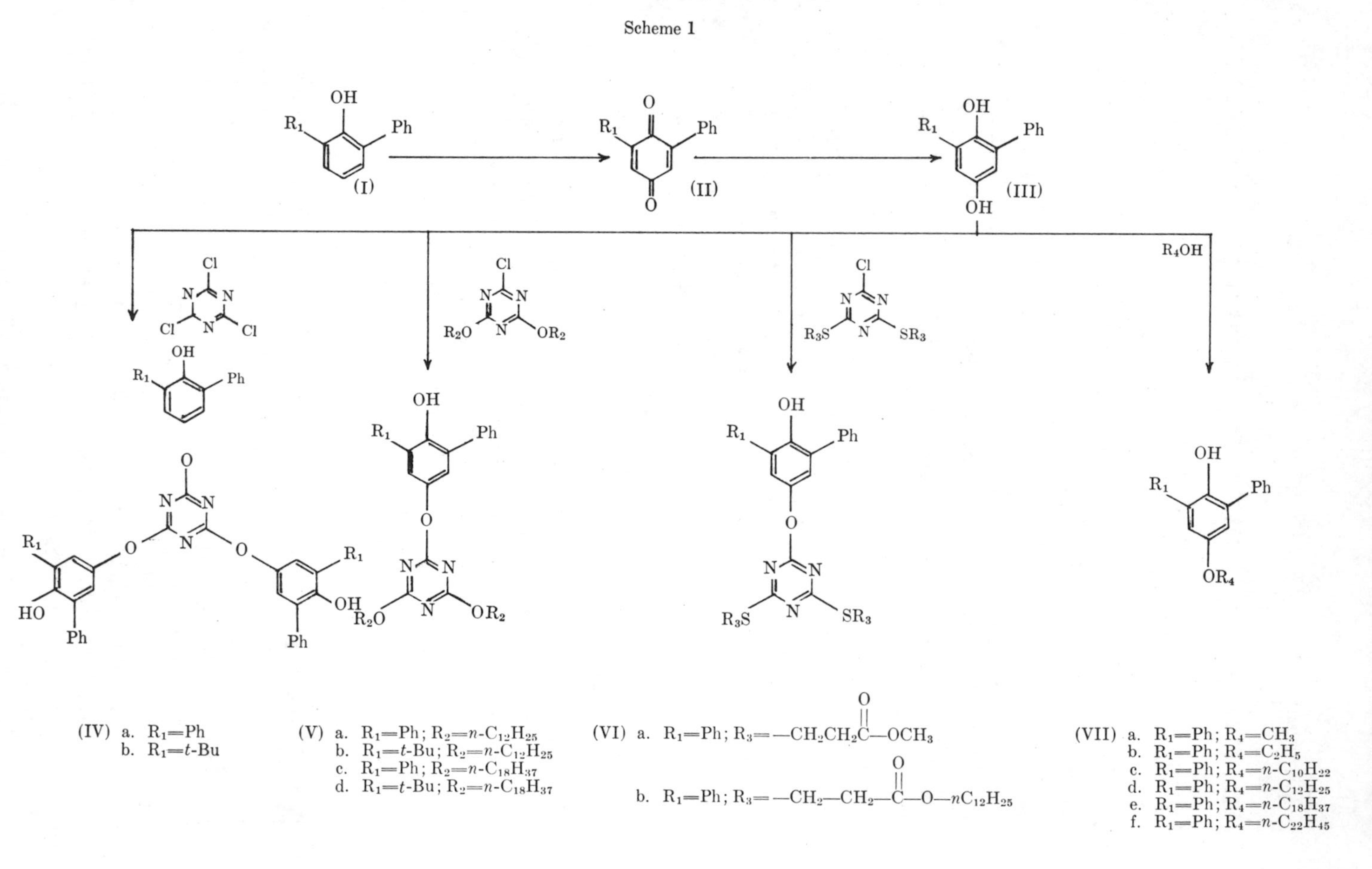
OH
R₁
Ph
(I)
O
(II)
(III)
R₄OH
Cl
N
R₂O
OR₂
R₃S
SR₃
OR₄
HO
(IV) a. R₁=Ph
b. R₁=t-Bu
(V) a. R₁=Ph; R₂=n-C₁₂H₂₅
b. R₁=t-Bu; R₂=n-C₁₂H₂₅
c. R₁=Ph; R₂=n-C₁₈H₃₇
d. R₁=t-Bu; R₂=n-C₁₈H₃₇
(VI) a. R₁=Ph; R₃=—CH₂CH₂C(=O)—OCH₃
b. R₁=Ph; R₃=—CH₂—CH₂—C(=O)—O—nC₁₂H₂₅
(VII) a. R₁=Ph; R₄=CH₃
b. R₁=Ph; R₄=C₂H₅
c. R₁=Ph; R₄=n-C₁₀H₂₂
d. R₁=Ph; R₄=n-C₁₂H₂₅
e. R₁=Ph; R₄=n-C₁₈H₃₇
f. R₁=Ph; R₄=n-C₂₂H₄₅

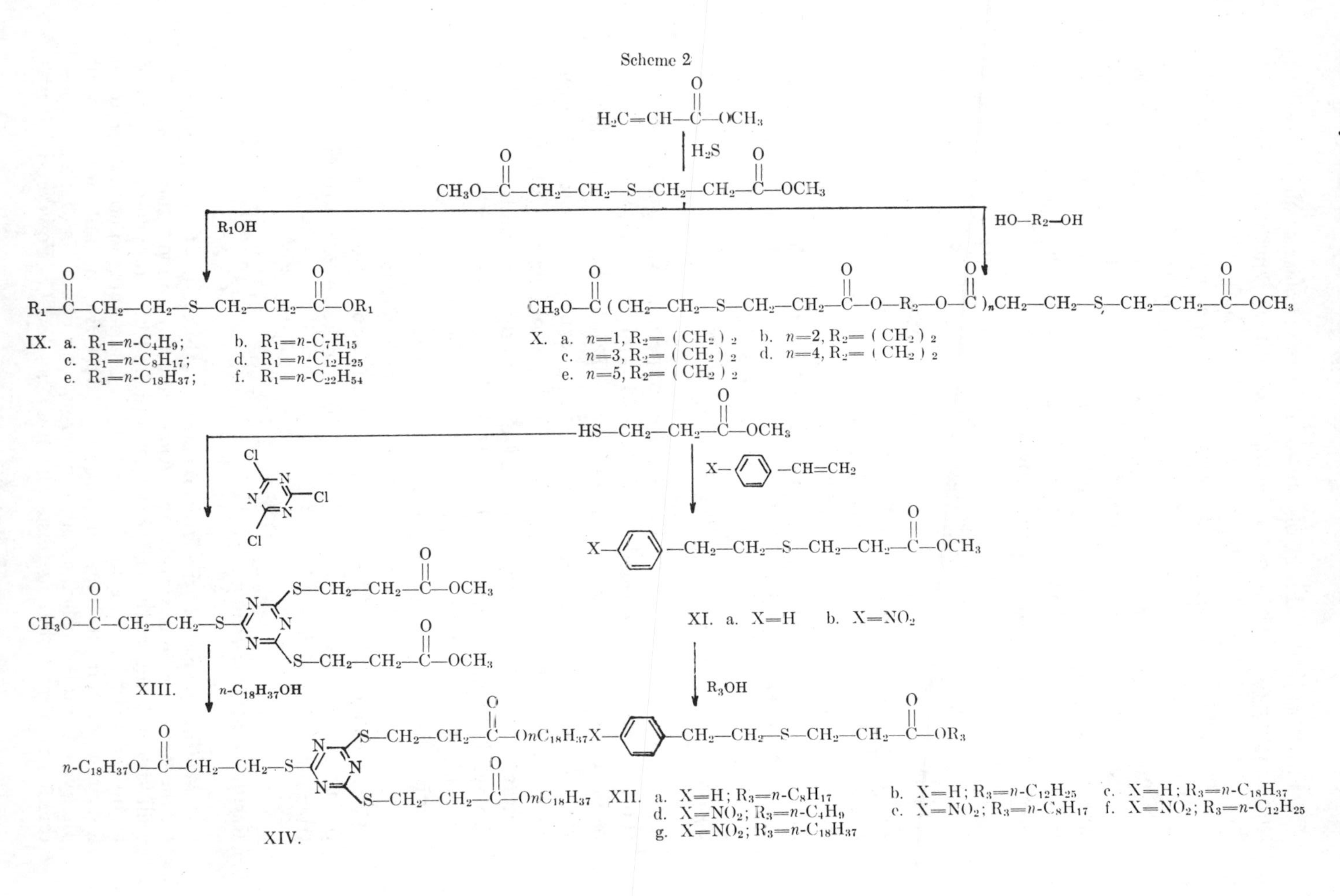
Scheme 2
H2S
R1OH
HO—R2—OH
IX. a. R1=n-C4H9; b. R1=n-C7H15
c. R1=n-C8H17; d. R1=n-C12H25
e. R1=n-C18H37; f. R1=n-C22H54
X. a. n=1, R2=(CH2)2 b. n=2, R2=(CH2)2
c. n=3, R2=(CH2)2 d. n=4, R2=(CH2)2
e. n=5, R2=(CH2)2
XI. a. X=H b. X=NO2
R3OH
XIII.
n-C18H37OH
XII. a. X=H; R3=n-C8H17 b. X=H; R3=n-C12H25 c. X=H; R3=n-C18H37
d. X=NO2; R3=n-C4H9 e. X=NO2; R3=n-C8H17 f. X=NO2; R3=n-C12H25
g. X=NO2; R3=n-C18H37
XIV.

Phenyl-substituted phenols constitute a novel class of chain-breaking antioxidants (*4*). Their action with and without β-activated thioethers has been investigated in polypropylene (PP) (*5, 6, 7*).

Stabilizer performance was assessed by oxygen-uptake measurements at 180°C and by oven-aging over a temperature range of 80°–150°C. In the long-term aging experiments the specificity of stabilizer performance is interpreted in terms of functionality, compatibility, and volatility. Model reactions with products originating from the action of the chain-breaking and preventive antioxidants were carried out to study the factors contributing to the synergistic action.

Synthesis of Chain-Breaking Antioxidants

The synthetic scheme adopted for the preparation of the chain-breaking phenolic antioxidants involves the oxidation of a phenyl-substituted phenol (**I**) to the corresponding *p*-benzoquinone (**II**) followed by reduction to the hydroquinone (**III**). The hydroquinone is allowed to react then with cyanuric chloride and its derivatives or with alcohols as shown in Scheme 1.

Synthesis of Preventive Antioxidants

In Scheme 2 the synthetic routes to some β-activated thioethers are shown. Starting from thiodipropionic acid dimethylester (VIII) a series of thiodipropionic dialkylesters (IX) were obtained by transesterification with the appropriate alcohols. Following the same procedure with diols, e.g., ethylene-glycol, it was possible to obtain oligomers (X). Chromatography of the oligomer mixture gave the pure compounds (a–e). For the synthesis of assymmetrical β-activated thioethers another approach was followed, viz. by using β-mercaptopropionic acid methylester as the starting material. Upon reaction with styrene and *p*-nitrostyrene, the β-activated thioethers of the type XI were obtained, which gave the higher homologues (XII) after transesterification. Finally β-mercaptopropionic acid methylester was used to synthesize 2, 4, 6-trisubstituted triazines XIII and XIV.

Results and Discussion

Measurements of the Antioxidant Activity of Combinations of Chain-Breaking and Preventive Antioxidants. Among the variety of well-known aging methods, the oxygen-uptake test and oven-aging proved to be the most versatile methods for high temperature and moderate to low temperature testing, respectively (*8*). Antioxidant activities of phenyl-substituted phenols and synergistic mixtures with β-activated thioethers were tested in unstabilized PP at 180°C. Powdered PP was mixed with the stabilizers and homogenized with acetone. After evaporation of the solvent, the sample was placed in a tube, and the oxygen-uptake was measured at regular time intervals.

With respect to the results obtained in a variety of studies (9), we used the oven-aging technique to measure the effectiveness of antioxidants in PP to predict the service life of the stabilized polymer. Times of failure of PP samples (6-mil thick) were measured in forced-air ovens at 80°–150°C. Oven-aging at 80°C was considered especially important in predicting the service life of PP stabilized with synergistic combinations of chain-breaking and preventive antioxidants. In practice the service life of stabilized polymers is predicted often by extrapolating results obtained with synergistic combinations of antioxidants at high temperatures (130°–160°C). However, these extrapolations are inconsequential, since other mechanisms are probably operative at higher temperatures. Before times of failure can be measured at 80°C, it will last several months. Therefore we used UV spectrophotometry to measure the decay of antioxidants in PP samples. Advantages of these UV measurements are the relatively short time during the induction in which it can be shown whether synergism takes place or not and the possibility of predicting the end of the induction period (time to failure) i.e., when the concentration of the phenolic antioxidant is $< 5\%$ of the original concentration. The predicted and measured induction periods are in good agreement with each other ($\pm 6\%$).

Antioxidant Activity of Some Phenyl-Substituted Phenols (VII) in PP at 180°C (Oxygen-Uptake Test). In Table I the results obtained with some phenyl-substituted phenols are shown and compared with some well-known *tert*-butyl-substituted phenols. A general conclusion from this table is the marked synergism between phenyl-substituted phenols and β-activated thioethers compared with the *tert*-butyl-substituted analog. Since the measurements are done in a closed tube, the influence of the volatility of the thiodipropionates is not reflected in the induction periods. An explanation for the anomalous behavior of phenyl-substituted phenols in synergistic mixtures can be found in the results of the model reactions, carried out with phenoxyl radicals and sulfides or sulfoxides (*see* "Model Reactions").

Antioxidant Activity of Triazine-Based, Phenyl-Substituted Phenols IV, V, and VI (Oven-Aging). Times of failure were measured on PP films (6 mil) stabilized with 0.1% (w/w) of the chain-breaking antioxidants IV, V, and VI with and without dilaurylthiodipropionate (DLTDP) or distearylthiodipropionate (DSTDP) (0.25%). In Table II the results obtained at 140°, 120°, and 80°C are summarized. Moreover the synergistic action (S.A.) was calculated.

$$\text{S.A.} = \frac{\begin{array}{c}\text{induction period}\\ \text{(chain-breaking a.o + preventive a.o)}\end{array}}{\begin{array}{c}\text{induction period chain-breaking a.o +}\\ \text{induction period preventive a.o}\end{array}}$$

Table I. Induction Periods[a] for Combinations of

β-Activated Thioether 05.%	*2,4,6-Trisubstituted Phenol 0.2%*	
		(2,6-diphenyl-4-methoxyphenol: OH, Ph, Ph, OCH_3)
	0.2	0.75
$S(CH_2{-}CH_2{-}\overset{O}{\overset{\|\|}{C}}{-}OCH_3)_2$	5	50
$S(CH_2{-}CH_2{-}\overset{O}{\overset{\|\|}{C}}{-}O{-}nC_{12}H_{25})_2$	2.5	32

[a] Time (hr) required for the consumption of 20 ml O_2/g unstabilized PP, stabilized

Table II. Times of Failure (hr) of PP Films[a] (6-mil Phenols and Synergistic

Chain-Breaking Antioxidant		*140°C*			
		0.25% DSTDP		*0.25% DLTDP*	
		(hr)	*S.A.*	*(hr)*	*S.A.*
(general structure: HO–phenyl(Ph, R_1)–O–triazine(X)(X))	5	18	—	15	—
IVa; R_1 = Ph, X = –O–(3,5-diphenyl-4-hydroxyphenyl)	8	600	(23)	1400	(61)
Va; R_1 = Ph, X = $-O{-}nC_{12}H_{25}$	15	480	(15)	1700	(51)
Vc; R_1 = Ph, X = $-O{-}nC_{18}H_{37}$	10	480	(17)	1600	(64)
VIa; R_1 = Ph, X = $-S{-}CH_2{-}CH_2COOCH_3$	15	780	(24)	1080	(36)
VIb; R_1 = Ph, X = $-S{-}CH_2{-}CH_2COO{-}nC_{12}H_{25}$	15	600	(18)	650	(22)
IVb; R_1 = *tert*-Bu, X = –O–(3-phenyl-5-tert-butyl-4-hydroxyphenyl)	15	980	(30)	2900	(97)
Vb; R_1 = *tert*-Bu, X = $-O{-}nC_{12}H_{25}$	15	700	(21)	2400	(80)
Vd; R_1 = *tert*-Bu, X = $-O{-}nC_{18}H_{37}$	16	640	(19)	2200	(71)
(HO–(3,5-di-tert-butylphenyl)–O–triazine($O{-}nC_{12}H_{25}$)($O{-}nC_{12}H_{25}$))	30	70	(1.5)	120	(2.7)

[a] The samples were prepared by compression molding for 2 min at 200°C. PP

Primary and Secondary Antioxidants at 180°C in PP

2,4,6-Trisubstituted Phenol 0.2%

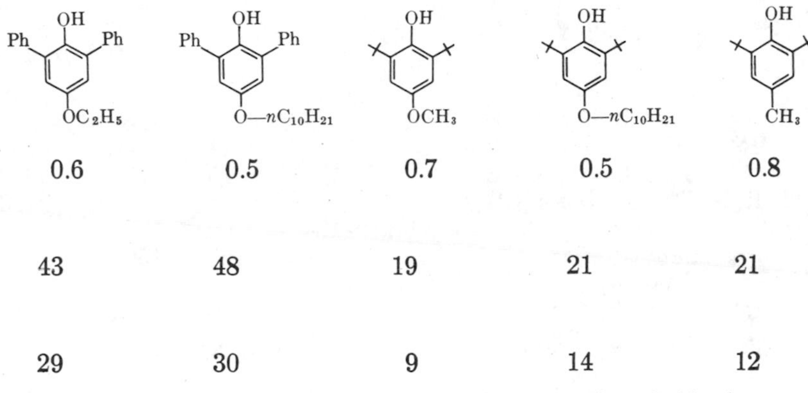

0.6	0.5	0.7	0.5	0.8
43	48	19	21	21
29	30	9	14	12

with 0.2% of a 2,4,6-trisubstituted phenol and 0.5% of a β-activated thioether.

thickness) Stabilized with Triazine-Based,Phenyl-Substituted Combinations at 140°, 120°, and 80°C

	120°C					*80°C*			
	0.25% DLTDP		*0.25% DSTDP*			*0.25% DLTDP*		*0.25% DSTDP*	
	(hr)	*S.A.*	*(hr)*	*S.A.*		*(hr)*	*S.A.*	*(hr)*	*S.A.*
20	280	—	210	—	300	3000	—	1100	—
80	1200	(3.3)		(8.6)					
			2500						
90	1320	(3.6)	4250	(14)					
30	1600	(5.2)	4200	(17)	475	7000	(2)	8000	(5.1)
60	—	—	3300	(12)					
60	2100	(6.2)	3290	(12)					
100	2000	(5.3)	5440	(18)					
105	1320	(3.4)	4900	(16)					
108	1350	(3.5)	4450	(14)					

type: experimental PP.

Table III. Times of Failure (hr) of PP Films[a] (6 mil Synergistic Combinations

Chain-Breaking Antioxidant	*140°C*		
Ph; HO; OR_4; R_1		*DLTDP (0.25%)*	*DSTDP (0.25%)*
	5	18	15
VIIc; R_1 = Ph; R_4 = n–$C_{10}H_{22}$	16	230	—
VIId; R_1 = Ph; R_4 = n–$C_{12}H_{25}$	14	—	580
VIIe; R_1 = Ph; R_4 = n–$C_{18}H_{37}$	15	780	2000
VIIf; R_1 = Ph; R_4 = n–$C_{22}H_{45}$	15	—	2100
Tetrakis methylene-3-(3′,5′-di-*tert*-butyl-4′-hydroxyphenyl)	1150	1500	1730
1,3,5-Trimethyl-2,4,6-tris(3′,5′-di-*tert*-butyl-4-hydroxybenzyl)benzene	800	1100	

[a] The samples were prepared by compression molding for 2 min at 200°C. PP

In general a higher degree of synergism is obtained with DSTDP than with DLTDP. The effect of compatibility of the chain-breaking antioxidants with PP on the induction periods can be well established by comparing the times of failure of (IVa) and (IVb).

From solubility measurements in squalane (*10*) it appears that 2% solutions of IVb, Va, and Vc can be obtained easily, whereas IVa is hardly soluble under the same conditions ($< 0.3\%$). In this case, where one phenyl group in IVa is replaced by a *tert*-butyl-group, the induction period of IVb is increased to a considerable extent in combination with DSTDP. Comparing IVa with Va and Vc, it is evident that, although the OH content of the chain-breaking antioxidant is decreased drastically, the induction periods of the latter increased.

These phenomena have been attributed to a better compatibility of these compounds compared with IVa, since no distinct differences in volatility of all of the compounds in Table II are observed (TGA measurements). It is noteworthy that no internal synergism was observed with VIa and VIb.

Antioxidant Activity of 4-Alkoxy-2,6-Diphenylphenols (VIII) (Oven-Aging). In Table III the results obtained with 4-alkoxy-2,6-diphenylphenols (VII) are summarized. With respect to the increasing alkyl chain in VIIc–VIIf the induction period increased, although the OH content decreased. Since all of the compounds were soluble in squalane (*10*) (2%) (a model compound for PP), which can be a measure of compatibility with PP, the increasing induction periods can be ascribed to decreasing volatility (TGA measurements); e.g., $T_{10\%\ wt\ loss}$ of VIId = 256°C and $T_{10\%\ wt\ loss}$ of VIIe = 331°C (heating rate, 5°C/min; air, 50 mL/min. The most promising results obtained thus far

thick) Stabilized with 4-Alkoxy-2,6-diphenylphenols and at 140°, 130°, and 80°C

120°C			*100°C*			*80°C*		
	DLTDP (0.25%)	*DSTDP (0.25%)*		*DLTDP (0.25%)*	*DSTDP (0.25%)*		*DLTDP (0.25%)*	*DSTDP (0.25%)*
20	280	210	70	1300	640	300	3000	1100
70	840	—	—	—	—	860	7100	8000
55	—	2250	—	—	—	—	—	—
60	2500	5800	360	5000	9500	2500	7600	14000
—	—	—	—	—	—	—	—	—
2700	3600	3400	—	—	—	6000	10000	8000
2600	2900					6200	11000	

type: experimental PP.

were those where VIIe and DSTDP were used as chain-breaking antioxidant and preventive antioxidant, respectively. Even at lower temperatures synergism is observed still; e.g., S.A. at 80°C of VIIe with DSTDP is ca. 4. This combination was investigated further with respect to the optimum ratio between chain-breaking antioxidant and hydroperoxide decomposer and the effect of processing on the induction periods at lower temperatures (oven-aging).

Influence of the Thioether Structure on the Synergism between Phenyl-Substituted Phenols and β-Activated Thioethers (Oven-Aging). From the preceding section it could be concluded that among the variety of phenyl-substituted phenols, 4-stearoxy-2,6-diphenylphenol (VIIe) proved to be the most powerful chain-breaking antioxidant when used in combination with β-activated thioethers. 4-Stearoxy-2,6-diphenylphenol is used now as the chain-breaking antioxidant and tested with a variety of β-activated thioethers.

To study the influence of volatility and compatibility, four distinct classes were tested viz., thiodipropionates, β-carboalkoxy-β′-phenyldiethylsulfides, β-carboalkoxy-β′-*p*-nitrophenyldiethylsulfides, and oligomers of the transesterification reaction of dimethylthiodipropionate and ethylene glycol. The results are shown in Table IV.

The TGA figures give the impression of the volatility and thermal stability. β-Activated thioethers with molecular weight < 400 do not show any synergism with VIIe, probably because of their extremely high volatility. β-Activated thioethers which are compatible with PP show a marked increase in synergistic action at increasing molecular weight (> 400). The oligomer series (Xa–Xe) do not show this effect, which is certainly caused by their lack of compatibility with PP.

Table IV. Times of Failure (hr) of PP Films[a] (6-mil Thickness), Stabilized with 4-Stearoxy-2,6-diphenylphenol (VIIe) and β-Activated Thioethers at 145° and 100°C

		TGA (5°C/min in air)		*145°C*		*100°C*	
β-Activated Thioether (0.25%)	*Mol Wt*	*10% Wt Loss*	*50% Wt Loss*		*0.1% VIIe*		*0.1% VIIe*
				5	55		1800
S (CH_2–CH_2–$COOCH_3$)$_2$	206			10	60	100	1820
S (CH_2–CH_2–COO–nC_4H_9)$_2$	290			10	60	120	1820
S (CH_2–CH_2–COO–nC_2H_{15})$_7$	374			10	60	190	1820
S (CH_2–CH_2–COO–nC_8H_{17})$_2$	402			10	60	310	2100
S (CH_2–CH_2–COO–$nC_{12}H_{25}$)$_2$	514	255°C	295°C	70	430	2600	4880
S (CH_2–CH_2–COO–$nC_{18}H_{37}$)$_2$	682	280°C	330°C	90	1000	2200	9750
Xa[b]	410	265	305	10	60	300	2450
Xb[b]	614	280	315	10	240	520	3240
Xc[b]	818	295	325	15	240	360	4010
Xd[b]	1022	300	325	10	240	300	4010
Xe[b]	1226	295	330	10	300	300	4050
XIIa[b]	322			10	60	190	2450
XIIb[b]	378			15	85	300	4220
XIIc[b]	462			35	300	820	8100
XIId[b]	311			10	60	300	2100
XIIe[b]	367			10	70	300	2500
XIIf[b]	423			60	160	1380	3740
XIIg[b]	507	270°C	330°C	80	360	2200	9400
XIV	1149			10	50	—	—

[a] The samples were prepared by compression molding for 2 min at 200°C. PP type: propathene HF–20 (ICI).
[b] See Scheme 2.

The triazine-based β-activated thioether XIV does not show any synergism with VIIe, although this compound is comparable with distearylthiodipropionate in hydroperoxide decomposition activity. Using the method of Hiatt et al. (*11*) to measure the rate of decomposition of *tert*-butylhydroperoxide, both compounds showed the same rate in *o*-dichlorobenzene at 80°C. In Figure 1 the effect of compatibility of β-activated thioethers on the induction periods of their synergistic mixtures with VIIe is shown. The best results are obtained again with the mixture VIIe–DSTDP.

Determination of the Optimum Ratio between VIIe and STDP. The optimum ratio was determined with a total antioxidant amount of 0.35%. PP films 6-mil thick were compression–molded and oven-aged at 140°C. During the induction period the decay of the phenolic antioxidant was measured with the aid of UV spectrophotometry. The measured induction periods are summarized in Figure 2. The results show that the best induction periods are obtained at the highest DSTDP/phenol ratio, viz, 4:1 or higher. The UV spectrophotometry results are given in Figure 3. At concentrations of VIIe $\leqslant$ 0.14% the decrease in absorption is proportional with time. At concentration levels > 0.14% the phenol concentration decreases very fast after a definite time, which probably is caused by the fact that no more DSTDP is present, which is essential for synergism.

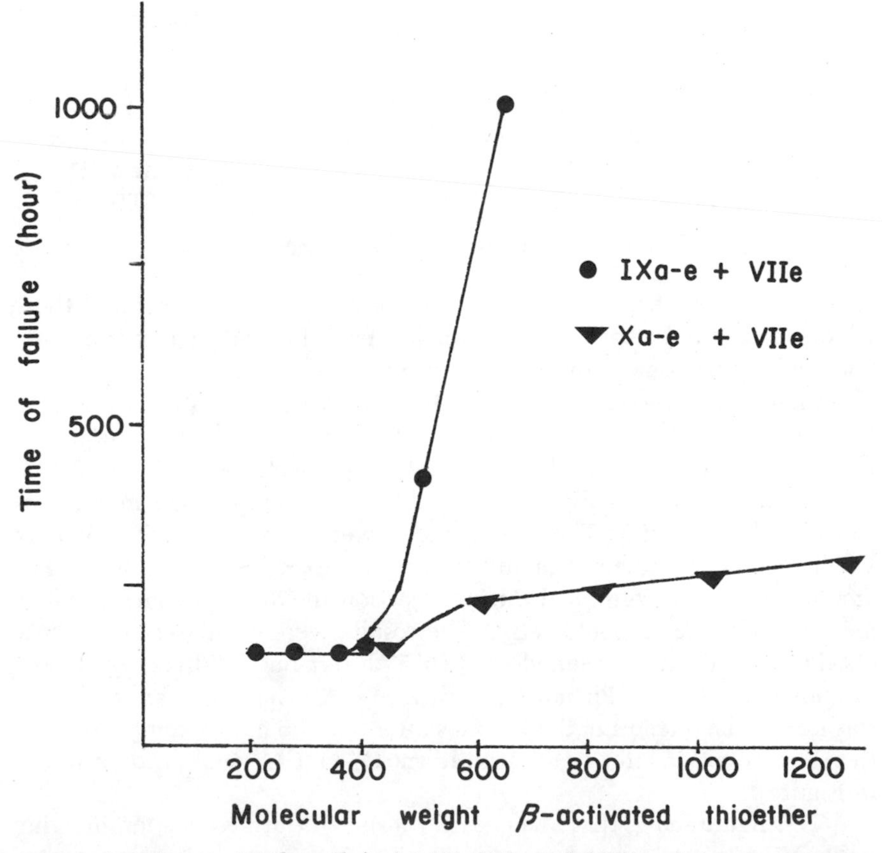

Figure 1. Times of failure (hr) of 0.1% VIIe and 0.25% of β-activated thioethers

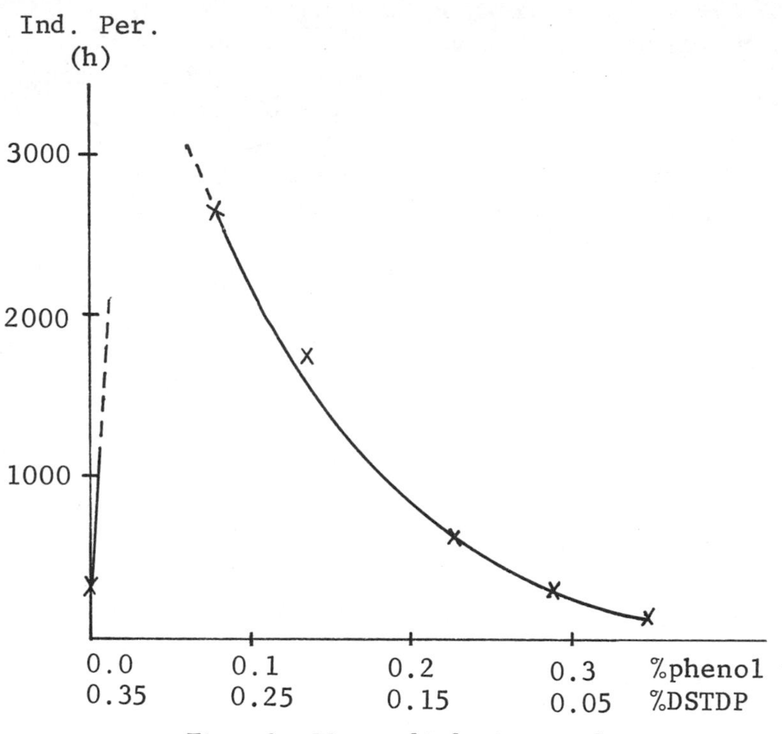

Figure 2. Measured induction periods

Comparison of the measured induction periods in Figure 2 and those which can be predicted (e.g., after 750 hr) from Figure 3 makes it evident that both failure times are in statement with each other.

Effect of Processing on the Induction Periods of (VIIe)/DSTDP Mixtures at 140°C. The effect of processing on the induction periods of (VIIe)/DSTDP mixtures was investigated by injection-molding at 260°C (1, 3, and 5 passes) of PP samples, followed by compression molding at 190°C of films (6 mil). These experiments were carried out with (VIIe)/DSTDP concentrations ranging from 0.03/0.32%–0.12/0.23%. The amount of phenol consumed during injection molding was measured by means of UV spectrophotometry. The results were compared with those obtained from tetrakis methylene-3-(3′,5′-di-*tert*-butyl-4′-hydroxyphenol) propionate methane (Phenol A) with DSTDP. The decrease in UV absorption of the phenols in the PP films after 3 and 5 passes compared with the UV absorption after 1 pass at different DSTDP/phenol ratios is shown in Figure 4.

This figure shows that Phenol A is consumed during injection molding while VIIe is regenerated completely. The latter result is corroborated by the fact that in a model experiment (*see* "Model Reactions") the phenyl-substituted phenol is still present (> 95%) (measured with GLC after methylene–chloride extraction of the PP after 5 passes).

Oven-aging of the 6-mil films was carried out at 140°C. The failure times are summarized in Table V. This table shows that at a DSTDP/phenol ratio of 4/1, (VIIe) gives better results than Phenol A. VIIe gives the best results at high DSTDP/phenol ratios, whereas the opposite is observed for Phenol A.

Model Reactions. In view of the marked synergism observed in measuring the antioxidant action of combinations of phenyl-substituted

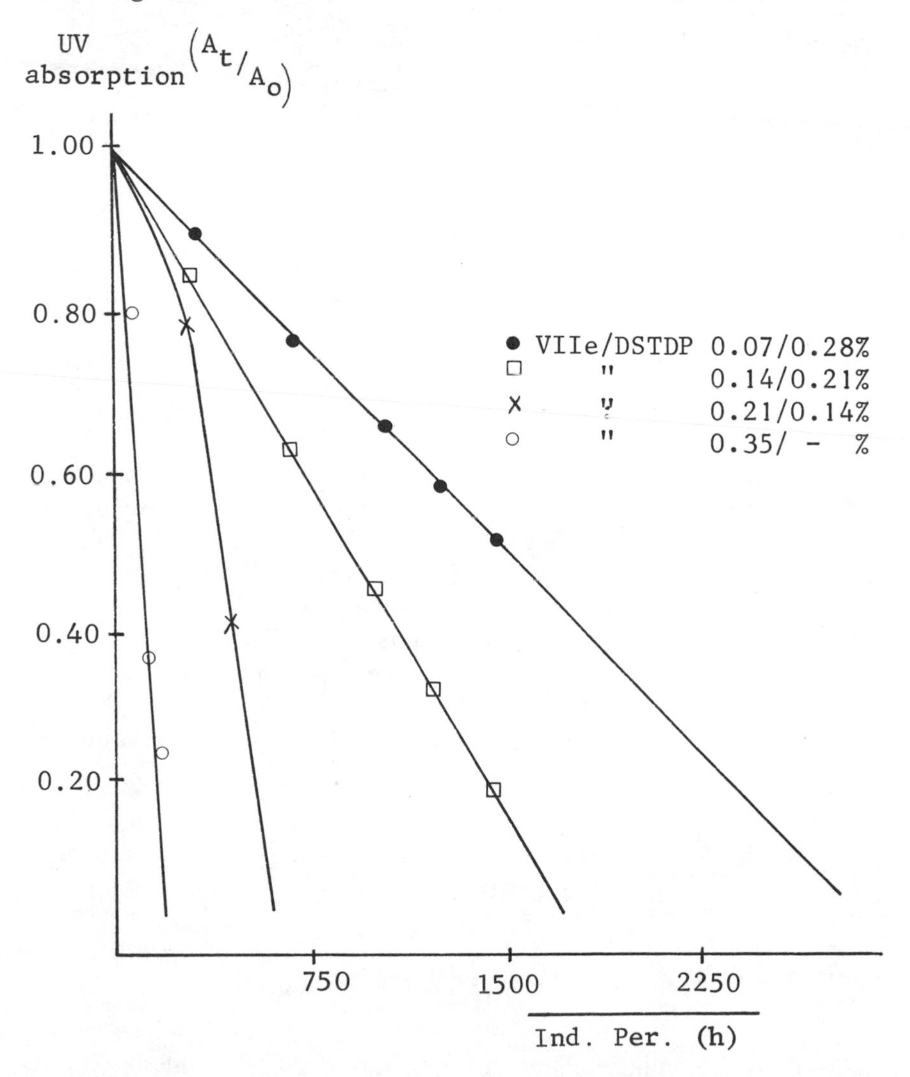

Figure 3. Predicted induction periods

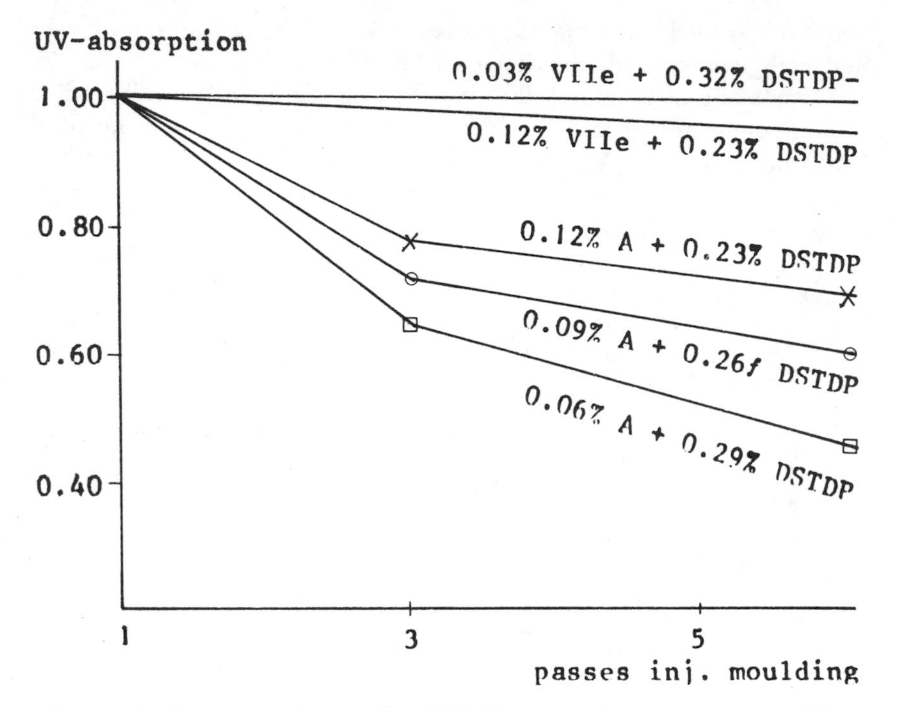

Figure 4. Decrease of antioxidant UV-absorption during injection molding

phenols and β-activated thioethers and considering the role of the chain-breaking antioxidant (continuous regeneration at high temperatures (demonstrated with UV spectrophotometry)), it was important to study the factors contributing to the synergistic action. Therefore model reactions were carried out with products originating from the action of the chain-breaking and preventive antioxidants (*12*). The primary products of the chain-breaking antioxidant (AH) and the preventive antioxidant ($X\text{–}CH_2\text{–}CH_2\text{–}S\text{–}CH_2\text{–}CH_2\text{–}X$) formed after inhibition are the phenoxyl radical (A·) and, after decomposition of hydroperoxide, the sulfoxide:

$$(X\text{–}CH_2\text{–}CH_2\text{–}\overset{\overset{\displaystyle O}{||}}{S}\text{–}CH_2\text{–}CH_2\text{–}X).$$

$$ROO\cdot + AH \rightarrow ROOH + A\cdot$$

$$ROOH + X\text{–}CH_2\text{–}CH_2\text{–}S\text{–}SH_2\text{–}CH_2\text{–}X \rightarrow ROH + X\text{–}CH_2\text{–}CH_2\text{–}\overset{\overset{\displaystyle O}{||}}{S}\text{–}CH_2\text{–}CH_2\text{–}X$$

4-Methoxy-2,6-diphenylphenoxyl (A·) and β,β′-diphenyldiethylsulfoxide were used as the model compounds. 4-Methoxy-2,6-diphenylphenoxyl was prepared by oxidation of the corresponding phenol and proved to be stable toward oxygen. The radical was obtained as its quinone–ketal dimer, which dissociates into radicals upon heating and recombines to the dimer again upon cooling.

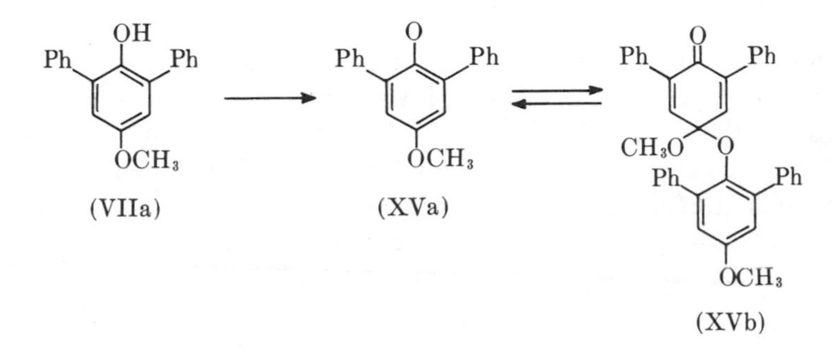

β,β′-Diphenyldiethylsulfoxide was obtained by oxidation of the corresponding sulfide with hydrogen peroxide and served as a model for the preventive antioxidant after its action as a hydroperoxide decomposer.

Reaction of 4-Methoxy-2,6-Diphenylphenoxyl (XVa) and β,β′-Diphenyldiethylsulfoxide. 4-Methoxy-2,6-diphenylphenoxyl (XVa) and β,β′-diphenyldiethylsulfoxide (molar ratio 2:1) were heated in *o*-dichlorobenzene at 150°C. From high-pressure liquid chromatographic (HPLC) analysis it was concluded that sulfinate (XVI) (*see* Scheme 3) arises and disappears within 5 min.

Following the formation of sulfinate (XVI), after a reaction time of 2.5 min, the concentration of XVI is maximal and XVI completcly dccomposed in 5 min. The products XVI, XVII, and VIIa were in a 1:1:1 ratio. Owing to the behavior of sulfinate (XVI) the consecutive reactions of sulfinate were studied.

Table V. Times of Failure (hr) of PP Films[a] (6 mil) Stabilized with Combinations of VIIe/DSTDP and Phenol A/DSTDP at 140°C

Sample	*1 Pass (260°C)*	*3 Passes (260°C)*	*5 Passes (260°C)*
0.12% Phenol A + 0.23% DSTDP	2660	1750	1450
0.03% VIIe + 0.32% DSTDP	2010	1500	1100
0.06% VIIe + 0.29% DSTDP	1500	1200	740
0.09% VIIe + 0.26% DSTDP	1300	740	400
0.12% VIIe + 0.23% DSTDP	1400	1150	580
0.03% Phenol A % 0.32% DSTDP	1600	1200	640
0.06% Phenol A + 0.29% DSTDP	1940	1500	1200
0.09% Phenol A + 0.26% DSTDP	2090	1930	1560

[a] The samples were prepared by compression-molding 2 min at 190°C. PP type: propathene HF-20 (I.C.I.).

It has been well established (*13*) that sulfenic acids and alkenes originate from thermolysis of sulfoxides. However, as shown in Scheme 3, it appears from kinetic measurements combined with analog computer simulation of the kinetic data (*14*) that sulfinate (XVI) was formed immediately. No induction period, indicating the formation of a sulfenic, was found.

From these observations it may be concluded that the sulfinate originates from the direct attack of the phenoxyl radicals on the β-hydrogens of the sulfoxide, followed by elimination of styrene (XVII) and subsequent recombination of the sulfinyl radical and the phenoxyl radical to form sulfinate (XVI). Under conditions where the sulfoxide is stable (80°C), the same reaction products were formed.

Scheme 3

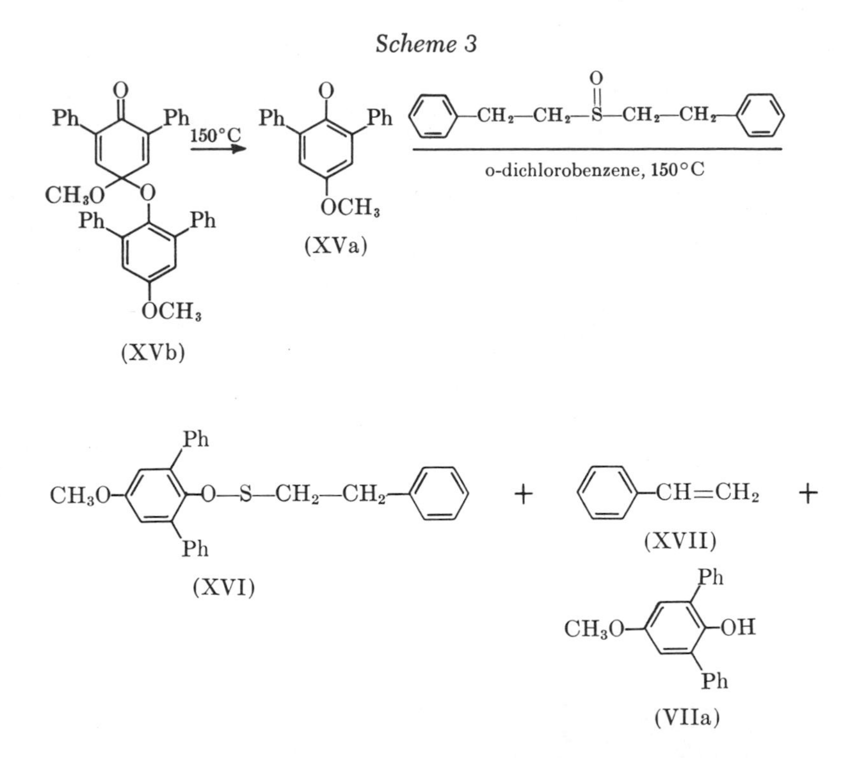

Thermal Decomposition of Sulfinate (XVI). Heating sulfinate (XVI) in *o*-dichlorobenzene at 150°C, XVI completely decomposed within 5 min. No styrene was formed. 4-Methoxy-2,6-diphenylphenol (VIIa) is isolated from the reaction mixture together with other products, which proved to be thiosulfonate (XVIII), sulfonate (XIX), and sulfide (XX). Reaction Scheme 4 gives possible routes for the formation of products XVIII, XIX, and XX.

Scheme 4

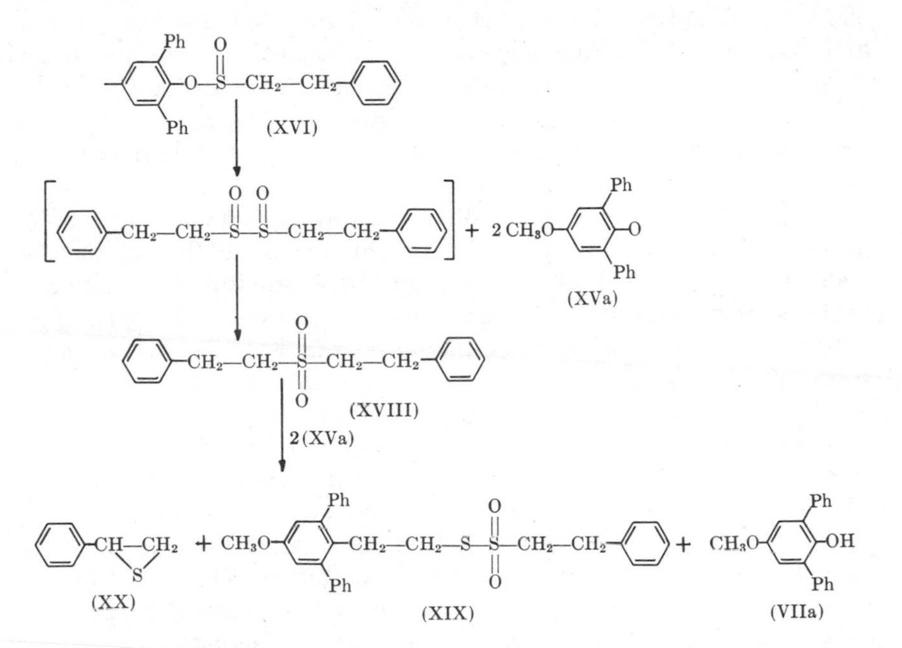

Thermal Decomposition of Sulfinate (XVI) in the Presence of an Excess of 4-Methoxy-2,6-Diphenylphenoxyl (XVa). Besides the products formed as indicated in Scheme 4 an increasing amount of 4-methoxy-2,6-diphenylphenol (VIIa) is formed. Moreover styrene is formed in equimolecular amount with the sulfite (XXI). In Scheme 5 a possible explanation is given for the formation of the above-mentioned products.

In summary, the reactions of phenoxyl radicals (XVa) and sulfoxides give a regeneration of the phenol (chain-breaking antioxidant) together with the formation of several sulfur compounds (potential preventive

Scheme 5

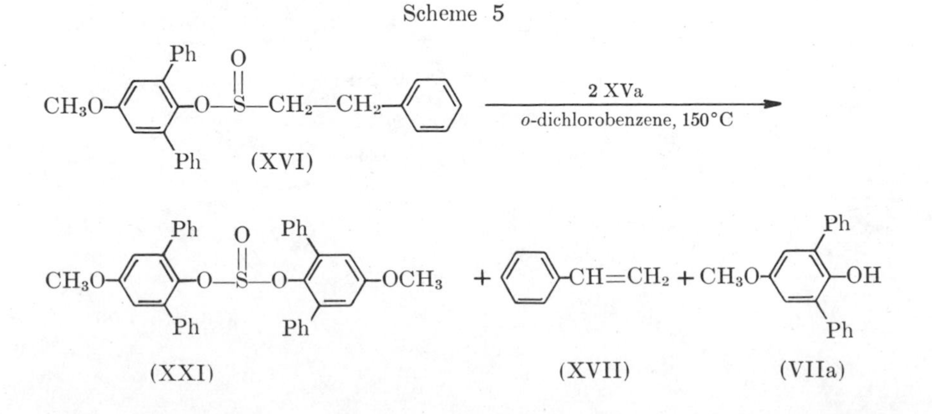

antioxidants). These observations can account for the marked synergistic effects of phenyl-substituted phenols and β-activated thioethers at high temperatures. Considering, however, synergistic effects with mixtures of chain-breaking and preventive antioxidants in ratios of 1:3 or even higher (*see* e.g., Table V), it seems more realistic, particularly for the initial stages of autoxidation, to study the interaction of phenoxyl radicals and sulfides.

Reaction of 4-Methoxy-2,6-Diphenylphenoxyl (XVa) and β,β'-Diphenyldiethylsulfide. At 150°C in *o*-dichlorobenzene the reaction between phenoxyl radical XVa and β,β'-diphenyldiethylsulfide (molar ratio 2:1) was completed within 15 min as concluded from HPLC. The main products proved to be the α,β-unsaturated sulfide XXII, the α- and β-substitution products XXIII and XXIV in a ratio 9:1, and phenol XII. These main products represent over 90% of all products in this reaction. The consecutive reactions of the α,β-unsaturated sulfide XXII and of α- and β-substitution products XXIII and XXIV were investigated also. The α,β-unsaturated sulfide XII gave no further reaction even with phenoxyl radical XVa at 150° in *o*-dichlorobenzene. Both the α- and β-substitution products XXIII and XXIV reacted further with phenoxyl radical XVa. Compound XXIV reacted much faster than XXIII. In the of phenoxyl radical XVa only XXIV gave further reactions.

From the results obtained it can be concluded that the phenol (VIIa) is regenerated by the H-donating action of the preventive antioxidant. A prerequisite for this regeneration seems to be the stability of the phenoxyl radical (XVa) or its homologs.

Scheme 6

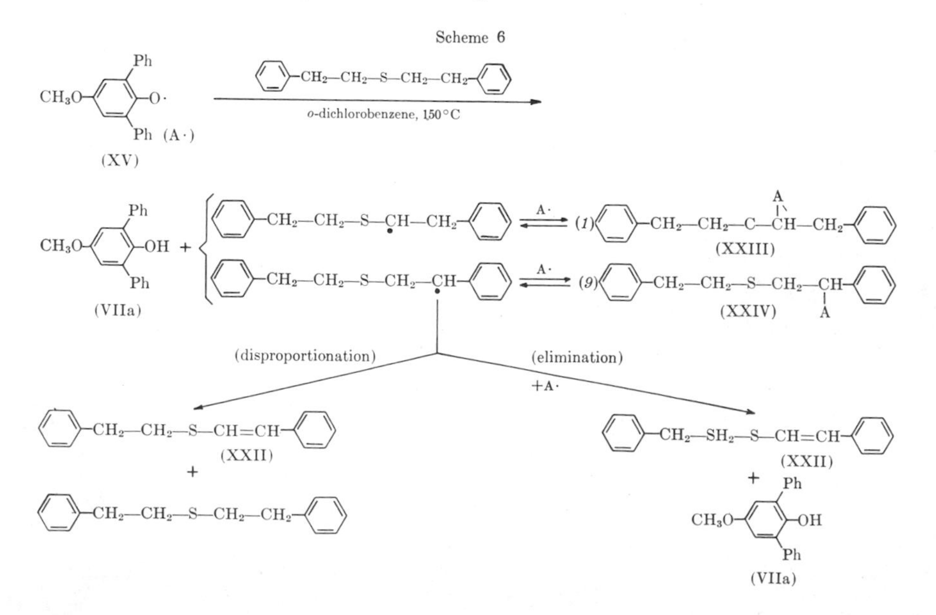

H-Donation of β-Activated Thioethers and Phenoxyl Radical (XVa). Representatives of different classes of β-activated thioethers, viz., IXe, Xc, XIIc, XIIg, and XIV (*see* Scheme 2) and phenoxyl radical XVa were heated in *o*-dichlorobenzene at 130°C in a 10:1 molar ratio. H-donation of the β-activated thioethers was followed visually. As soon as the green color of the phenoxyl radical had disappeared, samples were taken from the reaction mixture and here analyzed by TLC and GLC.

In all cases, except for XIV, the green color had disappeared in 8–10 min. When thioether XIV and phenoxyl radical XVa were heated in *o*-dichlorobenzene at 130°C, the green color, indicating the presence of phenoxyl radical (XVa), was still present after 120 min. In the case of IXe, Xc, XIIc, and XIIg, TLC showed the formation of phenol (VIIa). Quantitative GLC measurements confirmed the observations of TLC. In all four cases more than 90% of the theoretical amount of phenol (VIIa) was formed. Here again the importance of H-donation of preventive antioxidants for the synergistic action between phenyl-substituted phenols and β-activated thioethers is shown. The absence of H-donating ability of XIV explains the absence of internal synergism of VIa and VIb (*see* Table II) and the absence of synergism between VIIe and XIV (*see* Table IV.)

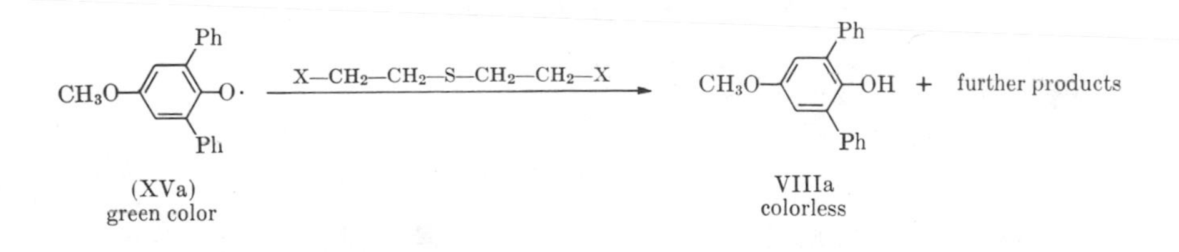

As an intermediate experiment between the model reactions in *o*-dichlorobenzene and the oven-aging tests of PP stabilized with combinations of phenyl-substituted phenols and β-activated thioethers, the behavior of 4-methoxy-2,6-diphenylphenol (VIIa) and β,β′-diphenyldiethylsulfide was studied in PP under realistic processing conditions and in aging of the sample.

PP, stabilized with 0.1% VIIa and 0.25% β,β′-diphenyldiethylsulfide was passed 1, 2,3, 4, and 5 times through an extruder at 200°C. Even after 5 passes the concentration of VIIa remain unchanged (0.099%—quantitative GLC measurements), whereas the concentration of sulfide had decreased to 0.21%. This PP sample was aged at 150°C under oxygen, and at regular time intervals samples were taken from the PP and the volatile products, which were trapped by cooling. The PP samples were extracted and the sublimates dissolved. In both solutions the amount of (VIIa) and the sulfide were determined quantitatively with

GLC. A blank experiment was carried out under nitrogen. The results are shown in Table VI.

From this table it can be concluded that after only 3 days aging at 150°C there is only 14% of the original amount of the thioether left, whereas after 8 days there is still 83% of the original amount of the phenol. In view of the regeneration mechanism outlined in the model reactions, other sulfur compounds have to be responsible for the regeneration of the phenol. Indeed, other reaction products in samples where no thioether was left (e.g., after 6 days aging in PP) were detected and isolated with the aid of HPLC. The unsaturated sulfide (XXII) and β,β′-diphenyldiethylsulfoxide proved to be present among the isolated reaction products.

Table VI. Concentrations of VIIa and β,β′-Diphenyldiethylsulfide in PP during Aging at 150°C in O_2

Antioxidant	*3 Days (O_2) in Sublimate*	*3 Days (O_2) in PP*	*6 Days (O_2) in Sublimate*	*6 Days (O_2) in PP*	*8 Days (O_2) in Sublimate*	*8 Days (O_2) in PP*	*8 Days (N_2) in Sublimate*	*8 Days (N_2) in PP*
VIIa	0.016	0.102	0.002	0.08	0.003	0.08	0.008	0.101
β,β-Diphenyl diethylsulfide	0.006	0.03	—	—	—	—	0.045	0.21

Conclusions

Phenyl-substituted phenols are excellent chain-breaking antioxidants in long-term aging when used in combination with β-activated thioethers, provided that prerequisites such as good compatibility and low volatility are fulfilled. From model experiments we learned that the antioxidant effectiveness of the synergistic combinations is determined greatly by the stability of the phenoxyl radicals, the H-donating ability of the β-activated thioethers, and the formation of the other preventive antioxidants, originating from the β-activated thioethers.

Experimental

Apparatus. IR spectra (KBr discs or neat) were recorded on a Hitachi EPI–G3 spectrophotometer. NMR spectra were taken with a Jeol C–60HL spectrometer with TMS ($\delta = 0$ ppm) as an internal standard. Mass spectra were recorded at 70 eV with a Varian MAT CH–5 spectrometer. UV spectra were taken with a Bausch and Lomb Spectronic 600E spectrophotometer. Oven-agings were carried out in a Heraeus Type-340 oven. All liquid chromatographic separations were carried out on a Hupe and Busch Liquid Chromatograph (Model UFC 1000). The melting points were determined with a melting point microscope (Leitz model 553215) and are corrected; the boiling points are uncorrected.

Syntheses. The solvents and reagents used for syntheses were of reagent grade (puriss.) and were not purified further.

2,6-Diphenyl-*p*-Benzoquinone II. In a 1-L oblong flask 50 g (0.2 mol) of 2,6-diphenylphenol were placed in 150 mL of dimethylformamide with 2 g (0.006 mol) of bis(salicylidene)-ethylene di-iminocobalt (II) (Salcomine) as a catalyst. The flask was connected with a gas burette, while the temperature was kept at 40°–45°C. The whole system then was flushed and filled with oxygen and shaken for 3 hr. The reaction was completed as shown by a sudden stop in the oxygen-uptake (4.81). The reaction mixture was poured onto 600 mL of crushed ice and 30 mL of 4*N* HCl. The immediately formed orange–red precipitate was filtered and washed three times with water. The crude 2,6-diphenyl-*p*-benzoquinone was recrystallized from 1-butanol, (48 g, 93%) mp 136.2°–136.8°C.

2,6-Diphenylhydroquinone (III). In a 250-mL stainless steel autoclave 40 g (0.15 mol) of 2,6-diphenyl-*p*-benzoquinone in 100 mL of methanol and 1 g of platinum on charcoal were placed. After closing the reaction vessel and removing the air, hydrogen was introduced until the pressure was ca. 50 atm. Agitation was started, and the temperature was maintained at 25°C. After 1 hr the hydrogenation was stopped, and the pressure was released. The reaction mixture was filtered and the methanol evaporated in vacuo. A sticky crystalline material was obtained. Recrystallization gave 2,6-diphenylhydroquinone(III), (38 g, 95%) mp 181°C. (Found: C, 82.5; H, 5.3; $C_{18}H_{14}O_2$ requires C, 82.4; H, 5.3%.)

A solution of 80 g of $Na_2S_2O_4$ in 400 mL of water was added to a solution of 40 g of 2,6-diphenyl-*p*-benzoquinone in 500 mL of chloroform. The mixture was stirred for 1 hr at 40°C. The initially formed, deep red color turned light yellow over this period. The reaction mixture was filtered and the residue washed with water and cold ethanol. Recrystallization from hexane gave 2,6-diphenylhydroquinone(III), (33 g, 82%) mp 180.2°–181°C).

Phenyl-Substituted Phenols (IV). To a solution of 3.7 g (0.02 mol) of cyanuric chloride in 50 mL of dry acetone, 15.6 g (0.06 mol) of 2,6-diphenylhydroquinone was added. Under N_2, 12 mL of a 5*N* NaOH solution was added dropwise, while stirring and cooling (20°C) for 1 hr. When the addition of base was complete, the reaction mixture was refluxed for 30 min, and acetone was distilled, while 50 mL of water was added, and the temperature was kept at 70°C for 1 hr, until pH 7.0. The reaction mixture was diluted with chloroform, washed with water, dried over $MgSO_4$, and the chloroform was removed in vacuo. The residual slightly colored syrup solidified when an acetone/hexane 50:50 mixture was added and gave 14.5 g of pure IVa, 84%, mp 239.0°–240.1°C. (Found: C, 79.2; H, 4.7; N, 4.9; $C_{57}H_{39}N_3O_6$ requires C, 79.4; H, 4.5; N, 4.9%.) The same procedure was followed for IVb. Yield—65%; mp 191.1°–192.5°C. The structure was confirmed by ir and NMR spectroscopy. In the case of IVb, the intermediates were prepared by the reaction of *o*-phenylphenol and isobutylene (*15*) which gave 2-*tert*-butyl-6-phenylphenol. Oxidation to the benzoquinone and subsequent reduction gave 2-*tert*-butyl-6-phenylhydroquinone, mp 100.8°–101.1°C.

Phenyl-Substituted Phenols (V). 2,4-Di-octadecanoxy-6-chloro-*s*-triazine (*16*). To a solution of 18.4 g of cyanuric chloride (0.1 mol) and 48.5 g of octadecanol (0.18 mol) in 300 mL of methylisobutylketone, 14.4 g (0.36 mol) of powdered NaOH was added while stirring in 10 min at 25°–30°C. The reaction mixture was stirred for 45 min and then an excess of a concentrated HCl solution was added. The reaction mixture then was heated to 80°C and filtered. On cooling to 15°C 2,4-di-octadecanoxy-6-chloro-*s*-triazine crystallized completely. The white solid product was filtered and dried. Yield, 57.8 g, 89%; mp 70.5°–71.5°C.)

The same procedure was followed for 2,4-di-dodecanoxy-6-chloro-*s*-triazine. (Yield, 90% mp 43.5°–44.9°C.)

Va (*17*). In a 250-mL, three-necked flask, 4.8 g (0.01 mol) of 2,4-didodecanoxy-6-chloro-*s*-triazine and 3.0 g of 2,6-diphenylhydroquinone(III) were placed in 150 mL of dry acetone. To this solution 0.4 g of NaOH dissolved in 7 mL of water was added dropwise under N_2. After the addition was complete, the reaction mixture was kept at reflux for 45 min. The orange–red color disappeared gradually. The reaction product was poured in ice water, extracted with diethylether, washed, dried over $MgSO_4$, and ether was removed in vacuo. The slightly colored, sticky product solidified at 15°C. Yield, 7 g; 98%.

The same was followed for Vb, Vc, and Vd. Vb yield, 84%, yellow syrup. Vc yield, 90%, yellow wax; mp 24°C. Vd yield, 93%, light brown wax; mp 22°C. The structures of Va, Vb, Vc, and Vd were confirmed by ir, NMR, and mass spectrometry.

Phenyl-Substituted Phenols VI. VIb. A solution of 1.85 g (0.01 mol) of cyanuric chloride in 6 mL of dry acetone was added to 6 mL of ice water, and the mixture was cooled (0°–5°C). Then 5.5 g (0.02 mol) of β-mercaptopropionic laurylester was added dropwise, while stirring and keeping the temperature at 0°–5°C. After adding the β-mercaptopropionic laurylester the temperature was raised to 20°C, and a solution of 1.1 g of Na_2CO_3 in 4 mL of water was added dropwise. Then the temperature was raised to 40°C ($CO_2\nearrow$). After 2 hr the reaction mixture was poured into ice water and was extracted with 3 x 50 mL of dichloromethane, washed with water, and dried over $MgSO_4$. After removing the solvent, a white solid was obtained. Recrystallization from hexane gave 6.5 g (98% yield) of 2,4-(dicarbododecanoxy)ethyl-thio-6-chloro-*s*-triazine, mp 56.7°–57.4°C. The structure of this product was confirmed by ir and NMR spectroscopy. The same procedure was followed for the preparation of VIb as for Va. A light yellow syrup was obtained in 90% yield (structure confirmed by ir, NMR, and mass spectrometry (M) *m/e* 821).

Phenyl-Substituted Phenols VII. General procedure. To a solution of 26.2 g (0.1 mol) of 2,6-diphenylhydroquinone(III) in 200 mL of the appropriate alcohol there was added, while stirring, 5 mL of concentrated sulfuric acid over 3 hr. After additional stirring for 1 hr at 65°–130°C, depending upon the alcohol used, the reaction mixture was diluted with 200 mL of chloroform and was washed with water. After evaporation of the chloroform, a light yellow syrup was obtained which gave the 4-alkoxy-2,6-diphenylphenol after crystallization from a hexane/toluene mixture (95:5). VIIa, yield 87%; mp 67.6°–68.1°C: VIIb, yield 72%; mp 89.1°–89.9°C: VIIc, yield 55%; mp 34.2°–35.3°C: VIIe, yield 84%; mp 56.9°–60.3°C. The identity and purity of the phenyl-substituted phenols followed from spectroscopic data (ir, NMR, and mass spectrometry), TLC, and as far as possible, GLC.

β-Activated Thioethers. DIALKYLTHIODIPROPIONATES IX. β,β′-Dicarbomethoxydiethylsulfide was prepared according to the method described for β,β′-diphenyldiethylsulfide (vide infra). The yield was 62%, bp 105°–106°C/0.4 mm (Found: C, 46.8; H, 7.1; S, 15.6; $C_8H_{14}O_4S$ requires C, 46.6; H, 6.8; S, 15.5%). The same product was obtained by introduction of hydrogen sulfide in methylacrylate using Triton B as a catalyst. The yield of β,β′-dicarbomethoxydiethylsulfide was 96%. β,β′-dicarbomethoxydiethylsulfide, transesterified with the appropriate alcohols, gave the dialgyl thiodipropionates IXa–IXf.

The general procedure for this transesterification is as follows: to a solution containing 25.0 g of β,β′-dicarbomethoxydiethylsulfide in 40 g of 1-butanol 0.5 g of concentrated sulfuric acid was added. While stirring at 115°C, methanol distilled. The reaction was followed with TLC. After 4 hr, the reaction product was diluted with dichloromethane and washed with water. After drying over $MgSO_4$ and distillation of the excess 1-butanol, the residue was distilled in vacuo. IXa was obtained in 81% yield, bp 148°–149°C/0.6 mm. IXb, 79%; bp 191°–192°C/0.5 mm: IXc, 74%; bp 208°–209°C/0.5 mm: IXd, 91%; mp 41°–42°C: IXe, 89%; mp 63.7°–64.4°C: IXf, 65%; mp 72.4°–73.7°C.

Oligomers X. In a 100-mL, 3-necked flask, 30.9 g (0.15 mol) of β,β′-dicarbomethoxydiethyl sulfide and 6.8 g (0.11 mol) ethylene glycol are placed. To this solution 0.5-mL of tetraisopropyl orthotitanate was added as the transesterification catalyst. The solution was heated 4 hr at 160°C under N_2. Methanol was distilled. Then the reaction mixture was cooled and diluted with chloroform, washed with water, and dried over $MgSO_4$. Then the chloroform was removed by evaporation in vacuo (12 mm). After 1 hr at 160°C/12 mm, the residue was heated at 200°C/0.4 mm for 1 hr. The yield was 20.4 g or 70%. TLC (ethyl acetate-*n*-hexane 150/55) showed the presence of(oligomers X ($n = 1$–12). Column chromatography of 10 g of the reaction product gave the products Xa–Xe, when ethyl acetate-*n*-hexane 55/45 was used as the elutent. This yielded 1.3 g of Xa, 0.45 g of Xb, 0.50 g of Xc, 0.3 g of Xd, and 0.12 g of Xe. NMR spectroscopy: integration of the signals δ($CDCl_3$): 4.30 ($O-CH_2-CH_2-O$), 3.73 (OCH_3), and 2.70 ($-\overset{O}{\overset{\|}{C}}-CH_2-CH_2-S-$) gave: Xa = 4:6:16, Xb = 8:6:24, Xc = 12:6:32, Xd = 16:6:40, and Xe = 20:6:48. Mass spectrometry: parent peaks *m/e* for Xa = 410, Xb = 614, and Xc = 818.

β-Aryl-β′-Carboalkoxydiethylsulfides XII. *β Phenyl-β′-Carbomethoxydiethylsulfide XIa.* In a 250-mL, three-necked flask containing 100-mL of MeOH (dried on mol sieves 3Å) 3.0 g (0.1 mol) of Na was added. After removing the methanol, 10.6 g (0.1 mol) of β-mercaptopropionic methyl ester was introduced slowly. Then 18.5 g (0.1 mol) of β-phenylethylbromide was added dropwise and the temperature was kept at 45°C. After 1 hr, the reaction proved to be completely TLC. Dilution with methylene chloride, washing with water, drying ($MgSO_4$), and removing the CH_2Cl_2 in vacuo gave 13.8 g, 62% of a colorless liquid, bp 124°–126°C/0.5 mm.

β-*p*-Nitrophenyl-β′-Carbomethoxy diethylsulfide XIb. β-(*p*-Nitrophenyl)ethyl bromide was prepared according to the method in Ref. 18. Then *p*-nitrostyrene was prepared according to the method in Ref. 19. To 60 g (0.5 mol) of β-mercaptopropionic methylester, 74.5 g (0.5 mol) of *p*-nitrostyrene was added. Then piperidine (catalyst) was added very slowly (exothermic reaction). The temperature was kept at 35°C. After stirring for 2 hr, the reaction mixture is diluted with chloroform, washed with water, dried over $MgSO_4$ and the chloroform removed in vacuo. 97 g of XIb (72%) was obtained, by 172°C/0.005 mm. Transesterification of XIa with the appropriate alcohols gave XIIa–c (liquids). Transesterification of XIb with alcohols gave XIId–g, XIId–f (liquids), and XIIg, mp 44.5°–44.7°C. The identity and purity of XIIa–g followed from IR, NMR, and mass spectrometry.

Triazines XIII and XIV. TRIAZINE XIII. In a 250-mL, three-necked flask 12.0 g (0.06 mol) of cyanuric chloride was placed with 50 mL dry methylisobutyl ketone and 22 g (0.2 mol) of sodium carbonate. While stirring, 24.0 g (0.2 mol) of β-mercaptopropionic methylester is introduced slowly. Then 10 mL of water is added and the reaction mixture is kept 1 hr at 50°C. Then 20 mL of acetone is introduced at 55°C and the reaction mixture is stirred for 1 hr. TLC showed that the reaction was complete and after cooling to 20°C, chloroform was added. The $CHCl_3$ solution is washed with water and dried over $MgSO_4$. Distillation at 150°C/0.7 mm gave 24.5 g of an almost colorless residue (XIII).

TRIAZIANE XIV. 10.0 g of residue XIII and 21.6 g of octadecanol were heated at 150°C, while tetraisopropyl-*o*-titanate was added as a catalyst. After destillation of the methanol, the residue was recrystallized from methylisobutyl ketone, which gave 23.1 g of triazine XIV, 87%, mp 68.8–69.7°C. Mass spectrometry: parent peak m/E 1149. The identity and purity of the β-activated thioethers followed from spectroscopic data (ir, NMR, and mass spectrometry).

4-Methoxy-2,6-Diphenylphenoxyl Dimer (XVb). To a solution of 4-methoxy-2,6-diphenylphenol VIIa (10 g) in 350 mL of methanol and 90 g of KOH in 21 g of water, a solution of 100 g $K_3Fe(CN)_6$ in 750 mL of water was added under nitrogen and while stirring over a period of 30 min. A gray-white precipitate was found, which was filtered and washed with water. Crystallization from acetone/hexane (9:1) gave the phenoxyl radical dimer (XVb). Yield, 9.6 g (97%); mp, 158.4°–159.2°C. (Found: C, 82.8; H, 5.6; $C_{38}H_{30}O_4$ requires C, 82.9; H, 5.3%3. The ν_{max} 2830 w; 1680 s, 1650 s, and 700 s cm^{-1}. The $\delta(CDCl_3)$ at −25°C. 7.10–7.00 (20H, m, aromatic), 6.97 (2H, s, aliphatic), 6.45 (2H, s, aromatic m − H), 3.95 (3H, s, aromatic OCH_3), and 2.95 (3H, s, aliphatic OCH_3).

β,β′-Diphenyldiethylsulfide. To a solution of 37 g (0.2 mol) of β-phenylethyl bromide in dimethylsulfoxide, 26.4 g (0.11 mol) of $Na_2C \cdot 9H_2O$ was added and the reaction mixture was stirred at 90°C for 2 hr. Then the reaction mixture was diluted with chloroform and washed with water. The chloroform solution was dried and after evaporation in vacuo at 30°C, the residue was distilled. Yield of β,β′-diphenyldiethyl sulfide 80%, bp, 151°C/0.2 mm. (Found: C, 79.2; H, 7.6; $C_{16}H_{18}S$ requires C, 79.3; H, 7.4%.)

β,β′-Diphenyldiethylsulfoxide. To a solution of 12.1 g (0.1 mol) of β,β′-diphenyldiethylsulfide in 100 mL of acetone, 20 mL of 30% hydrogen peroxide was added. After refluxing for 2 hr the reaction mixture was concentrated in vacuo at 30°C. Crystallization from hexane/acetone (95:5) gave 10.3 g (80%) of β,β'-diphenyldiethylsulfoxide, mp, 70.2°–70.4°C. (Found: C, 74.2; H, 7.2; S, 12.5, $C_{18}H_{18}OS$ requires C, 74.4; H, 7.0; S, 12.4%.) The ν_{max}, 1045 m (S = 0) and 700 s cm^{-1}. The $\delta(CDCl_3)$ 7.18 (10 H, s, aromatic) and 2.78 (8 H, s, aliphatic).

Oxygen-Uptake Measurements. Combinations of phenyl-substituted phenols and β-activated thioethers were tested in unstabilized isotactic PP according to the method described by Pospisil, Taimr, and Kotulak (*20*). Powdered PP (1.000 g) was mixed with 2 mg of the phenol and 5 mg of the thioether. This mixture was homogenized with 5 mL of acetone. The solvent was evaporated carefully in vacuo at 40°C and the sample was placed in a tube connected with a gas burette filled with oxygen. The oxygen uptake was measured at regular time intervals at 180°C (*see* Table I).

Oven-Aging. Powdered PP (experimental or propathene HF-20 (unstabilized) (I.C.I.) mixed with the stabilizer was compression-molded at 200°C for 2 min. From the 6-mil films, test pieces (15 mm × 50 mm) were cut and clamped in a circulating air oven. At regular time intervals the test pieces were controlled for embrittlement. The end of the induction period was shown by a rapid disintegration of the test pieces, starting at the edges. Moreover, the decay of the antioxidants in the films was measured during the induction period by means of UV spectrophotometry. Conditions for the UV measurements are: (1) the antioxidants must have an absorption at a wavelength 220 mm, (2) the reaction products must not disturb the measurement of the antioxidant, and (3) when combinations of antioxidants are used, they must absorb at different wavelengths.

Model Reactions. General procedure: All reactions were carried out by dissolving the reactants in *o*-dichlorobenzene and heating this solution in a thermostat-regulated oil bath. At proper time intervals samples were taken to follow the course of the reaction. These samples were quenched in *n*-hexane. From this solution samples were injected into the liquid chromatograph. The separated reaction products were quantified by peak area measurements, using external standards. All of the reaction products were isolated semipreparatively by liquid chromatography, using mixtures of spectrophotometrically pure solvents as the eluent and identified by ir, NMR, and mass spectrometry.

Reaction of 4-Methoxy-2,6-diphenylphenoxyl (XVa) and β,β'-Diphenyldiethylsulfoxide. A solution of 0.8 mmol of 4-methoxy-2,6-diphenylphenoxyl (XVa) and 0.78 mmol of β,β'-diphenyldiethylsulfoxide in 5 mL of *o*-dichlorobenzene was heated at 150°C. After 2.5-min reaction time the products were isolated by liquid chromatography using 5% of ethylacetate in *n*-hexane as the eluent. The products, styrene XVII and 4-methoxy-2,6-diphenylphenol (VIIa), were identified by their ir spectra as compared with standard spectra. Sulfinate XVI: ν_{max} 2820 w (OCH_3) and 140s cm^{-1} (sulfinate s = o); δ (acetone-d_6): 7.1–7.8 (15 H, m, aromatic), 7.06 (2 H, s, m aromatic H), 3.92 (3 H, s, OCH_3), 2.7 (4H, s, aliphatic). No parent peak (*m/e* 428) was found. However two strong molecule ion peaks with *m/e* 276 and 152 were present, which were assigned to 4-methoxy-2,6-diphenylphenol (VIIa) and sulfoxide

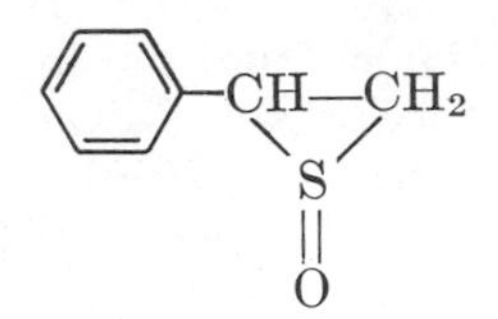

the decomposition products of sulfinate (XVI).

Thermal Decomposition of Sulfinate (XVI). A solution of 0.12 mmol of sulfinate (XVI) in 1 mL of *o*-dichlorobenzene was heated at 150°C. In the liquid chromatograms several peaks rose and disappeared during the reaction. Three main products were isolated: 4-methoxy-2,6-diphenylphenol (VIIa) was identified by its ir spectrum; thiosulfonate (XVIII), the ν_{max} 1200 s (SO_2), 750 s, and 700 s cm^{-1} (aromatic monosubstitution). (mass spectrometry: parent peak (M) *m/e* 306, peak *m/e* 242 (M–SO_2), *m/e* 105 ($C_6H_5C_2H_4$ fragment), and a molecule ion peak *m/e* 136, which was assigned to sulfide XX); and sulfonate XIX (the ν_{max} 2830 w (OCH_3), 1200 s (SO_2), and strong bands between 700 and 800 cm^{-1} (two types of aromatic substitution); mass spectrometry: parent peak (M) *m/e* 444, peak *m/e* 380 (M–SO_2)).

Thermal Decomposition of Sulfinate (XVI) in the Presence of an Excess of XVa. To a solution of 0.09 mmol of sulfinate (XVI), 0.11 mmol of phenoxylradical (XVa) was added and the solution was heated at 150°C. Styrene (XVII), phenol (VIIa), thiosulfonate (XVIII), and sulphonate XIX were identified by their ir and/or MS spectra. For sulfite XXI the ν_{max} 2830 w (OCH_3) strong bands between 1200–1250 (s = o of sulfite) and strong band between 700 and 800 cm^{-1} (two types of aromatic substitution). The δ (acetone-d_b) 7.1–7.8 (20H, m, aromatic), 6.8 (4H, s, aromatic m − H), 3.9 (6 H, aromatic OCH_3). MS: parent peak *m/e* 598.

Acknowledgment

The authors are indebted to M. E. F. Biemond, G. Hoentjen, and E. Verhulst for their technical assistance and to A. J. M. Weber and S. van der Werf for their advice in analytical problems.

Literature Cited

1. Hawkins, W. L., Sautter, H., Winslow, F. H., *Am .Chem. Soc., Div. Polym. Chem., Prepr.* (1963) **4**(1), 4331.
2. Scott, G., "Atmospheric Oxidation and Antioxidants," Elsevier, Amsterdam, 1965.
3. Shelton, J. R., "Polymer Stabilization," Hawkins, W. L., Ed., pp. 107–112, Wiley, New York, 1972.
4. de Jonge, C. R. H. I., Hageman, H. J., Mijs, W. J., Neth. Pat. Appl. **7,102,009.**
5. de Jonge, C. R. H. I., Hageman, H. J., Huysmans, W. G. B., Mijs, W. J., *J. Chem. Soc., Perkin Trans. 2* (1973) 1276.
6. de Jonge, C. R. H. I., Hageman, H. J., Huysmans, W. G. B., Mijs, W. J., *Am. Chem. Soc., Div. Org. Coat. Plast. Chem., Prepr.* (1974) **34**(2), 107.

7. de Jonge, C. R. H. I., van de Maeden, F. P. B., Biemond, M. E. F., Huysmans, W. G. B., Mijs, W. J., *J. Polym. Sci.*, in press.
8. Hawkins, W. L., "Polymer Stabilization," Hawkins, W. L., Ed., pp. 420–426, Wiley, New York, 1972.
9. Mayo, F. R., *Am. Chem. Soc., Div. Org. Coat. Plast. Chem., Prepr.* (1974) **34**(2), 117.
10. Roe, R. J., Bair, H. E., Gieniewski, C., *J. Appl. Polym. Sci.* (1974) **18**, 843.
11. Hiatt, R., Mill, T., Irwin, K. C., Castleman ,J. K., *J. Org. Chem.* (1968) **33**, 1415.
12. Emanuel, N. M., Denisov, E. T., Maizus, Z. K., "Liquid-Phase Oxidation of Hydrocarbons," Plenum, New York, 1967.
13. Shelton, J. R., Davis, K. E., *J. Am. Chem. Soc.* (1967) **89**, 718.
14. Smit, W., to be published.
15. Hay, A. S., *J. Org. Chem.* (1969) **34**, 1160.
16. Hoentjen, G., de Graaf, S. A. G., Bijkerk, A. H., de Jonge, C. R. H. I., Neth. Pat. Appl. **7,511,696.**
17. de Jonge, C. R. H. I., Giezen, E. A., Neth. Pat. Appl. **7,511,698.**
18. Foreman, E. L., McElvain, S. M., *J. Am. Chem. Soc.* (1940) **621**, 1435.
19. Strassburg, R. W., Gregg, R. A., Walling, C., *J. Am. Chem. Soc.* (1947) **69**, 2142.
20. Pospisil, J., Taimr, L., Kotulak, L., ADV. CHEM. SER. (1968) **85**, 170.

RECEIVED May 12, 1977.

INDEX

A

Absorption spectra of 2-butanone *tert*-butyl hydroperoxide 5
Acetone 84
  formation 87
  vapor-phase ir of 190
Aicd catalysis 232
Acids, Lewis 328
β-Activated thioethers 399, 421
  and phenoxyl radical (XVa), H-donation of 417
Activation energies for chemiluminescence from autoxidized, singlet-oxygenated polymers 22
Additive radicals 157
Additives in isotactic PP, effects of crystalline 261
Aging effect of irradiated PP in air 149
AIBN-initiated oxidation of cumene 235
Air-irradiated copolymer films, delayed-emission spectra of .. 106
Aldol-type reaction, reverse 79
Aliphatic polymer 3
Allylic hydroperoxides 11
4-Alkoxy-2,6-diphenylphenols, times of failure of PP films stabilized with 406, 407
4-Alkoxy-2,6-diphenylphenols (VIII), antioxidants activity of 406
Alkylperoxy radicals 31
Amorphous fluoroelastomers 294
Amorphous polypropylene 176
Anions in copper compounds, effects of 164
Annealing 270
Antioxidant(s)
  action, mechanism of 240
  activity
    of 4-alkoxy-2,6-diphenylphenols(VIII) 406
    of combinations of chain-breaking and preventive antioxidants, measurements of 402
    of some phenyl-substituted phenols 403
    of triazine-based, phenyl-substituted phenols 403
  disadvantage of 253
  effect of processing time at 150°C on carbonyl formation in LDPE containing 52

Antioxidant(s) (*continued*)
  in NBR vulcanizate, nonmigrating vs. conventional 259
  nonmigrating 253
  preventive 253
  on PP, grafting of 254
  in retarded antoxidation 223
  synthesis of chain-breaking 402
  UV absorption during injection molding, decrease of 412
Antioxidative properties of phenyl-substituted phenols 399
AO-1 132
AO-2 132
APP (amorphous polypropylene) . 176
  carbonyl groups during thermal oxidation of 168
  effect of
    tert-BuHPO on the thermal oxidation of 170
    copper stearate on the thermal oxidation of 164
    DTBP on the thermal oxidation of 171
    model compounds on the thermal oxidation of 173
    n-octanoic acid on the thermal oxidation of 172
    propionic acid on the thermal oxidation of 165
  gel permeation chromatograms of 169
  oxygen uptake curves 162
  preparation of oxidized samples of extracted 167
  weight loss during thermal oxidation of 168
Aromatic
  diketones 97
  nitroso compounds 253
  polyester polyurethane 119
Arrhenius plot of PP pyrolysis in air 187
β-Aryl-β′-carboalkoxydiethyl-sulfides XII 421
Atactic polypropylene (APP) 160
Autoradiography 262
Autoxidation 19
  chemiluminescence mechanism . 21
  of hydrocarbons, uninhibited ... 215
  of PP, inhibited 237
Autoxidized polymers, activation energies for chemiluminescence 22
Auxiliary stabilizers 317
Axion, degradation of polyurethanes by 212

B

Back diffusion coefficients 267
Barbender plasticorder 389
Barium coats 317
Benzene
inhibited oxidations of 9,10-dihydroanthracene in 244, 245
mechanism for the formulation of 313, 314
solution, inhibited oxidation of PP in 242, 243
solution, oxidation of PP in 240
Benzophenone LS-7 119
3,3'-4,4'-Benzophenonetetracarboxylic acid dianhydride (BTDA) 199
BHT-inhibition of PP oxidation ... 246
Biodegradable polymers 205
Biodegradability of step-growth polymers 205
Bis(8-oxyguinolate)zinc(II)
first-order rate constants for chain scission on PFAP(II) stabilized with 306
mechanism for the stabilization of PFAP(II) with 305
to PFAP(II), addition of 296
stabilization of PFAP(II) with . 302
Bis(2,2,6,6-tetramethyl-peperidinyl-4) sebacate (LS-1) ... 116
Blow-molded HDPE containers ... 122
Bond
fixation in heptatrienyl 358
fixation in hexatriene 358
scission 6
Bottle dispersion, metal carboxylates consumption in 381
Boundary cracking of the spherulites, radial 269, 270
($(Br(CH_2)_{10}CO_2)_2Cu$ into polyethylene, diffusion of 288, 289
Breaking elongation 144
Brittle snap 148
Bromination 72
BTDA (3,3',4,4'-benzophenonetetracarboxylic acid dianhydride) 199
DABP degradation 200
MDA degradation 200
Bulk carboxylate absorption 286
Bulk diffusion rates 279
Burning surface 176
1,3-Butanediol 88
2-Butanone *tert*-butyl hydroperoxide, absorption spectra of ... 5
2-Butanone, *tert*-butyl hydroperoxide, molar extinction coefficients of 5
tert-BuHPO on the thermal oxidation of APP, effect of 170
tert-Butyl alcohol 86
hydroperoxide
absorption spectra of 2-butanone 5
hydroperoxide (*continued*)
decomposition of 162
in hexane, quantum yields for photolysis of 5
molar extinction coefficients for 2-butanone 5
photolysis, quantum yields . 9
peroxide 244

C

C NMR spectrum of PVC reduced with tri-*n*-butyltin hydride .. 327
C-13 NMR spectra of 102-reacted CB 17
C_7-insoluble fractions 114
C_7-soluble PP fraction 114
Ca-Zn formulation 375
Cadmium soaps 317
Calcium stearate
consumption of 391
synergism, zinc and 389
transformation from chloride ions titration 391
transformation, viscosity evolution of plasticized PVC and 389
Calculated bond lengths for linear polyenes 335
Calendering 375
Carbenes 255
Carbon
arc fadeometer exposure, color as a function of 126
-13
chemical shifts 137
NMR 133
spectroscopy 12
spectra of branched polyethylene 138
Carbonyl
band of highly oxidized fractions 112
carbons, spin lattice relaxation times of the 368
chromophore, α,β-unsaturated .. 71
chromophores, phosphorescent .. 71
compounds, photochemistry of α,β-unsaturated 76
concentration relationship, embrittlement time formation . 48
in LDPE containing antioxidants, stabilizers, effect processing time at 150°C on 52
in LDPE, effect of processing time 40
in the thermal oxidation of LDPE films 52
groups 80
effect of photooxidation on the luminescent α,β-unsaturated 73
during thermal oxidation of APP 168

Carbonyl (*continued*)
impurities, α,β-unsaturated 74
region for the dibutuyltin diiso-octylthioglycolate/dibutyltin dichloride system, C NMR spectra of 369
Carboxamide group, hydrolysis of the 200
Carboxylate absorption, bulk 286
Carboxylic acids 235
potentiometric titration of 378
Chain
initiation 189
propagation 193
scission
in PFAP(II), first-order rate constants for 301
for PFAP(II), rate of 305
in PFAP(II) stabilized with bis(8-oxyquinolate)zinc-(II), first-order rate constants for 306
quantum yield(s) 7
for hydroperoxide decomposition 7
termination 191
transfer 189
intramolecular 189
breaking
antioxidants 216, 223
phenolic antioxidants as ... 399
synthesis of 402
Charge densities
polyenyl cation 340
STO-3G π-orbital 340
at the terminal carbon atoms ... 341
Charge-transfer absorption bands bands of tropenylium halides in methylene chloride 350
Chemical geometric nature of the monomer unit 202
Chemical modification, stabilization by 318
Chemiluminescence
from autoxidized, singlet-oxygenated polymers, activation energies of 22
emission 19
from a film of transpolypentenamer 24
from oxidized polymer films, Arrhenius plots 23
from polymers, low-temperature 21
Chlorides and stearic acid titration during heating of the PVC sheets 394
Chlorine ions, coulometric titration of 378
Chromophore(s) 56, 57
phosphorescent carbonyl 71
α,β-unsaturated carbonyl 71
Chromophoric groups 69
Cis elimination of HCl 310
Cold hexane extract of poly-4-methylpent-l-ene 72, 73
Cold-drawn filament 65
Color formation in PVC, effect of processing 33
Combinations of VIIe/DSTDP and phenol, times of failure of PP films stabilized with 413
Conjugated carbonyl concentration in PP, effect of processing time on 46
Conjugation energy(ies) for linear α,ω-dimethylpolyenyl cations ... 346
polyenes 337
polyenyl cations 342
(L_n+) with reference to 1-methylpolyenes (MeI_{n-1}) . 344
Constants for dimethyltin derivatives, ligand exchange effects on the tin–proton coupling ... 365
Consumption of calcium stearate .. 391
Copolymer
films, delayed-emission spectra of air-irradiated 106
films, delayed-emission spectra of vacuum-irradiated 106
fluorescence 106
quenching 107
Copper
carboxylates 274
into polyethylene, diffusion of 280
surface measurements, of diffusion of 282
catalysis 160
catalyzed oxidation 274
catalyzed oxidation of polyolefins 289
compounds, effects of anions in . 164
polyethylene interface 273
polymer interface 160
salts, ligands of 163
salts in polyethylene 274
stearate concentration, effect of . 165
stearate on the thermal oxidation of APP, effect of 164
Coulometric titration of chlorine ions 378
Cracking of the spherulites, radial boundary 269,270
Cross-linking of PVA, photooxidative 79
Crystalline additives in IPP, effects of 261
Crystalline polyolefin, factors affecting the oxidative degradation of 261
Crystallinity 59
variation 267
Crystallization, slow polymer 270
Cumene
AIBN-initiated oxidation of 235
hydroperoxide
decomposition 226
products from 221, 222

Cumene (*continued*)
effect of
base on decomposition of 220
n-butyl sulfide on the decomposition of 230
radical trapping agents on the decomposition of 233
sulfoxide on the decomposition of 230, 232
water on the decomposition of 229
polar decomposition of 222
radical decomposition of 222
$Cu(RCO_2)_2$, solubility vs. chain length for 278
Cyclic oligomers, depolymerization to 297
Cyclizations, Diels–Alder 313
Cycloelimination reaction 93
intramolecular 234
Cyclohexadiene structures, thermal scission of 313

D

Dart impact strength, kilolangley of 117
Dart impact strength, surface gloss of 119
Deactivation theory, defect site .. 316
Decay
second-order 155
under vacuum 155
in vacuum, ESR spectra of 154
Decontamination of the optimum ratio between VIIe and STDP 409
Deep surface regions 283
Defect
-site deactivation theory 316
sites 311
structures 24
Degradation:
nonoxidative thermal 309
of polymer samples, determination of degrees of 206
rate of polymers 203
Dehalogenation with tri-*n*-butyltin hydride, reductive 324
Dehydration, hydroxyketone 91
Dehydrochlorination 310
El mechanism for PVC 334
ion-pair mechanism for the thermal 351
mechanism of nonoxidative thermal 314
mechanism for PVC stabilization, implications of the ion-pair 353
of PVC 310, 333
rates of syndiotactic sequences .. 313
Delayed-emission spectra 101
of air-irradiated copolymer films 106
of ketonic products 102
of vacuum-irradiated copolymer films 106
Depolymerization (unzipping) .. 189
Depolymerization to cyclic oligomers 297
Deuterium isotope effects 217
Dialkyltin
compounds 364
dichlorides 363
dimercaptides 363
possible role of the ligand exchange reaction (I) in PVC stabilization by 372
Diaminobenzophenone (DABP) .. 199
Diamyl ethylene 42
1,1-Diamyl ethylene, effect of photo-initiators in the photo-oxidation of 45
Diazabicyclooctane 15
Dibenzanthrone 24
Dibenzyltin
di(benzylmercaptide)/dibenzyltin dichloride system, H NMR data for the 365
di(benzylmercaptide)/dibenzyl dichloride system, tin–proton coupling constants for the 366
dichloride 365
3,5-Di-*tert*-butyl-4-hydroxybenzyl 4-azidoxulfonylcarbanilate (DTBSC) 257
2-[3-(3,5-Di-*tert*-butyl-4-hydroxyphenyl)propionoxy] ethyl 4-aziodosulfonylcarbanilate (DTBPSC) 257
Dibutylchlorotin compounds 363
Dibutyltin
dialurate/dibutyltin dichloride system 369
dichloride 364
diisooctylthioglycolate/dibutyltin dichloride system
C NMR data for the 368
C NMR spectra of the carbonyl region for the 369
H NMR data for the 366
Dicarboxylates 363
Diels–Alder cyclizations 313
Diene fluorescence 103
Dienones, phosphorescent 73
Differential scanning calorimetry (DSC) 111
Diffusion
of $(Br(CH_2)_{10}CO_2)_2Cu$ into polyethylene 288,289
of copper carboxylates into polyethylene 280
of copper carboxylates, surface measurements of 282
equations 279
mechanisms 288
of metal salts in polymers 274
rates, bulk 279
samples, configuration of 275

9,10-Dihydroanthracene in benzene, inhibited oxidations of 244, 245
Diisooctylthiogylcolate system, ir spectra on the 369
Diketones, aromatic 97
Dilaurylthiodipropionate (DLTDP) 403
-dimethylpolyenyl cations, conjugation energies for linear 346
Dimethyltin derivatives, ligand exchange effects on the tin–proton coupling constants for . 365
Dimethyltin di(benzyl mercaptide)/ dimethyltin dichloride system 369
Dioctylphthalate degradation 392
Dione phosphorescence 103
2,6-Diphenyl-*p*-benzoquinone II .. 419
2,6-Diphenylhydroquinone (III) . 419
Disadvantages of $LiAlH_4$ reduction 325
Discoloration 142
Dissociation energies, heterolytic gas-phase 347
Dissociation energy of organic chlorides, heterolytic gas-phase 348
Distearylthiodipropionate (DSTDP) 403
Disulfides, oxidation of 226
DLTDP (dilaurylthiodipropionate) 403
Double-bond shifts 11
Draw
  atmosphere on PPH photostability, effect of 63
  conditions for PPH monofilament production 56–57
  oxidation 61
  speed on PPH photostability, effects of 63
  temperature on PPH photostability, effects of 63
DSC, (differential scanning calorimetry) 111
  data 113
  E fractions data 113
  R fractions data 113
  tracings for fractions O/E, O/E/I, and O/E, II 113
  tracings of R fractions 111
DSTDP (distearylthiodipropionate) 403
DTBP on the thermal oxidation of APP 171
DTBPSC (2-3-(3,5-di-*tert*-butyl-4-hydroxyphenyl)-propionoxy ethyl 4-azidosulfonylcarbanilate) 257
DTBSC (3,5-di-*tert*-butyl-4-hydroxybenzyl-4-azidosulfonylcarbanilate) 257
DTBSU (*N* 2-(3,5 di-*tert*-butyl-4-hydroxyphenyl) ethyl *N'*-(4-azidosulfonylphenyl) urea) .. 257

E

E fractions 111
El elimination reaction 352
El mechanism for PVC dehydrochlorination 334
Eicosane 137
Elastic modulus as a function of irradiation dose 145
Elastomers 20
  synthetic 254
Electron spin resonance (ESR) spectra
  of decay in vacuum 154
  of peroxy radical formation 153
  studies 151
Elemental oxygen 215
Elimination reaction, El 352
Embrittlement 62, 92, 142, 270
  PP post-irradiation 151
  time–carbonyl concentration relationship 48
  time for PP 46
Emission changes during irradiation of polystyrene 102
End-linking 136
"Ene"-type process 11
Enones, fluorescent 72
Enzymes
  degradation of hydroxy-substituted polyamides by 210
  degradation of methyl-substituted polyamides by 210
  polymer degradations by 207
Ester-linkage cleavage 211
Ethyl chloride 310
Ethylenic unsaturation in PVC containing additives 47
Exchange experiments 287
Excimer quenching 99, 100
Excimer sites, photoconversion of . 100
Exposed PP, cross-section profile of 127
Extrusion
  conditions for PPH monofilament production 56–57
  system, small-scale 57
  temperature on PPH photostability, effects on hopper blanketing and 62

F

Fick's law diffusion 288
Filament(s)
  cold-drawn 65
  hot-drawn 64
  photostability 61
Films from multi-film laminates, preoxidized nonpreoxidized .. 282
First-order
  chain scission in PFAP(II) stabilized with bis(8-oxyquinolate)zinc(II) 306
  kinetic plots for $\overline{M}_w$ loss in PFAP(II) 300
  kinetic plots for molecular weight loss in PFAP(II) 306
  plot of PP pyrolysis in He 181

First-order (*continued*)
rate constants for
chain scission in PFAP(II) .. 301
Flame retardants, effect of 176
Flange bending test curve 144
Fluorescence
excitation spectra of polypropylene 73
poly(4-methylpent-lene) 71
phosphorescence excitation 70
quenching, copolymer 107
spectra from a polystyrene film . 99
Fluorescent
enones 72
excitation spectrum of PP film after irradiation, intensity of 74
excitation spectrum of PP film before irradiation, intensity of 74
species, identification of 70
Fluoroelastomers, amorphous 294
Formaldehyde 194
Fourier transform ^{1}H NMR, pulsed- 313
Fractions O/E, O/E/I, O/E/II, DSC tracings for 113
Fragmentation patterns of pyrolysis products 182
Free amine, formation of *N*-oxyl from 129
Free-radical chain reaction 215
Functional group formation in HIPS, effect of processing at 200°C 37
Functional group formation in HIPS, effect of thermal oxidation at 98°C in air 38
Fungal growth on polyureas 210
Fungi, degradation of hydroxy-substituted polyamides by ... 210
Fungi, degradation of methyl-substituted polyamides by 210

G

G-values 134
Gamma irradiation 134
Gas-phase
dissociation energies, heterolytic 347
dissociation energy of organic chlorides, heterolytic 348
oxidation of hydrocarbons 193
GC pyrogram of IPP in
air, high-boiling 188
air, low-boiling 188
He, low-boiling 182
Gel permeation
chromatograms of APP 169
chromatograms of PFAP(II) ... 298
chromatography (GPC) 110
Geometric nature of the monomer unit, chemical 202
GPC (gel permeation chromatography)
of aged PFAP(II) 298
of unaged PFAP 300
Grafting of antioxidants on PP 254
Growth on polyureas, fungal 210
Growth rejection process 269

H

H-donation of β-activated thioethers and phenoxyl radical (XVa) 417
HALS LS-1 117
Haxacosane 137
HCl, cis elimination of 310
HDPE containers, blow-molded .. 122
Head-to-head PVC 311
Head-to-tail PVC 310
He (helium)
carrier gas 192
first-order plot of PP pyrolysis in 181
kinetics of pyrolysis of PP in ... 182
low-boiling GC pyrogram of IPP in 182
product distribution of pyrolysis of PP in 184
pyrolysis vs. time in 181
TGA of PP in 180
Heptane
extraction 110
-soluble and -insoluble fractions, photooxidative aging of PP: separation of 109
-soluble and -insoluble fractions, thermooxidative aging of PP: separation of 109
Heptatrienyl, bond fixation in 358
Heterolytic gas-phase dissociation energy 347
of organic chlorides 348
Hexane extraction 72
Hexatriene, bond fixation in 358
cis-3-Hexene 27
High-
crystalline plastics 293
density polyethylene 119, 122, 123
impact strength of 120
energy irradiation sterilization .. 142
energy oxidation on polyethylene model paraffins, effects of . 133
energy radiation on polyethylene model paraffins, effects of .. 133
Hindered-amine light stabilizers .. 116
Hindered-amine stabilizers, mechanism of 127
HIPS(rubber-modified polystyrene), effect of 36
processing at 200°C on functional group formation 37
thermal oxidation at 98°C in air on functional group formation 38

HIPS (*continued*)
thermal oxidation on the photo-oxidation ... 39
HMO (Huckel Molecular Orbital) theory
HMO treatment ... 338
HOBP (2-hydroxy-4 octyloxbenzophenone) ... 53
Homolytic decomposition of hydroperoxides ... 235
Homolytic processes ... 229
Homopolymers ... 294
Hopper blanketing extrusion temperature on PPH photostability, effects of ... 62
Hot-drawn filaments ... 64
Huckel Molecular Orbital (HMO) theory ... 356
Hydride reductions, tri-*n*-butyltin . 330
Hydrocarbon(s)
gas-phase oxidation of ... 193
resistance ... 293
uninhibited autoxidation of ... 215
Hydrogen
atom abstraction from PVA ... 91
denation mechanism ... 216
peroxide ... 88
Hydrolysis of the carboxamide group ... 200
Hydrolytic stability ... 293
Hydroperoxidation in *cis*-1,4-polybutadiene ... 14
Hydroperoxide(s)
allylic ... 11
decomposer, sulfur compounds as 399
decomposition ... 217
chain scission, quantum yields 7
homolytic decomposition of ... 235
induced-decomposition ... 3, 9
photolysis ... 5
2-Hydroxy-4 octylbenzophenone (HOBP) ... 53
Hydroxy-substituted polyamides by fungi, degradation of ... 210
Hydroxyketone dehydration ... 91
Hydroxyl formation ... 36

I

Impact polystyrene ... 125
Impact strength of high-density polyethylene ... 120
Impurities by growing spherulites, redistribution of ... 262
Impurities in IPP ... 261
Impurity
carbonyl groups ... 68
distribution curves ... 266
distributions on oxidation behavior, effects of ... 269
Induced-radical mechanism ... 4
Induction peroxides
for combinations of primary and secondary antioxidants at 180°C in PP ... 404, 405
Induction peroxides (*continued*)
measured ... 410
predicted ... 411
of (VIIe)/DSTDP mixtures at 140°C, effect of processing on ... 410
Inhibited
autoxidation of PP ... 237
oxidation
of 9,10-dihydroanthracene in benzene ... 244, 245
of PP ... 240
in benzene solution ... 242, 243
-rate dependence on
inhibitor concentration ... 247
initiator concentration ... 246
polymer concentration ... 247
Inhibition
mechanisms of ... 291
period, rate of oxygen consumption during ... 245
of PP oxidation ... 248
Inhibitor
concentration, inhibited-rate dependence on ... 247
properties ... 237
solutions ... 239
Initial PVA, ir spectra of ... 84
Initiation reaction ... 215
Initiation sites of nonoxidative thermal degradation ... 310
Initiator concentration, inhibited-rate dependence on ... 246
Injection molding, decrease of antioxidant UV absorption during 412
Insoluble fractions, separation of heptane-soluble and ... 109
Instron syringe flange bending device ... 143
Insulating cables, x-ray diffractogram of lead formulation for . 380
Interaction between zinc chloride or calcium chloride and epoxidized soya bean oil ... 396
Interfaced pyrolysis gas chromatographic peak identification system (IPGCS) ... 176
block diagram of ... 177
Interfibrillar cohesion ... 65
Internal reflection Ir measurements, summary of ... 284, 285
Intramolecular
chain transfer ... 189
cycloelimination reaction ... 234
transfer ... 191
Intrinsic viscosity ... 82
Ion-air dehydrochlorination mechanism for PVC stabilization, implications of the ... 353
Ion-pair mechanism for the thermal dehydrochlorination of PVC . 351
Ionic mechanism of nonoxidative thermal degradation of PVC .. 334

IPGCS (interfaced pyrolysis gas chromatographic peak identification system) 176
IPP (semicrystalline or isotactic polypropylene) 176
in air, high-boiling GC pyrogram of 188
in air, low-boiling GC pyrogram of 188
in He, low-boiling GC pyrogram of 182
oxygen uptake curves of 161
IR
internal reflection spectroscopy (IRS) 274
spectra of initial PVA 84
spectra of oxidized PVA 84
spectroscopy 12
spectrum of *cis*-1,4-polybutadiene (CB) after singlet oxygenation 12
Irradiation
embrittlement of PP syringe flange 147
of poly(styrene-alt-methyl methacrylate) films, prompt-emission changes during 106
of polystyrene, emission changes during 102
of polystyrene, prompt-emission changes during 100
undervacuum 155
IRS (ir internal reflection spectroscopy) 274
4-Isocyanatobenzenesulfonyl azide 256
Isotacophoresis analysis 81
Isotactic polypropylene (IPP) ..160, 261
effects of crystalline additives in 261
impurities in 261
y-irradiation sterilization of 151

K

Keto groups 2, 9
Ketone
distribution in the photooxidation of poly(vinyl alcohol) 92
photolysis 5
Ketonic products, delayed-emission spectra of 102
Kilolangley of dart impact strength 117
KLY-to-visual surface crazing 121

L

Laminate stacks 283
Laplace transform 279
LCAO–SCF molecular orbital theory, single-determinant ab initio 334
LDPD, photooxidation of 53
LDPE
containing antioxidants, effect of processing time at 150°C on carbonyl formation in 52
LDPE (*continued*)
effect of processing
on the molecular weight distribution 41, 42
time on
carbonyl formation 40
the formation of peroxides . 43
the melt-flow index (MFI) 40
rate of photooxidation 44
rate of vinylidene decay ... 44
films, carbonyl formation in the thermal oxidation of 52
thermal photooxidative reactions of vinylidene groups 43
Lead formation 375
for insulating cables, x-ray diffractogram of 380
x-ray diffraction for 379
x-ray diffraction results for 384
Lewis acids 328
$LiAlH_4$ (lithium aluminum hydride)
reduced polymers 328
reduced PVC, trans-alkene linkages in 327
reduction technique 325
Ligand(s)
of copper salts 163
exchange effects on the tin–proton coupling constants for dimethyltin derivatives 365
exchange reaction(I) in PVC stabilization by dialkyltin dimercaptides, possible role of 372
Light
-sensitive commercial polyolefins, luminescence photooxidation of 68
-sensitive polyolefins 69
stabilizers, hindered-amine 116
Linear
α,ω-dimethylpolyenyl cations, conjugation energies for ... 346
polyenes
calculated bond lengths for .. 335
conjugation energies for 337
observed bond lengths for ... 335
polyenyl cations (L_n+)
conjugation energy for 342
idealization of resonance forms found for 340
polyenes, numbering system nomenclature used for .. 336
with reference to 1-methylpolyenes(MeL_{n-1}), energy conjugation for 344
STO-3G calculated bond lengths for 339
Liquid
chromatography analysis of the organic compounds 378
-phase solubilities 276
Lithium aluminum hydride ($LiAlH_4$) 324

Low-temperature chemiluminescence from polymers 21
LS-1, (bis(2,2,6,6-tetramethylperidinyl-4)sebacate) ..116, 130
  performance of 117
LS-2 ....119, 130
LS-3 130
LS-4 131
LS-5 131
LS-6 131
LS-7 131
LS-8 131
Luminescence 68
  measurements 69
  photooxidation of light-sensitive commercial polyolefins 68
Luminescent α,β-unsaturated carbonyl groups, effect of photooxidation on 73

M

Maleate stabilizer during the thermal induction period 51
Maleic anhydride UV absorption, effect of processing time on the growth of 50
Measured induction periods 410
Mechanical
  mixing of PFAP(II) compounds 296
  properties 142
    during PPH photooxidation, changes in 64
Mechanism for the formulation of benzene 313, 314
Mechanism of nonoxidative thermal dehydrochlorination 314
Mechanooxidation of
  polymers during processing 31
  PVC 32
  rubber 31
Mechanoscission of the PVC chain 34
Melt
  degradation 61
  diffusion rates 266
  flow index (MFI) 38
  of LDPE, effect of processing time 40
  oxidation 50
  surface 176
Mercaptides vs. carboxylates, implications for PVC stabilization in the behavior of 370
Metal
  carboxylates consumption in bottle dispersion 381
  deactivators 274
  salts in polymers, diffusion of ... 274
  salts in polymers, solution of ... 274
  soaps 386
    stabilization with 317
  stearates, effect of 161
  surfaces 273
Metallic catalyst, functions of 163
Methanol charcateristics of PVA photooxidation 83
4-Methoxy-2,6-diphenylphenoxyl dimer (XVb) 422
  (XVa)
    and β,β'-diphenyldiethylsulfide, reaction of 416
    and β,β'-diphenyldiethylsulfoxide, reaction of 413, 423
Methyl *tert*-butyl ketone 9
4-Methyl-2-pentene 15
Methyl-substituted polyamides by fungi, degradation of 210
Methylene
  blue 12
  chloride, charge-transfer absorption bands of tropenylium halides in 350
  dianiline (MDA) 199
    basicity 200
  resonances 364
Methylperoxy radicals 85
MFI, (melt flow index) 38
  of LDPE, effect of processing time on 40
Microorganisms, polymer degradation by 207
Microscopic mechanisms of oxidative degradation 273
Microstructure investigation, poly(vinyl chloride) 324
Microwave discharge 11
Model
  compounds, photooxidations of ..87, 89
  compounds, values for polymer . 25
  extinction coefficients for 2-butanone *tert*-butyl hydroperoxide 5
  paraffins, effects of 133
Molecular
  elimination 314
  orbital theory of polyenes 333
  orbital theory, single-determinant ab initio LCAO–SCF 334
  weight
    distribution of LDPE, effect of processing 41, 42
    distribution of oxidized PP ... 170
    loss in PFAP(II), first-order kinetic plots for 306
Monomer unit, chemical geometric nature of the 202
Multi-film laminates, preoxidized non-preoxidized films from ... 282

N

N-[2-(3,5-di-*tert*-butyl-4-hydroxyphenyl)] ethyl (*N*′-4-azidosulfonyl) urea (DTBSU) 257
N-oxyl from free amine, formation of 129
NBR vulcanizate, nonmigrating vs. conventional antioxidant in .. 259

Neutral polyenes 333
Nickel chelates 127
Nickel diethylthiocarbamate
(NiDEC) 53
Nitrosocompounds, aromatic 253
Nitroxyls 129
Nomenclature used for the linear
polyenyl cations polyenes,
numbering system 336
Nonmigrating vs. conventional antioxidant in NBR vulcanizate .. 259
Nonoxidative thermal
degradation 309
dehydrochlorination, mechanism of 314
initiation sites of 310
of PVC 310
ionic mechanism of 334
pathways for 333
Non-preoxidized films from multifilm laminates preoxidized
Norcon 201 rapid-scan, vapor-phase
spectrophotometer 178
Norrish Type I reaction 76, 92
Norrish Type II reaction 76
NR vulcanizate, nonmigrating vs.
conventional antioxidant in .. 259
Nucleophile, water as a 228
Nucleophiles, sulfur 228
Numbering system nomenclature
used for the linear polyenyl
cations polyenes 336

O

O_2 absorption of polymers 258
O_2 absorption study 258
O_2/polyethylene/copper system ... 273
Observed bond lengths for linear
polyenes 335
n-Octanoic acid on the thermal oxidation of APP, effect of 172
O/E, DSC tracings for Fractions . 113
O/E/I, DSC tracings for Fractions 113
O/E/II, DSC tracings for Fractions 113
Olefinic unsaturation 312
development during PVC processing, peroxide 32
Oligomers, depolymerization to
cyclic 297
Oligomers X 421
One-pass organotin hydride treatment 328
Optimum ratio between VIIe and
STDP, decontamination of ... 409
Organic
chlorides, heterolytic gas-phase
dissociation energy of materials, oxidative degradation
of 215
sulfur compounds P 226
Organotin stabilizers 363
Organotins, stabilization with 316
Orientation 59
Oven-aging 406, 423
of PP 251
Oxidation
behavior, effects of impurity
distributions on 269
BHT-inhibited PP 246
copper-catalyzed 274
of cumene, AIBN-initiated 235
of 9,10-dihydroanthracene in
benzene 244, 245
of disulfides 226
mechanism, radical-chain 3
morphology on 262
on polyethylene model paraffins,
effects of high-energy 133
polyisoprene 217
of polyolefins, copper-catalyzed . 289
of PP
in benzene solution 240
inhibited 242-243
inhibited 240
products during PPH photooxidation, changes in 64
scheme 139
of the sulfenic acid 221
sunlight-induced 68
technique, solution 238
Oxidative
bleaching of colored polyenes
in PVC 315
chemiluminescence 20
cleavage 140
degradation 399
of a crystalline polyolefin,
factors affecting 261
microscopic mechanisms of ... 273
or organic materials 215
of polymers 78
pyrolysis 182, 193
of PP 175, 187
procedures for 180
polymer films, Arrhenius plots of
chemiluminescence 23
PVA, ir spectra of 84
4,4′-Oxydiphthalic anhydride
(ODPA) 199
Oxygen
-aging of vulcanizates 258
consumption during the inhibition period, rate of 245
elemental 215
uptake curves of APP 162
uptake curevs of IPP 161

P

Partition coefficient 269
2,4-Pentanediol 88
Peroxide(s)
concentration in PP, effect of
processing time on 46
decomposer, formation of the
active 227

Peroxides (*continued*)
decomposers as UV stabilizers .. 51
decomposition 217
formation in PVC containing additives 47
in LDPE, effect of processing time on the formation of .. 43
in LDPE, effect of temperature on the formation of olefinic unsaturation development during PVC processing 32
Peroxy
gel formation in PVC 35
radical 152
formation, ESR spectra of 153
PFAP (poly(fluoroalkoxyphosphazene))
elastomer (II), mechanism of the thermal degradation of 295
formulations 306
GPC of unaged 300
(I) 294
(II) 294
addition of bis(8-oxyquinolate)zinc(II) to 296
with bis(8-oxyquinolate)zinc(II), stabilization of 302
change in solution viscosity of 303
compounds, mechanical mixing of 296
containing 0 wt % bis(8-oxyquinolate)zinc(II) 303
containing 1 wt % bis(8-oxyquinolate)zinc(II) 303
containing 2 wt % bis(8-oxyquinolate)zinc(II) 303
containing 3 wt % bis(8-oxyquinolate)zinc(II) 303
first-order kinetic plots for $\overline{M}_w$ loss in 300
first-order kinetic plots for molecular weight loss in . 306
gel permeation chromatograms in 298
GPC of aged 298
investigation of the thermal degradation of 296
O-ring seal compounds, mechanical testing of 296
rate of chain scission for pure 305
solution viscosity of 297
stabilized with bis(8-oxyquinolate)zinc(II), first-order rate constants for chain scission in 306
thermal stability of 302
wt % of 298
Phenolic antioxidants as chain-breaking antioxidants 399
Phenols
antioxidant activity of triazine-based, phenyl-substituted .. 403
monofunctional 250
phenyl-substituted 402
Phenols (*continued*)
singly hindered 238, 250
times of failure of PP films 405
Phenoxy radical products, doubly hindered 251
Phenyl-substituted phenols 402
antioxidant activity of some 403
antioxidative properties of 399
(IV) 419
(V) 420
(VI) 420
(VII) 420
α-Phenylindole 395
Phosphorescence
dione 103
excitation 70
excitation spectra of polypropylene poly(4-methylpent-l-ene) 72
Phosphorescent
carbonyl chromophores 71
dienones 73
species, identification of 70
Phosphorus nitrogen chain 293
Photochemistry of α,β-unsaturated carbonyl compounds 76
Photoconversion of excimer sites .. 100
Photodegradation of polyolefins, fundamental processes of 1
Photoengraving processes 79
Photo-initiators 59
in the photooxidation of 1,1-diamylethylene, effect of 45
Photolysis
of *tert*-butyl hydroperoxide in hexane, quantum yields ... 5
hydroperoxide 5
ketone 5
of polyethylene 6
of *cis*-polyisprene hydroperoxide 8
PVA 79
Photooxidation(s) 2
in aqueous solution, PVA 80
of 1,1-diamylethylene, effect of photo-initiators on 45
of HIPS, effect of thermal oxidation on 39
of LDPE, effect of processing time on the rates of 44
of light-sensitive commercial polyolefins, luminescence .. 68
on the luminescent α,β-unsaturated carbonyl groups, effect of 73
methanol characteristics of PVA 83
of mildly LDPD 53
of model compounds 87, 89
of polymers 30
of PVC 32, 78, 82, 83
at 70°C 81
ketone distribution in the 92
ketone distribution statistical chain scissions in the ... 92

Photooxidation(s) (*continued*)
of PVC containing a tin maleate stabilizer, effect of processing time on the rate of 48
of PVC, effect of processing time on the rate of 36
of rubber-modified polymers ... 36
sensitization 2
Photooxidative aging of polypropylene: separation of heptane-soluble and -insoluble fractions 109
cross-linking of PVA 79
reactions of vinylidene groups in LDPE 43
results 110
Photooxidized filaments 60
Photosensitivity of PP fibers, effects of production conditions on the 56
Photosensitivity, strain-induced ... 62
Photosensitization 11
Photosensitized oxidation of 1,4-polyisprene 12
Photosensitized oxidation of polymer films 22
Photosensitizer residual 14
Photostability 61
Photoyellowing 104
in films 98
formation 97
of polystyrene poly(styrene-alt-methyl methacrylate) 96
Piperidene derivatives, substituted . 116
Plasticized PVC and
calcium stearate transformation, viscosity evolution of 389
zinc and calcium stearate transformation, viscosity evolution of 390
zinc stearate transformation, viscosity evolution of 390
Plastics, high-crystalline 293
Polar decomposition of cumene hydroperoxide 222
Polar processes 228
Poly(alkylene D-tartrates), extent of growth of A. Niger on 208
Poly(amic-acid) resin 198
Polyamide(s) 209
precursor resins 198
Poly(amide urethanes) by subtilisin, degradation of 212
Poly[bis(*m*-chlorophenoxy)phosphazene], thermal degradation of 295
Poly[bis(trifluoroethoxy)phosphazene], thermal degradation of 295
1,2-Polybutadiene, autoxidized ... 21
cis-1,4-Polybutadiene (CB) 11
extent of hydroperoxidation 14
after singlet oxygenation, ir spectrum 12
β-values 26
Polycaprolactone 208
Polyene(s)
calculated bond lengths for linear 335
conjugation energies for linear . 337
energies 336
geometries 336
growth, effects of tacticity on ... 313
molecular orbital theory of 333
neutral 333
numbering system nomenclature used for the linear polyenyl cations 336
Polyenes, observed bond lengths for linear 335
Polyenyl
cation(s) 333, 338
charge densities 340
conjugation energy for linear . 342
geometries 338
idealization of resonance forms found for linear 340
polyenes, numbering system nomenclature used for the linear 336
with reference to 1-methylpolyenes(MeI_{n-1}), energy conjugation for linear ... 344
STO-3G calculated bond lengths for linear 339
STO-3G energies of conjugation for linear 345
framework, β_2-type distortion of . 338
radical 156
Polyenylic chloride, concerted migration of 353
Polyester(s) 207
derived from aromatic diacids .. 207
polyurethane, aromatic 119
Polyethylene
Carbon-13 spectra of branched . 138
copper salts in 274
diffusion of $(Br(CH_2)_{10}CO_2)_2Cu$ into 288, 289
esters 207
films
analysis of 276
effect of processing conditions on the UV lifetime of ... 54
solubilities in 277
high-density 119, 122, 123
impact strength of high-density . 120
model paraffins, effects of 133
high-energy oxidation on 133
oxidation products, distribution of 140
photolysis of 6
sebacate 207
tartrate 208
thermal oxidation of 137
Poly(fluoroalkoxyphosphazene) (PFAP) elastomer 293
Polyisoprene, oxidation 217
cis-1,4-polyisoprene, singlet-oxygenated 21
cis-1,4-polyisoprene, β-values for 26

1,4-Polyisprene, photosensitized oxidation of 12
cis-1,4-Polyisprene 20
Polymer(s)
biodegradability of step-growth 205
biodegradable 205
concentration, inhibited-rated dependence on 247
cross-linking 392
crystallization, slow 270
degradation
by enzymes 207
by microorganisms 207
rate of 203
diffusion of metal salts in 274
discoloration 392
films, photosensitized oxidation of 22
gelation 392
$LiAlH_4$-reduced 328
luminescence spectra 69
O_2 absorption of 258
oxidative degradation of 78
photooxidation 30
of rubber-modified 36
during processing, mechanooxidation of 31
samples, determination of degrees of degradation of 206
shear 31
solution of metal salts in 274
stabilization 254
fundamentals in thermal autooxidation of 215
styrenic 124
UV stabilization of 116
β-values, double reciprocal plot for determination of 27
Polymerizations, step-growth 206
Poly(4-methylpent-l-ene) 69
cold hexane extract of 72, 73
fluorescence excitation spectra of 71
phosphorescence excitation spectra of PP 72
Polyolefin(s) 30, 68
copper-catalyzed oxidation of .. 289
effect on processing 38
factors affecting the oxidative degradation of a crystalline 261
fundamental processes in the photodegradation of 1
light-sensitive 69
luminescence photooxidation of light-sensitive commercial . 68
semi-crystalline 269
Polypropylene(PP) 69
in air, aging effect on irradiated 149
amorphous 176
atactic(APP) 160
in benzene solution, inhibited-oxidation of 242, 243
in benzene solution, oxidation of 240
cross-section profile of exposed . 127
effects of crystalline additives in isotactic 261

PP (*continued*)
effect of processing time on the peroxide 46
embrittlement time 46
fibers, effects of production conditions on the photosensitivity of 56
film after irradiation, intensity of fluorescent excitation spectrum of 74
fluorescence excitation spectra of 72, 73
grafting of antioxidants on 254
in He
kinetics of pyrolysis of 182
product distribution of pyrolysis of 184
TGA of 180
identification of pyrolysis products of 184
impurities in isotactic 261
inhibited autoxidation of 237
inhibited oxidation of 240
irradiated in air, electron spin resonance spectra of 154
isotactic 261
PPH 56
isotactic (IPP) 160
mechanical properties of modified 157
mechanisms for pyrolysis of 189
molecular weight distribution of oxidized 170
oven-aging of 251
oxidation, BHT inhibition of ... 246
oxidation, inhibition of 248
oxidative pyrolysis of 175
photooxidative degradation of .. 2
poly(4-methylpent-l-ene), phosphorescence excitation spectra of 72
post-irradiation embrittlement .. 151
pyrolysis 175, 180, 184
in air
in air, Arrhenius plot of 187
in air, first-order plot of ... 186
in He, first-order plot of ... 181
products of 183
rate of volatilization of 185
semicrystalline (IPP) 176
separation of heptane-soluble and insoluble fractions, photooxidative aging of 109
separation of heptane-soluble and insoluble fractions, thermooxidative aging of 109
stability of y-irradiated 142, 151
stress–strain curve of 145
syrings barrel flange, post- irradiation effect on the bending angle of 148
syringe flange, irradiation embrittlement of 147
tensile bar, percent extension of . 146
thermal oxidative degradation of 159

PP (*continued*)
γ-iniated oxidations of 60
γ-irradiation sterilization of isotactic 151
Polystyrene(s) 98
emission changes during irradiation of 102
film, fluorescence spectra from .. 99
impact 125
prompt-emission changes during irradiation of 100
weathering stability of 124
Poly(styrene-alt-methyl methacrylate) 104
films, prompt-emission changes during irradiation of 106
photoyellowing of polystyrene .. 96
Polyureas 211
fungal growth on 210
Polyurethane(s) 211
aromatic polyester 119
by axion degradation of 212
Poly(vinyl alcohol) (*see* PVA)
Poly(vinyl chloride) (*see also* PVC)
degradation, recent fundamental developments in the chemistry of 309
dehydrochlorination of 310, 333
ion-pair mechanism for the thermal dehydrochlorination of . 351
ionic mechanism of nonoxidative thermal degradation of 334
microstructure investigations ... 324
nonoxidative thermal degradation of 310
oxidation 315
pathways for nonoxidative thermal degradation of 333
peroxy gel formation in 35
stabilization, recent fundamental developments in the chemistry of 309
thermal stability of 324
Post-
embrittlement 148
irradiation effect on the bending angle of PP syringe barrel flange 148
irradiation embrittlement, PP .. 151
Potentiometric titration of carboxylic acids 378
PPH
fiber samples, γ-initiated oxidation of 59
monofilament production, draw conditions for 56, 57
photooxidation, changes in mechanical properties 64
photooxidation, oxidation products during 64
photostability, effect(s) of
draw
atmosphere on 63
speed on 63
PPH (*continued*)
temperature on 63
hopper blanketing and extrusion temperature on 62
γ-initiated oxidation, role of sample thickness in 58
Predicted induction periods 411
Preoxidized non-preoxidized films from multi-film laminites 282
Preventive antioxidants 216, 218, 222
syythesis of 402
Primary radical 190
Processing
on color formation in PVC, effect of 33
conditions on the UV lifetime of polyethylene films, effect of 54
at 200°C on functional group formation in HIPS, effect of 37
on the molecular weight distribution of LDPE, effect of 41, 42
peroxide olefinic unsaturation development during PVC . 32
on the polyolefins, effect of 38
of PVC 31
time
at 150°C on carbonyl formation in LDPE containing antioxidants and stabilizers, effect of 52
on carbonyl formation in LDPE, effect of 40
on the decay of the tin maleate carboxylate ir absorption, effect of 49
on the formation of peroxides in LDPE, effect of 43
on the growth of ester in PVC containing a tin maleate stabilizer, effect of 50
on the growth of maleic anhydride UV absorption, effect of 50
on the melt-flow index (MFI) of LDPE, effect of 40
on the peroxide conjugated carbonyl concentration in polypropylene, effect of . 46
on the rate of
photooxidation of
LDPE, effect of 44
PVC 36
PVC containing a tin maleate stabilizer, effect of 48
vinylidene decay in LDPE, effect of 44
at 170°C on UV absorbance maxima, effect of 34
Product
distribution of pyrolysis of PP in air 188
distribution of pyrolysis of PP in He 184

Product (*continued*)
I prompt-emission spectra 99
II prompt-emission spectra 104
III 104
Production conditions, effects of .. 59
Production conditions on the photosensitivity of PP fibers, effects of 56
Prompt-emission
changes during irradiation of polystyrene 100
changes during irradiation of poly(styrene-alt-methylmethacrylate) films 106
spectra, Product I 99
spectra, Product II 105
Pro-oxidant effects 235
Propagation reaction 216
2-Propanol 88
Propionic acid in solvent, effect of . 167
Propionic acid on the thermal oxidation of APP, effect of 165
Proton NMR spectroscopy 12
Pulsed Fourier-transform ^{1}H NMR 313
PVA
effect of time on UV spectra of photooxidized 85
hydrogen atom abstraction from . 91
ir spectra of initial 84
ir spectra of oxidized 84
oxidation 78
photolysis 79
at 70°C, photooxidation of 81
photooxidation(s)
in aqueous solution 80
methanol characteristics of ... 83
reaction pathways in 88
photooxidative cross-linking of .. 79
UV spectra of oxidized 86
PVC
allylic groups 35
chain, mechanoscission of 34
containing
additives, ethylenic unsaturation of 47
additives, peroxide formation of 47
a tin maleate stabilizer, effect of processing on the growth of ester in 50
a tin maleate stabilizer, effect of processing time on the rate of photooxidation of . 48
effect of processing on color formation in 33
effect of processing on the rate of photooxidation of 36
head-to-head 311
head-to-tail 310
peroxy gel formation in 35
processing 31
peroxide olefinic unsaturation development during 32

PVC (*continued*)
reduced with tri-*n*-butyltin hydride, C NMR spectrum of 327
stabilization
implications of the ion-pair dehydrogenation mechanism for 353
in the behavior of mercaptides vs. carboxylates, implications for 370
by dialkyltin dimercaptides, possible role of the ligand exchange reaction I in ... 372
synergistic mechanisms between zinc, calcium soaps, and organo compounds in 386
stabilizer consumption during processing of 374
stabilizer formulations 375
Pyrolysis
oxidative 182
of PP 175, 180, 184
in air 186
product distribution of 188
in He, kinetics of 182
mechanism for 189
in He, product distribtuion of 184
procedures for 179
products
fragmentation patterns of 182
of PP 183
identification of 184
vs. time in He 181

Q

Quantum yield for chain scission .. 7
Quantum yields for hydroperoxide decomposition chain scission . 7
Quenched melt regions 264
Quenching
copolymer fluorescence 107
efficiency, singlet-oxygen 128
excimer 99, 100
singlet-oxygen 128
triplet-state 127

R

R fractions, DSC tracings of 111
Radiation degradation 142
Radical(s)
additive 157
boundary cracking of the spherulites 269, 270
build-up 155
chain oxidation mechanism 3
combination 4
decay in vacuum 156
decomposition of cumene hydroperoxide 222
inhibitors 233
methylperoxy 85

Radical(s) (*continued*)
migration 136
recombination 156
scavenging 129
trapping agents on the decomposition of cumene hydroperoxide, effect of 233
Random scission 300
RBS (Rutherford Back Scattering) 282
Reaction pathways in PVA photooxidations 88
Redox potential 164
Reductive dehalogenation with tri-*n*-butyltin hydride 324
Relative work as a function of irradiation for tensile bars 146
Residual photosensitizer 14
Resonance forms found for linear polyenyl cations, idealization of 340
Retarded antioxidation, antioxidants in 223
Retention of tenacity 118
Reverse Aldol-type reaction 79
Rubber
mechanooxidation of 31
modified polymers, photooxidation of 36
modified polystyrene (HIPS) ... 36
Rutherford Back Scattering (RBS) 282
Rubrene consumption 25

S

Salt/polymer interface 279
Scanning electron microscopy (SEM) 58
Scission, random 300
Second-order decay 155
Secondary radical 191
Secondary stabilizers, zinc stearate and organo compounds as 395
SEM, (scanning electron microscopy) 58
Semicrystalline polyolefins 269
Semicrystalline PP (IPP) 176
Sensitization mechanisms 2
Shear conditions 61
Shear of the polymer 31
Single-determinant ab initio LCAO–SCF molecular orbital theory . 334
Singlet
molecular oxygen 19
-oxygen 11
addition 6
adduct 6
-oxygen quenching 128
efficiency 128
-oxygenated *cis*-1,4-polyisoprene 21
-oxygenated polymers, activation energies for chemiluminescence 22
oxygenation 12, 15
ir spectrum of *cis*-1,4- polybutadiene (CB) 12
Soil burial testing 206
Solubilities of polyethylene films .. 277
Solubility
direct measurement of 277
vs. chain length for $Cu(RCO_2)_2$. 278
vs. (molar volume of solvent)$^{-1}$ for $Cu(RCO_2)_2$ salts 277
Solubilization mechanism 287
Solution
of metal salts in polymers 274
oxidation technique 238
phase ionization of tropenyl chloride 349
viscosity of PFAP(II) 297
change in 303
Spectral shifts 231
Spectrophotometer, Norcon 201 rapid-scan vapor-phase 178
Sperulite 265
Spex Industries MC-2 mass chromatograph 178
Spherulites, radial boundary cracking of 269, 270
Spin lattice relaxation times of the carbonyl carbons 368
Spin-trapping technique 91
Stabilization
by chemical modification 318
of PFAP(II) with bis(8-oxyquinolate)zinc(II) 302
mechanism for 305
with metal soaps 317
with miscellaneous compounds . 317
with organotins 316
Stabilizer(s) 56, 399
commercial 65
consumption 375
in bottles according to their processing condition ..376, 377
during processing of PVC 374
effects of 48, 65
processing time at 150°C on carbonyl formation in LDPE containing antioxidants 52
Standard STO-3G bond angles 335
Standard STO-3G bond lengths ... 335
Statistical chain scissions in the photooxidation of PVC 92
4-Stearoxy-2,6-diphenylphenol, times of failure of PP films stabilized with 408
Step-growth polymers, biodegradability of 205
Step-growth polymerizations 206
STO-3G
calculated bond lengths for linear polyenyl cations 339
energies of conjugation for linear polyenyl cations 345
energies for tropenyl chloride .. 348
π-orbital charge densities 339, 340

Strain
-induced photosensitivity 62
rate 144
Stress–strain curve of PP 145
Styrenic polymers 124
Substituted piperidene derivatives . 116
Sulfenic acid(s) 223, 235
oxidation of 221
Sulfenyl free radical 223
Sulfinate (XVI)
in the presence of an excess of 4-methoxy-2,6-diphenylphenoxyl (XVa), thermal decomposition of 415
in the presence of an excess of XVa, thermal decomposition of 424
thermal decomposition of ...414, 424
Sulfonyl
axide intermediates 253
nitrene(s) 255
insertion mechanism 255
Sulfoxides on the decomposition of cumene hydroperoxide, effect of 230, 232
Sulfur compounds as hydroperoxide decomposers 399
Sulfur nucleophiles 228
Sunlight-induced oxidation 68
Surface
cracking 64
crazing, KLY-to-visual 121
gloss of dart impact strength ... 119
measurements of diffusion of copper carboxylates 282
region measurements 282
-to-surface gradients 282
Syn transition state 310
Syndiotacticity 313
Syndiotactic sequences, dehydrochlorination rates of 313
Synergism between zinc and cal - cium soaps 387
Synergism, zinc and calcium stearate 389
Synergistic mechanisms between zinc, calcium soaps, and organo compounds in PVC stabilization 386
Synergistic UV stabilizers 54
Synthetic elastomers 254

T

Tacticity on polyene growth, effects of 313
Temperature on the formation of peroxides in LDPE, effect of . 43
Tenacity, retention of 118
Termination reaction 216
Tetraphenylporphine 14
TGA of PP in air 185
TGA of PP in He 180
Thermal
autoxidation of polymers, stabilization fundamentals in 215
decomposition
of sulfinate (XVI) 414, 424
in the presence of an excess of 4-methoxy-2,6-diphenylphenoxyl(XVa) 415
in the presence of an excess of XVa 424
degradation 293
initiation sites of nonoxidative 310
mechanism of 295
nonoxidative 309
of a PFAP elastomer (II), mechanism of 295
of PFAP(II), investigation of . 296
of poly[bis(*m*-chlorophenoxy)-phosphazene] 295
of poly[bis(trifluoroethoxy)-phosphazene] 295
of PVC
pathways for nonoxidative . 333
ionic mechanism of nonoxidative 334
nonoxidative 310
dehydrochlorination, mechanism of nonoxidative 314
dehydrochlorination of PVC, ion-pair mechanism for 351
oxidation
of APP
effect of
tert-BuHPO on 170
copper stearte 164
DTBP on 171
model compounds on ... 173
n-octanoic acid on 172
propionic acid on 165
weight loss 168
of LDPE films, carbonyl formation in the 52
on the photooxidation of HIPS, effect of 39
of 1,4-polybutadiene 12
of polyethylene 137
of PVC 32
oxidative degradation of polypropylene 159
processing history 30
reactions of vinylidene groups in LDPE 43
scission of cyclohexadiene structures 313
stability of PFAP(II) 302
stability of PVC 324
Thermooxidative aging of polypropylene: separation of heptane-soluble and insoluble fractions 109
Thermooxidative results 110
Thioethers, β-activated 399
Thiolsulfinate(s) 219, 226
decomposition 222

Thiosulfuric acid 234
Tin
ligand bonding 371
maleate
carboxylate ir absorption, effect of processing time on the decay of 49
stabilizer, effect of processing time on the growth of ester in PVC containing . 50
stabilizer, effect of processing time on the rate of photooxidation of PVC containing 48
proton coupling constants for the dibenzyltin di(benzyl-meccaptide)/debenzyltin dichloride system 366
proton coupling constants for dimethyltin derivatives, ligand exchange effects on . 365
stabilizer 375
Toepler pump 80
Transition metal compounds 159
Transition metal stearates 160
Triazine(s)
-based, phenyl-substituted phenols, antioxidant activity of 403
XIII 422
XIV 422
Tri-n-butyltin hydride 326
C NMR spectrum of PVC reduced with 327
reductions 330
redutive dehalogenation with .. 324
Triene photolysis 107
Trigonal bipyramidal configuration 369
Triphenlphosphine method 7
Triplet-state quenching 127
2,4,6-Trisubstituted phenol 404
Tropenyl chloride 348, 350
STO-3G energies for 348
Tropenylium halides in methylene chloride, charge-transfer absorption bands of 350
Triphenylmethyl chloride 349
Two-pass n-Bu_3SnH reduction 327

U

α,β-Unsaturated carbonyl
chromophore 71
compounds, photochemistry of . 76
groups, effect of photooxidation on the luminescent 73
impurities 74
Unsaturation, conjugated 34
Unsaturation, olefiinc 312
Unzipping (depolymerization) ... 189
UV
absorbance maxima, effect of processing at 170°C on 34
absorbers 126
degradation 126

UV (*continued*)
radiation 126
screening 271
spectra of oxidized PVA 86
spectra of photooxidized PVA, effect of time on 85
stabilization of polymers 116
stabilizers, peroxide decomposers as 51
stabilizers, synergistic 54
transmission microscopy 262
Uvitex OB 263

V

Vacuum-irradiated copolymer films, delayed-emission spectra of .. 106
Vapor-phase ir of acetone 190
Vinylidene decay in LDPE, effect of processing time on the rate of 44
Vinylidene groups in LDPE, thermal photooxidative reactions of 43
Viscosity evolution of plasticized PVC and
calcium stearate transformation . 389
zinc and calcium stearate transformation 390
zinc stearate transformation 390
Volatilization of polypropylene, rate of 185
Vulcanizates, oxygen-aging of 258

W

Water on the decomposition of cumene hydroperoxide, effect of 229
Water as a nucleophile 228
Weak-link concentration in PFAP-(II), first-order rate constants for 301
Weight loss during thermal oxidation of APP 168
Weight percent of PFAP(II) 298

X

X-ray diffraction for lead formulation 379, 384
for insulating cables 380

Y

Y-initiated oxidation(s) 57
of polypropylene 60
of PPH fiber samples 59
Y-irradiated polypropylene, stability of 142, 151
Y-irradiation sterilization of IPP .. 151